Chemistry of Terpenes and Terpenoids

Chemistry of Terpenes and Terpenoids

Edited by

A. A. NEWMAN

Honorary Research Assistant,
University College, London, England

1972

ACADEMIC PRESS
London and New York

ACADEMIC PRESS INC. (LONDON) LTD.
24/28 Oval Road,
London NW1

United States Edition published by
ACADEMIC PRESS INC.
111 Fifth Avenue
New York, New York, 10003

Library of Congress Catalog Card Number: 75-185198
ISBN: 0-12-517950-2

PRINTED BY J. W. ARROWSMITH LTD., BRISTOL, ENGLAND

Contributors

D. V. BANTHORPE, *Chemistry Department, University College, London, England.*

B. V. CHARLWOOD, *Department of Pharmacognosy, The School of Pharmacy, University of London, England.*

J. D. CONNOLLY, *Chemistry Department, The University of Glasgow, Glasgow, Scotland.*

J. R. HANSON, *The Chemical Laboratory, University of Sussex, Brighton, England.*

K. H. OVERTON, *Chemistry Department, The University of Glasgow, Glasgow, Scotland.*

R. RAMAGE, *The Robert Robinson Laboratories, University of Liverpool, Liverpool, England.*

J. S. ROBERTS, *Chemistry Department, University of Stirling, Stirling, Scotland.*

D. WHITTAKER, *Chemistry Department, University of Liverpool, Liverpool, England.*

Preface

The book presented here is conceived as an advanced survey of terpene chemistry for the use of graduates and research workers. Being a collaborative effort, a certain amount of overlapping was inevitable. No contributor can be asked to give less than a well rounded account of development in his own field just because another colleague has already touched on a particular aspect common to other types of terpenoids. I should also like to explain that owing to the multitude of skeletal types in the sesquiterpenoid field, it was felt that the inclusion of a tabulated list in the abbreviated form would serve no useful purpose.

It is hardly necessary to mention that each chapter was written by an acknowledged specialist in the field, and this being so, it is permissible to hope that the book will be found helpful to all interested in terpene chemistry. Readers may wish to know that the subject matter of one, excellent chapter is more restricted than that of the rest. The responsibility for this is entirely mine.

To my colleagues who have helped to realize the book, and to the publishers who have accepted the commercial risk inherent in such an undertaking, I express my sincere gratitude.

London
February, 1972

A. A. NEWMAN

CONTENTS

Chapter 4

The Di- and Sesterterpenes

Part 1: The Diterpenes

J. R. HANSON

Part 2: The Sesterterpenes

Chapter 5

The Triterpenoids

J. D. CONNOLLY and K. H. OVERTON

Chapter 6

Carotenoid Chemistry

R. RAMAGE

Chapter 7

Biogenesis of Terpenes

D. V. BANTHORPE and B. V. CHARLWOOD

1

INTRODUCTION AND NOMENCLATURE

J. R. HANSON

Lecturer in Chemistry, University of Sussex

I. Introduction

The terpenes are amongst the most widespread and chemically interesting groups of natural product. However, despite their structural diversity, they have a simple unifying feature which we shall use as their definition. Terpenoid compounds may be defined as a group of natural product whose structure may be divided into isoprene units. This immediately leads to a rational classification of the terpenes depending upon the number of such isoprene units.

Monoterpenes	C_{10}
Sesquiterpenes	C_{15}
Diterpenes	H_{20}
Sesterterpenes	C_{25}
Triterpenes	C_{30}
Carotenoids	C_4O
Rubber	$(C_5)_n$

In practice it is possible to discern a further subdivision of the terpenes. The monoterpenes, sesquiterpenes, diterpenes and sesterterpenes contain the isoprene units linked in a head to tail fashion. The triterpenes and carotenoids are made up of two C_{15} and C_{20} units respectively linked in the middle head to head. This structural unity is exemplified in the structures shown in Fig. 1. There are a number of apparent exceptions to this regular isoprenoid backbone. These, however, can be seen to originate in the rearrangement of regular isoprenoid precursors during biosynthesis. In addition, isoprenoid units are often found within the framework of other natural products. For example, the indole alkaloids such as vindoline have been shown to contain a monoterpenoid fragment whilst the lysergic acid group of indole alkaloids

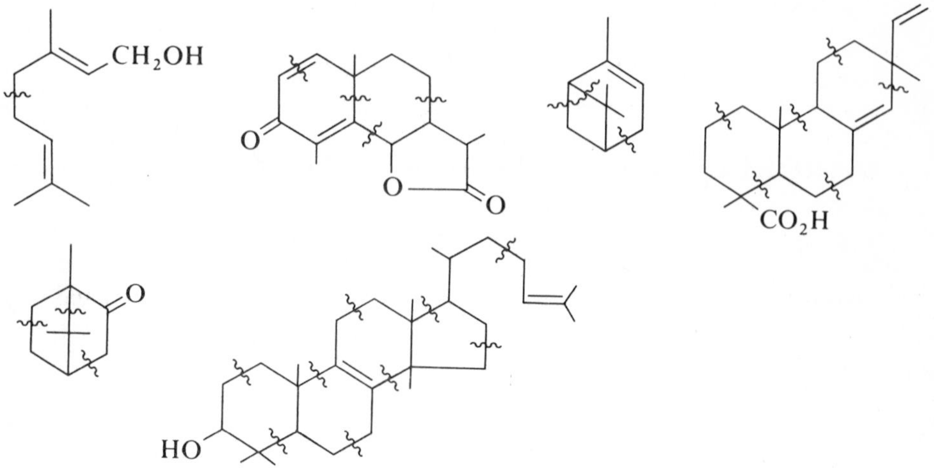

FIG. 1. Structural unity of the terpenes.

contain a single isoprene unit. A number of quinones [such as vitamin K and the pigment lapachol] also contain terpenoid fragments as do the humulone resins from hops and cannabinol from hashish.

Terpene chemistry has provided an immense fund of problems whose solutions have, in many instances, provided the stimulus for theories that are now fundamental to organic chemistry. The Wagner–Meerwein rearrangement of α-pinene, the various descriptions of the non-classical carbonium ion and the theories of conformational analysis have firm foundations in terpene chemistry. The same applies to the Woodward–Hoffman rules of cycloaddition reactions. The photochemistry of terpenoid substances such as santonin has been a point of considerable interest. Indeed, the juxtaposition of diverse functionality on a terpenoid framework permits the chemist to study the interaction between functional groups within molecules. These interactions are often reflected in the spectroscopic properties of terpenoid substances. The subtle variations in structure that are often available within a closely related group of terpene can afford the organic chemist an insight into the interplay between electronic and stereochemical features that control the outcome of organic reactions.

The structures of the terpenes, particularly the polycyclic compounds, have provided a challenging source of synthetic objectives. It is often possible to discern a series of increasingly complex structures, as in the diterpenes, which are of nicely graded difficulty. Thus, as our control over the stereochemical consequences of reactions has improved, so the synthetic objectives often involving the construction of ring junctions of known stereochemistry have become sophisticated.

In recent years the origin of the ubiquitous isoprene unit and its conversion into various terpenes has been successfully studied. Fifteen years ago Sir Robert Robinson wrote that "Organic chemists have often been tempted to leave the security of their own proper pastures and to graze, albeit speculatively, in the attractive fields of biochemistry". In the intervening years, terpene biochemistry has moved from placing compounds in a possible order of biogenesis to providing experimental biosynthetic evidence for such schemes. Indeed, terpene enzymology is now becoming an area of investigation.

The history of the terpenes spans the centuries of civilization. Essential oils, particularly oil of turpentine, were known to the Ancient Egyptians whilst they also receive mention in Dioscorides' "De Materia Medica". Astringent and toxic properties of sesquiterpenoid and diterpenoid bitter principles figure in many folk medicines. Thus, the toxic nature of honey from bees feeding on rhododendrons figures in the writing of Xenophon and in agricultural records ever since. Camphor was introduced to Europe from the East by the Arabs and is recorded in several eleventh-century manuscripts. The process of obtaining oils by fatty extraction, the so-called "enfleurage" process, was known by the early Middle Ages. Arnald de Villanova, who died in 1311, describes distilled oils from rosemary and sage. His "oleum mirabile" was an alcoholic solution of oil of turpentine and rosemary. However, the development of the art of distillation belongs to the sixteenth century and is recorded in such books as Brunschwig's "Liber de Arte Distilland" and Lonicer's "Krauterbuch". Whilst oil of turpentine, juniper, rosemary and spike were known at the beginning of that century, by 1592 when the Nuremburg edition of "Dispensatorium Valerii Cordi" was written, some 61 oils were described. There is evidence from the writings of Walter Ryff of Strasburg in 1550 and of Quercetanus in 1607 that a vestigial essential oil industry existed in France producing oil of lavender and oil of juniper. Eau de Cologne appears to date from the early eighteenth century.

The terpenes have been the subject of chemical study from the dawn of modern chemistry. Analyses of oil of turpentine were recorded in 1818 by J. J. Houton de la Billardière who showed that the carbon–hydrogen ratio was five to eight. Physical and chemical data on many oils were recorded in the work of Dumas and Berthelot from 1830 onwards. Thus, Dumas in 1833 determined the formula $C_{10}H_{16}O$ for camphor. The first records of the isolation of the sesquiterpene santonin from wormseed date from 1830 whilst crude diterpene resin acids were isolated from colophony about the same time. The name, terpene, derived from turpentine, appears to have originated in the writings of Kékulé in 1866. For many years, however, the term "camphors" was used to describe not only camphor itself but also other crystalline oxygenated substances, such as thyme camphor (thymol) and peppermint camphor (menthol). The latter half of the nineteenth century was dominated by the structural work of chemists such as Tilden, Wallach, Semmler, Tiemann and Harries. Thus, in 1877 Tilden introduced nitrosyl chloride as a suitable reagent for obtaining crystalline derivatives. The lack of adequate proof of homogeneity was a confusing feature of much of the chemistry of this period. It is, nevertheless, an amazing feature of nineteenth-century chemistry that it was capable of developing and tackling problems of considerable complexity in the absence of the criteria of purity afforded by modern physical methods. One may contrast the controversy that raged over a problem such as the isopropenyl–isopropylidene isomerism with its present facile solution by nuclear magnetic resonance techniques.

In 1887 Wallach proposed an isoprene rule to distinguish the monoterpenes and the sesquiterpenes, and within a few years the structures of quite a number of compounds were known.

For example, the structure of camphor was finalized by Bredt in 1893, the structure of α-pinene was proposed by Wagner in 1894 and that of citral by Tiemann and Semmler in 1895. Barbier and Bouveault recorded the first synthesis of citral in 1896. Camphoric acid which had previously been reconverted to camphor, was synthesized by Komppa in 1903 and by Perkin and Thorpe in 1904, thus completing a formal total synthesis of camphor.

Considerable insight was obtained during this period into the complex chemistry of the sesquiterpene santonin, particularly by Italian workers such as Cannizzaro, Francesconi and Andrescocci. Indeed, the first records of its photochemistry date from this period. However, the correct structure for santonin was not proposed until 1929, and its photochemistry remained in doubt until the work of Barton in 1957. Although sesquiterpene hydrocarbons such as cadinene, isolated in 1840 from oil of cubebs and cedrene, and in 1841 from cedar wood oil, were studied through the nineteenth century, clarification of the sesquiterpene structures developed through the work of Ruzicka during the 1920s. The introduction of dehydrogenation with sulphur, and later selenium, produced naphthalenes such as cadalene from sesquiterpenes such as cadinene. In 1921 Ruzicka demonstrated the isoprenoid structure of cadalene (1,6-dimethyl-4-isopropylnaphthalene). The following year he observed that eudesmol and selinene gave another naphthalene, eudalene, 1-methyl-7-isopropylnaphthalene. Again, its isoprenoid carbon skeleton was rationalized, this time with the extrusion of an angular methyl group. The structures of the azulene-based sesquiterpenes such as guaiazulene (1,4-dimethyl-7-isopropylazulene) and vetivazulene (2-isopropyl-4,8-dimethylazulene), were first formulated during the 1930s.

The importance of dehydrogenation as a structural tool can be clearly seen in the development of diterpene chemistry during this inter-war period. Although the first application of dehydrogenation was made by Vesterberg in 1903 who isolated retene (1-methyl-7-isopropylphenanthrene) from abietic acid, it was not until the work of Ruzicka in the 1920s that the generality of this technique was established. Pimanthrene (1,7-dimethylphenanthrene) was also isolated from a number of diterpenoid substances, and it was synthesized by Haworth in 1932. Ruzicka made an extensive study of the oxidation products of abietic acid during this period leading to a classification of the structure by 1931.

The carotenoids had been recognized as plant pigments during the nineteenth century. For example, crocetin was known as early as 1818 whilst bixin was isolated by Boussingault in 1825 from commercial orleans pigment. β-Carotene was isolated by Wackenrodder from carrots in 1837 whilst Willstätter determined the correct molecular form in 1907. However, it was not until the 1930s that progress in determining their structure was made by Karrer, Kuhn and Zeichmeister. Thus, the structure of β-carotene was demonstrated in 1930. This field contains some of the first applications of chromatography to natural product chemistry, a technique which was to revolutionize natural product chemistry in the post-war period. The chemistry of vitamin A, which is related to the carotenoids, was investigated by Heilbron and Jones during this period. These compounds become the object of a number of interesting syntheses using acetylenic methods during the following decade.

The period since 1945 has witnessed an immense explosion in natural product chemistry stimulated by the advent of spectroscopic techniques, more refined criteria of purity and chromatographic methods. Thus, although the triterpenes had been studied for many years, lanosterol from wool fat was known as early as 1872, the structure was not finally clarified until 1952 by Ruzicka and Jeger in Zurich and Barton and McGhie in London. The same year saw a complete elucidation of the stereochemistry of the pentacyclic triterpene, β-amyrin. The theories of conformational analysis put forward by Barton in 1950 led to the rationalization of the stereochemistry of many reactions amongst the polycyclic terpenes. Indeed, this period in terpene chemistry is dominated, on the one hand, by the interrelationship of various terpenes with compounds of known structure and the recognition of various skeletal types, and on the other hand, by the extensive application of conformational analysis.

The increasing use of spectroscopic techniques, firstly ultra-violet spectra, then infra-red spectra followed by nuclear magnetic resonance spectra and mass spectra, has meant that many of the more highly oxygenated structures could be elucidated. Indeed, the approach to natural product chemistry has shown a subtle change. During the interwar period the techniques involved removal of functional groups and the determination of the underlying carbon skeleton by dehydrogenation. Present-day physical methods with their many correlations between functionality, ring-size and stereochemical environment, thrive on the more highly oxygenated compounds.

The synthetic organic chemist has turned his attention to the higher terpenes, particularly over the last fifteen years. The synthesis of lanosterol was recorded in 1954 by Woodward and Barton. Since then many syntheses of sesquiterpenes, diterpenes and triterpenes have been completed illustrating the advent of novel reagents, as in the syntheses from Corey's laboratory, or in the control of stereochemical problems from Ireland's work.

In 1956 mevalonic acid was shown to act as an irreversible biosynthetic precursor of cholesterol. Subsequently, its incorporation into a number of terpenoid substances has been demonstrated. Many of the techniques developed in the study of steroid biosynthesis have been extended to the fungal terpenes. For example, over the last few years most of the pathway to the fungal diterpenes which include the gibberellin plant growth hormones, has been clarified.

Recently, an increasing number of terpenoid substances have been shown to be responsible for the biological activity of various systems. There are a number of antibiotics amongst the sesquiterpenes and the diterpenes, whilst the triterpene fusidic acid is available commercially. The insect juvenile hormones appear to be derived from sesquiterpenes, whilst the gibberellin plant growth hormones are diterpenes. A number of sesquiterpenes are active against experimental tumours.

A. Nomenclature

The systematic naming of a terpene contains the following elements. Firstly a prefix indicates to which enantiomeric series a compound belongs. Secondly, there comes a group of prefixes arranged in alphabetical order to define the location and stereochemistry of substituents. Thirdly, there are a series of operators which may designate structural modifications to the parent stem (e.g. ring expansion or ring contraction). These immediately precede the stem of the parent hydrocarbon. This stem is modified by the degree and position of unsaturation and is followed by a suffix designating the principal functional group according to the I.U.P.A.C. rules of organic nomenclature.

In practice, trivial names have been given to many natural products, often because they were isolated and described in the literature long before their structure was known. In many cases it is easier to refer to a compound by a diagram or by a simple modification of a trivial name.

The rules for nomenclature are the subject of revision and at the time of writing a set of tentative proposals on Diterpene Nomenclature have been made. The common skeleta and their numbering are set out in the following figures.

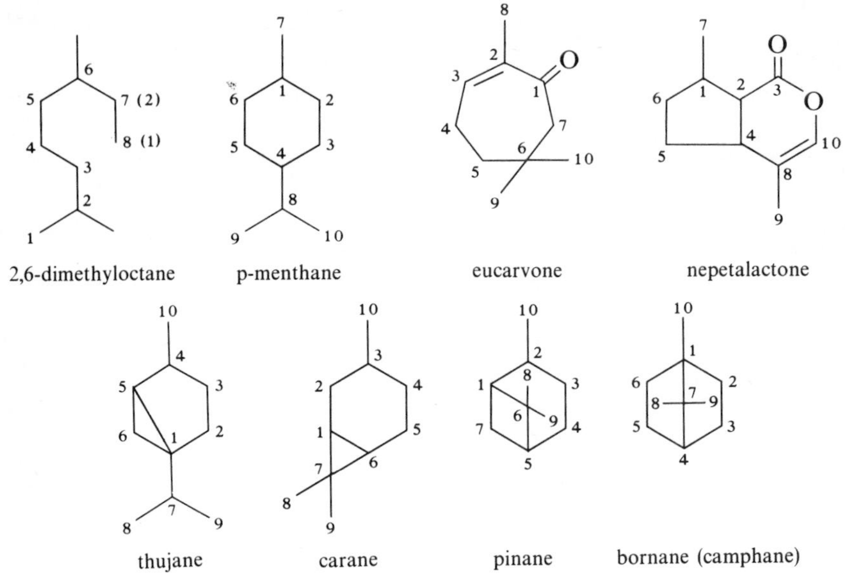

FIG. 2. Typical monoterpene skeleta.

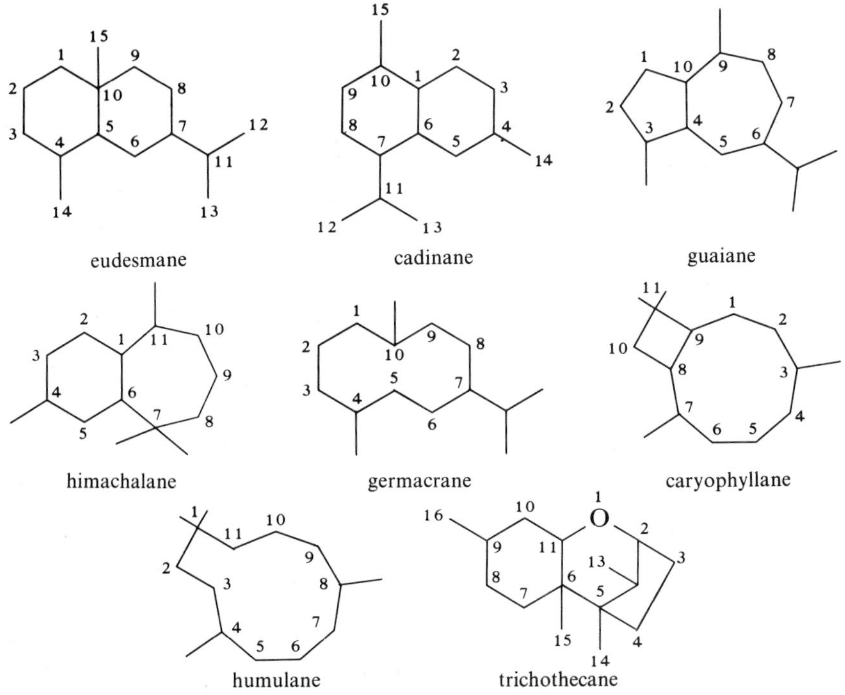

FIG. 3. Typical sesquiterpene skeleta.

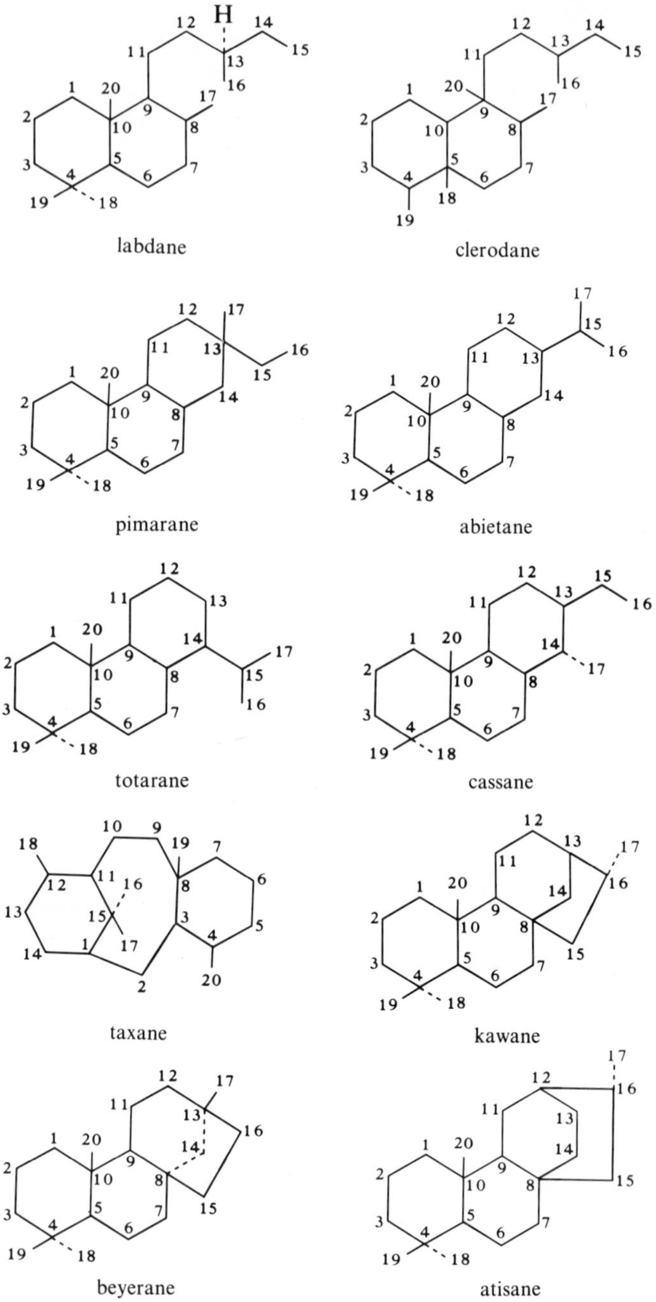

FIG. 4. Typical diterpene skeleta.

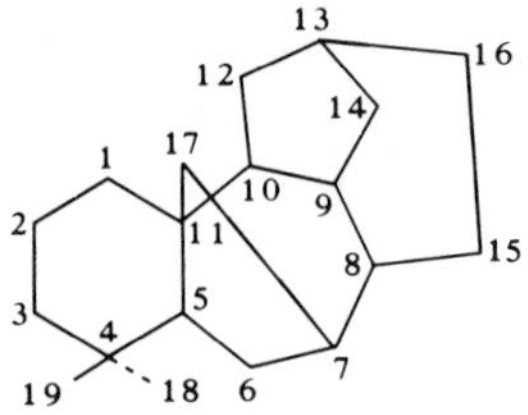

aconane

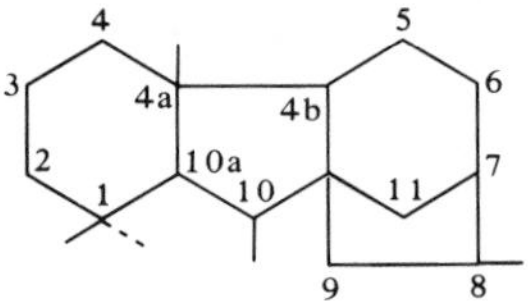

gibbane

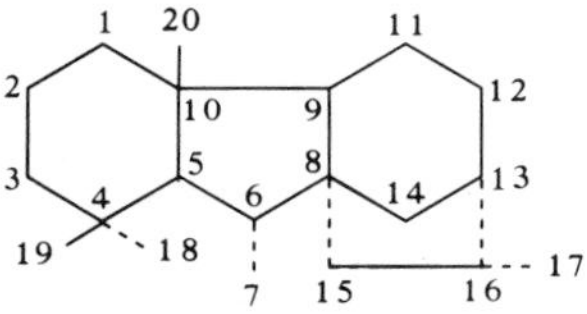

gibberellane

FIG. 4 (continued).

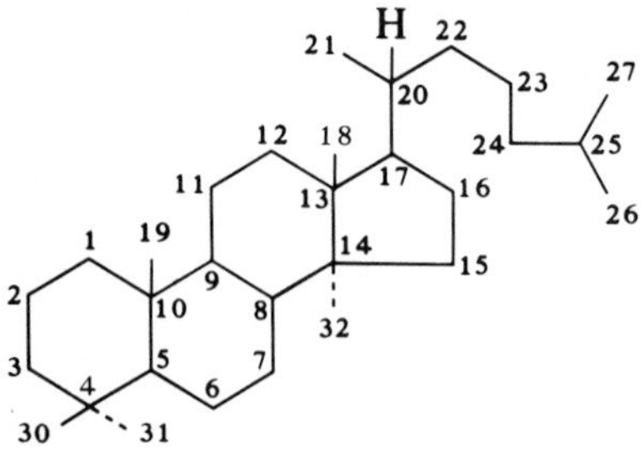

lanostane

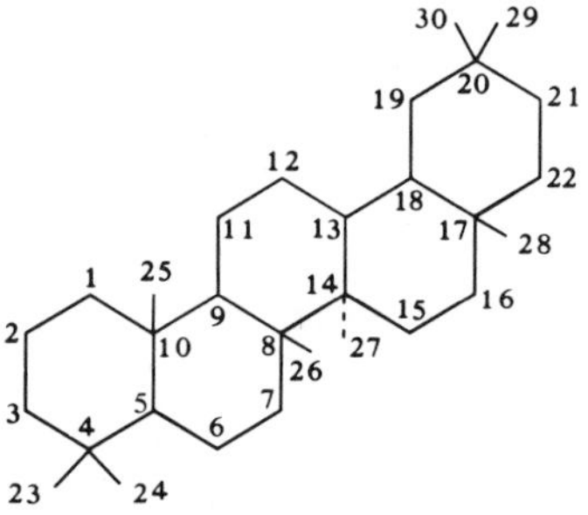

oleanane

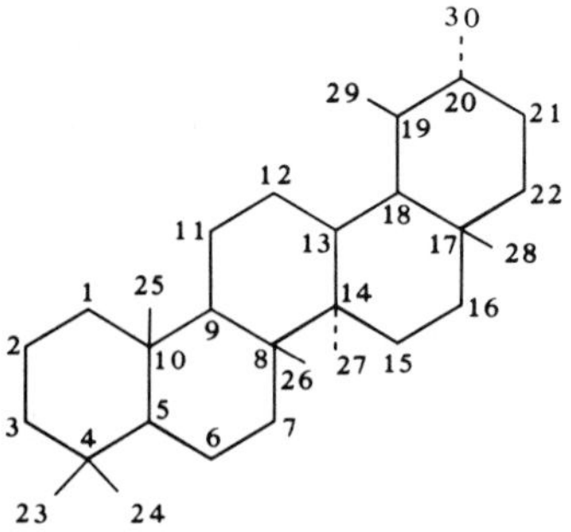

ursane

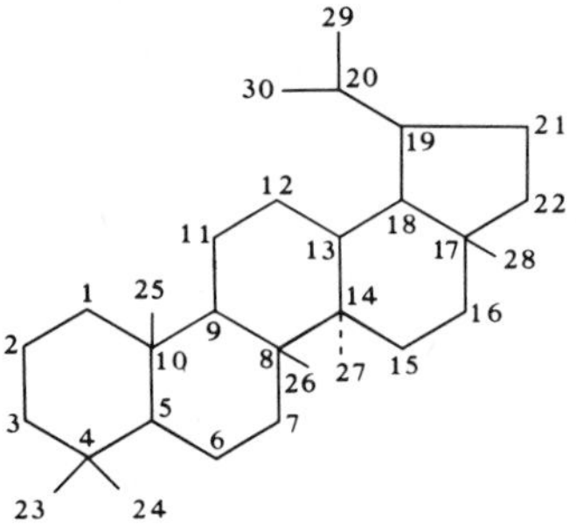

lupane

FIG. 5. Typical triterpene skeleta.

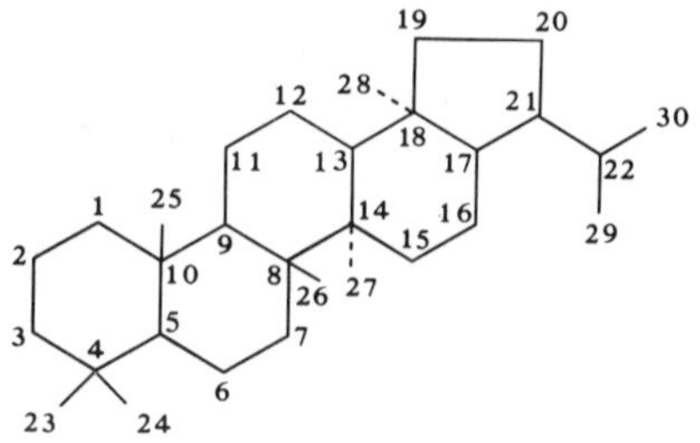

hopane

FIG. 5 (continued)

Some of the more commonly used operators to modify these skeleta are as follows. "Cyclo" is used to describe the formation of a single bond between two atoms of the parent stem leading to the formation of an additional ring and naturally requires the specification of these centres. "Homo" is used to indicate the addition of a methylene to the parent stem whilst "nor" refers to the subtraction of a methylene. The numbering is modified by the addition of xa where x is the lower number of the two carbon atoms adjacent to the insertion. The number preceding the prefix "nor" is the number of the atom eliminated. The term "seco" is used to designate the cleavage of a bond between two atoms of the parent structure. Two terms are used to cover rearrangements. When a bond joining one carbon atom (x) to a second carbon atom (y) has been broken and a new bond formed between the first and third carbon atom (z) then the "abeo" operator is used preceded by numbering of the form x(y → z). This terminology has been introduced to cover modifications in ring size whilst the other term "friedo" designates one or more consecutive 1,2-migrations of alkyl groups and/or hydrogen atoms around a ring system. Again this must be preceded by the numbers referring to the start and end of the rearrangement. The use of these terms is not yet widespread and it remains to be seen how successful they will be.

In studying the older literature the reader will come across many different numbering systems and considerable caution must be exercised in their interpretation.

2

THE MONOTERPENES

D. WHITTAKER

Lecturer, Department of Chemistry, University of Liverpool

I. Introduction

The monoterpenes have been known for hundreds of years as components of the fragrant oils which can be obtained from the leaves, flowers and fruits of many plants. Although a few, such as camphor, occur in a sufficiently pure form to allow crystallization of a single substance, most occur as complex mixtures, often of isomeric substances which are difficult to separate. Since they are complex, and readily available, the naturally occurring monoterpenes have occupied chemists since the middle of the nineteenth century. Early workers were handicapped by the complicated rearrangements which they underwent with great ease, but the majority have by now been obtained in pure form, their structures determined, and then confirmed by synthesis. This chapter, then, will not be concerned with the determination of structure or complete synthesis or simple reactions of the monoterpenes, for details of which the reader is referred to the classic works of Simonsen (1947, 1949a), but will be concerned with the rearrangements which these compounds undergo, and with those features of their chemistry in which their structures cause reactions to take unusual courses.

It is difficult, if not impossible, to understand the rearrangements of the terpenes without having some preliminary understanding of carbonium ion rearrangements. The simplest type of carbonium ion rearrangement is the example shown below, in which the shift of a methyl group converts a secondary carbonium ion to the more stable tertiary ion which then decomposes to a rearranged product.

$$CH_3-\underset{\underset{CH_3}{|}}{\overset{\overset{CH_3}{|}}{C}}-\underset{\underset{H}{|}}{\overset{\overset{CH_3}{|}}{C}}-Cl \rightarrow CH_3-\underset{\underset{CH_3}{|}}{\overset{\overset{CH_3}{|}}{C}}-\underset{\underset{H}{|}}{\overset{\overset{CH_3}{|}}{C^+}} \rightarrow CH_3-\underset{\underset{+}{}}{\overset{\overset{CH_3}{|}}{C}}-\underset{\underset{CH_3}{|}}{\overset{\overset{CH_3}{|}}{C}}-H$$

$$\xrightarrow{H_2O} CH_3-\underset{\underset{OH}{|}}{\overset{\overset{CH_3}{|}}{C}}-\underset{\underset{CH_3}{|}}{\overset{\overset{CH_3}{|}}{C}}-H$$

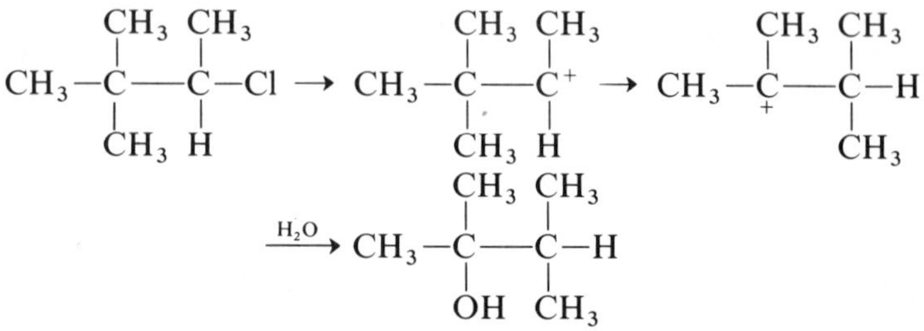

In a bicyclic system, such a bond shift would result in a change in the skeleton of the molecule, i.e.,

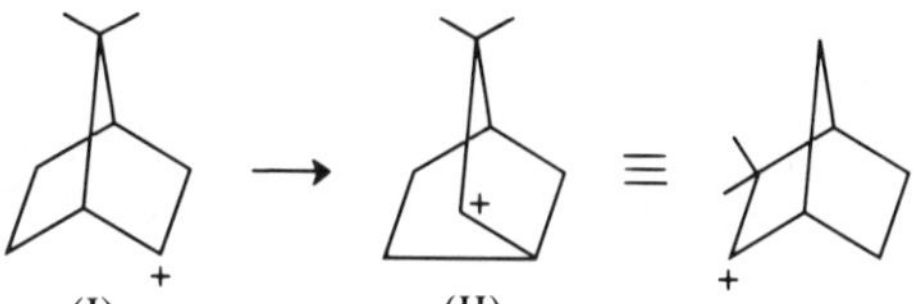

(I)   (II)

Reactions involving either of the ions (I) or (II) have been found experimentally to give a mixture of the products which would be expected to be derived from each ion, so that either the ions must be undergoing interconversion at a rate which is rapid compared with the rate of capture of the ion, or instead of two ions, we must have single ion which is capable of reacting to give products of the type expected from either ion. Such an ion is known as a bridged ion, and may be represented by (III).

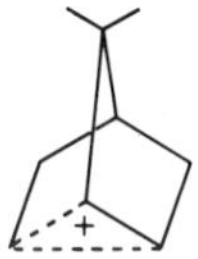

(III)

Throughout the last decade, controversy has raged over whether ions such as (III) have a finite existence, or merely represent a transition state in the rapid interconversion of (I) and (II). This problem has been adequately discussed elsewhere (Brown, 1962, 1966; Berson, 1963; Bartlett, 1965) and need not be repeated here. Most rearrangements mentioned in this chapter can be understood equally well using either representation of the intermediate, though the bridged ion will, in fact, be used.

A bridged ion can be formed during the ionization of a molecule, provided the geometry of the molecule is such that it has a bond in a position for its electrons to assist, and hence accelerate the departure of the leaving group, such as in the ionization of *exo*-norbornyl *p*-toluenesulphonate shown below (Winstein and Trifan, 1952b).

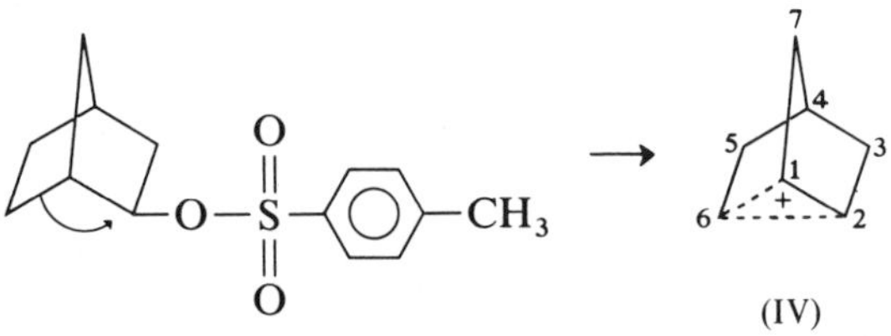

(IV)

If the geometry of the molecule is such that no bond has electrons in a position to assist departure of the ionizing group, or has suitable bonds only in such a position that electron shift would be sterically unfavourable, ionization proceeds to give a normal, unbridged ion, which may, however, subsequently form a bridged ion, as in the ionization of *endo*-norbornyl *p*-toluenesulphonate, shown below (Winstein and Trifan, 1952a).

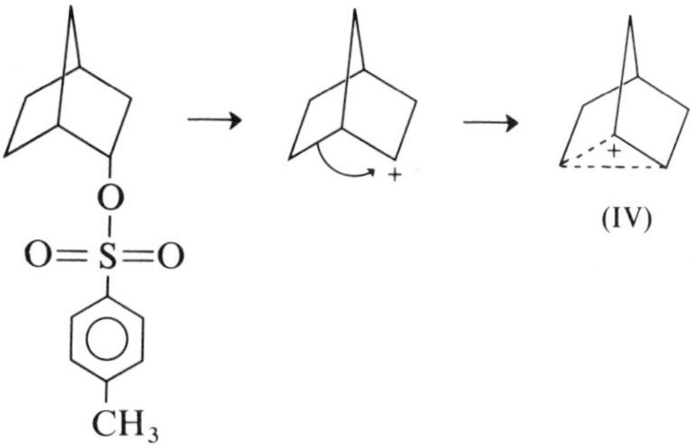

(IV)

Bridged ions such as the norbornyl cation (IV) decompose by a route which is the reverse of their formation, the attacking nucleophile approaching along the path of the departing group. Thus, these ions exert a rigid control over the stereochemistry of the product, and in acetic acid, the ion (IV) gives a product which is at least 99·5% *exo*-norbornyl acetate (Goering and Schwene, 1965).

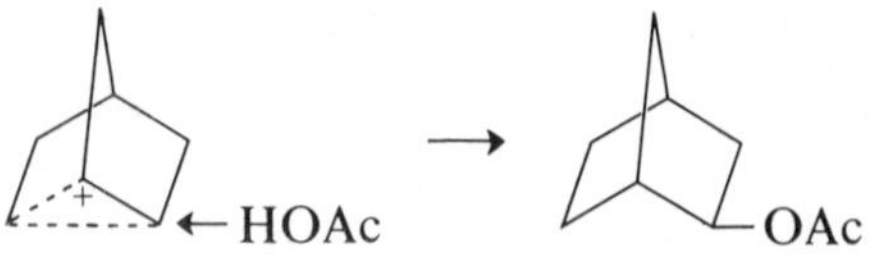

However, ion (IV) can also react at the bridgehead carbon atom, the stereochemistry of reaction being equally rigidly controlled by the geometry of the ion. The product, once again, is pure *exo*-norbornyl acetate, but in this case, the mirror image of that generated by reaction at the carbon atom from which the *p*-toluenesulphonate group departed.

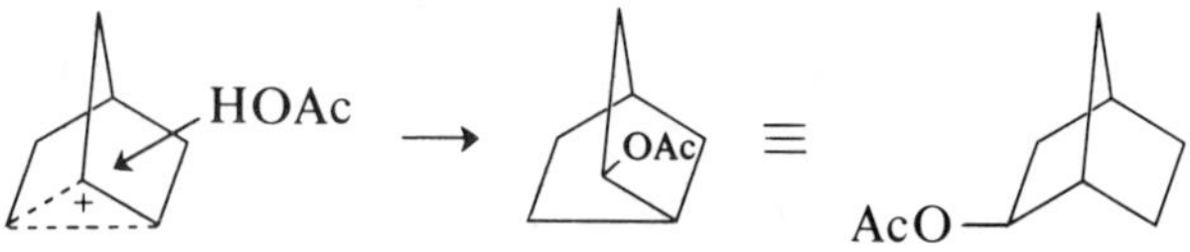

In a substance of less symmetry than norborneol, reaction at the two possible places can give rise to two different substances with different carbon skeletons, as can be seen in the reaction of *apo*-isobornyl bromobenzene-sulphonate to give a mixture of *apo*-isoborneol and *exo*-camphenilol, plus products involving further rearrangement (Winstein, 1955).

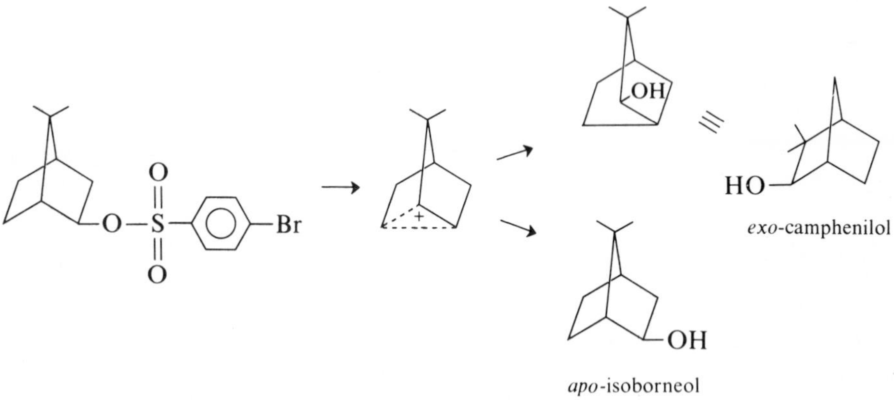

If the bridged ion is intermediate between a secondary and a tertiary ion, reaction takes place more rapidly at the more substituted carbon atom, so that the methanolysis of isobornyl chloride gives a mixture of isobornyl methyl ether and camphene hydrate methyl ether in which the latter predominates by a factor of approximately 50 (Beltramé *et al.*, 1960, 1964).

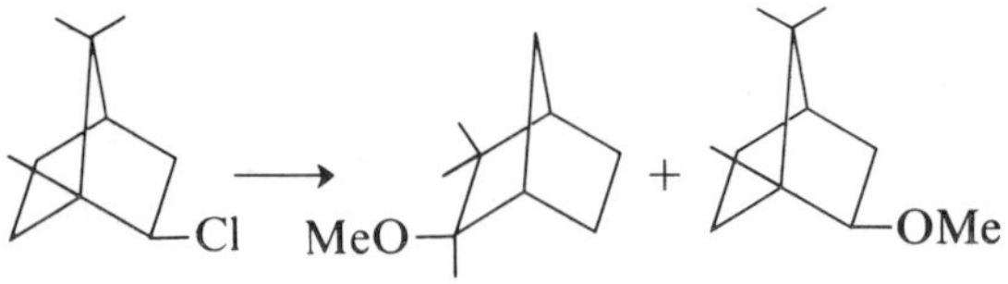

isobornyl chloride — camphene hydrate methyl ether — isobornyl methyl ether

An ion such as (IV) can undergo rearrangements, the most common being hydride shift. In the case of (IV), the most rapid hydride shift, the 2,6 shift, produces a mirror image ion, so that this affords another route to racemization of the products. In a less symmetrical compound, however, such a shift can result in a change in the carbon skeleton of the compound.

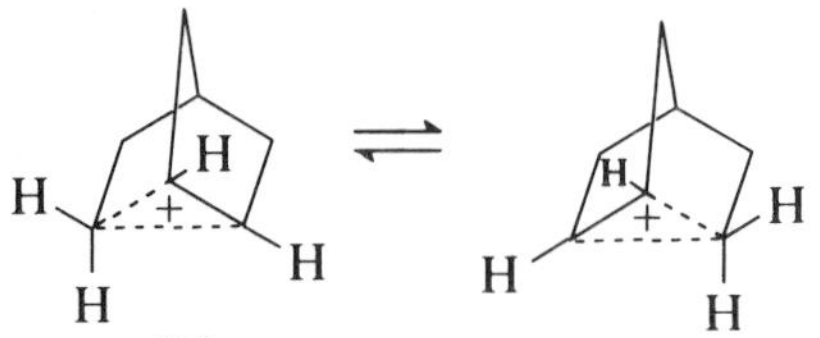

(IV)

Rearrangement, including ring formation and ring fission, can also take place readily in the excited states of terpene molecules produced by irradiation with ultra-violet or visible radiation. Consideration of the fundamental principles of photochemistry is beyond the scope of this chapter, but the reader is referred to one of several excellent texts which have appeared on this subject (Turro, 1967).

II. Acyclic Monoterpenes

The acyclic monoterpenes can be regarded as derivatives of the saturated hydrocarbon, 2,6-dimethyloctane (V). Associated with the four carbon atoms of the isopropyl group of virtually all acyclic monoterpenes is a double bond, the position of which has been much disputed in the past. Early attempts to locate it, based on ozonolysis experiments, gave products which were mixtures of those expected from compounds containing an isopropylidene group (VI) and an isopropenyl group (VII). It has since been shown by infra-red spectroscopy that most natural terpenes are in the isopropylidene form (the β-isomer) though in many cases, the isopropenyl form, often known as the α-isomer, can be prepared. These double bonds play an important part in the cyclization reactions which many acyclic terpenes undergo, the ring closing to give either derivatives of either *p*-menthane (VIII) or 1,1,3-

trimethylcyclohexane (IX). Most of the naturally occurring acyclic monoterpenes have pleasant odours, so have been synthesized on a large scale for use in the perfumery industry (Fordham, 1968; Ansari, 1970).

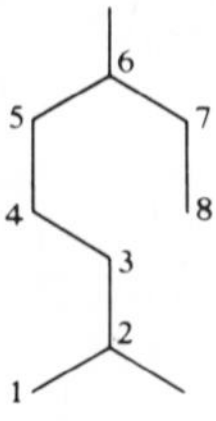

(V)

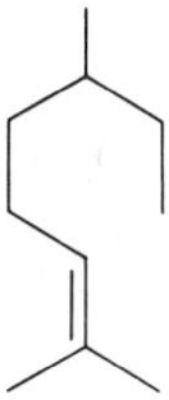

(VI)

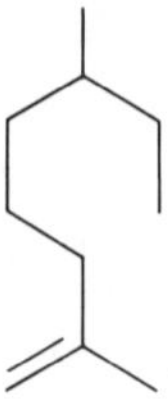

(VII)

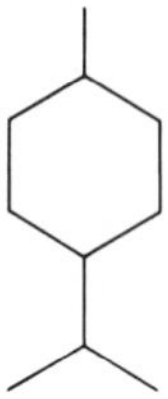

(VIII)

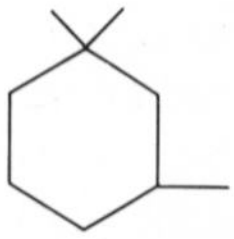

(IX)

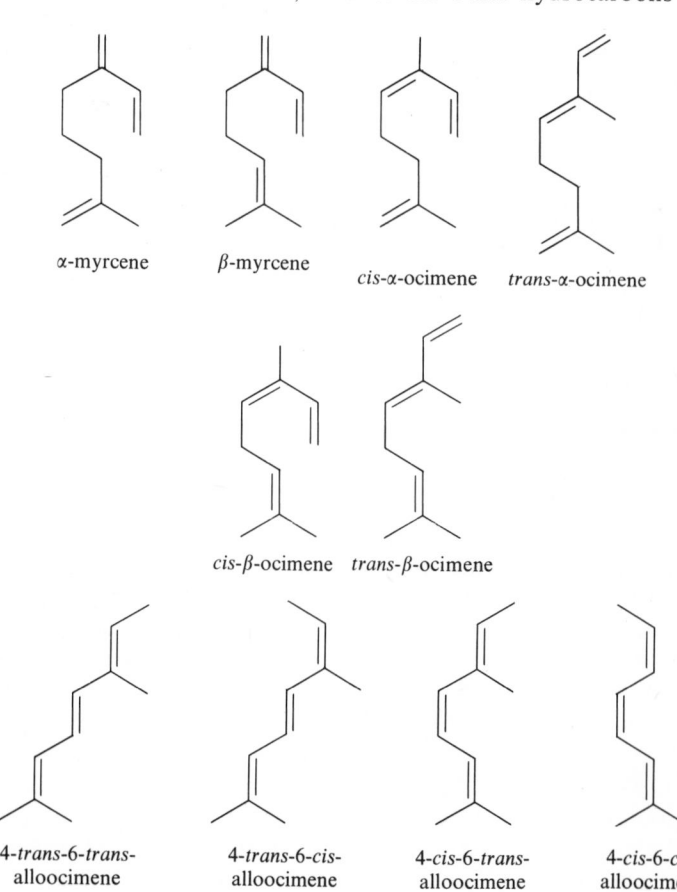

Three acyclic trienes occur naturally, ocimene, myrcene and alloocimene. Myrcene exists as two isomers; each of the other hydrocarbons as four.

α-myrcene

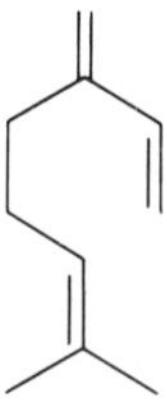

β-myrcene

cis-α-ocimene

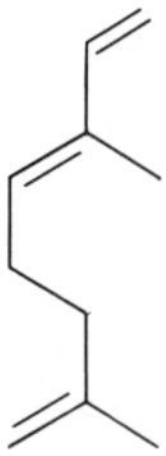

trans-α-ocimene

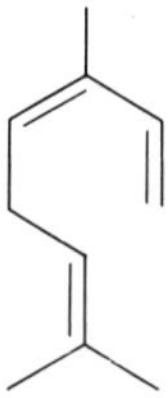

cis-β-ocimene

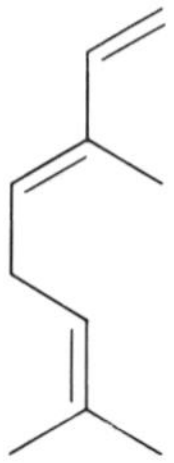

trans-β-ocimene

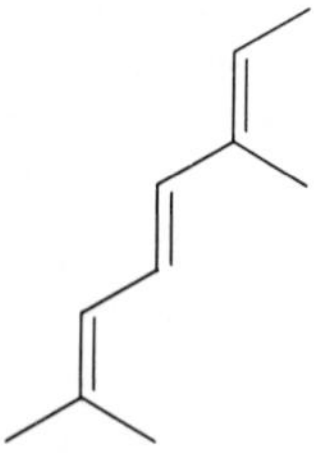

4-*trans*-6-*trans*-alloocimene

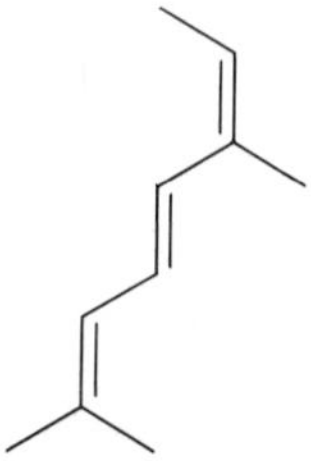

4-*trans*-6-*cis*-alloocimene

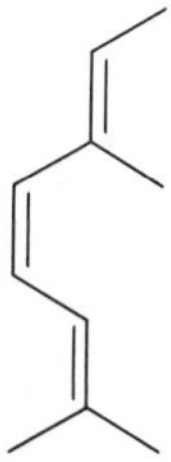

4-*cis*-6-*trans*-alloocimene

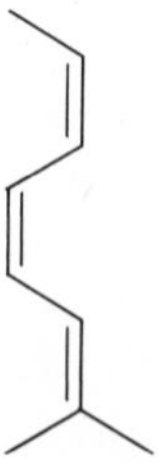

4-*cis*-6-*cis*-alloocimene

Myrcene is readily obtained on a large scale by the thermal cracking of β-pinene, a reaction which gives up to 90 % yield of a product which is almost entirely the β-isomer (Patton, 1950). Thermal cracking of α-pinene yields ocimene, almost exclusively the *cis*-β-isomer (Hawkins and Burnis, 1959) but under the conditions of reaction this readily isomerizes to alloocimene, giving a mixture of the two 4-*trans* isomers (Wolinsky *et al.*, 1962). On irradiation, this mixture isomerizes to yield all four isomers (Crowley, 1968). The β-ocimenes can be obtained by pyrolysis of linalyl acetate, the reaction taking place over a surface, permitting reaction conditions mild enough to prevent isomerization to alloocimene (Mitzner *et al.*, 1966a). The α-isomers of myrcene and the ocimenes have also been obtained from ester pyrolysis reactions (Mitzner *et al.*, 1965a; Ohloff *et al.*, 1964).

On heating, β-myrcene dimerizes to give α-camphorene (Goldblatt and Palkin, 1941), and *cis*-β-ocimene isomerizes by a 1,5-hydrogen migration to yield 4-*trans*-6-*cis*-alloocimene. (Sasaki *et al.*, 1971, Hawkins and Hunt, 1951). The *trans*-β-isomer of ocimene is, however, reported to be thermally stable (Wolinsky *et al.*, 1962). Alloocimene isomers are all reported to undergo a cyclization reaction on heating giving mainly β-pyronene (Wolinsky *et al.*, 1962), but as this is readily formed from one of the other products, α-pyronene, under reaction conditions, it may well be a secondary product. The pyrolysis of alloocimene also gives 4 % of α-terpinene (Parker and Goldblatt, 1950).

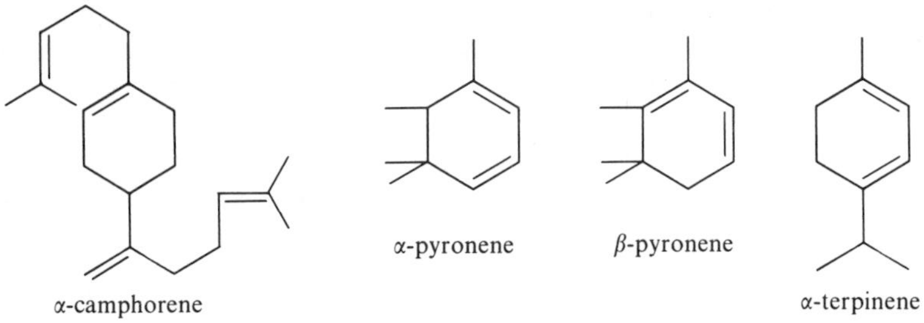

On irradiation, β-myrcene cyclizes to a mixture of the three products shown below (Dauben, 1966). This has been suggested as a biogenetic route to β-pinene (Crowley, 1962). In the presence of a photosensitizer, the bicyclo[2.1.1]hexane is the main product (Liu and Hammond, 1964).

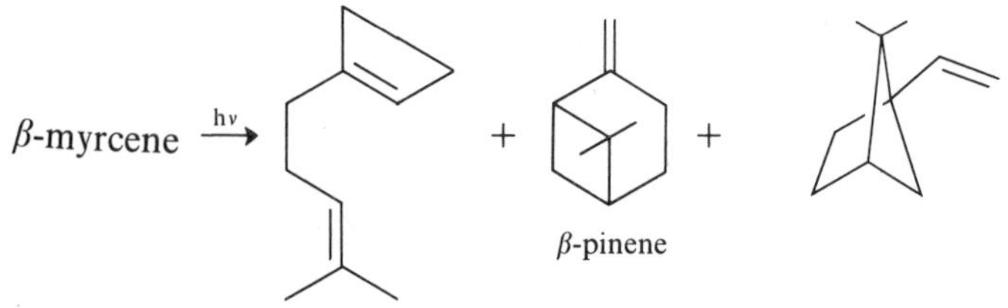

In contrast, the main photochemical reaction of *cis*-β-ocimene is formation of a mixture with the *trans* isomer (Frank, 1968). Alloocimene behaves similarly in initial formation of a mixture of the four possible isomers, but the mixture subsequently undergoes further reaction to give an allene, which readily rearranges to α-pyronene, together with β-pyronene and a bicyclic compound (Crowley, 1968).

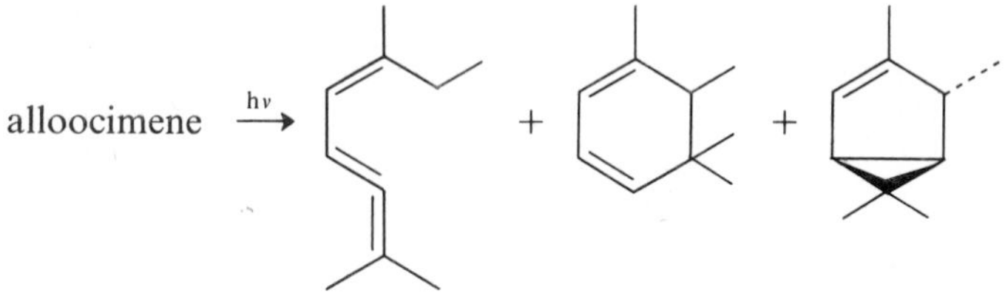

The addition reactions of ocimene and myrcene can take two different courses, depending on whether addition takes place across the conjugated double bonds, or across the isolated double bond. Thus, both β-ocimene (Enklaar, 1907) and β-myrcene (Semmler, 1901) on hydrogenation, yield dihydromyrcene by addition of hydrogen to the conjugated double bonds, while alloocimene yields a mixture of isomers. The addition of diisoamylborane to myrcene proceeds by a 1,2 addition to the conjugated diene system, giving a product which on oxidation followed by hydrolysis, gives myrcen-8-ol (Brown *et al.*, 1963). Addition of acetic acid to β-myrcene in

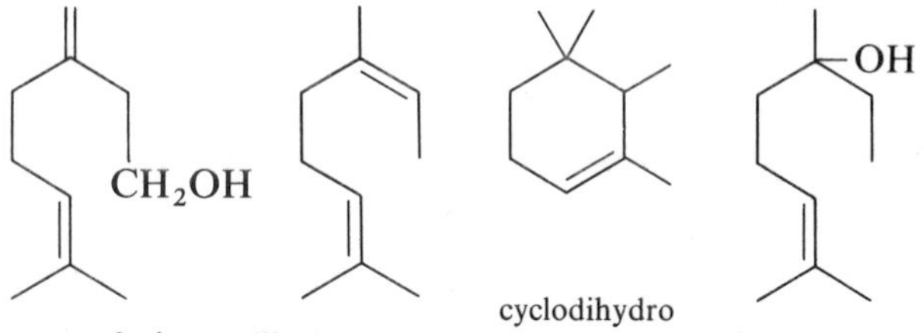

myrcen-8-ol dihydromyrcene cyclodihydro myrcene dihydrolinalol

the presence of sulphuric acid gives the acetate of myrcen-2-ol (Houlihan *et al.*, 1959), reaction occurring on the unconjugated double bond; the corresponding reaction of dihydromyrcene gives the cyclic product, cyclodihydromyrcene, although under mild conditions dihydrolinalyl acetate is formed (Semmler, 1894). Myrcene adds hydrogen chloride to give the chloride of myrcen-2-ol, together with the chlorides of nerol, geraniol and terpineol (Ansari, 1970). However, in the presence of cuprous ions, preferably in the form of cuprous chloride, hydrogen chloride can be added across the conjugated double bonds of myrcene to give a mixture of only geranyl chloride and neryl chloride. Reaction of this mixture of chlorides with sodium acetate in a unimolecular reaction gives linalyl acetate, while in the presence of triethylamine, a bimolecular reaction yields geranyl and neryl acetates. The process is of industrial importance, the three alcohols being of considerable use in perfumery, though the exact function of cuprous

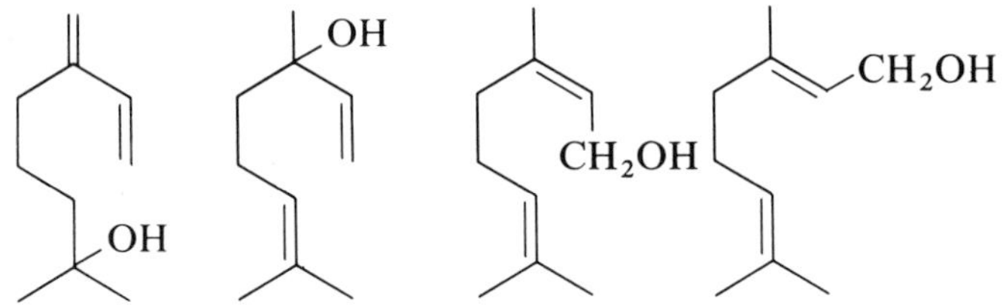

myrcen-2-ol linalol nerol geraniol

ions is not yet clear (Sully, 1964; Ansari, 1970). A similar reaction is undergone by α-myrcene, which adds hydrogen chloride in the presence of copper to give a mixture of the chlorides of α-nerol and α-geraniol. The chlorides, on treatment with sodium acetate give the acetates, which can be hydrolysed to the alcohols. There is no evidence that these α-isomers occur naturally (Mitzner *et al.*, 1966d).

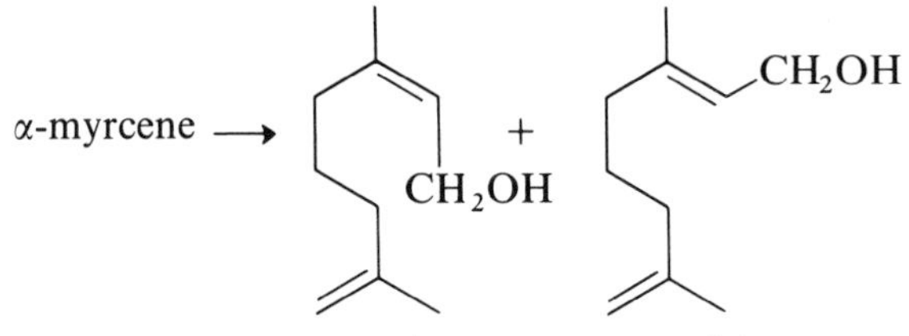

α-nerol α-geraniol

The three naturally-occurring acyclic allylic alcohols, linalol, geraniol and nerol, interconvert in acids via an allylic rearrangement. A kinetic study of the reaction has shown that geraniol and nerol react at approximately the same rate, but that geraniol is converted mainly to linalol, while nerol forms α-terpineol via a cyclization reaction approximately 18 times faster than does geraniol, no doubt because the stereochemistry of nerol is favourable to this cyclization (Valenzuela and Cori, 1967). When the hydroxyl group of geraniol is protected by, for example, acetylation, then geraniol cyclizes in acids by a reaction which probably involves protonation of one double bond followed by attack of the carbonium ion thus produced on the other double bond to give cyclogeranyl acetate, as a mixture of the α- and β-isomers (Smit *et al.*, 1962). There is no report of a similar reaction of nerol, though such a reaction would be expected to be slower on account of less favourable stereochemistry. The different carbon skeleton of the cyclization products formed under different conditions results from the tendency of the carbonium ion centre, generated in different places in the different reactions, to react to give

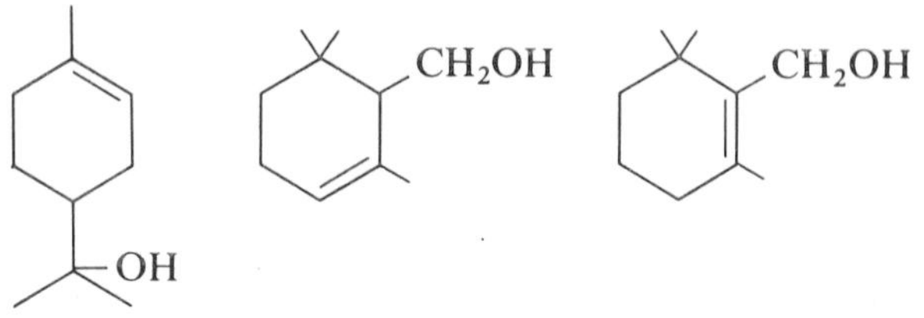

α-terpineol α-cyclogeraniol β-cyclogeraniol

the unstrained six membered ring. Geranyl acetate also undergoes a free radical oxidative cyclization in the presence of an initiator such as benzoyl peroxide (Breslow *et al.*, 1966).

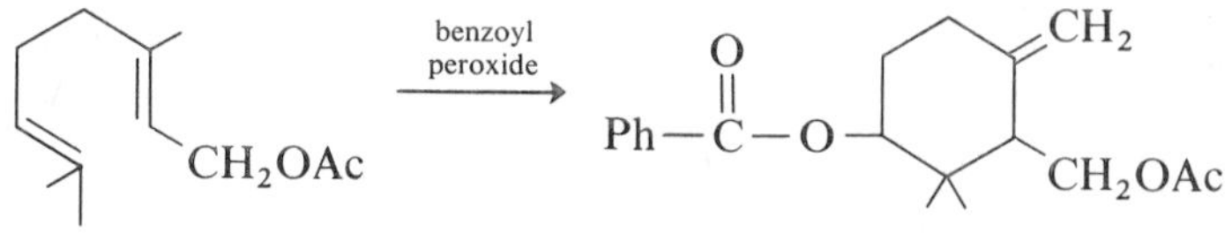

Linalol undergoes some cyclization during acid catalysed dehydration, but in this case, generation of a primary allylic carbonium ion favours cyclization to the *p*-menthane skeleton (Mitzner *et al.*, 1966a). Linalol also cyclizes thermally, in this case giving rise to the four isomeric plinols, which have a five-membered ring (Strickler *et al.*, 1967).

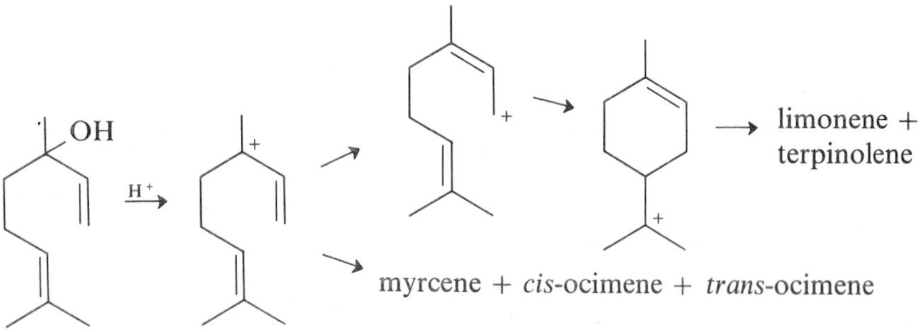

On partial hydrogenation of geraniol, the monounsaturated alcohol, citronellol, is obtained. This alcohol is of industrial importance because of its occurrence in rose oil, and strong rose smell. In the older literature, it was also called rhodinol, and this was at one time thought to be an isomer of citronellol. Industrially, it is prepared by hydration of citronellene, using the Zeigler process to obtain anti-Markovnikov addition (Sully, 1964). It does not cyclize readily in the presence of acids, but in the presence of an initiator such as di-tertiary butyl peroxide, it undergoes a free radical cyclization to give a mixture of the menthol stereoisomers (Van Bruggen, 1968). On oxidation, citronellol gives citronellal, also known as rhodinal.

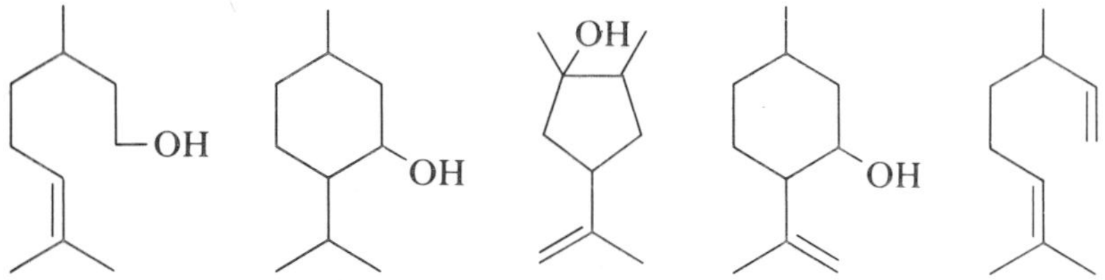

citronellol menthol plinol isopulegol citronellene

Citronellal cyclizes much more readily than its alcohol, giving isopulegol in good yield with sulphuric acid. This process can be reversed, pyrolysis of isopulegol giving citronellal (Sully, 1964).

Condensation of citronellal with acetone gives, via an aldol-type condensation, dihydropseudoionone, which readily cyclizes to dihydroionone, a substance with a fresh smell of flowers (Rupe, 1900). The corresponding

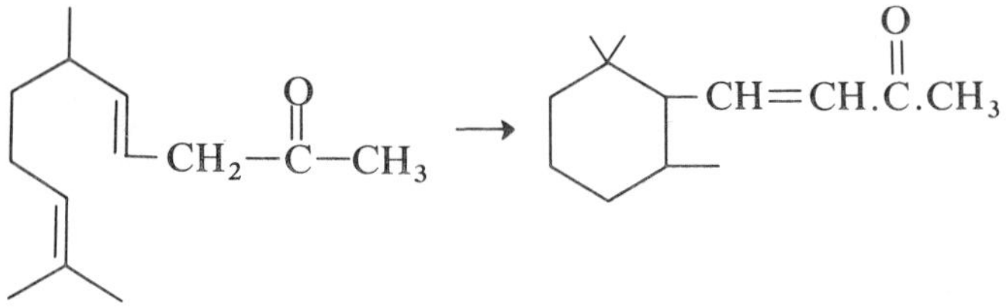

dihydropseudoionone dihydroionone

diunsaturated aldehydes, neral and geranial, are readily obtained by oxidation of the corresponding alcohols. They are difficult to separate, and their mixture was once thought to be a single substance, which was named citral. They are sometimes known as citral A and citral B. They show similar reactions to citronellal, cyclizing in the presence of aqueous acids to give 3,8-*p*-menthane diol (Whalley, 1958). They cannot be cyclized to the cyclocitrals, since they readily undergo dehydrative cyclization with strong acids to give *p*-cymene, though the cyclocitrals have been obtained by oxidation of the cyclogeraniols. Both isomers condense with acetone, giving pseudoionone A from citral A and pseudoionone B from citral B; the pseudoionones cyclize readily in acids, giving a mixture of the α- and β-ionones (Hibbert

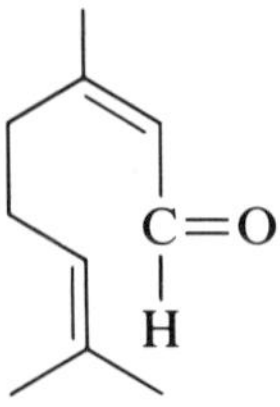

neral
(citral B)

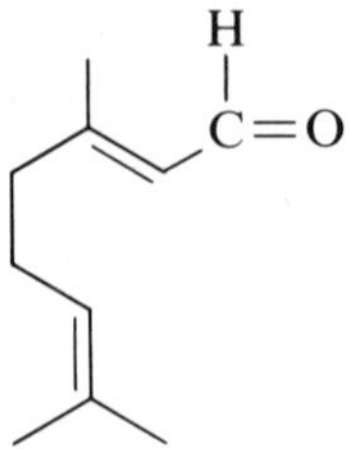

geranial
(citral A)

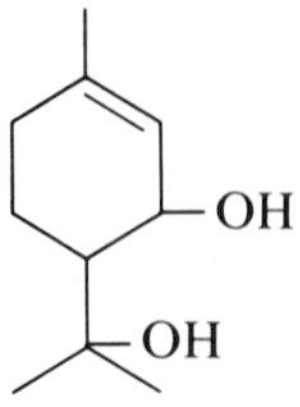

3,8-*p*-menthane-diol

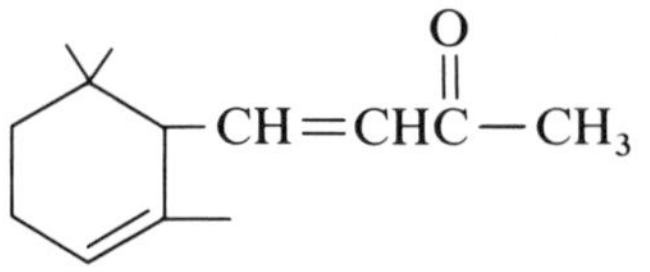

α-ionone

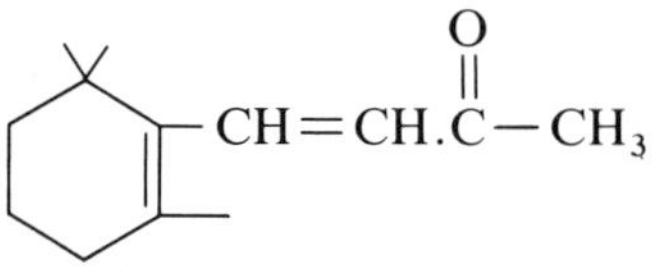

β-ionone

and Cannon, 1924). These compounds, in high dilution, have an odour resembling that of violets.

On irradiation, the mixture of the citral isomers cyclizes to a mixture of a monocyclic and a bicyclic product (Cookson *et al.*, 1963).

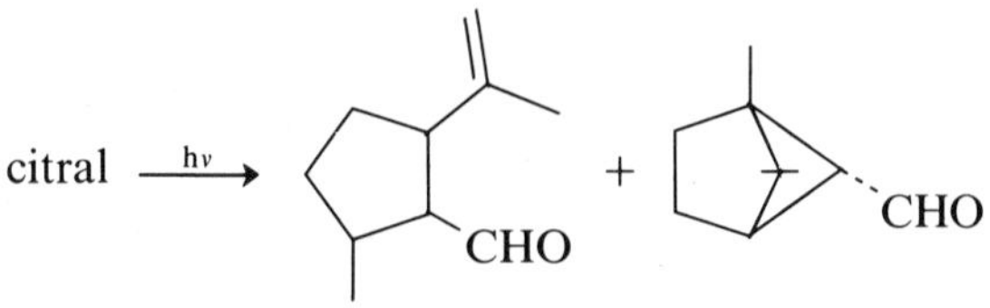

photocitral A photocitral B

III. Monocyclic Monoterpenes

The monocyclic monoterpenes may be regarded conformationally as disubstituted cyclohexanes in which the isopropyl substituent, being bulkier than the methyl, assumes the equatorial position. Saturated *p*-menthane derivatives should then have a chair conformation for the ring, which is consistent with the observations of reaction stereochemistry, except in the cases of the oxygen bridged rings in compounds such as the cineoles and ascaridole. Compounds having one double bond in the ring would be expected to have a half chair conformation (Barton *et al.*, 1954), while two double bonds within the ring would be expected to produce a nearly flat ring.

Substituents such as carbonyl groups or exocyclic double bonds could produce some flattening of the ring. To date, there have been no systematic n.m.r. studies of the monocyclic monoterpenes, so that these generalizations remain unproved. The infra-red spectra of many derivatives have been published (Mitzner *et al.*, 1965b, 1968, 1969).

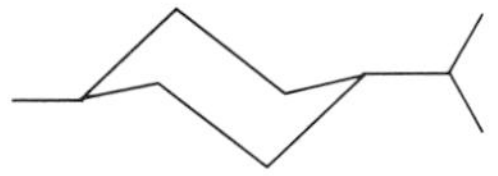

p-menthane

The numbering of the *p*-menthane ring follows the general rules of nomenclature,

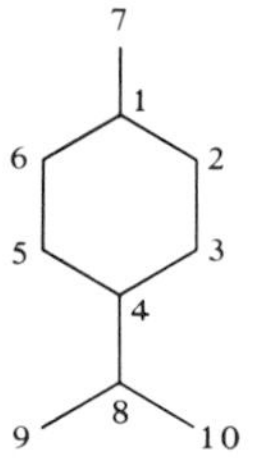

(VIII)

Fourteen dienes having the *p*-menthane skeleton can theoretically exist, and all have been synthesized. Only six definitely occur in nature, these being limonene (X), terpinolene (XI), α-terpinene (XIII), γ-terpinene (XII), α-phellandrene (XV) and β-phellandrene (XIV). All these six dienes can rearrange as shown in Fig. 1 in the presence of acids (Bardyshev, 1966), in the presence of bases (Bates *et al.*, 1969), or on surfaces such as silica gel (Hunter and Brogden, 1963) or by γ-irradiation (Buechi and Iconomou, 1965). α-Terpinene itself, on prolonged treatment with acid, is converted to iso-terpinolene (Williams and Whittaker, 1971a). This isomer, together with 3,8(9) *p*-menthadiene, is also obtained when limonene is treated with sodium metal in the presence of a promoter such as *o*-chlorotoluene, though on prolonged reaction both isomers are converted to *p*-cymene (Pines and Eschinazi, 1955). The interconversions of the *p*-menthadienes have recently been reviewed (Verghese, 1967) and several reviews on the chemistry of limonene also exist (Verghese, 1968, 1969).

The acid catalysed hydration of limonene proceeds mainly on the vinyl double bond, giving α-terpineol, though some reaction proceeds on the endocyclic double bond, giving β-terpineol (Williams and Whittaker, 1971a). The addition of HCl proceeds *via* the same route, though in this case the

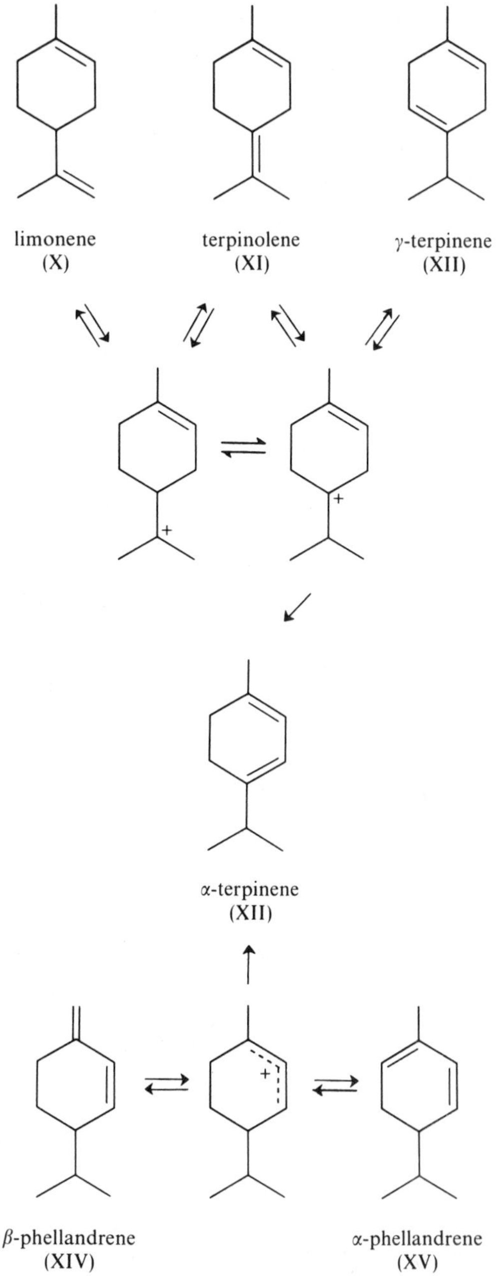

FIG. 1. Interconversions of *p*-menthadienes.

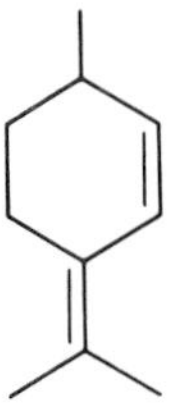

iso-terpinolene (XVI)

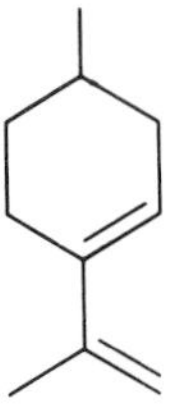

3,8(9) *p*-menthadiene (XVII)

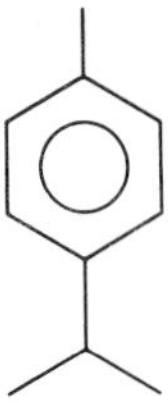

p-cymene

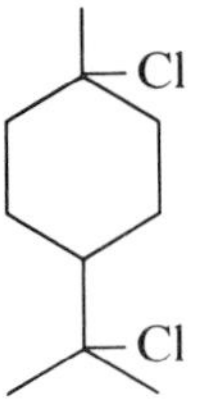

limonene dihydrochloride

dihydrochloride is formed readily. The phellandrenes behave in a different manner, as would be expected for a conjugated system, α-phellandrene reacting to give the same product as β-phellandrene (Schenck *et al.*, 1964). The chloride produced, on hydrolysis, gives a mixture of allylic alcohols which all give piperitone on oxidation with chromic acid/sulphuric acid, the tertiary alcohols presumably rearranging to piperitol under the influence of the sulphuric acid, then being oxidized. Although the *p*-menthane skeleton

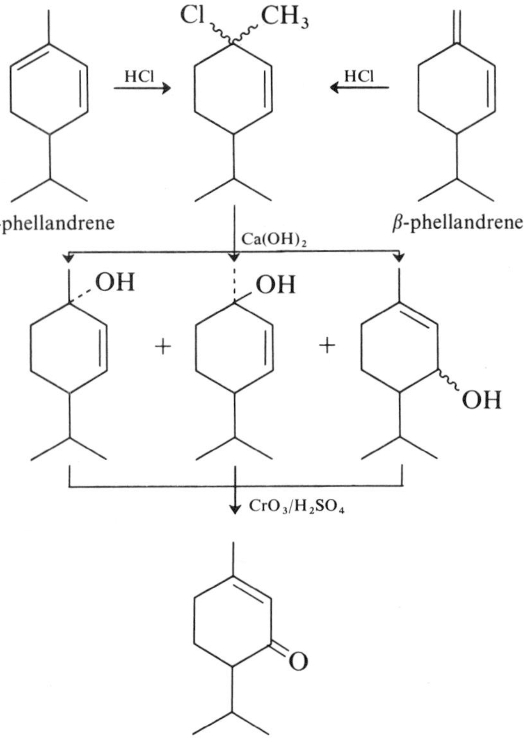

piperitone

is readily formed by cyclization of acyclic terpenes, and can be formed by ring opening of any of the five classes of bicyclic monoterpenes in ionic or pyrolytic reactions, these reactions are not readily reversed. The *p*-menthane skeleton appears to represent the most stable monoterpene structure. Ring fission is possible only in relatively high energy reactions, such as thermally, or on photochemical irradiation, and even then the product often recyclizes. Cyclization reactions take place only when the molecule contains a suitable activating group, such as a carbonyl function, or under very vigorous conditions, such as the cycloisomerization of limonene in the vapour phase at 200° over a silico-phosphoric acid catalyst. The product in this case is a dimethyl bicyclooctane, a system which does not occur naturally (Ipatieff *et al.*, 1947, 1952).

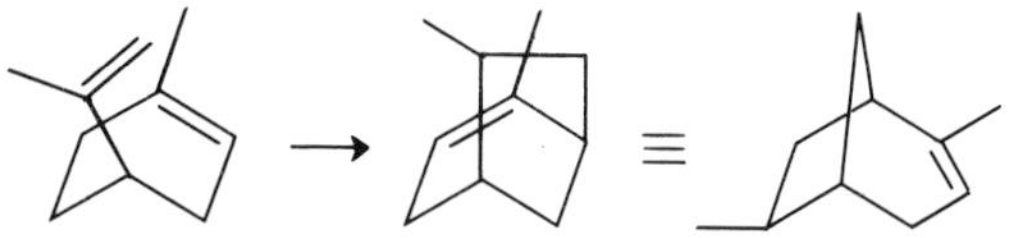

limonene

The photosensitized oxidation of limonene (Schenck *et al.*, 1964) proceeds normally to give a mixture of hydrogen peroxides which, on reduction, give the six alcohols shown below.

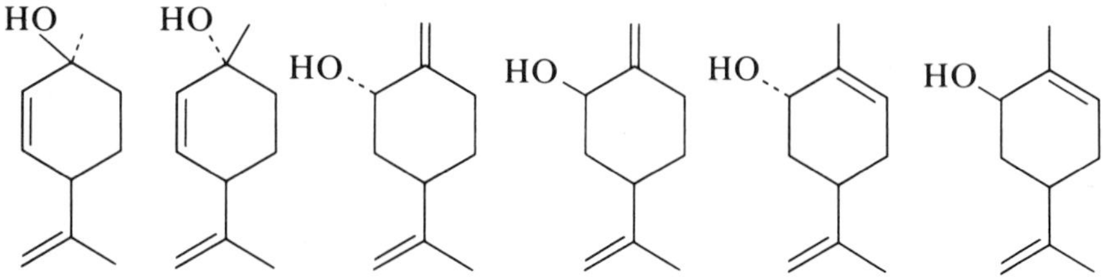

The formation of these alcohols is consistent with the general reaction scheme:–

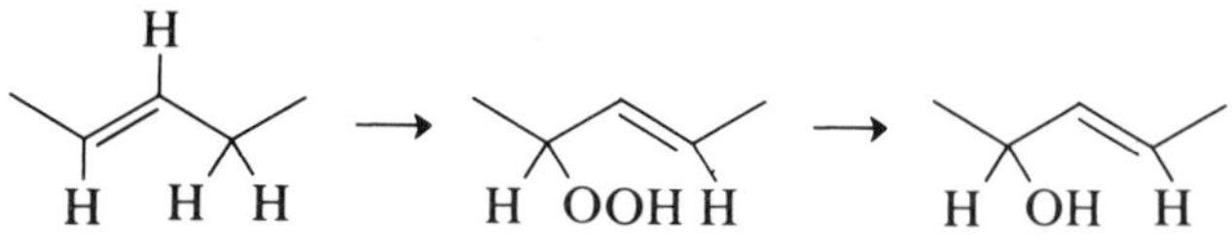

In the absence of a photosensitizer, the reaction follows a different course, in that the two alcohols having an *exo*-cyclic double bond are not formed, and the two carveols are obtained as racemates. It is suggested that this reaction proceeds via the allylic radicals,

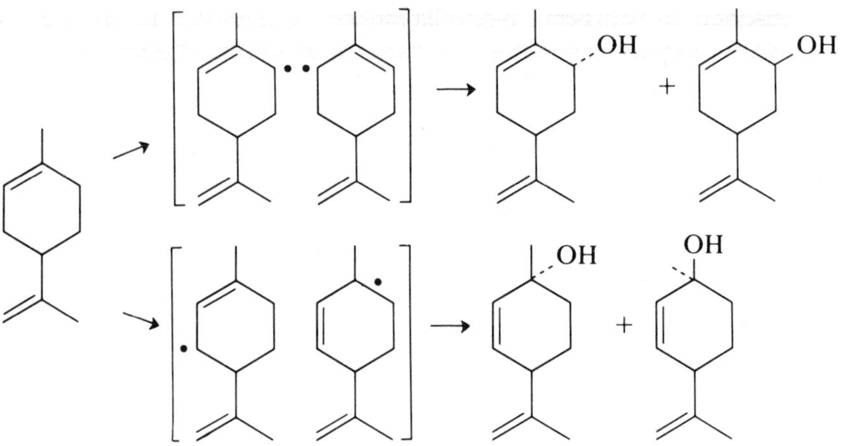

The conjugated dienes, on photosensitized oxidation, form epidioxides. In this way, ascaridole has been synthesized from α-terpinene (Schenck and Ziegler, 1944), while α-phellandrene yields phellandrene *trans*-epidioxide in a similar reaction (Schenck *et al.*, 1953b). Ascaridole also occurs naturally in oil of chenopodium, and is the only naturally-occurring terpene peroxide.

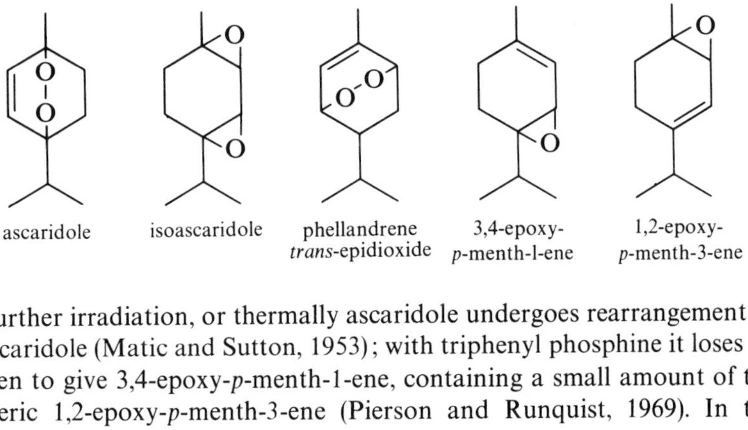

ascaridole | isoascaridole | phellandrene *trans*-epidioxide | 3,4-epoxy-*p*-menth-l-ene | 1,2-epoxy-*p*-menth-3-ene

On further irradiation, or thermally ascaridole undergoes rearrangement to isoascaridole (Matic and Sutton, 1953); with triphenyl phosphine it loses an oxygen to give 3,4-epoxy-*p*-menth-1-ene, containing a small amount of the isomeric 1,2-epoxy-*p*-menth-3-ene (Pierson and Runquist, 1969). In the

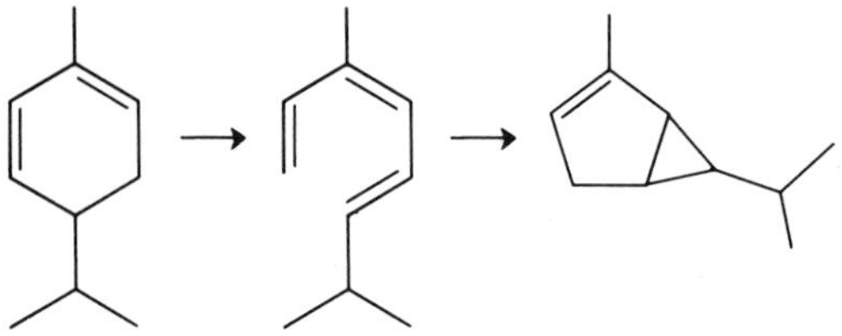

absence of oxygen, α-phellandrene is cleaved to an acyclic triene, which subsequently recyclizes to a bicyclic product (Meinwald *et al.*, 1965; Baldwin and Kreuger, 1969).

Hydration of the *p*-menthadienes gives monounsaturated alcohols, eight of which, known as the terpineols, can be obtained by dehydration of terpin hydrate. This process is of considerable industrial value, since terpin hydrate is readily obtained by hydration of α-pinene, and is relatively insoluble in organic solvents, so that it can be separated from the bornyl and fenchyl impurities which this process produces (Watson and Gladden, 1964). Any one of the two pairs of diols, the 1,8- and 1,4-terpins could serve as starting material, since they interconvert as shown in Fig. 2. Under dehydrating conditions, the reactions shown in Figs. 2 and 3 proceed to give a mixture of alcohols and cineoles, the major product being α-terpineol. The exocyclic double bond isomers, δ-terpineol and 1(7)-terpinene-4-ol occur to less than 0·5 % each (Mitzner *et al.*, 1966c). A similar equilibrium mixture can be obtained by treating α-terpineol with dilute acid, the reaction presumably proceeding via dienes or diols.

On dehydration with acids (Von Rudloff, 1961) α-terpineol forms a mixture of 1;8-cineole plus dienes consistent with formation of the ions,

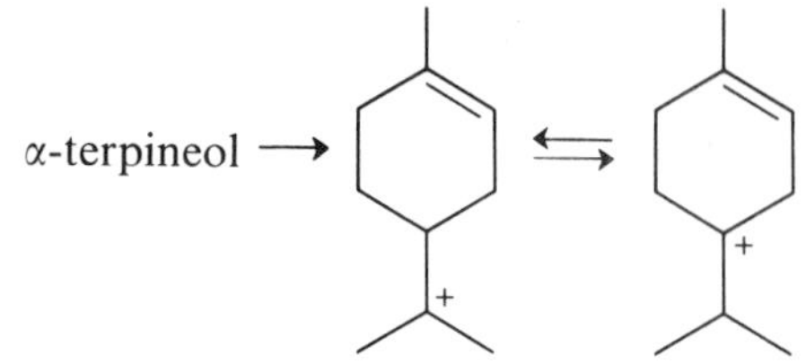

The other unsaturated alcohols are relatively uncommon, except for the isopulegols, which can be obtained either from the cyclization of β-citronellal to give isopulegol itself, or by reduction of *iso*-isopulegone. All four possible isomers have been characterized (Schulte-Elte and Ohloff, 1967).

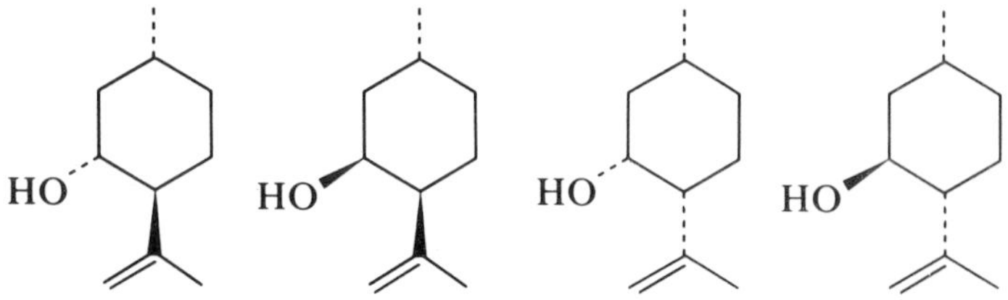

isopulegol neo-isopulegol neo-iso-isopulegol iso-isopulegol

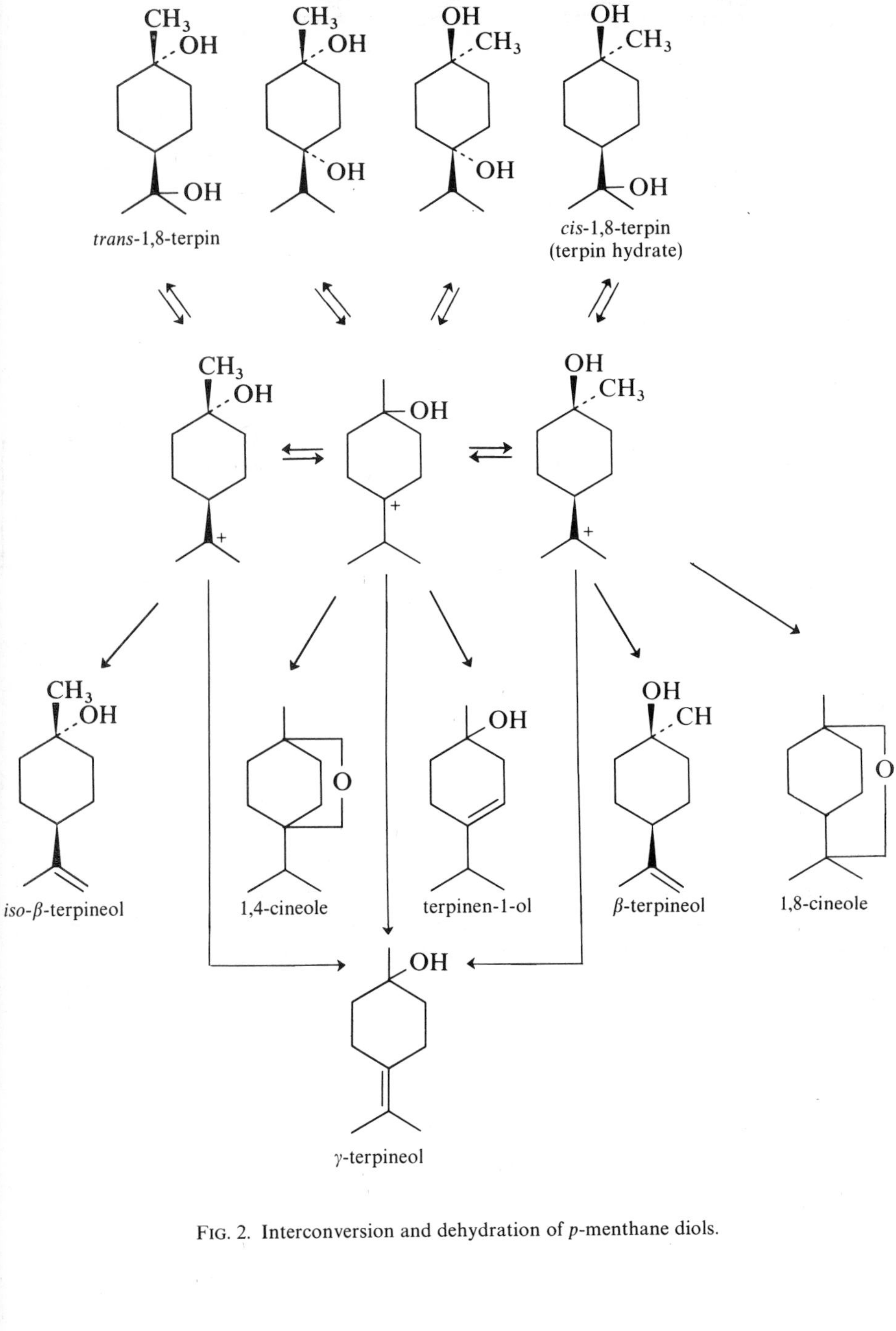

FIG. 2. Interconversion and dehydration of *p*-menthane diols.

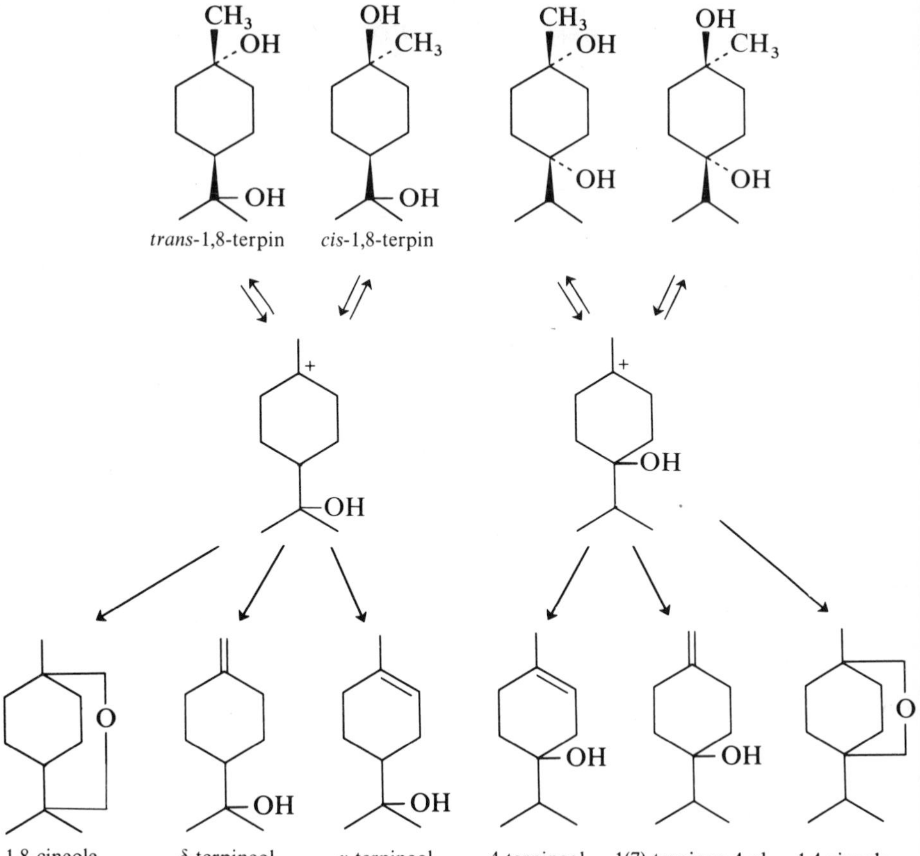

FIG. 3. Dehydration of *p*-menthanediols.

The most common diunsaturated ketone, carvone, occurs naturally in many essential oils, being the chief ketonic constituent of dill and caraway oils. With hydrogen bromide, it gives the hydrobromide which, on treatment with alcoholic potash, cyclizes to give the ketone, carenone, which rearranges to eucarvone (van Tamelen *et al.*, 1957).

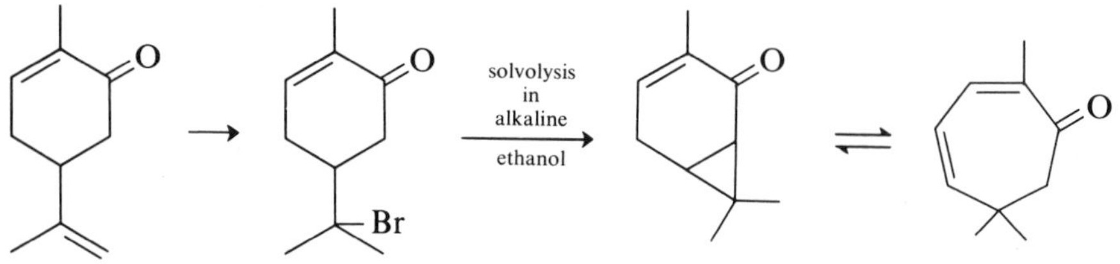

On irradiation of carvone, in alcoholic solution, with either sunlight or radiation from an ultra-violet blacklight tube, it undergoes the unusual cyclization reaction to give carvonecamphor. This reaction was first observed in 1908 (Ciamician and Silber, 1908) when the correct structure of the product was put forward. More recently, carvone camphor has been shown to rearrange on alumina to give isocarvone camphor (Buchi and Goldman, 1957), and to decompose on irradiation with higher frequency radiation to give the ketene, which subsequently reacts with solvent, giving an acid in the presence of water or an ester in the presence of alcohols (Meinwald, 1965).

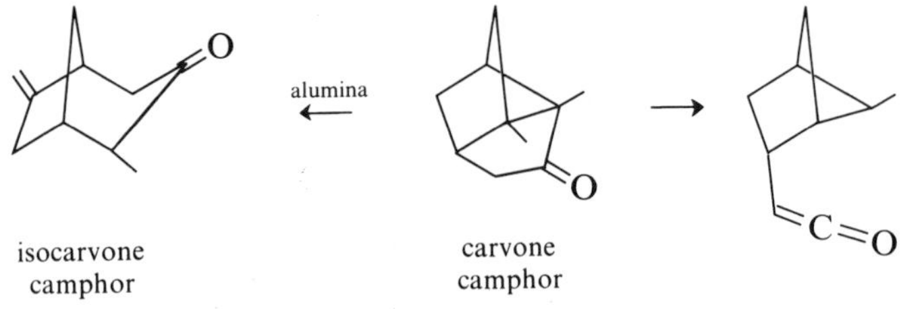

Two other diolefinic ketones exist, piperitenone and isopiperitenone. The alcohol derived from the latter is a compound of some industrial importance, being obtained by pyrolysis of *trans*-verbenol, itself obtained by oxidation of α-pinene, and being hydrogenated to give isomenthol (Sully, 1964). The ketone can be obtained, together with carvone by oxidation of limonene with a chromium trioxide–pyridine complex (Dauben *et al.*, 1969).

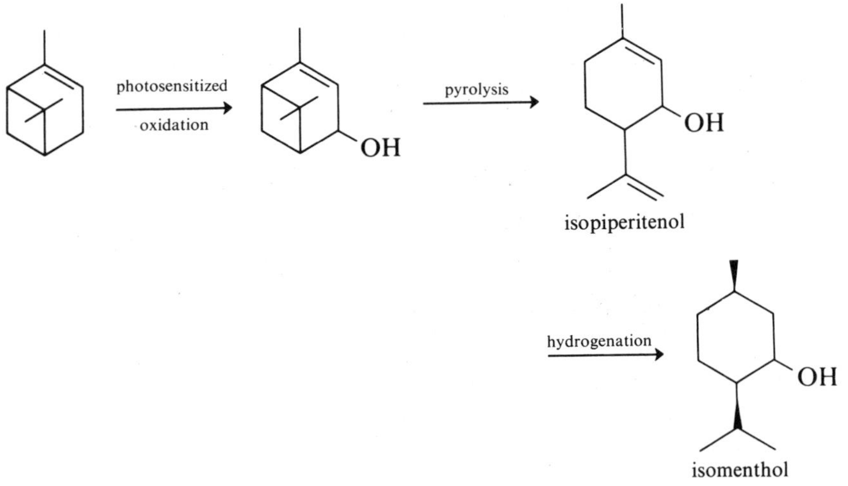

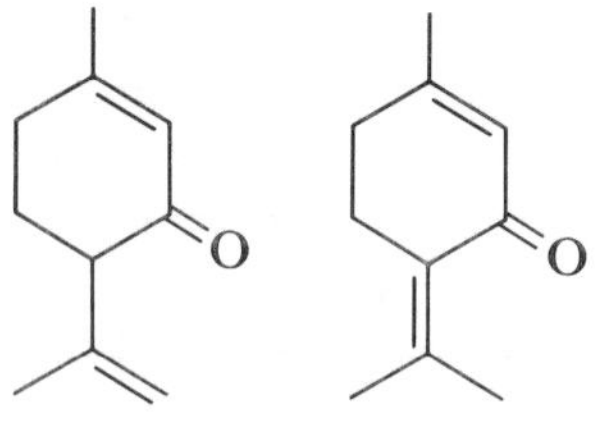

isopiperitenone piperitenone

Recently, isopiperitenone has been shown to undergo, on irradiation, a similar reaction to that of carvone to give 1,2-dimethyl bicyclo[3.3.0.0.^{2,7}]-octan-6-one. This compound has also been obtained by the irradiation of verbenone, so the authors suggest that isopiperitenone may be an intermediate on this route (Erman and Gibson, 1969).

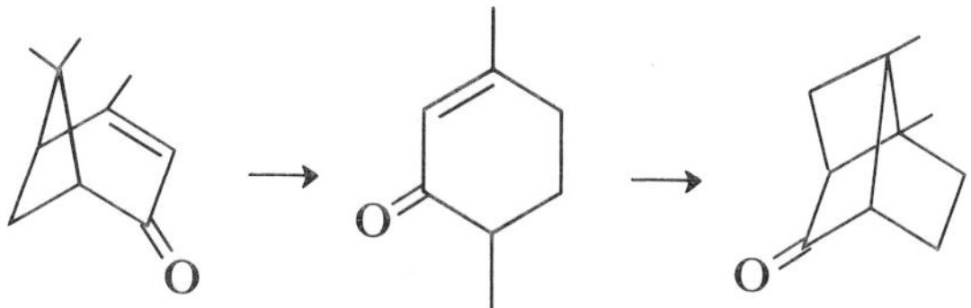

Verbenone isopiperitenone 1,2-dimethyl bicyclo (3.3.0.0.^{2,7}) octan-6-one

Investigation of the monounsaturated ketones appears to be governed mainly by their availability, the three best known being pulegone, isopulegone and piperitone, all of which have the oxygen function at C_3.

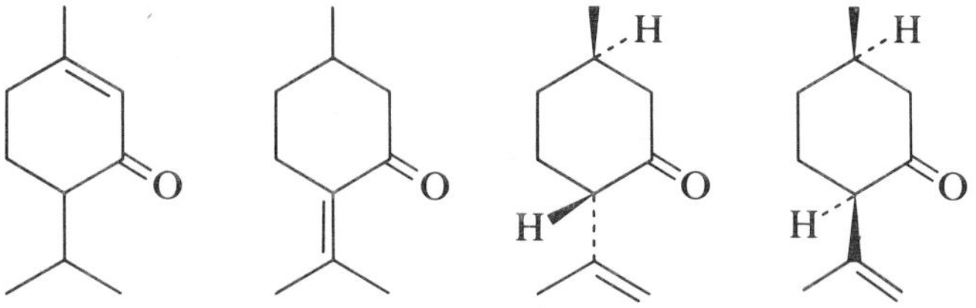

piperitone pulegone isopulegone iso-isopulegone

Piperitone is made from α-phellandrene by the route described earlier, pulegone is the main component of pennyroyal oil, and isopulegone is made by cyclization of citronellal. The other isomer of isopulegone is prepared by the thermal rearrangement of pulegone (Ohloff *et al.*, 1962).

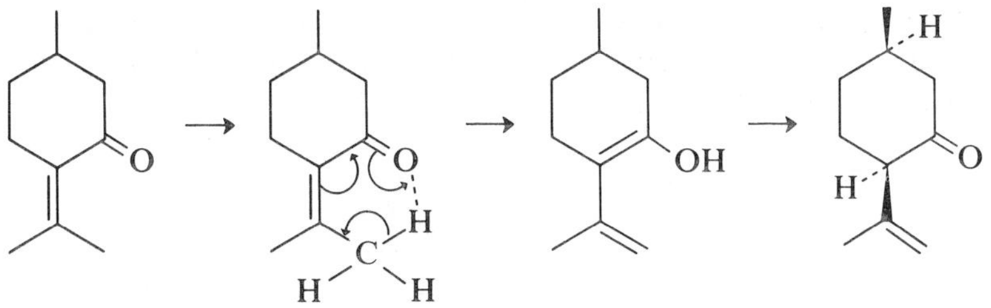

Pulegone oxide undergoes unusual rearrangements on heating, reaction in the gas phase occurring on surfaces to give a ring expanded product (Rousch *et al.*, 1968).

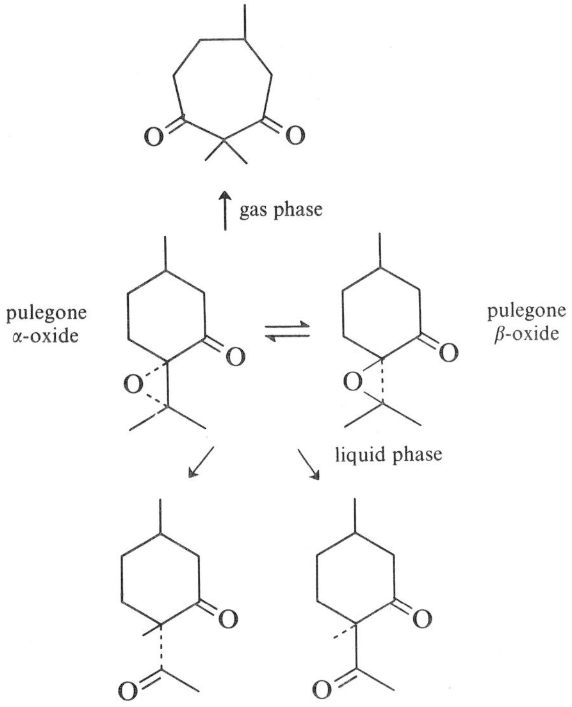

The interconversion of the α- and β-oxides, plus conversion to the products of liquid phase thermolysis, also takes place on irradiation (Johnson *et al.*, 1963).

Saturated ketones exist with the oxygen function on C_2, the carvomenthones, or on C_3, the menthones. The carvomenthones have been relatively little studied, but the menthone to isomenthone interconversion, by removal and replacement of the hydrogen on C_4, has been used as a convenient vehicle for studies of the effect of replacement of this hydrogen by deuterium or tritium (Bell and Cox, 1970).

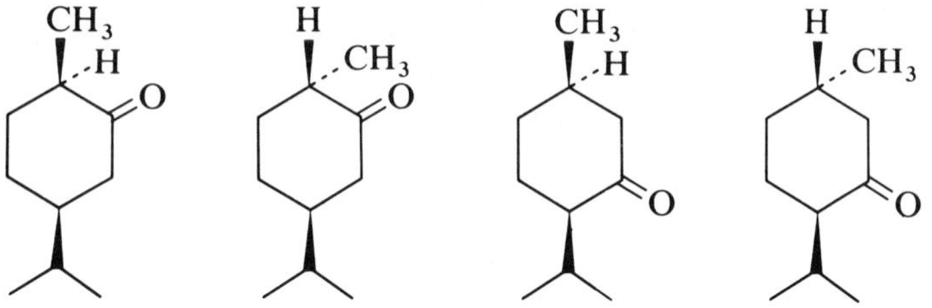

isocarvomenthone carvomenthone isomenthone menthone

Reduction of each of the carvomenthones gives rise to two alcohols, so that four carvomenthols exist in all. The stereochemistry of these isomers has been clarified only recently (Schroeter and Eliel, 1964), so that they have been used in relatively little mechanistic work.

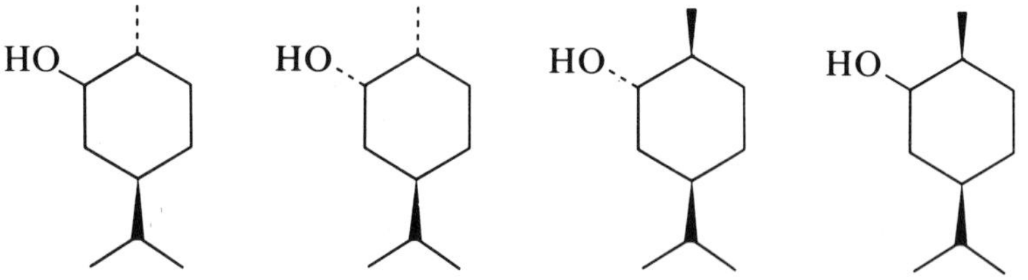

carvomenthol neocarvomenthol isocarvomenthol neoisocarvomenthol

In contrast, the menthols have been used extensively in work on the influence of conformation on the direction of elimination. The four menthols have been assigned the stereochemistry

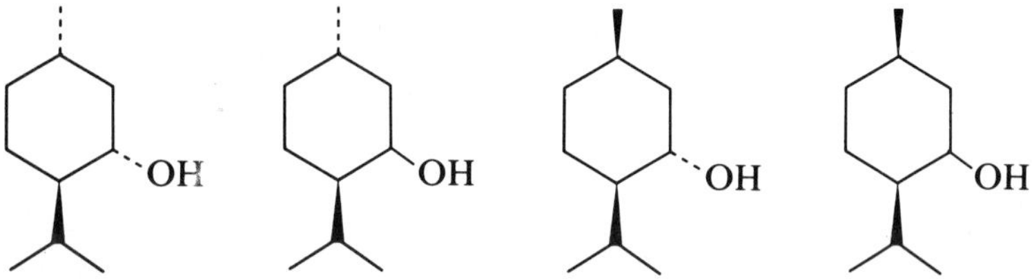

menthol neomenthol isomenthol neoisomenthol

In these structures of restricted geometry, the mechanism of the reaction is often framed to fit the structure, so that the need for *trans* diaxial orientation of the atoms eliminated in an E2 elimination can overrule the laws governing the direction of elimination derived by Hofmann and Saytzeff from consideration of acyclic systems. Thus, neomenthyl chloride takes up the conformation in which the methyl and isopropyl groups are equatorial and the chlorine is axial. Elimination of HCl can then take place to give either the menth-2-ene or the menth-3-ene; since both products are permitted sterically the reaction follows the Saytzeff rule, and gives mainly menth-3-ene (Hughes *et al.*, 1953).

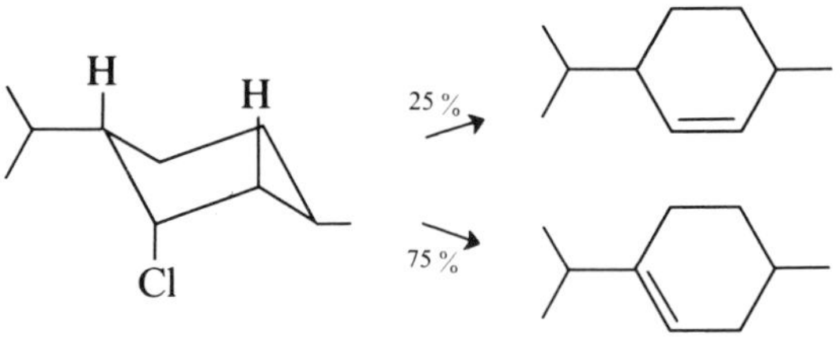

Menthyl chloride, however, has an equatorial chlorine atom, so that reaction can take place only if the molecule inverts itself into the less stable conformation, with all groups axial. For this reason the reaction is much slower than that of the neomenthyl isomer, and since, in the inverted form, C_2 has an axial hydrogen while C_4 does not, reaction takes place entirely by elimination to give the menth-2-ene. In this case, reaction goes 100% in the direction opposite to that predicted from consideration of acyclic systems.

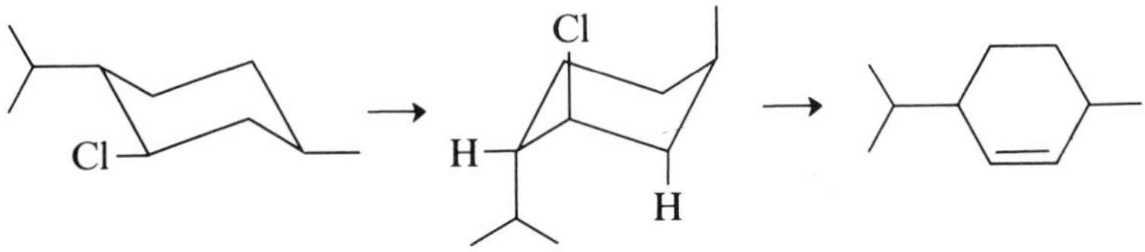

For a full account of the eliminations from menthyl derivatives, specialized texts should be consulted (Banthorpe, 1963; Banthorpe *et al.*, 1967).

IV. Bicyclic Monoterpenes

A. Carane Series

The carane skeleton undergoes a very wide range of rearrangements, since it possesses a structure which is relatively unsymmetrical, and can react by opening the cyclopropane ring in addition to reactions taking place at any other functional groups. The parent hydrocarbons, *cis*- and *trans*-carane, do not occur in nature, and are not readily prepared by hydrogenation of the olefins. The best route to *cis*-carane appears to be by hydroboration of car-3-ene to give *cis*-caran-*trans*-4-ol, conversion of the alcohol to its *p*-toluenesulphonate, followed by reduction with lithium aluminium hydride to give *cis*-carane. The *trans* isomer is prepared by Wolff–Kishner reduction of 2-caranone, itself prepared from carvone (Cocker *et al.*, 1966b).

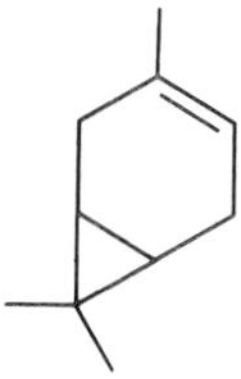

car-3-ene

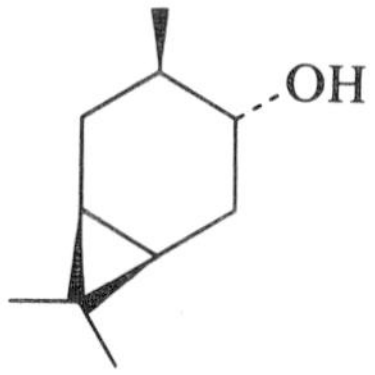

cis-caran-*trans*-4-ol

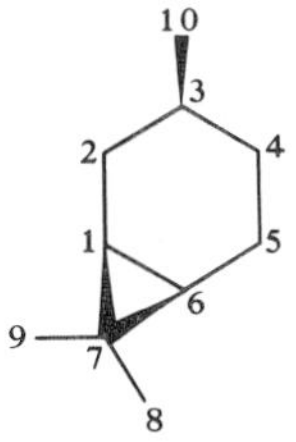

cis-carane

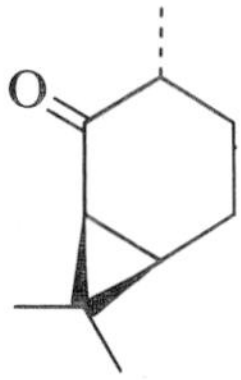

2-caranone

The two hydrocarbons are similar in spectroscopic properties, and very difficult to separate. It is suggested that they probably have the conformations

cis-carane

trans-carane

The caranes can undergo typical reactions of the cyclopropane ring, so that they add hydrogen chloride to give a mixture of chlorides by opening the ring in two possible ways (Borowiecki *et al.*, 1964):

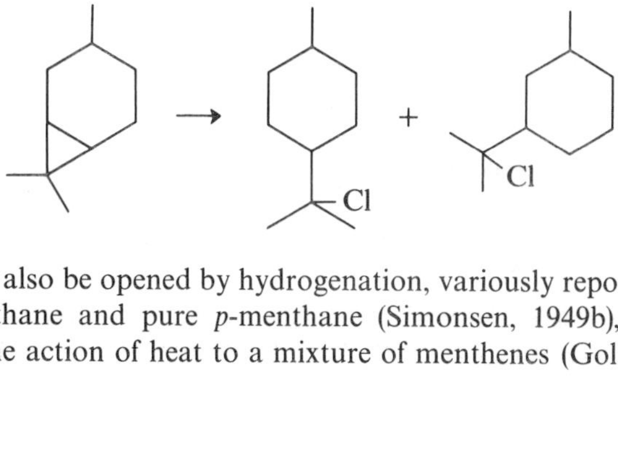

The ring can also be opened by hydrogenation, variously reported as giving pure *m*-menthane and pure *p*-menthane (Simonsen, 1949b), and can be opened by the action of heat to a mixture of menthenes (Gollnik, 1966b).

Five carenes can theoretically exist, and all are known. They are numbered in the normal manner, following the numbering of the carane ring, but an older numbering system exists still, and these names are given below, together with those generally used. The most common carene is car-3-ene, which occurs in a wide variety of plants, and is readily obtained by distillation from most turpentine oils. A review of its chemistry was published in 1965, but much new information has been obtained since then (Verghese, 1965).

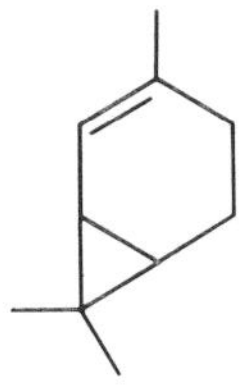

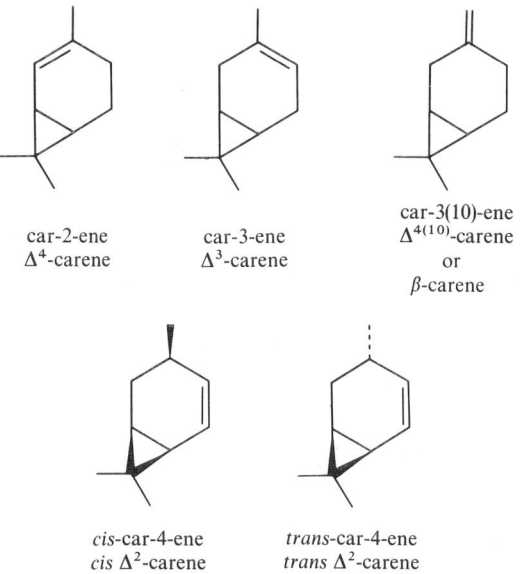

car-2-ene
Δ^4-carene

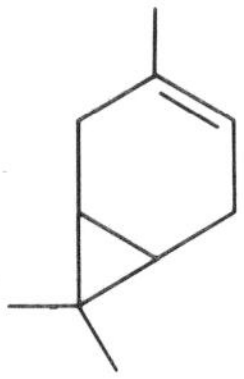

car-3-ene
Δ^3-carene

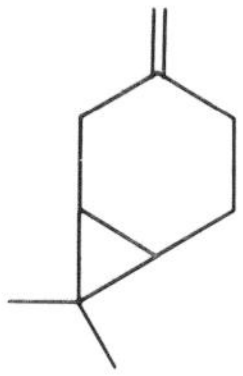

car-3(10)-ene
$\Delta^{4(10)}$-carene
or
β-carene

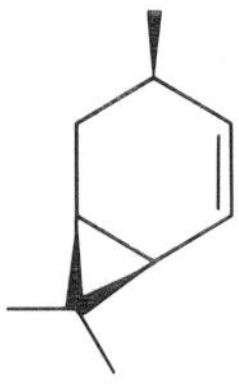

cis-car-4-ene
cis Δ^2-carene

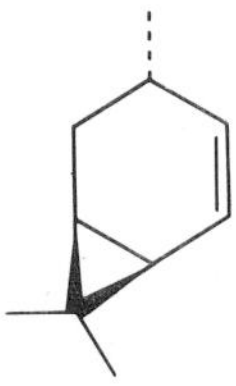

trans-car-4-ene
trans Δ^2-carene

Car-3-ene is believed to exist as a mixture of conformations (Cocker *et al.*, 1967a; Acharya and Brown, 1967):

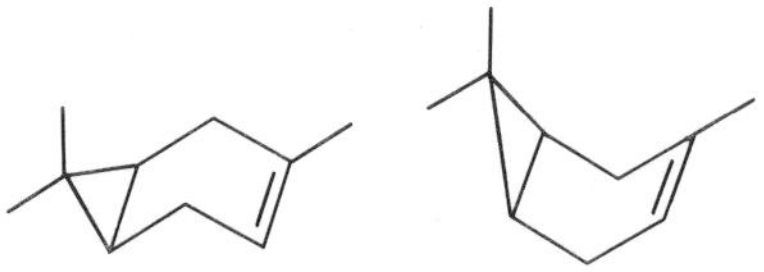

Both conformations favour addition reactions taking place by attack from the opposite side of the molecule to the *gem*-dimethyl group, so that hydrogenation gives *cis*-carane (Cocker *et al.*, 1966a). However, this reaction is less simple than would be expected, and has been shown to take place by preliminary equilibration to a mixture of 60 % car-3-ene and 40 % car-2-ene, together with traces of two cycloheptenes. The final products of hydrogenation are *cis*-carane and 1,1,4-trimethylcycloheptane. The authors suggest that hydrogen adds 1,4 to the conjugated double bond and cyclopropane ring of car-2-ene, giving 1,1,4-trimethylcyclohept-2-ene, which isomerizes to the observed -3-ene and -4-ene isomers.

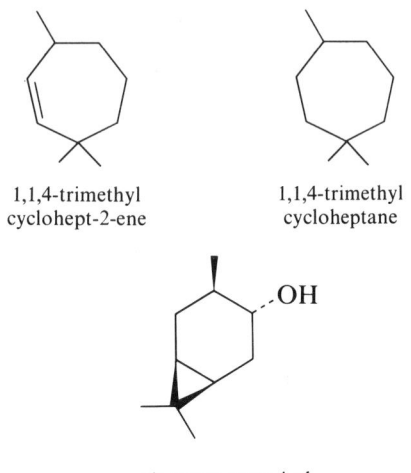

1,1,4-trimethyl cyclohept-2-ene

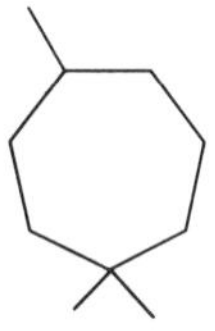
1,1,4-trimethyl cycloheptane

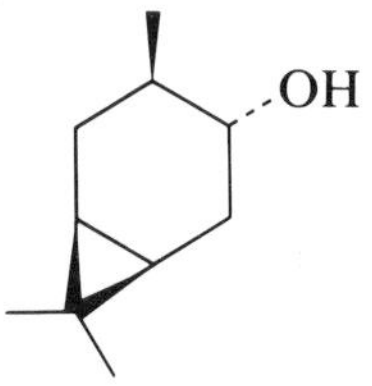

cis-caran-*trans*-4-ol

The hydroboration of car-3-ene proceeds normally to give the alcohol, *cis*-caran-*trans*-4-ol, which would be expected from addition of the borane to the less hindered side of the molecule (Cocker *et al.*, 1967a; Brown and Suzuki, 1967). If the borane complex is heated, rearrangement occurs to give a mixture of the five possible boranes which could be derived from the five carenes, plus a further borane obtained by opening the cyclopropane ring (Cocker *et al.*, 1969b). These boranes are believed to arise from shifts of the double bond resulting from successive addition and elimination of the borane.

Car-3-ene rearranges readily to an equilibrium mixture with car-2-ene, either during hydrogenation, on treatment with acid (Rudakov and Marchevskii, 1953), or with base (Ohloff *et al.*, 1965; Acharya and Brown, 1967). On irradiation in the presence of an aromatic hydrocarbon sensitizer, such as benzene, toluene, or xylene, it gives car-3(10)-ene (Kropp, 1966a), but no reports of isomerization to the other carenes have appeared. More vigorous treatment with sulphuric acid results in isomerization to dipentene, γ-terpinene and sylvestrene (Rudakov and Koratov, 1937) while the addition of hydrogen chloride opens the ring in either of two ways, and, if present in excess, subsequently adds to the double bond (Gollnik, 1966b). On surfaces, the cyclopropane ring is readily opened, so that with activated clay (Rudakov and Artamenov, 1945) car-3-ene yields α-terpinene, dipentene, terpinolene

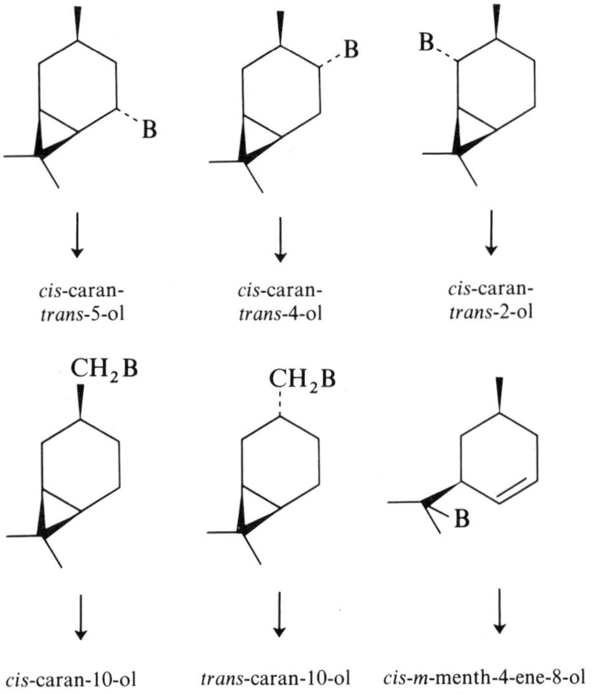

and sylvestrene, and with silica gel gives dipentene, and terpinolene (Arbuzov and Isaeva, 1953).

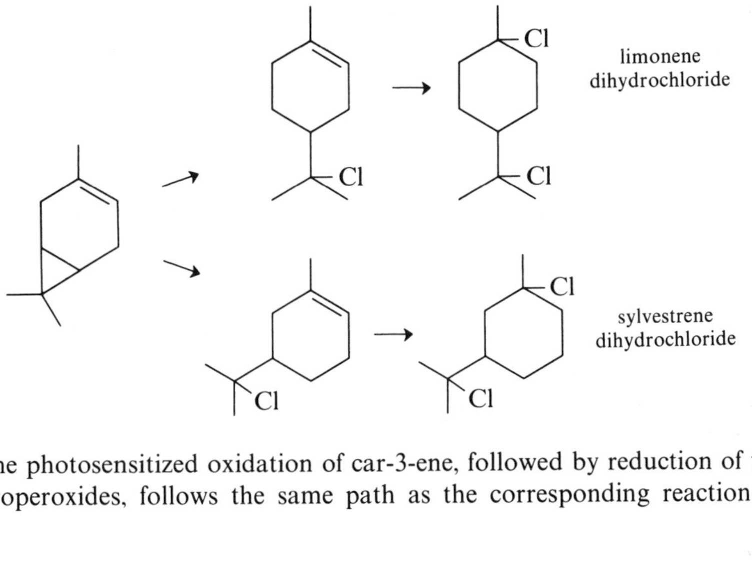

The photosensitized oxidation of car-3-ene, followed by reduction of the hydroperoxides, follows the same path as the corresponding reaction of limonene, giving the expected mixture of unsaturated alcohols, and has proved a convenient method of synthesis (Gollnik *et al.*, 1965).

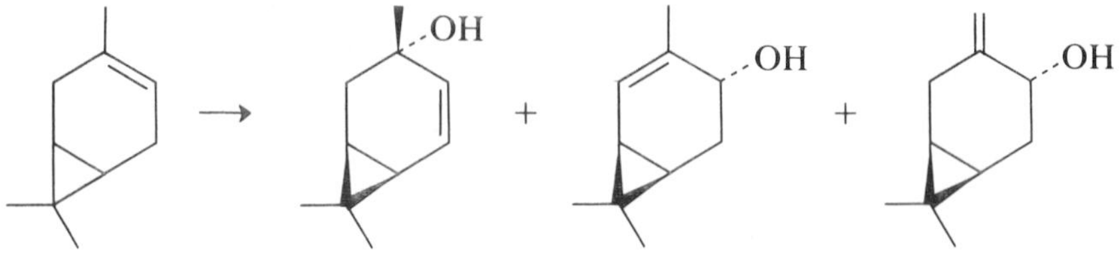

It is notable that the conformation *of* the carene again favours attack by the reagent on the face away from the *gem*-dimethyl group. This situation also prevails when the ring is attacked by electrophiles, as in the Prins and Friedel–Crafts acetylation reactions (Kropp *et al.*, 1968).

Car-3-ene shows much greater thermal stability than α-pinene, needing a temperature about 200° higher to promote thermal decomposition. It has been suggested that this is because the radicals which would be produced by ring opening would not be in a position to be stabilized by the double bonds, in contrast to those produced by α-pinene. Cracking of either of the bonds allylic to the double bond is also inhibited by their pseudo-vinylic character, produced by the cyclopropane ring (Cocker, 1968b).

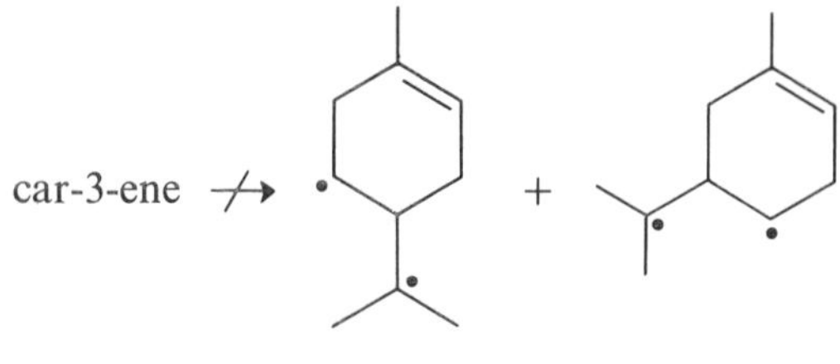

When pyrolysis does take place, the conditions are sufficiently vigorous to preclude the formation of any but aromatic compounds.

Car-2-ene is best prepared by isomerization of car-3-ene, basic conditions being reported (Acharya and Brown, 1967) to minimize side reactions. Separation is then possible by either distillation or chromatography (Simonsen, 1949a). It can also be obtained by pyrolysis of the phenylurethane (Piatkowski *et al.*, 1966) acetate or benzoate (Gollnik and Schade, 1966a) of either *cis*- or *trans*-3-caranol. These latter methods give a product contaminated with *trans*-2,8(9)-menthadiene, which is readily produced by pyrolysis of car-2-ene, a reaction which is suggested to go via a cyclic transition state.

The conformation in the following diagram is that suggested by Acharya and Brown (1967) pointing out that this conformation appears to offer the least steric strain, though it minimized conjugation between the cyclopropane

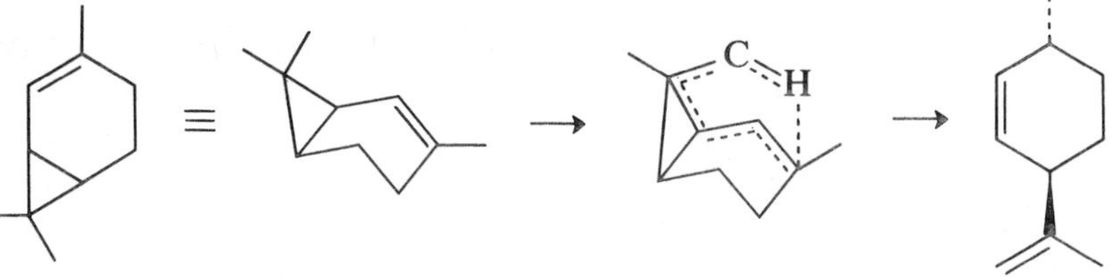

ring and the double bond, since the π-orbitals of the double bond now make an angle of 15° with the plane of the cyclopropane ring. This suggestion of diminished conjugation is also consistent with the equilibrium mixture of the two olefins favouring the unconjugated over the conjugated.

The conformation of car-2-ene, like that of the 3-isomer, favours attack on the double bond from the side opposite to the *gem*-dimethyl bridge, so that hydroboronation yields *cis*-caran-*trans*-2-ol (Cocker *et al.*, 1967a; Acharya and Brown, 1967).

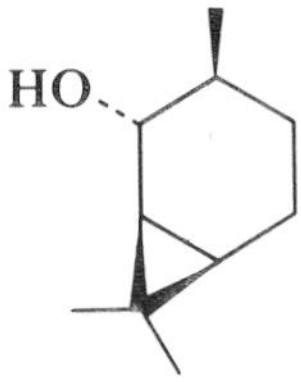

On photosensitized oxidation the same steric control of direction of attack dominates the reaction (Gollnik and Schade, 1966c). The exocyclic double bond isomer, car-3(10)-ene, has been relatively little studied, since it is not readily accessible. It has been reported as a product of irradiation of car-3-ene. and as a result of dehydration of the 3-caranols, the latter reaction yielding only around 10% of this isomer. It is probably best obtained from one of the products of photosensitized oxidation of car-3-ene, by chromic oxidation (presumably following acid catalysed rearrangement of car-3-ene-10-ol) and subsequent reduction of the aldehyde (Gollnik and Schade, 1966b).

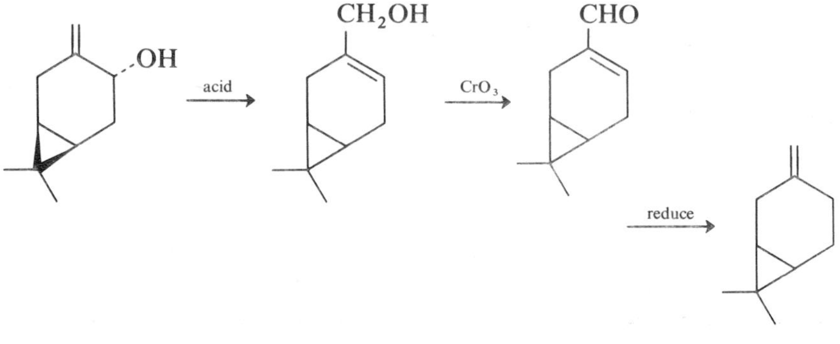

The car-4-enes can both be obtained by pyrolysis of the appropriate xanthate esters, *cis*-caran-*trans*-4-ol giving the *cis* isomer (Gollnik, 1966a) while the *trans* isomer is obtained from *trans*-caran-*trans*-4-ol. When prepared this way, the *trans* isomer undergoes some thermal decomposition to *m*-mentha-3(8),4-diene in a reaction analogous to the ring opening of car-2-ene, from which it differs only in the position of the methyl group.

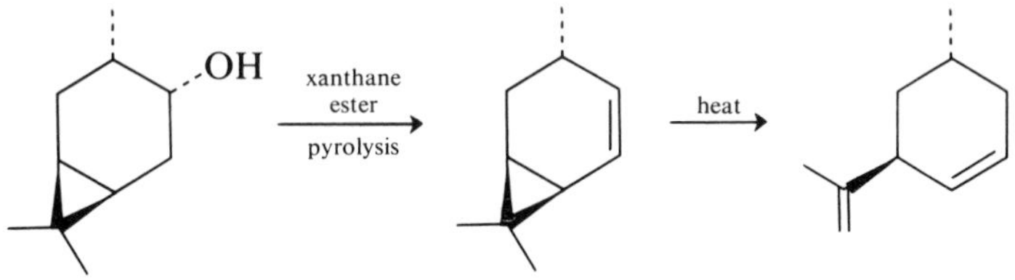

Recently (Carson *et al.*, 1970), the *cis* isomer has been prepared from *cis*-caran-4-one by treatment of the toluenesulphonylhydrazone with methyl lithium. With *cis*-caran-5-one, *cis*-car-4-ene is obtained, together with some *m*-mentha-3(8)4-diene.

The *cis* isomer, on hydroboration, gives a mixture of the isomeric alcohols, *cis*-caran-*trans*-4-ol and *cis*-caran-*trans*-5-ol, the latter being favoured by a factor of four.

The chemistry of the carenones is dominated by their ring expansion to cycloheptane derivatives. This expansion was first discovered when the cyclization of carvone by solvolysis of its hydrobromide gave not car-3-ene-2-one but eucarvone (Baeyer, 1894, 1898). Investigation of the chemistry of

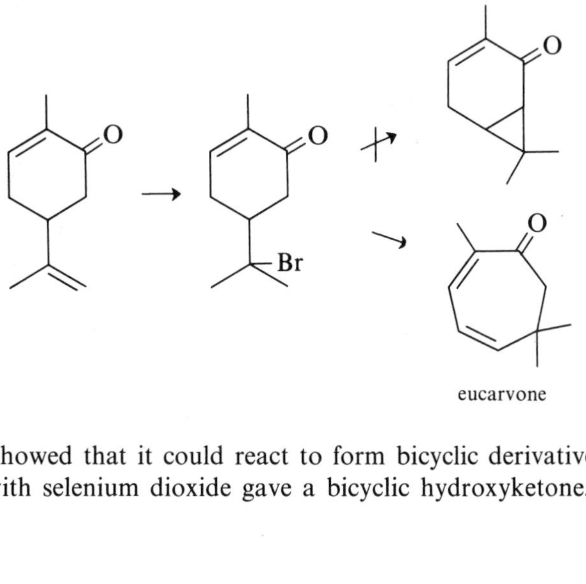

eucarvone showed that it could react to form bicyclic derivatives, so that oxidation with selenium dioxide gave a bicyclic hydroxyketone, and with sodium amide in dioxan it gave a sodium derivative, which reacted with methyl iodide to give 3-methyl car-4-en-2-one (Corey and Burke, 1956).

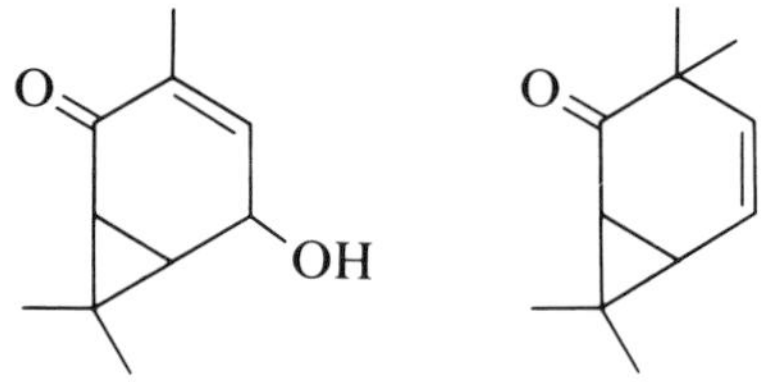

3-methyl car-4-ene-2-one

Preparation of the bicyclic car-3-en-2-one by solvolysis of the hydrobromide of carvone under conditions of carefully controlled pH has been claimed, though the product could not be purified. In the presence of base, it readily rearranged to eucarvone (van Tamelen *et al.*, 1957). The bicyclic ketone has since been reported as a product of the oxidation of car-3-ene (Burns *et al.*, 1968; Carson *et al.*, 1969). Isolation of the bicyclic ketone favours the theory that the formation of eucarvone from carvone hydrobromide involves this ketone, and is probably not a concerted process. The original mechanism put forward for the ring expansion (Wallach, 1899, 1905), that it proceeded by attack of hydroxide ion on the bridgehead carbon, opening the ring to give a hydroxyketone which subsequently dehydrated to eucarvone, has been disproved by preparing norcarenone (XVIII), and showing that it rearranges to a cycloheptadienone (XIX) (van Tamelen *et al.*,

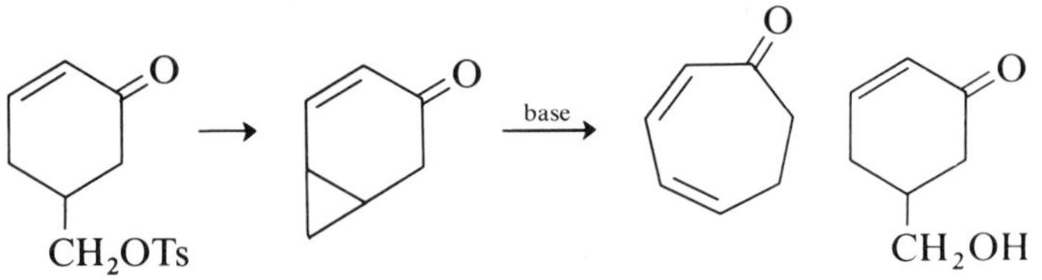

(XVIII) (XIX) (XX)

1957). In this case, attack by hydroxide on the ring would be expected to take place on the least hindered carbon atom, and would give the alcohol (XX), which has been shown to be stable to base. The ring expansion appears to result from removal of an active hydrogen from C_5 triggering an anionic shift to the anion of eucarvone. The presence of an equilibrium between the monocyclic and bicyclic ketones in the presence of base has been established by showing that eucarvone, in EtOD, exchanges three hydrogen atoms (Corey *et al.*, 1956). Exchange of the third hydrogen is possible only if the bicyclic ketone is formed from exchanged eucarvone.

When car-3-ene is oxidized with permanganate in neutral conditions, some car-3-ene-2-one is produced, as mentioned earlier, but the main product is car-3-en-5-one, the main reaction occurring on the less hindered side of the double bond (Burns *et al.*, 1968). This compound, which constitutes 75% of the neutral volatile products of the reaction, was characterized by spectroscopic studies, semicarbazone formation in the cold, and by hydrogenation to the known *cis*-caran-5-one. On treatment with base or borontrifluoride-ether complex, it was converted to the eucarvone isomer, 1,1,4-trimethylcyclohepta-2,4-dien-6-one (XXI). The mechanism of this latter reaction is

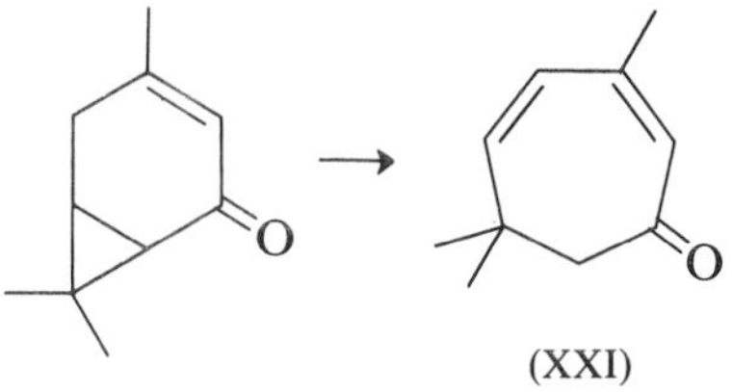

(XXI)

presumably similar to that of the car-3-en-2-one to eucarvone transformation, since the two bicyclic isomers can be regarded as differing only in position of the methyl group.

The oxidation of car-3-ene with permanganate also gave the other isomeric unsaturated ketone, car-2-en-4-one, which did not isomerize to a eucarvone derivative, presumably on account of the absence of an activated methylene group next to the cyclopropane ring. When the oxidation of

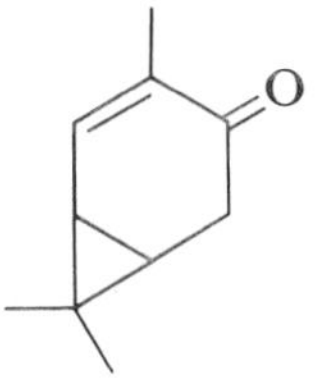

car-2-en-4-one

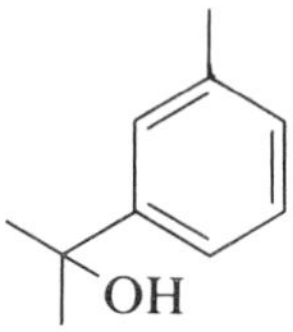

8-hydroxy-*m*-cymene

car-3-ene was carried out with chromium trioxide, both eucarvone and its isomer were obtained, but the main product was 8-hydroxy-*m*-cymene, probably formed by oxidation of products of acid catalysed ring opening of carene.

On account of their ready ring expansion to cycloheptadiene derivatives, the photochemical reactions of the carenones have been studied only with 3-methyl car-4-en-2-one, in which the extra methyl group stabilizes the bicyclic structure. This compound undergoes an interesting ring fission to the ketene; deuteration of the extra methyl group showed that the two methyl groups on C_3 remained distinguishable so that the stereochemistry of the product was determined by the ground state of the reactant (Baldwin and

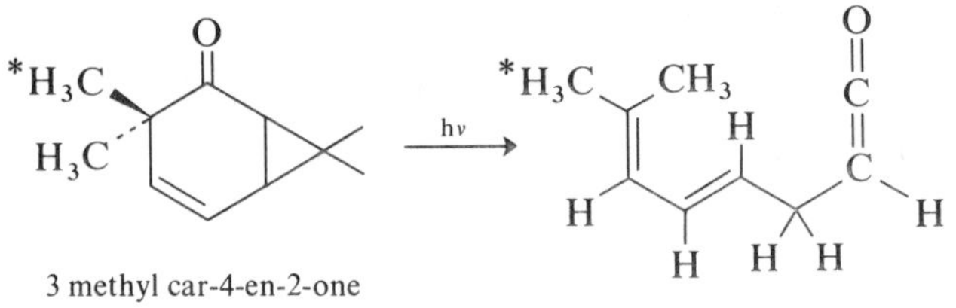

3 methyl car-4-en-2-one

Kreuger, 1969a; Bellamy and Whitham, 1964). On irradiation of eucarvone itself, it undergoes cyclization to a mixture of two bicyclic products (Hurst and Whitham, 1963).

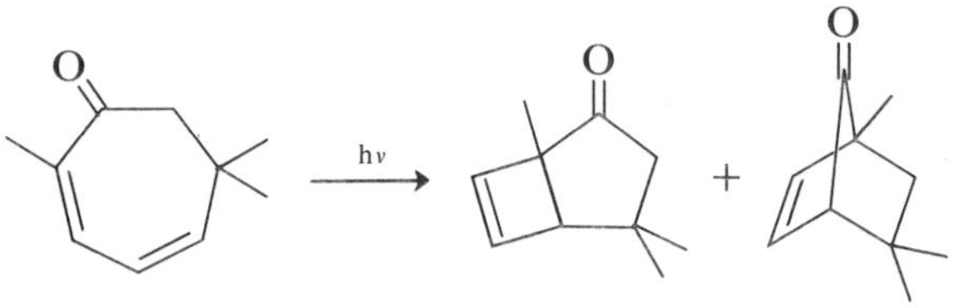

All of the six possible saturated ketones, the caranones, have been prepared. Hydroboration of car-2-ene, followed by oxidation of the alcohol produced, gives 2-isocaranone, while cyclization of dihydrocarvone by base catalysed hydrolysis gives 2-caranone. In the presence of sodium ethoxide, either ketone isomerizes to a mixture of 83% caranone and 17% isocaranone. The conformations suggested below have been shown to be consistent with the

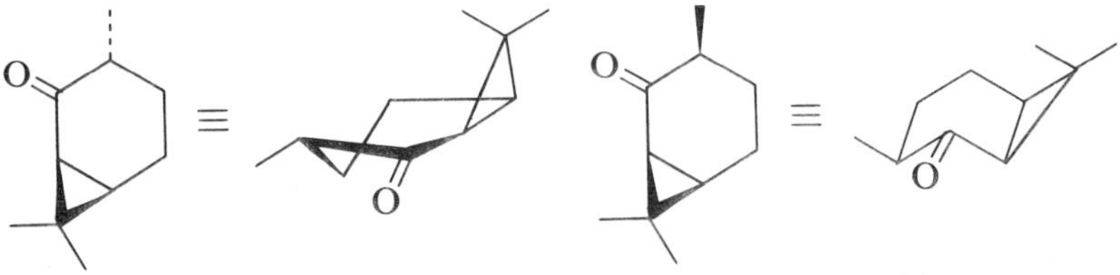

2-caranone 2-isocaranone

spectroscopic properties of the ketones (Acharya and Brown, 1967). In contrast to this result, 4-isocaranone, obtained by hydroboronation of car-3-ene, followed by oxidation of the alcohol, isomerizes in base to give an equilibrium mixture of 90% 4-isocaranone and 10% 4-caranone. This unexpected result has been rationalized on the basis of the conformations shown below (Brown and Suzuki, 1967). Thus, 4-caranone has a quasi-equatorial methyl group, while the different conformation of the ring of 4-isocaranone permits the methyl group to be fully equatorial, enhancing the stability of this isomer.

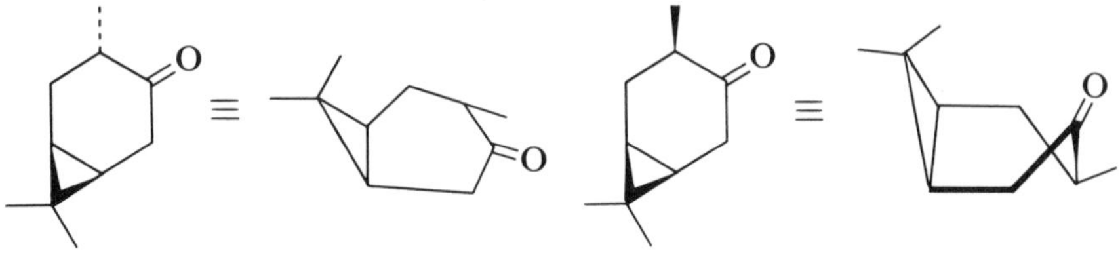

4-caranone 4-isocaranone

On irradiation, 4-isocaranone (also referred to as *cis*-4-caranone or *cis*-caran-4-one) departs from the normal pattern of decomposition of β,γ-cyclopropyl ketones (double α-cleavage leading to decarbonylation) but behaves in a manner similar, though not identical, to β,γ-unsaturated ketones. There is no evidence of epimerization of the ketones, and the overall reaction may be summarized as shown below (Heckert and Kropp, 1968; Carson *et al.*, 1968b).

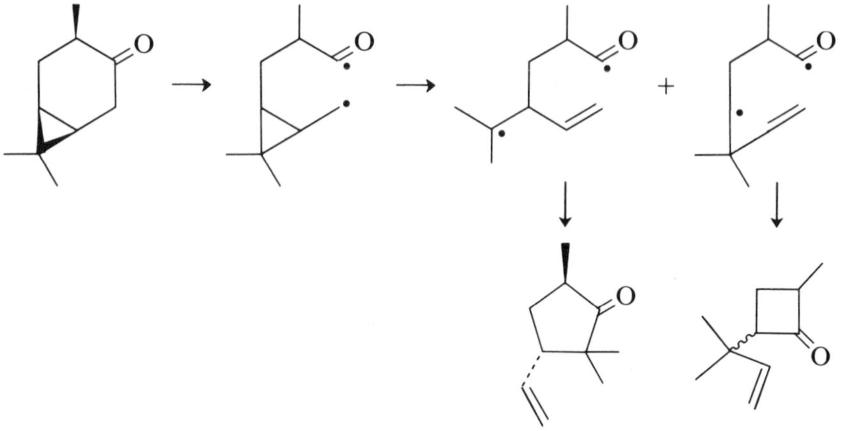

cis and *trans*-isomers

The other isomer, 4-caranone, behaves similarly, except that it gives the *cis* isomer of the cyclopentanone.

On reduction of the saturated ketones, mixtures of alcohols are formed whose composition is consistent with the conformations of the ketones, and the known stereospecificities of the reducing agents. Conformations of some alcohols have been suggested, on the basis of n.m.r. data (Carson *et al.*, 1968a). Rearrangement of the alcohols with OH at C_2 and C_5 with acid has shown expected influences over the direction of ring opening (Cocker *et al.*, 1968b).

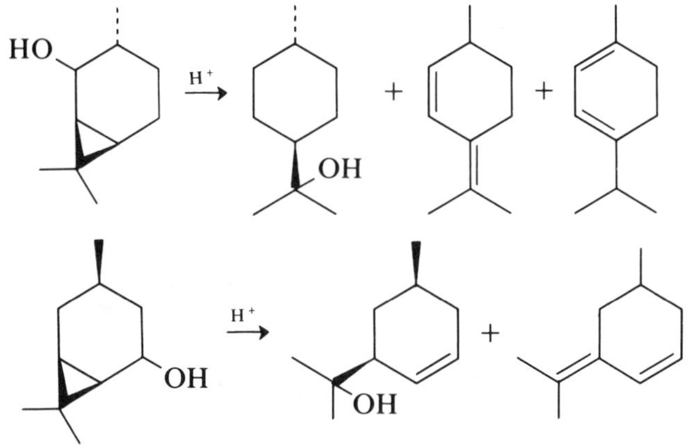

It has been suggested that the preference for reaction to occur *trans* to the cyclopropyl ring may be not entirely steric in origin, but in some cases may be reinforced by participation of the cyclopropyl ring (Kropp, 1966b.) In support of this, it has been shown that (XXII), on treatment with phosphorus oxychloride and pyridine, gave a mixture of products, one of which affords the first example of rearrangement of carane to another bicyclic system (XXIII). However, reaction does not take place when the acetoxy group is

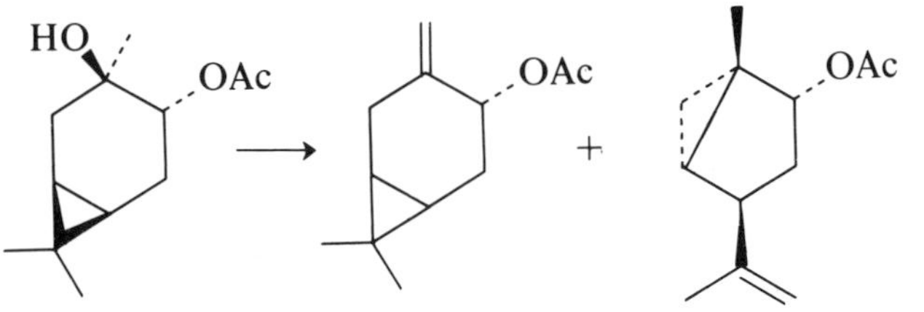

(XXII) (XXIII)

absent or when the hydroxyl group is *cis* to the acetoxyl, so that if participation is responsible for this result, it must be a result of the particularly favourable circumstances for existence of an ion whose extreme forms may be represented as below, rather than observation of a general phenomenon.

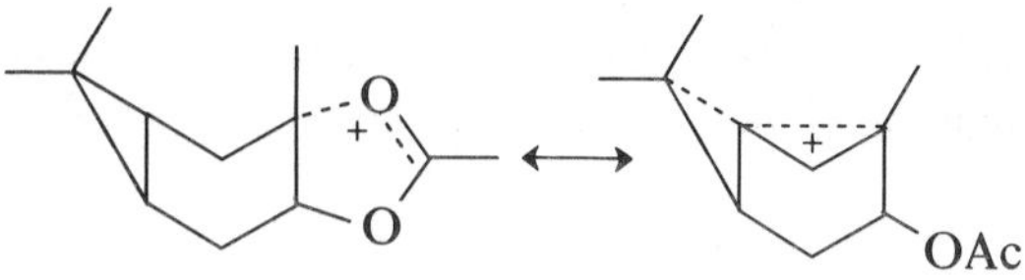

A number of caranamines have been prepared, and conformations proposed on the basis of n.m.r. studies (Cocker *et al.*, 1968a). On deamination, the 2- and 5-amines give a mixture of unrearranged and ring opened products, the ring opening reactions following closely the paths of similar reactions observed in the acid catalysed rearrangement of the corresponding alcohols. The 4-amines, however, give a mixture of the 3-ols, the 4-ols, and two rearranged products, which could arise from the diazonium ion, as shown, or the carbonium ion (Cocker *et al.*, 1969).

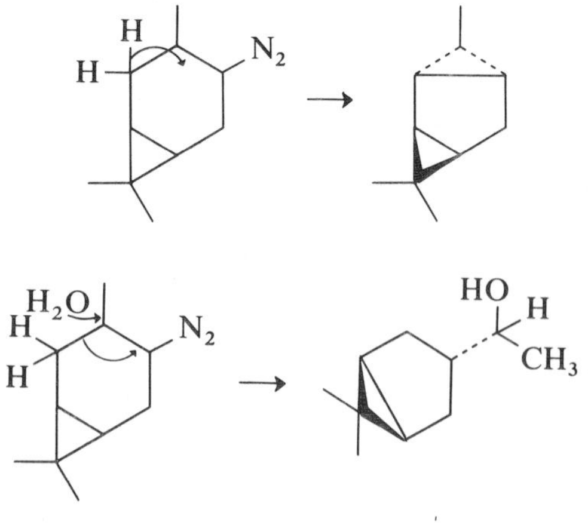

B. Thujane Series

The parent hydrocarbon, thujane, exists as the *cis* and *trans* forms, these having the absolute configurations shown below (Ohloff *et al.*, 1966).

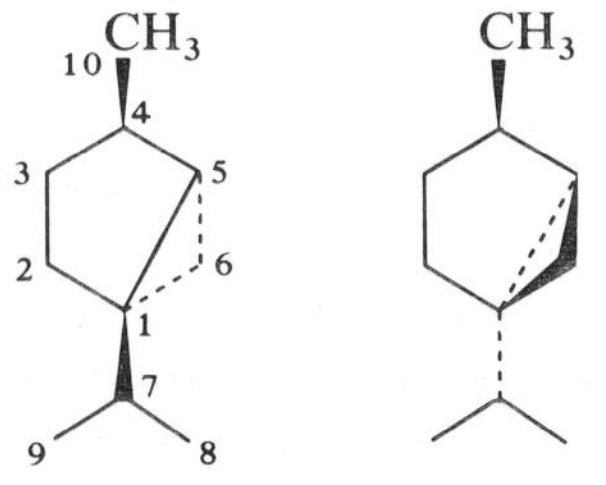

(+)-*cis*-thujane (+)-*trans*-thujane

Studies of the thujanes by n.m.r. have shown the conformation of *cis*-thujane to be a half chair, while the *trans* isomer prefers a boat-like conformation (Dieffenbacher and von Philipsborn, 1966).

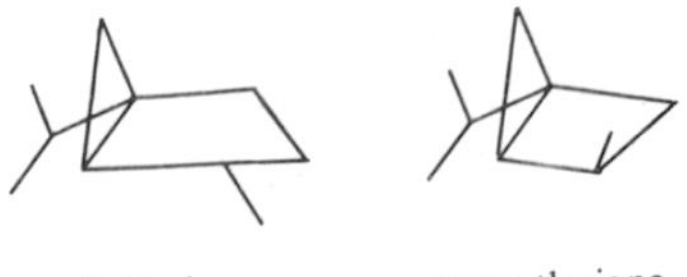

cis-thujane *trans*-thujane

Four isomeric thujenes exist; studies of α-thujene by n.m.r. have shown it to adopt an L-shaped conformation (Cooper and Whittaker, 1968), with the 5-membered ring planar. The β-thujenes, which also have their double bond in the ring, are presumably similar, but the conformation of sabinene has not been studied.

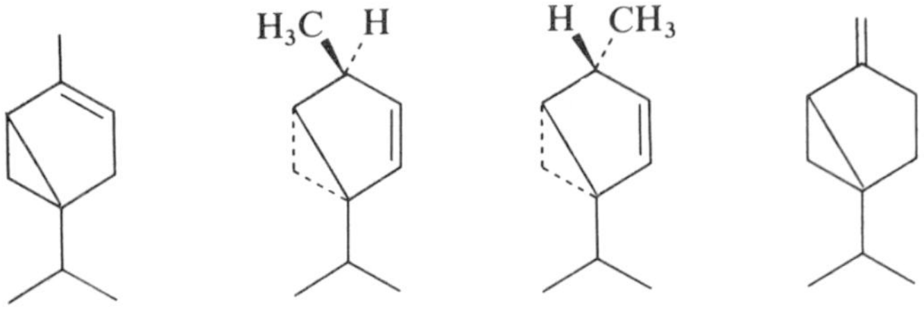

α-thujene β-thujene β-isothujene sabinene

Sabinene occurs naturally in Oil of Savin, while α-thujene is obtained fr n *Eucalyptus dives* oil. The β-thujenes have been prepared by pyrolysis o e xanthate esters of the corresponding thujyl alcohols (Banthorpe and D s, 1968b).

eith quilibrium between α-thujene and sabinene, attained by refluxing cal th potassium *t*-butoxide in dimethylsulphoxide, a procedure identi- 9% s used to equilibrate the 2- and 3-carenes, is at 91% α-thujene and Sine (Acharya *et al.*, 1969).

the do ene has an L-shaped conformation, it would be expected that propan d would undergo attack from the side opposite the cyclo- gives 85 is is in fact observed; hydrogenation at low temperature cyclopro ujane and 15% *cis*-thujane; at higher temperatures, the the side opened. On hydroboration, α-thujene adds borane on *trans*-thuja he cyclopropyl ring, and gives an alcohol with the

In contrast to this, sabinene, which may be presumed to have approximately the same L-shaped conformation as thujene, can be hydrogenated under identical conditions to give 86% *cis*-thujane. However, hydroboration of sabinene yields a mixture of the 10-thujanols in which the alcohol with CH_2OH and cyclopropyl ring *cis*- occurs to 35%. It has been suggested that sabinene has the cyclopropyl ring and double bond in a better alignment for conjugation than α-thujene, and that this partially conjugated system is more likely to attach itself to the platinum catalyst used in the hydrogenation reaction so as to attach both double bond and cyclopropane ring, whereas in α-thujene only the double bond will attach to the catalyst, and the molecule aligns itself with the cyclopropane ring away from the catalyst to reduce steric hindrance. Conjugation of double bond and cyclopropane ring may also provide an explanation of the unusual amount of the exocyclic isomer in the equilibrium mixture, but this could also arise from the greater strain of accommodating a double bond in what is essentially a five-membered ring (Acharya *et al.*, 1969).

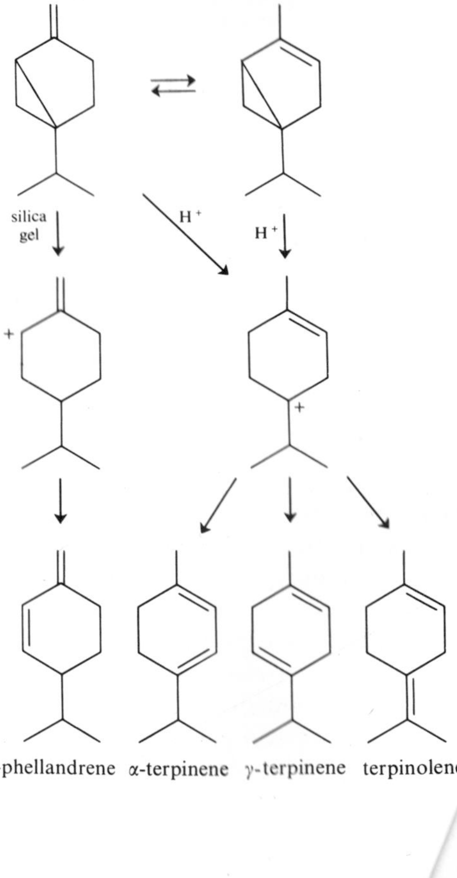

β-phellandrene α-terpinene γ-terpinene terpinolene

The rearrangements of both α-thujene and sabinene, catalysed by acids and by surfaces, have been studied by several authors (Goryaev *et al.*, 1964; Wrolstad and Jennings, 1965). The results obtained can be represented by the reaction scheme set out on page 50.

On heating α-thujene to 250°, it rapidly loses optical activity. Investigation of this reaction has shown that it is a true racemization, rather than decomposition to racemic products, and deuterium labelling experiments have shown that the reaction is an example of the vinyl cyclopropane type of rearrangement (Doering and Lambert, 1963).

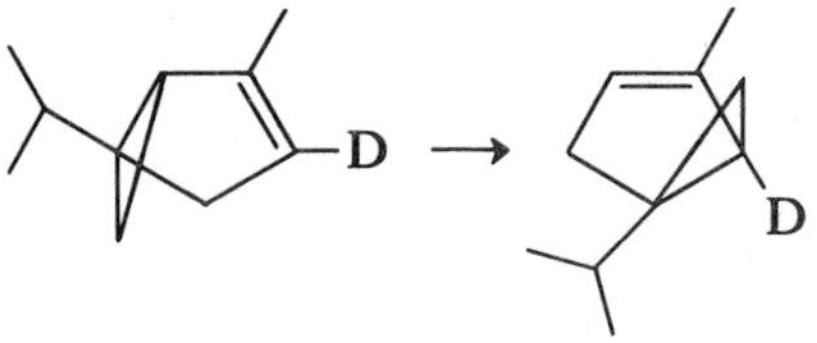

A similar rearrangement of β-thujene has been observed, but in this case the product undergoes further rearrangement.

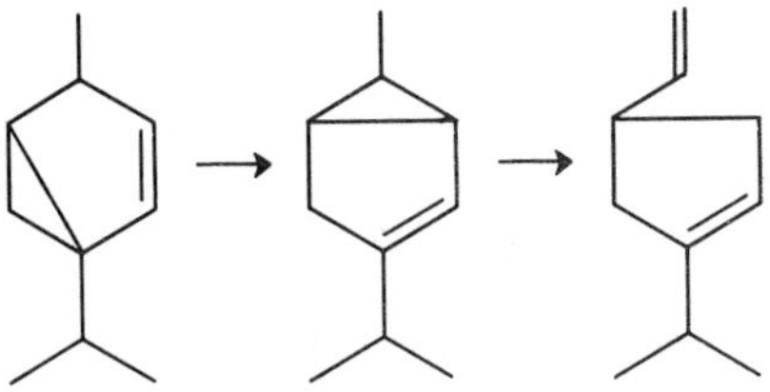

Sabinene, although it is also a vinyl cyclopropane system, does not react by this route because it would yield a strained product, but it decomposes at 600° by a reaction which is believed to involve a biradical (Mitzner and Theimer, 1962).

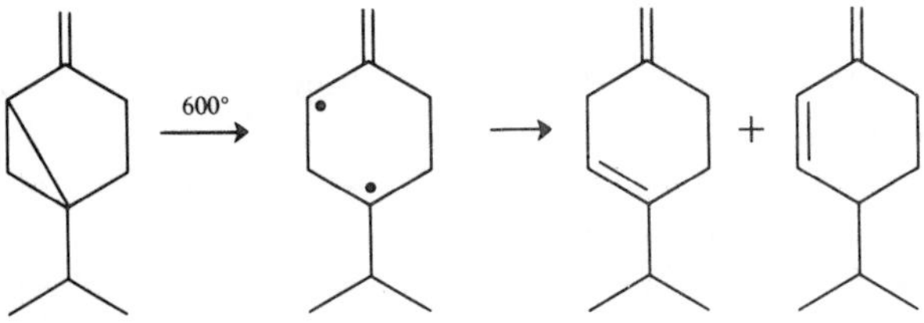

The saturated ketones, thujone and isothujone, have been assigned a boat conformation on the basis of n.m.r. studies (Tori, 1964), but since equilibration with sodium ethoxide shows thujone to be favoured in the equilibrium by a factor of two, it seems possible that the boat of thujone may be distorted to bring the methyl group equatorial. This is consistent with the results of Banthorpe and Davies (1968a) who studied the reduction of the ketones with a variety of reagents, and concluded that thujone probably has a pseudo chair conformation during the transition state for reduction.

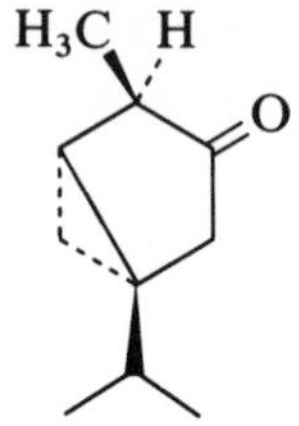

thujone

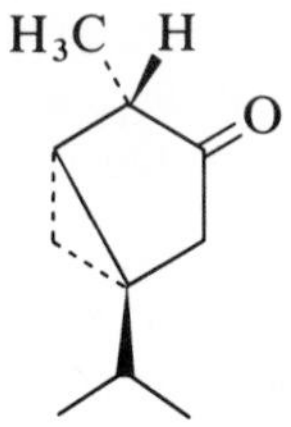

isothujone

On heating, thujone rearranges by a vinyl cyclopropane mechanism similar to that of α-thujene.

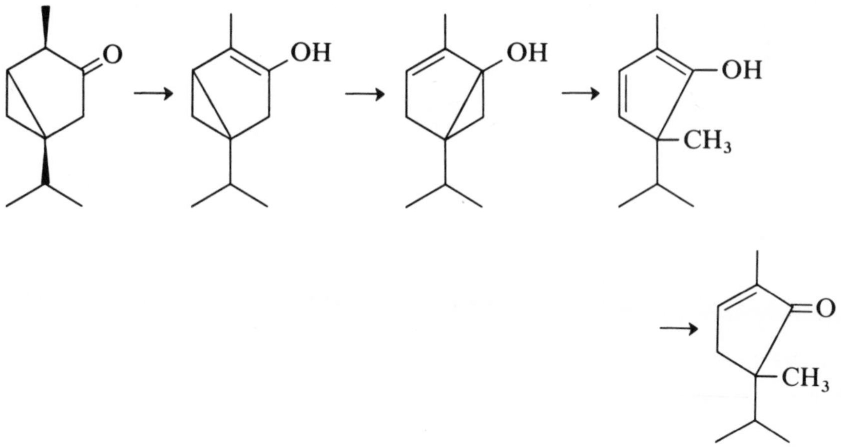

On reduction, thujone gives the alcohols thujol and neothujol, while isothujone gives isothujol and neoisothujol. Acharya *et al.* (1969) have pointed out that the accepted nomenclature of the thujols is complex and not completely consistent with the other thujanes, and has proposed an alternative nomenclature based on that of Schroeter and Eliel (1965) for the carvomenthol and related terpenes. In this, the prefixes iso or *cis* are used to denote that methyl and isopropyl are *cis* to each other, and neo is used to indicate that the methyl and hydroxyl are *cis* to each other. The revised names are in brackets below the old names.

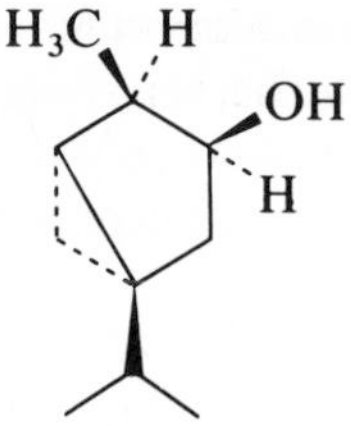

thujol
(XXIV)
(3-neoisothujanol)

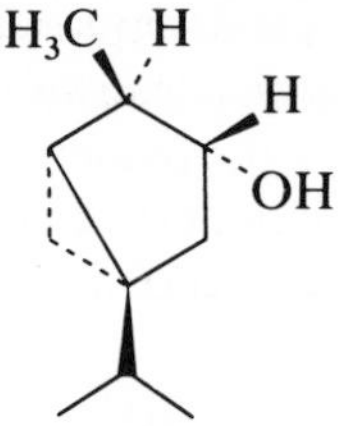

neothujol
(XXV)
(3-isothujanol)

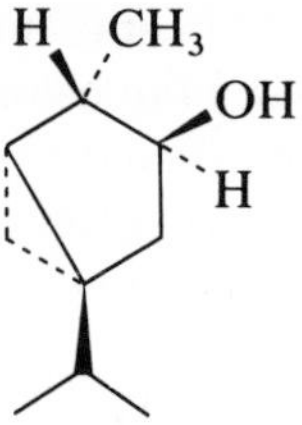

isothujol
(XXVI)
(3-thujanol)

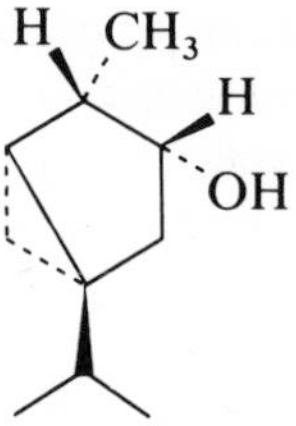

neoisothujol
(XXVII)
(3-neothujanol)

The acetolysis of the thujyl *p*-toluene sulphonates has been the subject of an important paper by Norin (1964). He found that optically active (XXV) *p*-toluene sulphonate gave, on acetolysis, the acetate of racemic (XXV); similarly, optically active (XXVII) *p*-toluene sulphonate gave, on acetolysis, the acetate of racemic (XXVII), while the *p*-toluene sulphonates of the other thujols gave a complex mixture of products. These results are consistent with a reaction scheme in which the esters of the equatorial alcohols, (XXV) and (XXVII), are assisted in solvolysis by participation of the cyclopropane ring, reaction involving ions of the type,

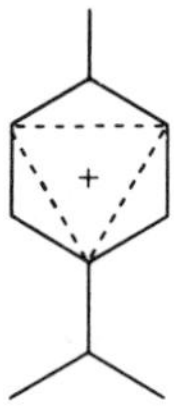

The esters of the axial alcohols, (XXIV) and (XXVI), are not assisted in this way, and hence react *via* an undelocalized ion.

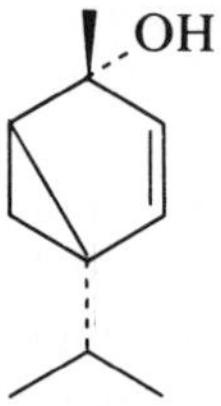

trans-4-hydroxy-
β-thujene

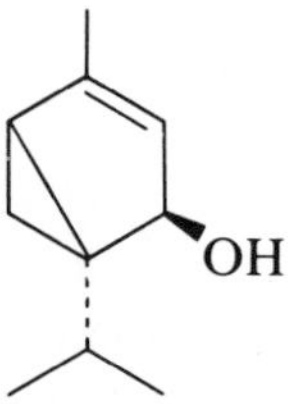

umbellulol

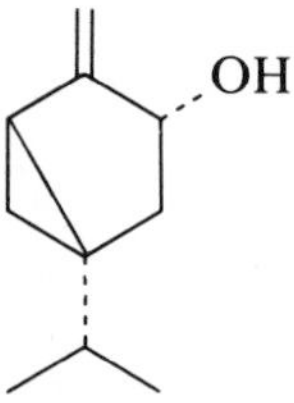

cis-sabinol

Three unsaturated thujyl alcohols have been obtained from the photosensitized oxidation of α-thujene (Klein and Rojahn, 1965). Umbellulol, on oxidation, gives the ketone umbellulone, which also occurs naturally in the oil of the California Mountain Bay Laurel. On irradiation, umbellulone rearranges to thymol by the reaction sequence shown below (Barber *et al.*, 1968).

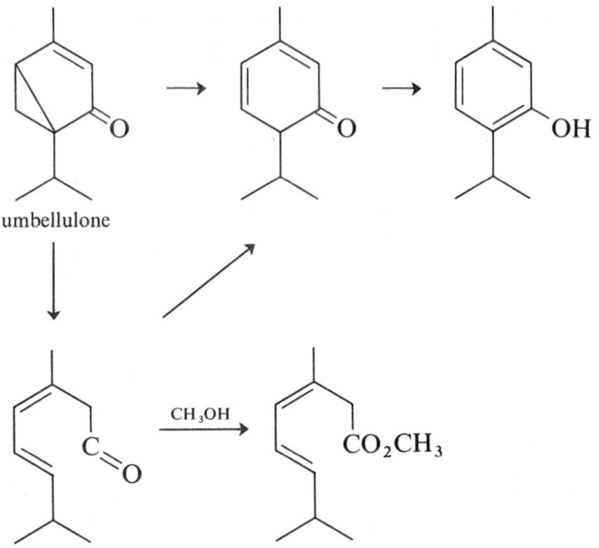

C. Pinane Series

The pinanes have a cyclobutane ring, which can readily undergo ring expansion or ring opening reactions, and these reactions dominate pinane chemistry (Banthorpe and Whittaker, 1966a). The source of almost all pinane derivatives is turpentine; it contains α-pinene and β-pinene, and from these almost all others are synthesized (Banthorpe and Whittaker, 1966b). These two olefins are of considerable industrial importance, as a good deal of the synthetic perfumery industry is based on reactions involving them as starting materials (Sully, 1964, Ansari, 1970).

Hydrogenation of either α-pinene or β-pinene, proceeds on the less hindered side of the double bond, i.e. away from the *gem*-dimethyl bridge, giving a mixture of *cis*- and *trans*-pinanes in which the former predominates (Cocker *et al.*, 1967a). Pure *cis*-pinane is better prepared by reduction of either olefin with diimide (van Tamelen and Timmons, 1962) this giving 99% of the *cis* isomer. The *trans* isomer can be obtained by hydroboration of β-pinene, followed by thermal equilibration of the boron complex to the *trans* isomer, and acidolysis (Zweifel and Brown 1964). This reaction involves normal addition of borane from the less hindered side of the double bond, but on heating isomerization occurs to put the bulky borane group on the side away from the *gem*-dimethyl group. Hydroboration of α-pinene also proceeds from the less hindered side of the double bond, but in this case equilibration cannot take place, and acidolysis gives the *cis* isomer.

Both isomeric pinanes have also been obtained by reduction of the δ-pinenes (Schmidt, 1947).

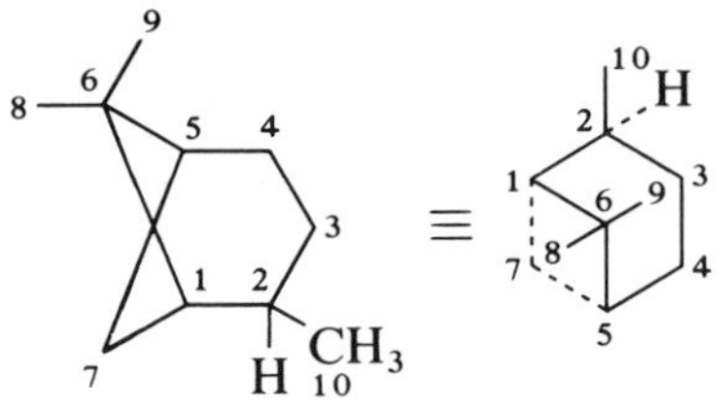

cis-pinane

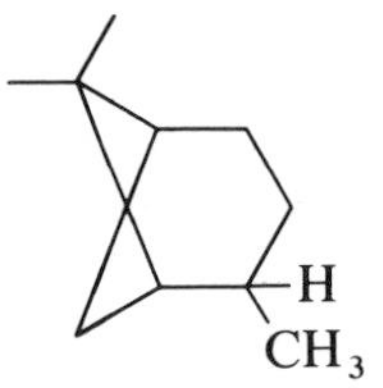

trans-pinane

Because of the restrictions of Bredt's rule, which forbids the existence of a double bond at a ring junction, and is rigidly obeyed in small ring systems such as this, only four mono-olefins can exist. Two of these occur naturally in a wide range of plants, mainly conifers; the two δ-pinenes are of synthetic origin. One diene, verbenene, can also be prepared synthetically by the pyrolysis of verbenyl acetate, a reaction involving a shift of the double bond in the starting material, which is unusual in this type of reaction (Ohloff and Klein, 1962). The δ-pinenes can be prepared by a number of routes, the most convenient being the pyrolysis of the methyl xanthate esters of the verbenols (Schmidt, 1947), and hydroboration of the verbenene produced, followed by acidolysis of the boron complex formed (Zweifel and Whitney, 1966).

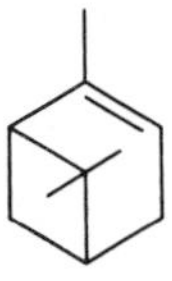

α-pinene

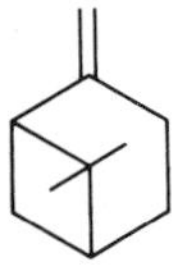

β-pinene

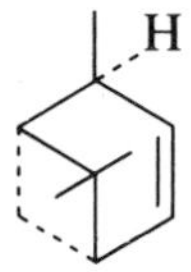

cis-δ-pinene

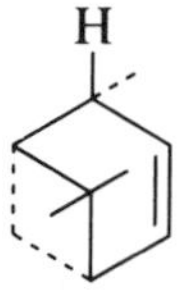

trans-δ-pinene

verbenene

On account of their availability and industrial importance, the chemistry of α-pinene and β-pinene has been thoroughly studied. They interconvert fairly readily, the simplest methods being with hydrogen on a platinum catalyst (Cocker *et al.*, 1967a) and with the homogeneous catalyst, iron pentacarbonyl (Spanninger and von Rosenberg, 1969). The equilibrium mixture consists of 98 % of the α-isomer and 2 % of the β-isomer. Both olefins readily undergo ring opening reactions, high energy reactions such as thermal cracking or irradiation giving a mixture of monocyclic and acyclic products, while lower energy catalytic isomerizations give ring opening to monocyclic products together with ring expansion to other bicyclic systems. All the high energy reactions proceed by initial fission of the cyclobutane ring to give a *p*-menthane structure containing two reactive sites, one of which can be stabilized by interaction with the double bond already present. The type of intermediate, and hence the products, depends on the means used to supply the energy for ring opening.

The thermal reactions of both olefins thus proceed through a biradical (Burwell, 1951). α-Pinene, on decomposition at 375°, gives a mixture of dipentene (racemic limonene), alloocimene, α-pyronene and β-pyronene (Hawkins and Hunt, 1951). The last three products almost certainly arise from decomposition of *cis*-β-ocimene formed directly from pinene, as this substance can be obtained in good yield by reaction at low temperature (175°) involving a short time of heating (Hawkins and Burnis, 1959).

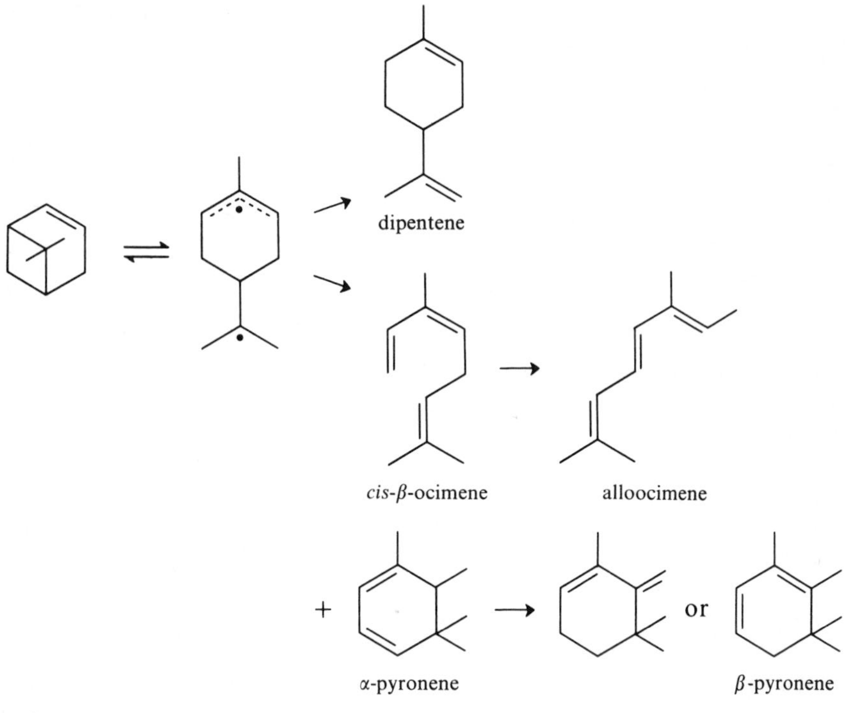

The thermal decomposition of β-pinene proceeds via a similar radical, with the important difference that this intermediate does not have a plane of symmetry, so should give rise to optically active products (Goldblatt and

Palkin, 1941). At 160°, thermal decomposition of β-pinene gives β-myrcene, limonene, myrcene dimer, (α-camphorene) 1,(7),8-*p*-menthadiene (Steinbach *et al.*, 1964) and α-myrcene (Steinbach and Walt, 1965) though by using high temperatures and low heating times, the reaction can be forced to give myrcene in almost quantitative yield (Patton, 1950).

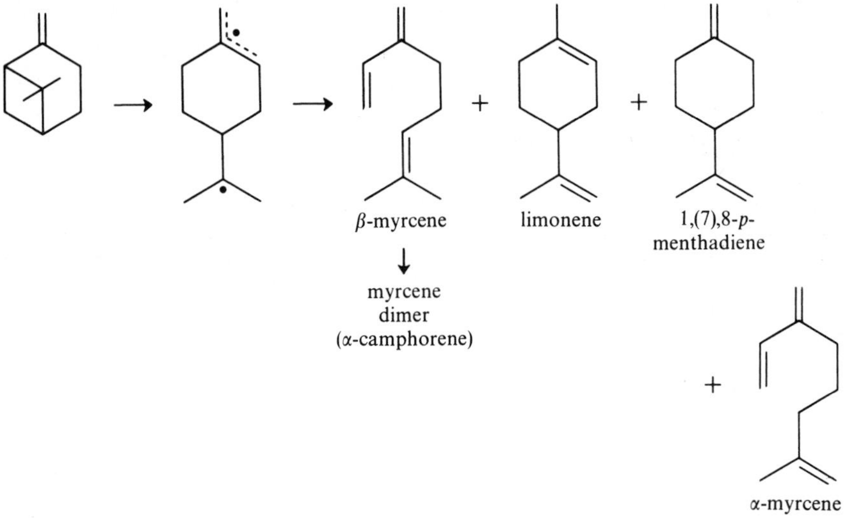

On γ-irradiation, both pinenes decompose in a manner which is generally similar to the thermal decomposition, but differs in sufficient detail for zwitterionic intermediates to be suggested (Bates *et al.*, 1962a, 1962b).

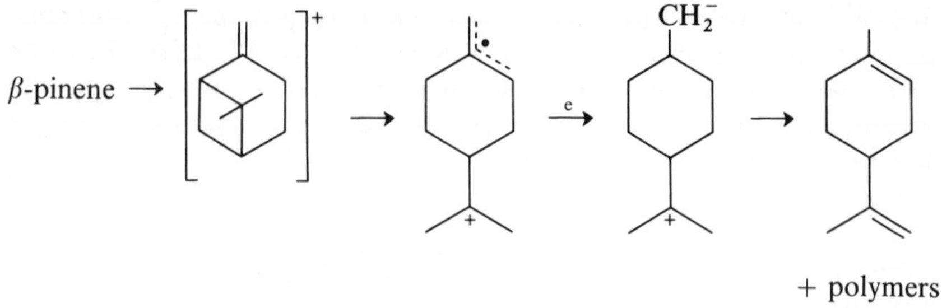

Direct photolysis of α-pinene in diethyl ether gives a mixture *cis*-β-ocimene, *trans*-β-ocimene, dipentene and cyclofenchene (Mayer *et al.*, 1964); in the presence of a photosensitizer such as acetophenone or propiophenone, the cyclic products are not formed (Frank, 1968). The *trans*-β-ocimene may well be a secondary product, since the *cis* isomer isomerizes under these reaction conditions.

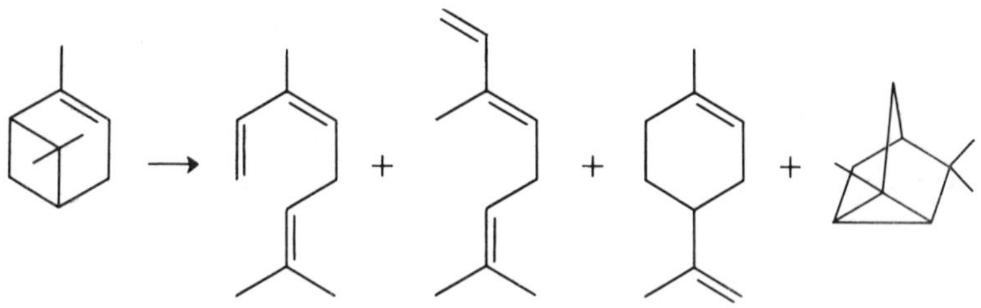

If irradiation is carried out in the presence of oxygen, α-pinene gives sobrerol, probably by decomposition of the α-pinene epoxide formed by oxidation of pinene with verbenyl hydroperoxide, which is itself formed by oxidation of pinene by oxygen alone (Schenck *et al.*, 1953b). In the presence of a photosensitizer, both α-pinene and β-pinene react by migration of the double bond, in the same manner as limonene, giving the hydroperoxides of *trans*-pinocarveol and myrtenol respectively (Hope and Mitchell, 1954).

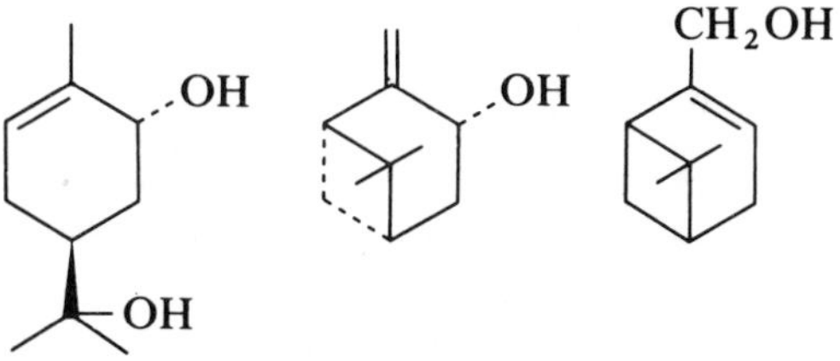

trans-sobrerol *trans*-pinocarveol myrtenol

Other rearrangements of the pinenes proceed on surfaces or result from addition reactions; this type of reaction proceeds at lower energy than the reactions discussed above, and consequently the main processes involved are enlargement of the cyclobutane ring to a cyclopentane ring by rearrangement to bornane or fenchane derivatives and ring opening to monocyclic products. Ring opening to acyclic products is not observed under these conditions. Both α-pinene and β-pinene isomerize readily in the liquid phase at temperatures in the 20–80° range over catalysts such as clays, minerals, and salts, yielding camphene as the main product, together with limonene (Wystrach *et al.*, 1957). On prolonged reaction, camphene slowly forms limonene, and the limonene is isomerized to a mixture of *p*-menthadienes. The process is of considerable industrial importance, forming an essential step of the route from pinene to camphor, and is the subject of an immense patent literature, consisting of several hundred variations on the basic catalyst types. Some of these catalysts, such as silica gel, titanic acid and alumina, are proton carriers, but others, such as clays, silicate minerals, and salts such as magnesium sulphate cannot act in this way (Swann, 1948), and the catalytic activity in these, and possibly in others too, depends on the formation of a π-complex between the substrate and a metal ion in the lattice of the catalyst. The stereochemistry of absorption of the pinene is of importance in determining the course of the reaction, absorption with the *gem*-dimethyl bridge away from the catalyst, which is favoured on steric grounds, leading to camphene formation, while absorption with the *gem*-dimethyl group towards the catalyst leads to ring opening and limonene formation (Banthorpe and Whittaker, 1966a).

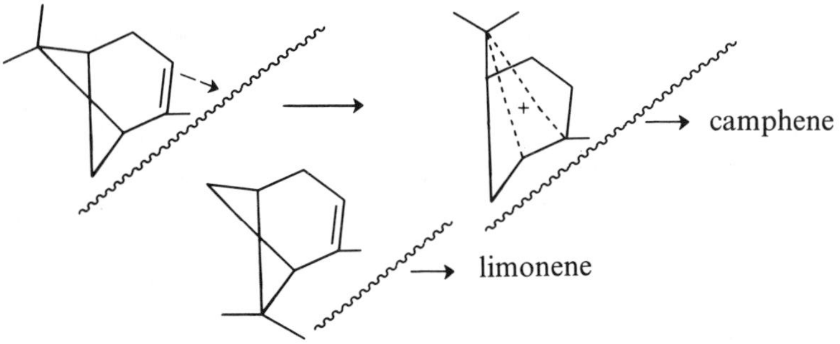

At low temperatures, camphene formation is favoured, but at higher temperatures, limonene formation dominates (Kosima, 1961).

In the presence of a nucleophile, capture of the intermediate complex is possible, so that refluxing α-pinene with a suspension of alumina in acetic acid gives limonene, together with a mixture of the acetates of α-terpineol, borneol and α-fenchol (Kuwata, 1936).

Acid catalysed reactions of the pinenes proceed in much the same way as the surface catalysed reactions, and the rearrangements involved in the reactions may be summarized, using hydration as an example (Valkanas and Iconomou, 1963; Williams and Whittaker, 1971a).

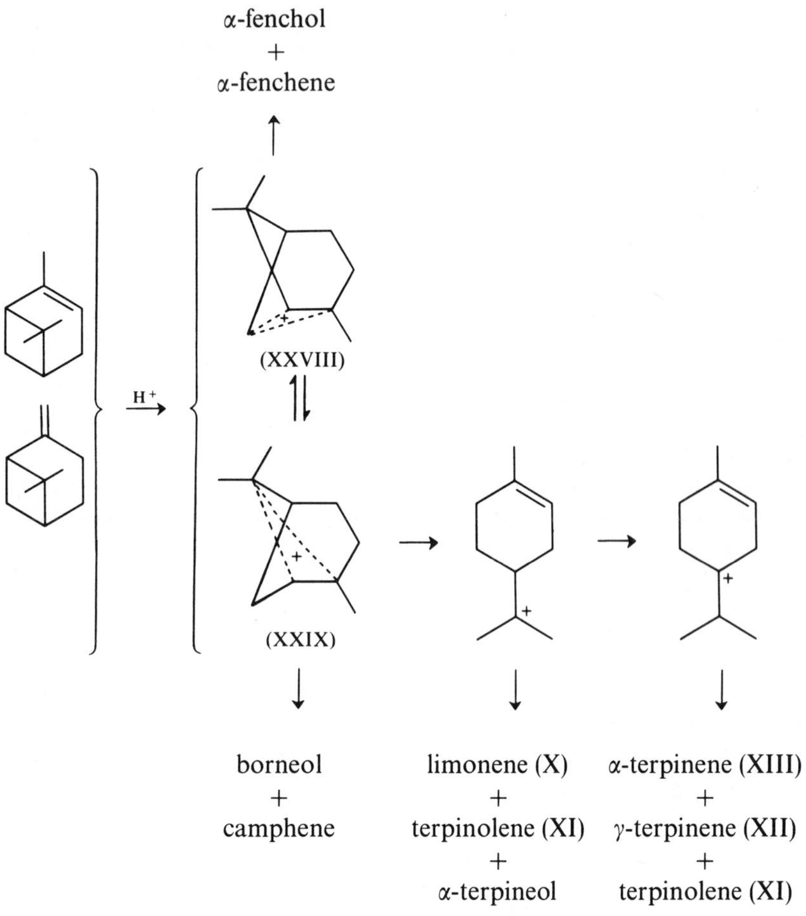

FIG. 4. Hydration of α-pinene and β-pinene.

Both α-pinene and β-pinene hydrate to give identical product mixtures in dilute acid, the β-isomer reacting more rapidly. With the Bertram–Walbaum reagent, that is sulphuric acid catalyst in acetic acid, the reaction gives mainly α-terpinyl acetate, but there is evidence here that the main reaction is isomerization of pinene to a *p*-menthadiene mixture, which subsequently adds acetic acid (Williams and Whittaker, 1970, 1971b).

If the hydration is carried out with concentrated aqueous acid, the main product is 1,8-terpin, formed by hydration of the double bond of α-terpineol.

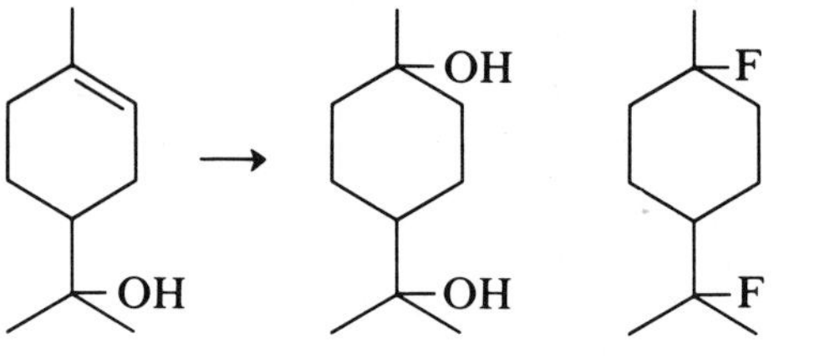

limonene hydrofluoride

The addition of HCl to α-pinene proceeds in a different manner, in that ring opening does not occur, the products being bornyl chloride and α-fenchyl chloride; the probable explanation for this is that the acid adds in the unionized form, giving an intimate ion pair which collapses more rapidly than ring opening. In contrast, the addition of HF results in complete ring opening, giving only limonene hydrofluoride (Hanack, 1960).

The addition of HBr to cis-δ-pinene results in ring expansion by shift of the other end of the bridge, to give some previously unreported isomers of the fenchanes (Barthélémy *et al.*, 1968).

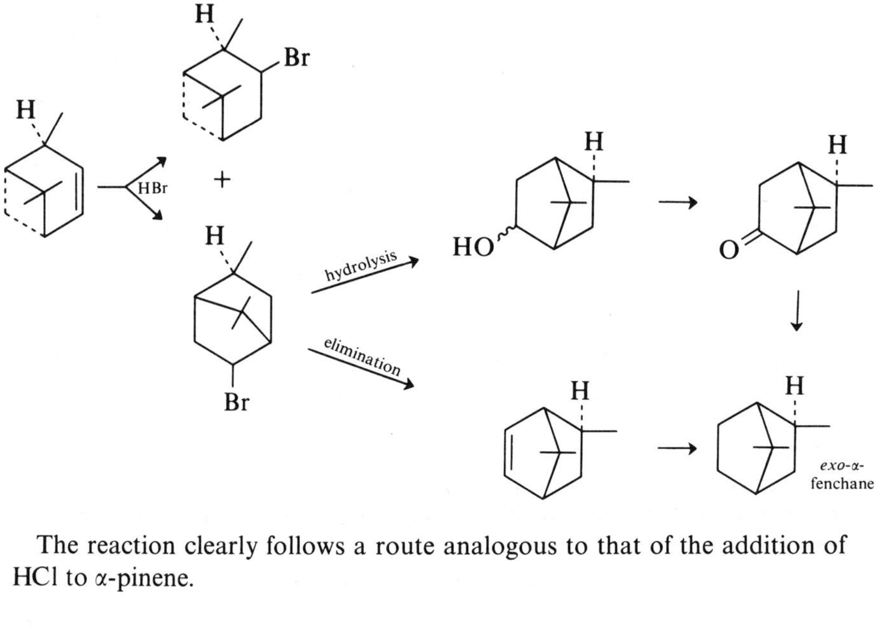

The reaction clearly follows a route analogous to that of the addition of HCl to α-pinene.

In contrast to the above, the addition of halogens to the pinenes can proceed with preservation of the pinane skeleton, though some ring opening does occur (Kergomard, 1953). The mechanism of the reaction has not been investigated, but the results are consistent with a concerted reaction involving addition and elimination within a cyclic transition state.

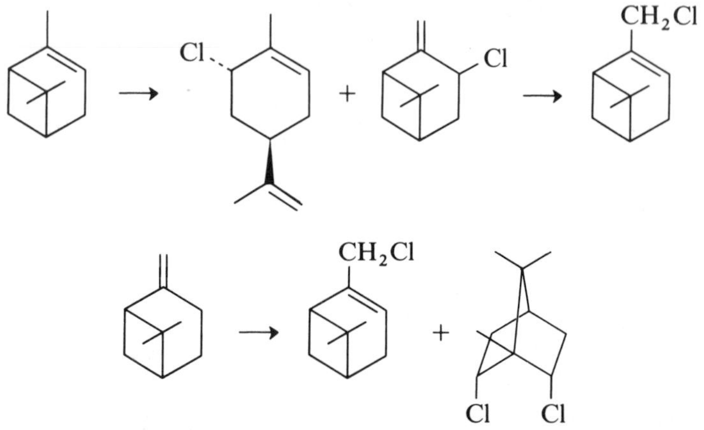

Three unsaturated ketones exist, pinocarvone, chrysanthenone and verbenone. The carbonyl group of pinocarvone completely blocks rearrangement taking place during the addition of HBr, the partial positive charge on C-3 forcing the carbonium ion centre to C-10, the product being the 10-bromo-ketone (Hartshorn and Wallis, 1964).

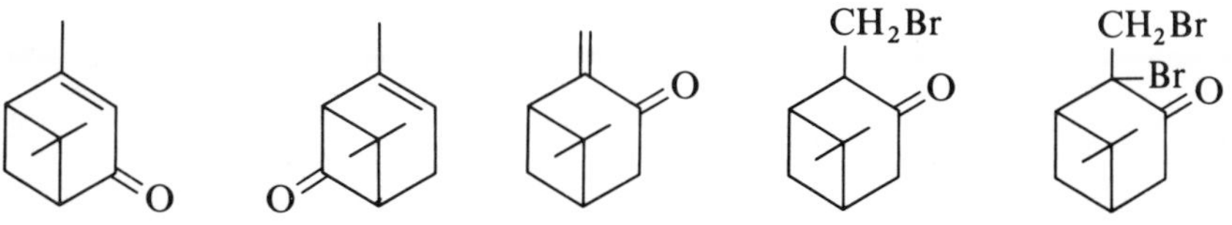

verbenone | chrysanthenone | pinocarvone | 10-bromo-pinocarvone | 2,10-dibromo pinocarvone

When bromine is added, reaction also takes place without rearrangement, yielding the 2,10-dibromo-ketone.

On irradiation with light of wavelength above 300 mμ, verbenone is converted into chrysanthenone, and forms a convenient method of preparing this otherwise inaccessible ketone (Hurst and Whitham, 1960). When the irradiation is carried out with a broad spectrum mercury lamp, a more complex reaction takes place, summarized in the diagram below (Erman, 1967). Reduction of the carbonyl groups of each of these ketones gives rise

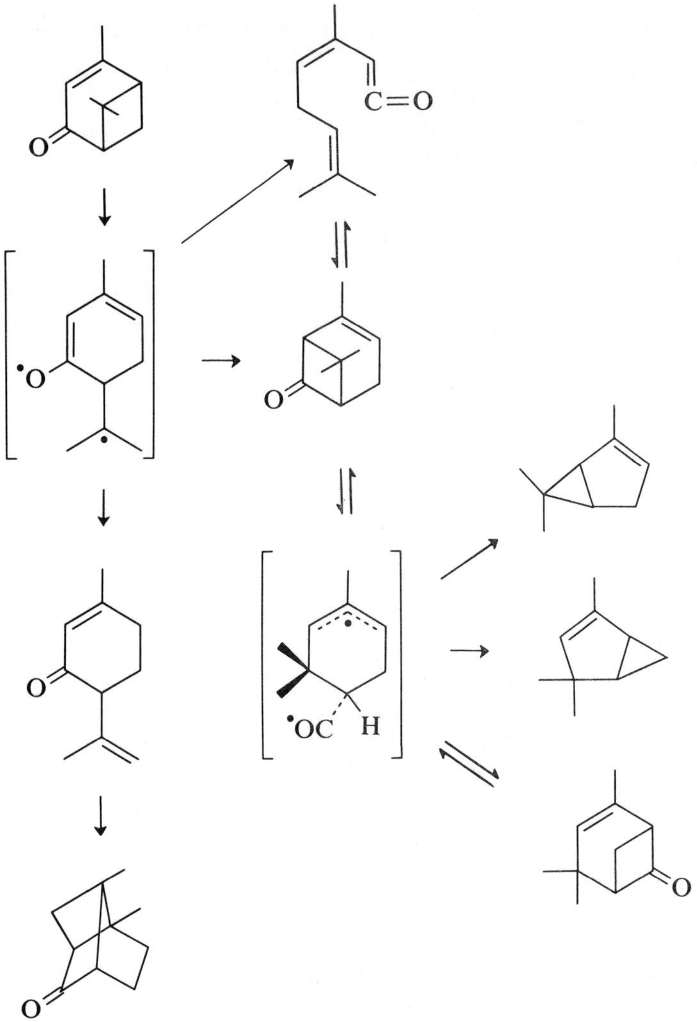

FIG. 5. Products of verbenone irradiation.

to two unsaturated alcohols, but only the verbenols (Cooper *et al.*, 1967) and the pinocarveols (Hartshorn and Wallis, 1964) have been satisfactorily characterized.

Trans-verbenyl acetate readily interconverts via the allylic ion with *cis*-pin-3-en-2-yl acetate (Whitham, 1961) while *trans*-pinocarvyl acetate similarly interconverts with myrtenyl acetate (Hartshorn and Wallis, 1964). In both cases, reaction on the side away from the *gem*-dimethyl group is

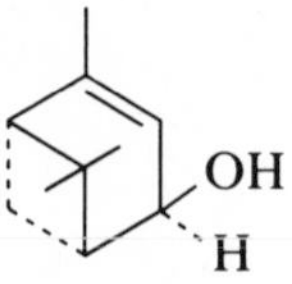

cis-verbenol

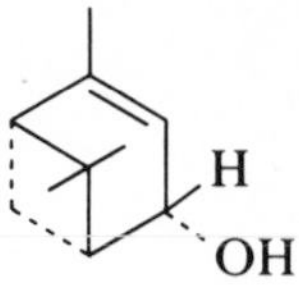

trans-verbenol

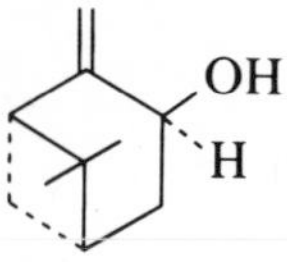

cis-pinocarveol

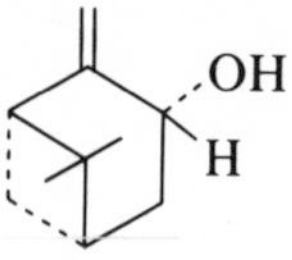

trans-pinocarveol

favoured, the *cis* isomer of pin-3-en-2-ol being that in which the methyl group is *cis* to the *gem*-dimethyl bridge. There is no evidence of any derivatives of the pin-3-en-2-ols rearranging to bornylene derivatives, as the ion centre

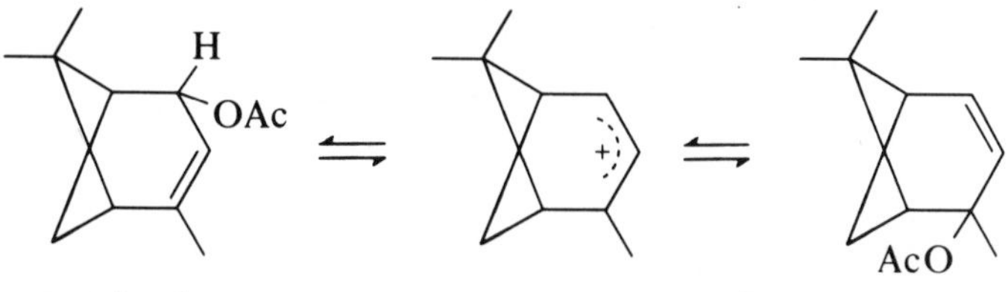

trans-verbenyl acetate　　　*cis*-pin-3-en-2yl acetate

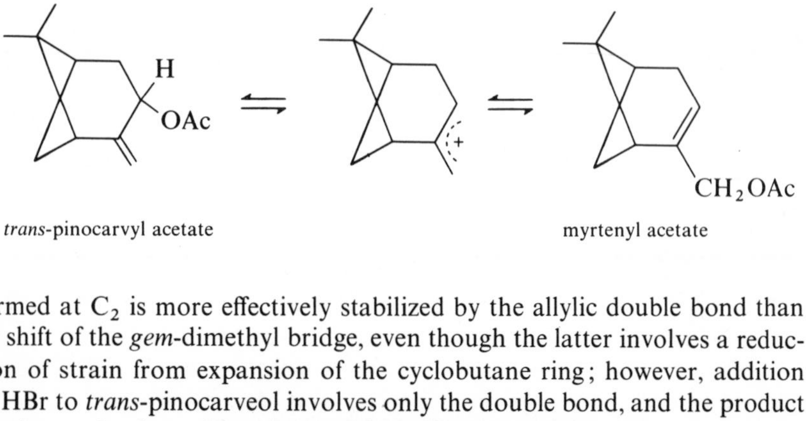

trans-pinocarvyl acetate　　　myrtenyl acetate

formed at C_2 is more effectively stabilized by the allylic double bond than by shift of the *gem*-dimethyl bridge, even though the latter involves a reduction of strain from expansion of the cyclobutane ring; however, addition of HBr to *trans*-pinocarveol involves only the double bond, and the product is a bromofenchane (Hartshorn and Wallis, 1964). Reduction of the double bond of each of the unsaturated ketones should give the two saturated ketones, though the chrysanthanones have not been prepared. The pinocamphones and the verbanones are all known, isoverbanone being one of the very few pinanes whose n.m.r. spectrum has been fully assigned (Abraham *et al.*, 1969), hence permitting determination of its conformation, which is midway between the half chair and Y-shape conformations; this reduces the eclipsing strains in the latter and the repulsion of the methyl group, C_2 hydrogens and *gem*-dimethyl bridge in the former.

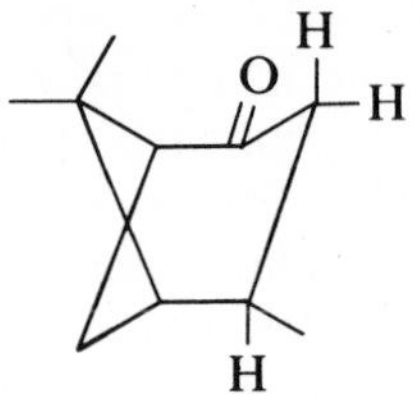

Reduction of the carbonyl group of the verbanones gives rise to the four verbanols, whose stereochemistry is as shown below (Regan, 1969). Similarly,

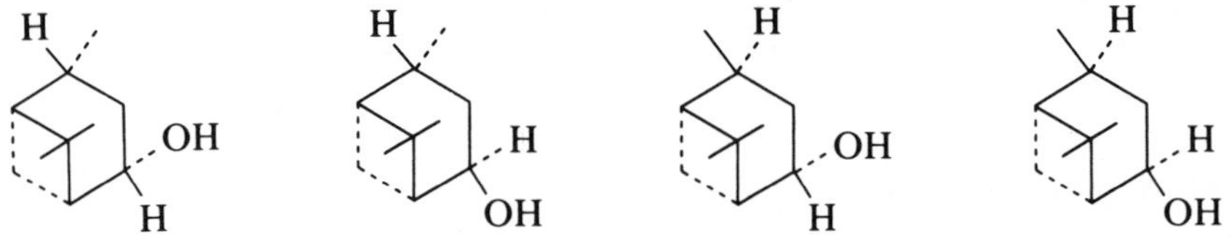

verbanol — neoverbanol — isoverbanol — neoisoverbanol

the pinocampheols are prepared from the pinocamphones, but do not follow quite the same nomenclature pattern as the verbanols. The other known

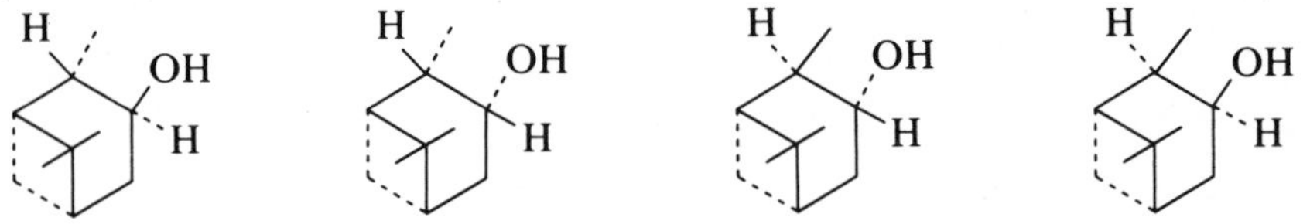

pinocampheol — neopinocampheol — isopinocampheol — neoisopinocampheol

saturated pinane alcohols are the 2-pinanols (Burrows and Eastman, 1959) and the myrtanols (Zweifel and Brown, 1964) whose stereochemistry is that shown below.

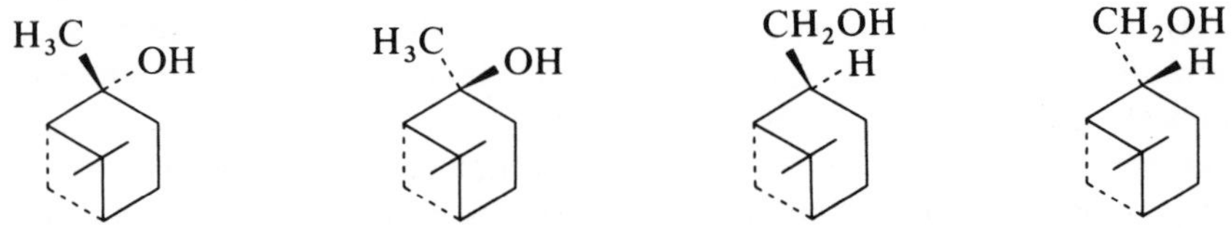

cis-2-pinanol (pinene hydrate) — *trans*-2-pinanol — *cis*-myrtanol — *trans*-myrtanol

Methanolysis of the *p*-nitrobenzoate esters of the 2-pinanols gives a mixture of products consistent with the scheme outlined in Fig. 5. The esters react to give the ions (XXVII) and (XXIX), but when the reaction is carried out in the presence of methoxide ion, kinetic control is imposed and the main products are the acid-labile 2-pinanyl methyl ethers; the presence of methoxide also shortens the life of the monocyclic ion sufficiently to suppress migration of the ionic centre (Salmon and Whittaker, 1971). Esters of the myrtanols (Salmon and Whittaker, 1967) and the pinocampheols (Huckel and Holzwarth, 1966), while giving some products of substitution at C_{10} and C_3 respectively, react mainly by shift of the carbonium ion centre to C_2, where delocalization gives rise to a mixture of (XXVII) and (XXIX).

The verbanyl esters have not been studied, but would be expected to react to give a mixture of unrearranged products and ring expanded products, the ring expansion involved being similar to that observed in the addition of hydrogen bromide to *cis*-δ-pinene.

D. Camphane Series

The parent hydrocarbons, camphane and the iso-camphanes, have in fact different carbon skeletons, but they interconvert readily, and are usually considered together. Camphane is readily prepared by the hydrogenation of bornylene (Forster, 1902) while hydrogenation of camphene (Sabatier and Senderens, 1901) or reduction with lithium aluminium hydride (Lalande and Ducasse, 1961) gives the *endo* isomer of isocamphane. Both *exo* and *endo* isomers of isocamphane have been prepared by hydrogenation of the

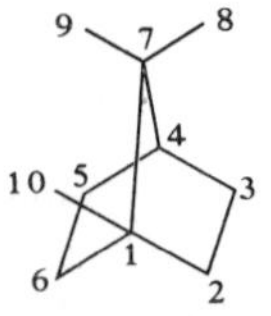

camphane

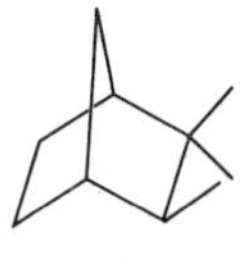

exo-isocamphane

endo-isocamphane

corresponding isocamphenes (Beckmann and Geiger, 1961). Six olefins can exist, and all have been reported. Isotricyclene (Hanack and Hahnle, 1962) and the isocamphenes are relatively difficult to synthesize (Beckmann and Geiger, 1961), but the other isomers are readily prepared. These latter olefins interconvert on refluxing over a titanium oxide catalyst (Swann and Cripwell, 1948) or over alumina (Watanabe *et al.*, 1962), the latter conditions giving the equilibrium mixture,

Camphene	73%
Tricyclene	24%
Bornylene	3%

camphene

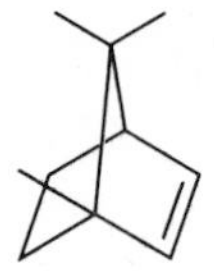

bornylene

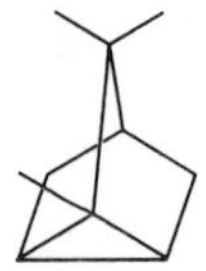

tricyclene

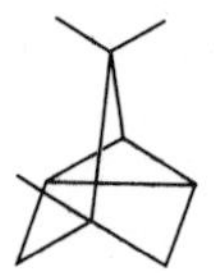

iso-tricyclene

exo-isocamphene

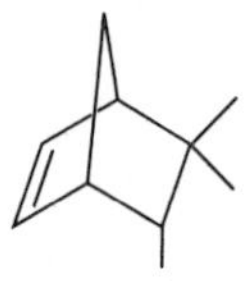

endo-isocamphene

Camphene is obtained industrially by rearrangement of α-pinene over a catalyst, and in the laboratory by dehydration of isoborneol. The former route yields a product contaminated with tricyclene and limonene, the latter tricyclene only. Both methods give partially racemized camphene (see below). Camphene has been obtained, free from tricyclene, and of high optical purity by pyrolysis of the methyl xanthate (Bunton *et al.*, 1961) or the phenylurethane (Vaughan *et al.*, 1963) of isoborneol; either reaction must involve a seven-membered ring intermediate in an ester pyrolysis reaction, but inspection of models shows this to be readily attainable. The methyl xanthate pyrolysis also yields bornylene, which can be obtained in better

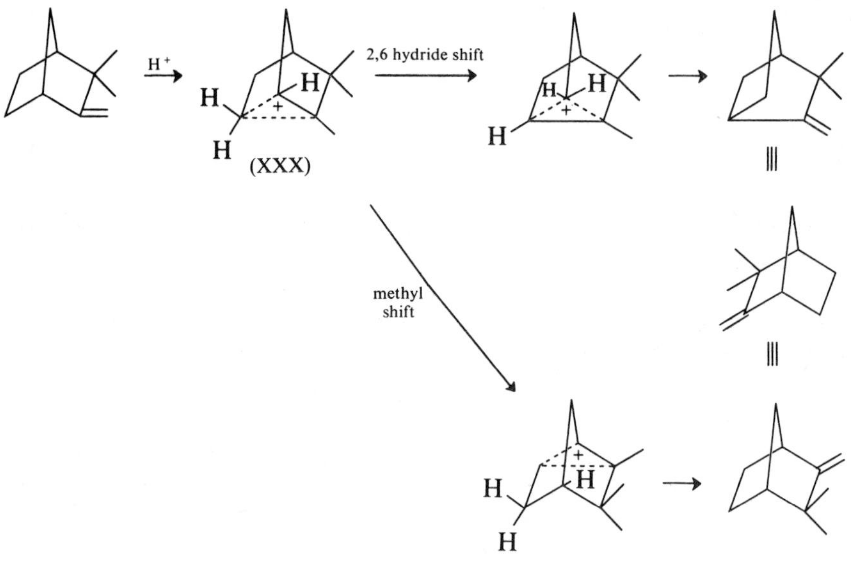

yield by pyrolysis of the bornyl ester, or by treatment of bornyl iodide with potassium (Meerwein and Joussen, 1922). Tricyclene is obtained by treatment of camphor hydrazone with mercuric oxide (Meerwein and van Emster, 1920), a reaction presumably going via the diazo compound; other routes to the diazo compound also give tricyclene as the end product (Clarke *et al.*, 1962).

Many of the acid catalysed rearrangements of camphene and its derivatives proceed via the ion (XXX). This ion can react at either C_1 or C_2, or can undergo either a 2,6-hydride shift or a methyl (Nametkin) shift. Both the latter processes lead, in camphene, to inversion, so both have been postulated as means of camphene racemization. Both in fact occur, the 2,6-shift being the faster (Vaughan *et al.*, 1963).

The addition of acids to camphene, and the acid catalysed hydration, also proceed through the ion involved in the racemization reaction (XXX). Thus,

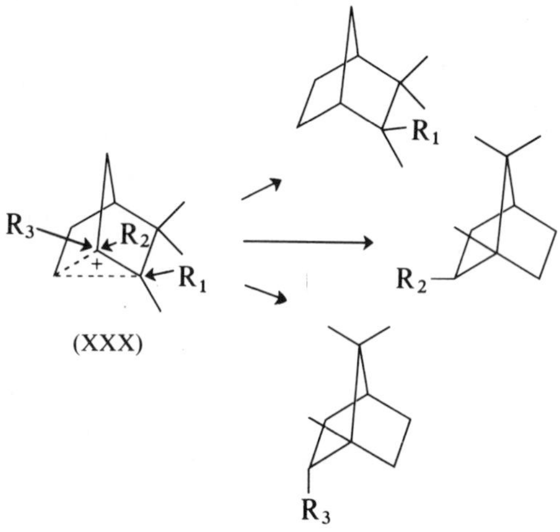

(XXX)

attack of nucleophile as shown by R_1 gives camphene hydro derivatives, R_1 = Cl, camphene hydrochloride (Nevell *et al.*, 1939), R_1 = OH, camphene hydrate (Bunton *et al.*, 1963). These products are not thermodynamically stable, however, so that prolonged reaction gives rise to products arising from attack of nucleophile as shown by R_2, giving isobornyl derivatives, R_2 = Cl, isobornyl chloride (Nevell *et al.*, 1939), R_2 = OH, isoborneol (Bunton *et al.*, 1963), R_2 = $OCOCH_3$, isobornyl acetate (Bertram and Walbaum, 1894). Prolonged reaction under vigorous conditions gives products arising from attack as shown by R_3, giving bornyl derivatives, R_3 = Cl, bornyl chloride (Meerwein and van Emster, 1922), R_3 = $OCOCH_3$, bornyl acetate (Beltrame *et al.*, 1960). The addition of phenols to camphene, catalysed by BF_3, does not obey these general rules; mild conditions give isobornyl phenyl ether, but under vigorous conditions, the *o*- and *p*-isomers of the phenols shown below were obtained (Erman, 1964), presumably from the ions (XXX) and (XXXI). The addition of chlorine to camphene (Jennings

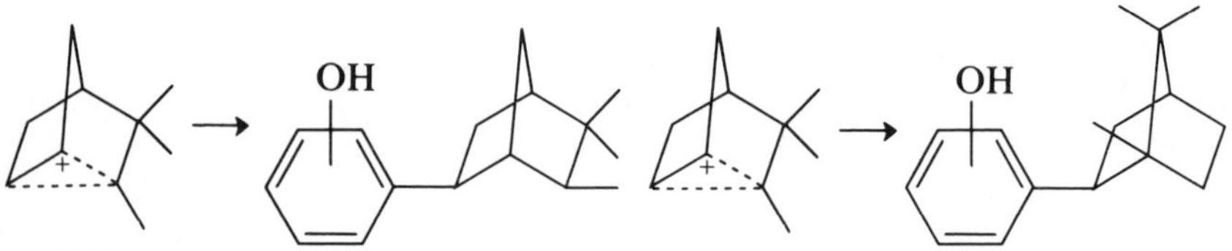

(XXXI) (XXX)

and Herschbach, 1965; Richey *et al.*, 1965) under conditions to preclude radical reactions gives a mixture of products which are consistent with initial addition of Cl^+ to give an ion which is a chlorinated derivative of (XXX).

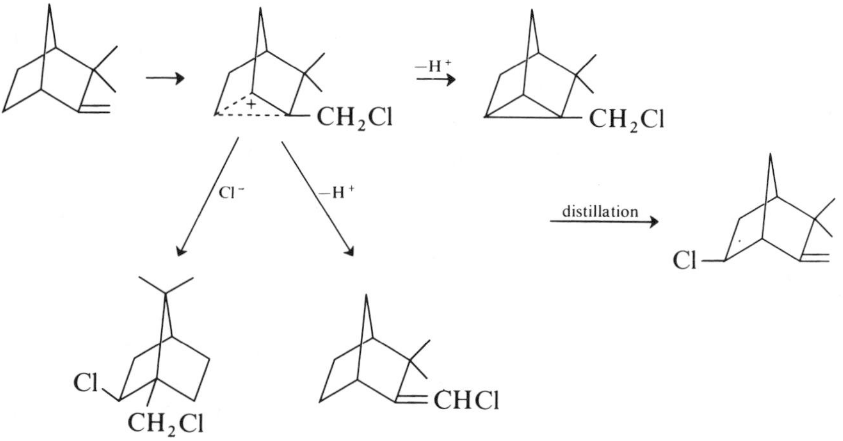

Tricyclene behaves like camphene in addition reactions, adding HCl more slowly than camphene to give camphene hydrochloride (Lipp, 1921; Meerwein and van Emster, 1920). Bornylene adds hydrogen fluoride (Hanack and Hahnle, 1962) to give camphene hydrofluoride, isobornyl fluoride, and the two fluorides produced by adding hydrogen fluoride to isotricyclene, these presumably having the fluorine on either C_4 or C_3. The isocamphenes also undergo hydration in dilute aqueous acid, giving as the main product isoborneol, together with the 5-hydroxy-iso-camphanes. The reaction is suggested to proceed via the route shown below (Beckmann and Geiger, 1962); though the ion would be expected to give camphene hydrate, this alcohol is not stable in acid solutions, and the observed product is isoborneol.

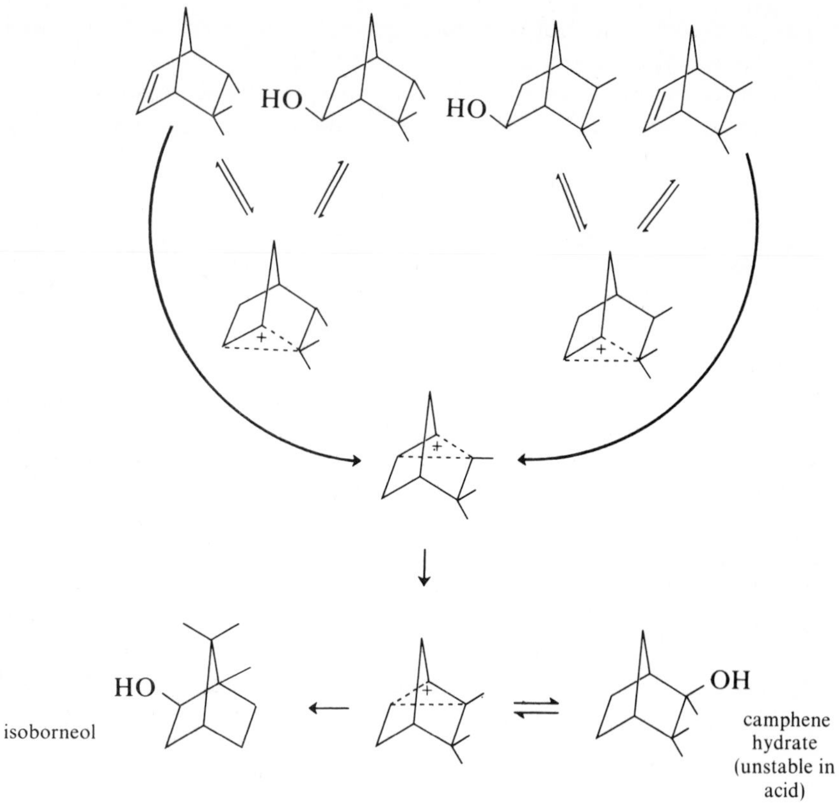

Three saturated ketones exist, camphor and epicamphor (Bredt and Perkin, 1913) having the full camphane skeleton, while camphenilone, which is formed by oxidative cleavage of the double bond of camphene, has one fewer carbon atom (Komppa and Hintikka, 1912). Camphenilone displays an

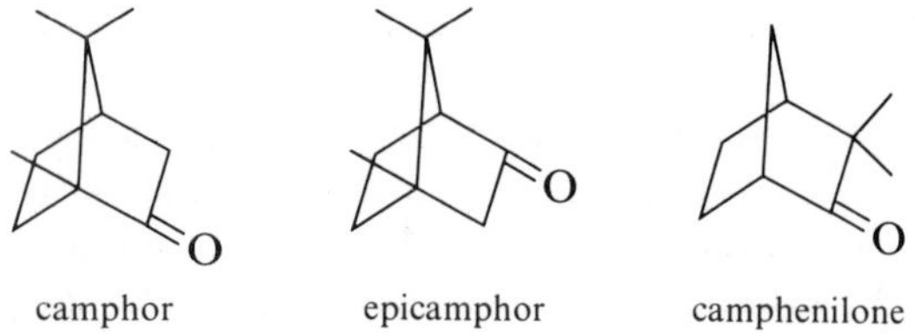

unusual reaction with base. Though lacking a hydrogen α- to the carbonyl group, it nonetheless incorporates deuterium when treated with base in solvents with exchangeable deuterium, and simultaneously racemizes. The reaction is believed to involve the homoenolate ion (Nickon *et al.*, 1966).

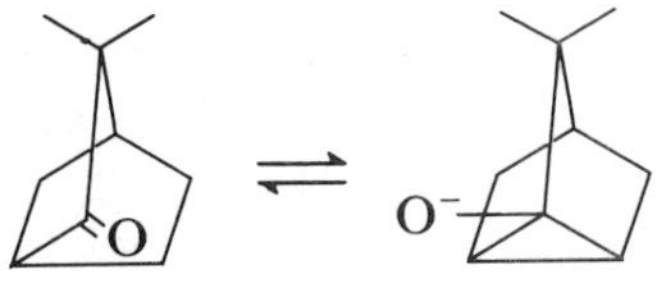

camphenilone homoenolate ion

Camphor occurs naturally in a fairly pure state, and has been known since the earliest days of chemistry (Libavius, 1595). It is of industrial importance, and is obtained industrially via the route,

pinene ⟶ camphene ⟶ isoborneol ⟶ camphor

Camphor undergoes a number of substitution reactions, the simplest being exchange of an α-hydrogen atom for a deuterium atom in the presence of base. Unexpectedly, it was found that the *exo*-α-hydrogen atom was replaced much more readily than the *endo*, though the latter did exchange on prolonged reaction (Thomas *et al.*, 1967). Similar observations of exchange of only the *exo*-hydrogen atom were made on norcamphor and isofenchone, showing that steric effects of the *gem*-dimethyl group did not influence the reaction. An ingenious explanation of this phenomenon has been put forward, based on strains in the transition state from the enol to the ketone resulting from eclipsing of the bridgehead hydrogen (Schleyer, 1964), but it has been pointed out that dehydrocamphor exchanges both α-hydrogen atoms (Jerkunica *et al.*, 1965) suggesting that it is the *endo*-hydrogen on C_6 which inhibits approach to the enol from the *endo* position.

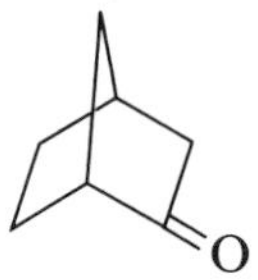

norcamphor

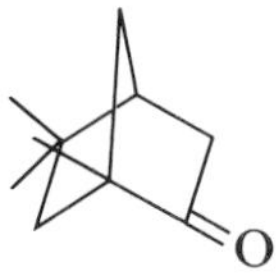

isofenchone

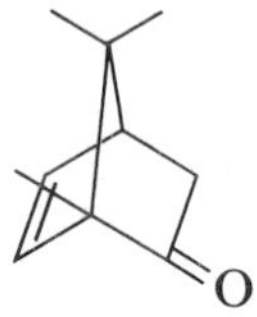

dehydrocamphor

Camphor readily undergoes halogenation on the α-carbon atom, C_3, via the enol (Marsh, 1890). Because of the lability of the halogen, its orientation has not been conclusively proved, but it is believed to be *exo*; methylation via the enol gives almost pure *exo* product, though, like the halogen derivative, it rearranges in the presence of base to a mixture of the *exo* and *endo* isomers (Corey, 1962). Under more vigorous conditions, dihalogenation on C_3 can be accomplished (Schiff, 1880), and a halogen can be substituted in the bridgehead methyl, C_{10} (Kachler and Spitzer, 1863). Halogenation on C_{10} can be accomplished more readily by reaction of the C_{10} alcohol with the halogen (Forster, 1902). Halogenation on C_9 does not take place directly, but C_9 halides can be prepared by the action of heat on the halogenosulphonic acids. The C_8 halogenosulphonic acids cannot be prepared directly, but the C_8 halides can be prepared by the ingenious route shown below (Corey *et al.*, 1957).

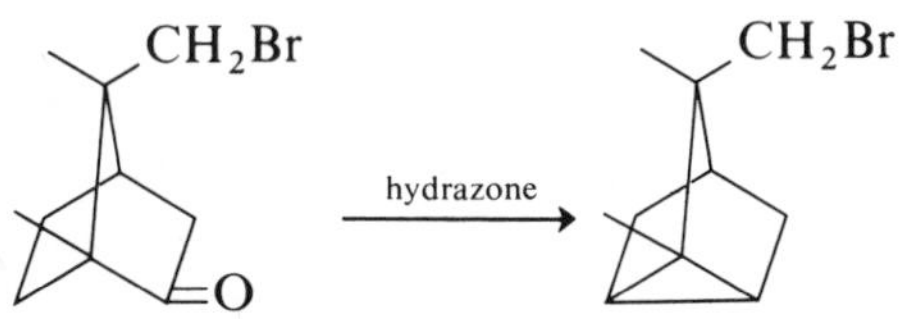

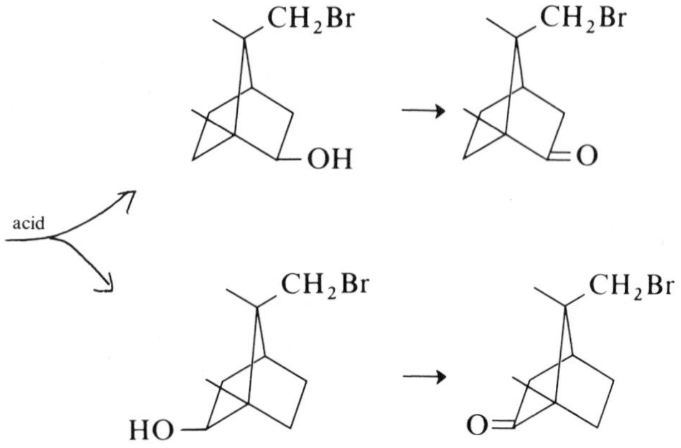

Selective removal of halogen is also possible, so that almost any combination of substituents in the three positions can be prepared.

In concentrated sulphuric acid, camphor racemizes; experiments in which C_8 is labelled with ^{14}C have shown that interchange of C_8 and C_{10} forms part of the racemization process, a mechanism of which is set out in Fig. 6 (Finch and Vaughan, 1969; Miki *et al.*, 1955). This process is consistent with the observation that the label shifts completely from C_8 to C_{10} rather than equilibrates between the two positions, this resulting from the preference shown to 2,3, *exo* shifts over 2,3 *endo* shifts in the bornanes and norbornanes, a phenomenon elegantly demonstrated by Berson *et al.* (1965).

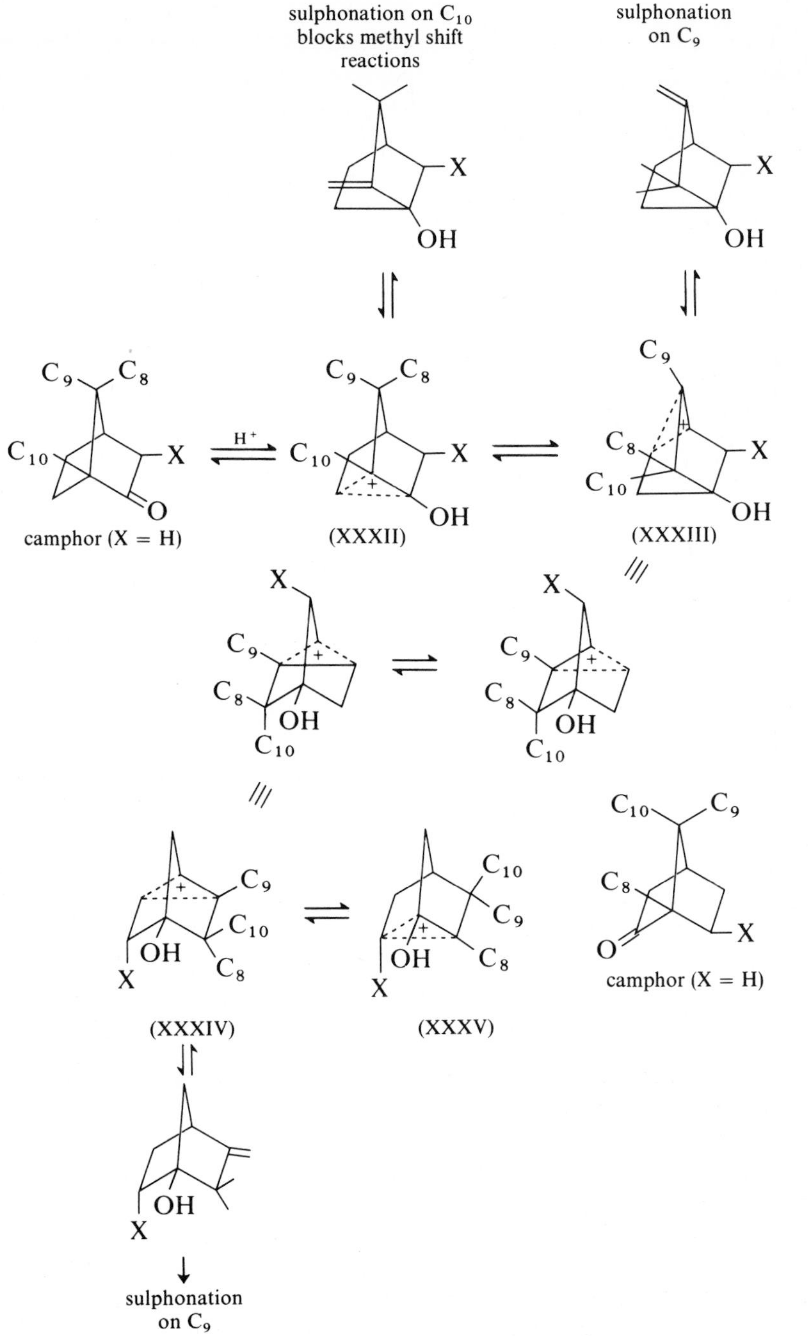

FIG. 6. Racemization and sulphonation of camphor.

Under similar conditions to those required for the racemization of camphor, 3,9-dibromocamphor rearranges to 6,9-dibromocamphor (Miki *et al.*, 1955). This process is also consistent with the scheme set out in Fig. 6. In the case of camphor, X = H, so that starting material and product are optical antipodes; in the second case, X = Br, and there is a second bromine on C_9, so that the process leads to a rearranged product.

When camphor is treated with oleum, or chlorosulphonic acid in concentrated sulphuric acid, racemization is accompanied by sulphonation. Under milder conditions (XXXII) is formed in low concentrations, so that (XXXIII) is formed very slowly, and sulphonation takes place on C_{10} by attack on the vinyl olefin in equilibrium with (XXXII). Under more vigorous conditions (XXXIII) is formed in higher concentration, so that reaction now takes place mainly with the vinyl olefin in equilibrium with this ion, giving sulphonation on C_9. Similarly, sulphonation can take place with the olefin in equilibrium with (XXXIV) which is the mirror image of (XXXIII). Sulphonation of (XXXIII) evidently favours the hydride transfer reaction required to convert it to (XXXIV) over the methyl shift required to convert it to (XXXII), since the sulphonic acid obtained is always completely racemic.

When the sulphonation reaction is carried out on α-bromocamphor, the 3-bromo-9-sulphonic acid obtained is optically active (Corey *et al.*, 1957), and does not contain any of the 6-bromo acid which would be obtained by a process similar to that involved in the racemization during camphor sulphonation. Bromine must destabilize the ion (XXXIV) by its electron withdrawing properties, thus favouring reaction of (XXXIII) by a methyl shift to give (XXXII) rather than by the hydride shift, which would lead to formation of the 6-bromo isomer. On prolonged reaction, however, rearrangement to the 6-bromo isomer does take place (Miki *et al.*, 1955) this being the thermodynamically favoured product.

Reduction of camphor gives a mixture of the secondary alcohols borneol and isoborneol, the proportions of each depending on the method of reduction used (Ourisson and Rassat, 1960). The *gem*-dimethyl bridge inhibits approach to the carbonyl function from the *exo* side, so that reagents such as lithium aluminium hydride give product mixtures in which isoborneol predominates. Hydration of camphene gives one of the corresponding

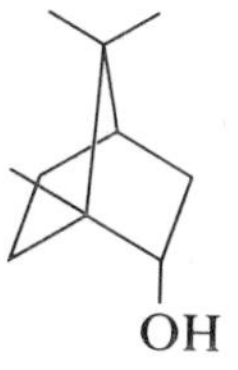

borneol

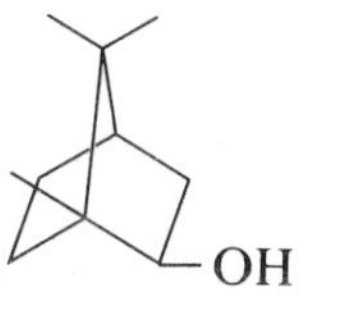

isoborneol

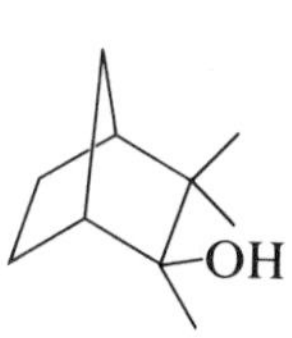

camphene hydrate

methyl camphenilol

tertiary alcohols camphene hydrate, while the other, methyl camphenilol, can be obtained indirectly. The corresponding chlorides are not usually obtained from the alcohols, as this type of reaction gives a mixture of products. Isobornyl chloride and camphene hydrochloride can be obtained from camphene, as noted earlier, while bornyl chloride is obtained by adding hydrogen chloride to α-pinene. Methyl camphenilyl chloride has not been obtained.

Reactions of these alcohols and their derivatives which involve fission of the bond between the ring and the derivative show marked rate differences within the secondary and tertiary pairs.

	Chloride methanolysis (Beltramé *et al.*, 1964)	Heterolysis of alcohol in acid (Bunton *et al.*, 1963)	Solvolysis of *p*-nitrobenzoate (Bunton *et al.*, 1967)
Borneol	7×10^{-6}	$3{\cdot}8 \times 10^{-6}$	—
Isoborneol	1	0·75	—
Methyl camphenilol	—	1	1
Camphene hydrate	2×10^{3}	$1{\cdot}3 \times 10^{3}$	$1{\cdot}7 \times 10^{3}$

These rate differences arise in part from enhancement of the rates of reaction of derivatives of the *exo* alcohols (camphene hydrate and isoborneol) by participation of electrons shifting to give the ion (XXX) in ionization of the substituent, and in isoborneol derivatives, in part from steric acceleration by the *gem*-dimethyl bridge. The *endo*-alcohol derivatives react at rates similar to the corresponding secondary and tertiary butyl alcohol derivatives,

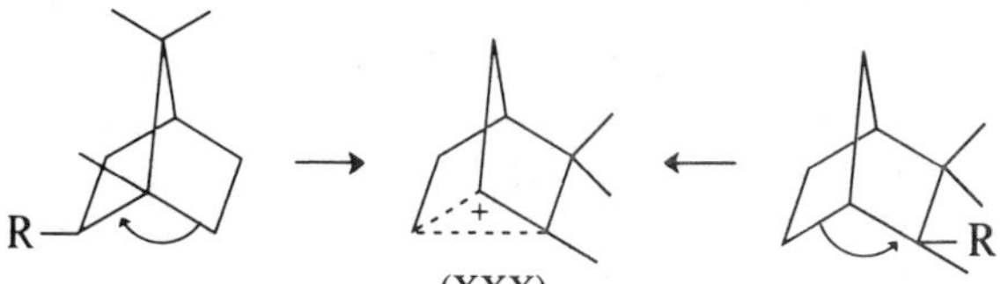

(XXX)

but presumably give an ion which subsequently rearranges to (XXX), since all three chlorides, on solvolysis in alkaline methanol (Beltramé *et al.*, 1964), give similar mixtures of camphene hydrate methyl ether, isobornyl methyl ether and camphene. The *p*-toluene sulphonate of borneol gives a product mixture under these conditions which is similar, except for a much higher yield of camphene, which may arise by a competing direct elimination reaction (Hückel *et al.*, 1968).

On deamination of the amines of borneol and isoborneol, the latter gives products similar to those obtained from the chloride, except for the unexplained production of 10% of camphor, but bornylamine deaminates to give approximately 40% products of ring opening, a reaction in which the high energy of the carbonium ion permits delocalization to give an ion (XXIX) similar to that obtained from the pinanols (Hückel and Kern, 1969).

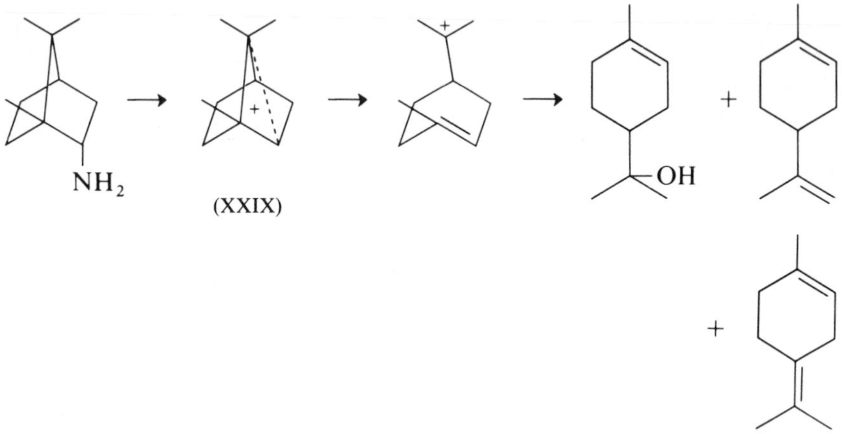

E. Fenchane Series

The fenchanes differ from the camphanes in the position of one methyl group; this has, however, a marked effect on the symmetry of the molecule, so that there are many more isomers of the fenchanes than the camphanes though the basic skeletons of the two systems are identical and both react by similar mechanisms. This increased complexity probably accounts for the lack of interest in these compounds, since *endo*-fenchol is a product of α-pinene hydration, and is readily available. The development of gas chromatography has made analysis of mixtures of the fenchenes and their derivatives possible, but even so, their study does not appear to have caught up from its early neglect.

Like the camphanes, the parent fenchanes have more than one basic carbon skeleton. In this case, there are three basic skeletons, all readily interconvertible, and a total of five fenchanes exist, all of which are known. Fenchane itself can be prepared by the action of sodium ethoxide on the hydrazone of either fenchone (Wolff, 1911) or isofenchone (Komppa and Hasselstrom, 1932); the *endo* isomers of α- and β-fenchanes can be obtained by hydrogenation of α-fenchene and β-fenchene (Zelinski, 1904), while the *exo* isomers can be obtained by hydroboration, equilibration of the boron complex and acidolysis on the same olefins (Barthélémy *et al.*, 1968).

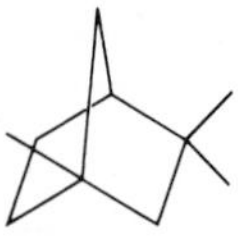
fenchane

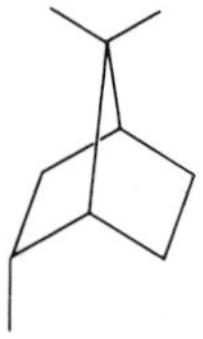
endo-α-fenchane (isobornylane)

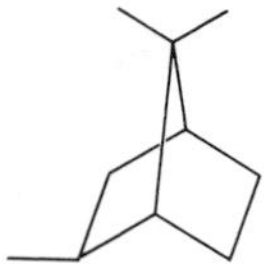
exo-α-fenchane

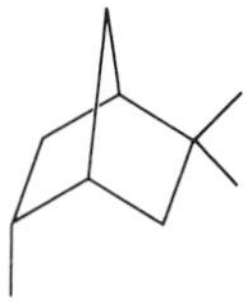
endo-β-fenchane

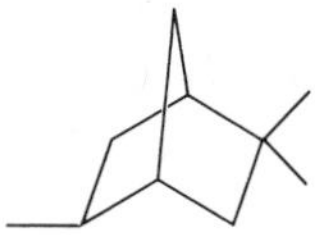
exo-β-fenchane

Six olefins are known and characterized out of the ten which would be expected from the above alkanes (omitting methyl santenes), and a seventh has recently been obtained by isomerization of *cis*-δ-pinene (Bartélémy *et al.*,

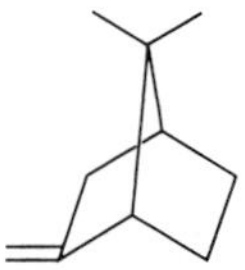
α-fenchene

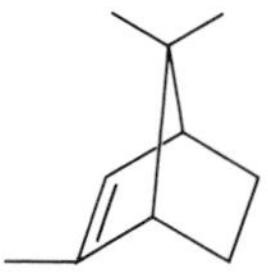
ξ-fenchene

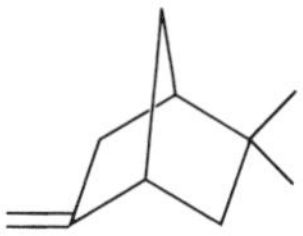
β-fenchene

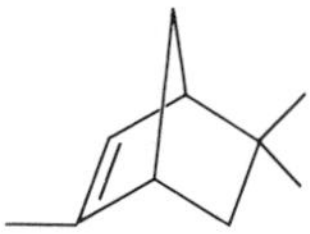
γ-fenchene

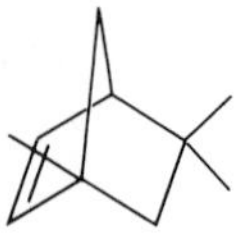
δ-fenchene

ε-fenchene (1-methyl santene)

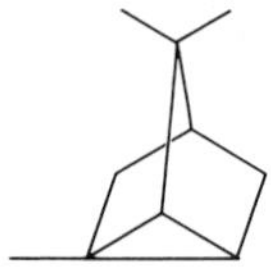
cyclofenchene

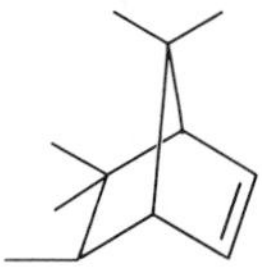
fenchene from *cis*-δ-pinene (could be named η-fenchene)

1968). A mixture of seven isomeric fenchenes can be obtained by the dehydration of *endo*-fenchol with potassium hydrogen sulphate (Hückel and Kern, 1965). It has the composition,

$$\alpha,\ 17{\cdot}3\%;\ \beta,\ 27{\cdot}0\%;\ \gamma + \xi,\ 19{\cdot}0\%;\ \delta,\ 5{\cdot}1\%;\ \varepsilon,\ 9{\cdot}4\%;\ \text{cyclo},\ 22{\cdot}2\%$$

which probably represents the equilibrium composition of the fenchenes. This reaction has been studied further by carrying out the dehydration of the O-deuterated alcohol with potassium deuterium sulphate (Doering and Wolfe, 1951), which showed that for the principal products of the reaction, α-, β-, and γ-fenchenes, and cyclofenchene, most of the deuterium is incorporated at the olefin or bridgehead carbon atoms, a result which rules out cyclofenchene as a possible intermediate in the formation of β- and γ-fenchenes from *endo*-fenchol, the reaction probably proceeding through a 2,6-hydride shift. The two ions involved correspond to the single ion obtained by

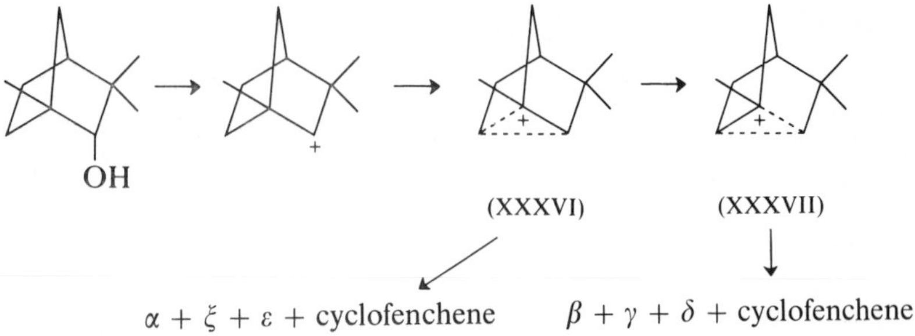

protonation of camphene (XXX), and have similarly important roles in the rearrangements of the fenchenes and their derivatives, though the details of many reactions remain to be studied. On account of their ready interconversions, the fenchenes are very difficult to obtain pure, with the exception of cyclofenchene, which is formed in good yield and about 95% purity by pyrolysis of the acetates of either *endo*-fenchol or *endo*-isofenchol (Hückel and Kern, 1965). The other fenchenes have been obtained to a high degree of purity, and their infra-red spectra recorded, in a comprehensive investigation by Pulkkinen (1965, 1957).

A detailed study of the interconversions of the fenchenes in acid has not been made; the acid catalysed hydration has also been neglected, though the addition of acetic acid, catalysed by sulphuric acid, to either α-fenchene or cyclofenchene, gives *exo*-isofenchyl acetate, probably the less sterically hindered, and therefore the most stable, of the secondary acetates obtainable from the ions (Bertram and Helle, 1900). Three other fenchenes, ξ, β, and γ would probably give the same product. The reaction of hydrogen chloride with the fenchenes has been more fully investigated (Hückel and Volkmann,

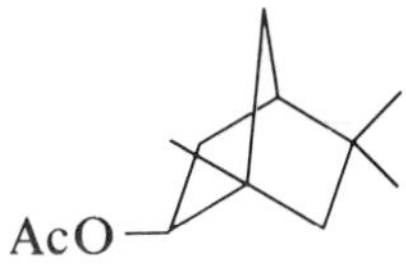

exo-isofenchyl acetate

1963). α- and ξ-fenchenes, on treatment with HCl, give a mixture of the tertiary *exo*- and secondary *exo*-chlorides, which slowly rearranges to the more stable tertiary *endo*-chloride. Under vigorous conditions, i.e. with phosphorus pentachloride, this further rearranges to give a mixture of the *exo* tertiary and *exo* secondary chlorides related to β-fenchene. These latter chlorides can also be obtained by the addition of HCl to β- or γ-fenchene. Two fenchones exist, fenchone being prepared by oxidation of either fenchol,

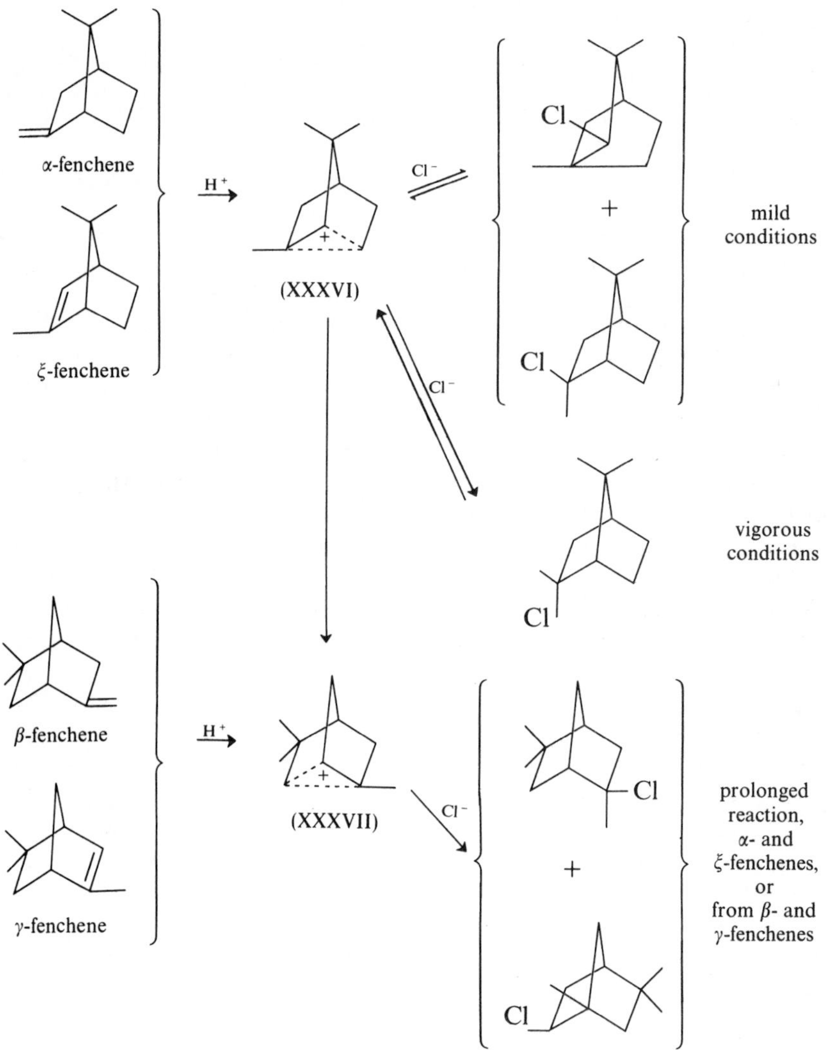

FIG. 7. Addition of HCl to fenchenes.

while isofenchone can be obtained from *endo*-fenchol or α-fenchene, by treatment with sulphuric acid in acetic acid to give *exo*-isofenchyl acetate, which can be saponified and oxidized. The fenchocamphorones, which, like camphenilone, contain one carbon atom fewer than the parent hydrocarbon, are prepared by oxidative cleavage of the double bond in α- and β-fenchenes. A feature of these ketones is that the fenchones and β-fenchocamphorone lack

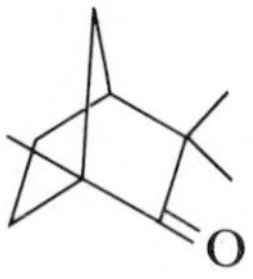
fenchone

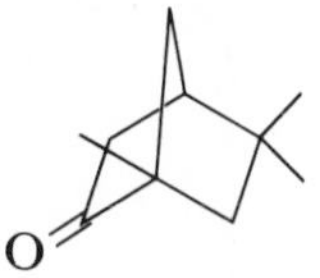
isofenchone

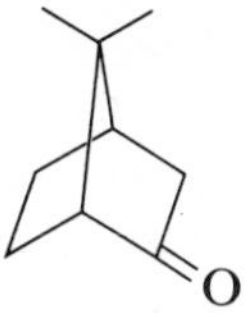
α-fenchocamphorone

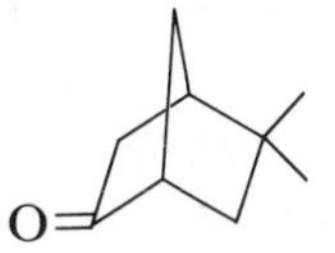
β-fenchocamphorone

the *gem*-dimethyl bridge of camphor; consequently, in contrast to camphor, the carbonyl group is more readily attacked from the bridge side, and reduction with reagents such as lithium aluminium hydride gives *endo* alcohols, while reduction with aluminium isopropoxide gives *exo* isomers. The reverse is true with α-fenchocamphorone, since here the *gem*-dimethyl group makes the bridge side of the ketone the more hindered.

Eight fenchyl alcohols are known, consisting of two sets of four, each set analogous to the camphane alcohols. The two sets are related by a 2,6-hydrogen shift. The older nomenclature, regarding the alcohols as α and β rather than *exo* and *endo* is particularly confusing, but these old names are included in brackets. Evidence that the *exo* and *endo* assignments are correct has been provided by spectroscopic studies (Hirstarvi, 1956).

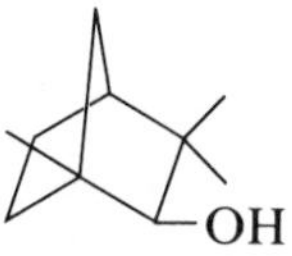

exo-fenchol
(β-fenchol)

endo-fenchol
(α-fenchol)

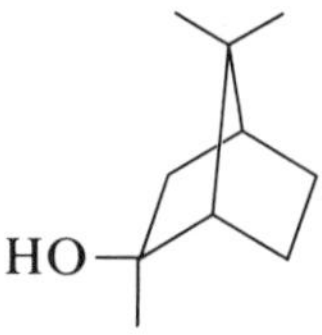

exo-α-fenchene
hydrate

endo-α-fenchene
hydrate

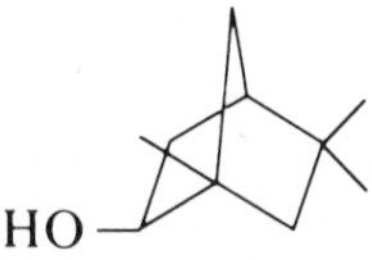

exo-isofenchol
(α-isofenchol)

endo-isofenchol
(β-isofenchol)

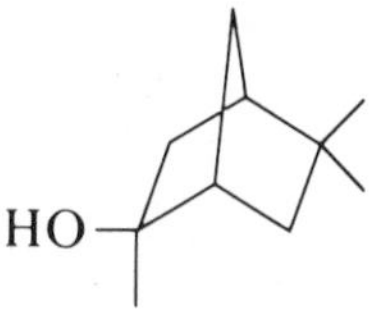

exo-β-fenchene
hydrate

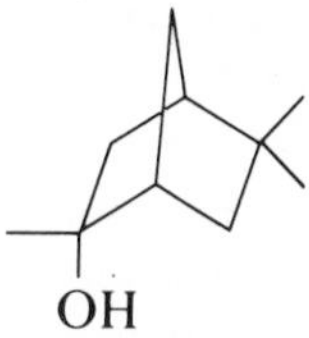

endo-β-fenchene
hydrate

All four secondary alcohols have been converted to their *p*-toluene sulphonates, and the solvolyses of these esters studied (Hückel and Kern, 1965). The rates of ethanolysis show acceleration of solvolysis of the *exo* isomers by electron participation, the *exo* isomer solvolysing faster than the *endo* by a factor of 3700 in the fenchols and 2800 in the isofenchols, a considerable reduction over the rate differences observed between the borneol isomers. This probably results from the lack of further acceleration to the *exo* isomer by the steric effect of the *gem*-dimethyl bridge, since the rate difference is similar to that observed in derivatives of the tertiary alcohols having the isocamphane skeleton, i.e. camphene hydrate and methylcamphenilol.

Methanolysis of the esters in alkaline methanol shows that the reaction products are in line with those expected for this type of reaction, kinetic control of the reaction preventing the 2,6-hydride shift needed to interconvert the two systems.

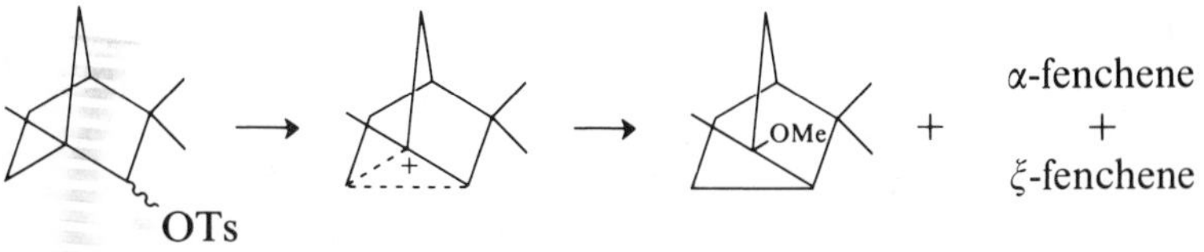

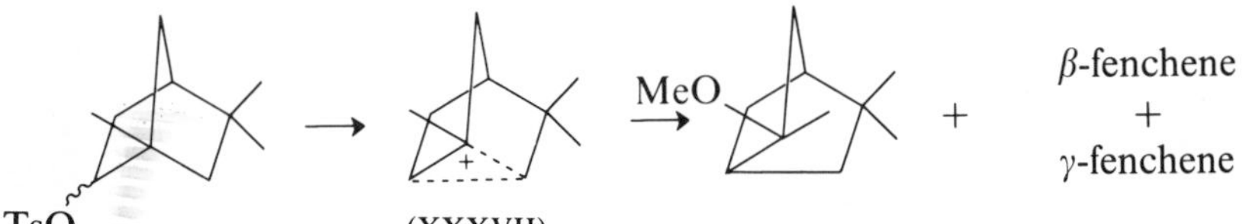

(XXXVII)

An exception is the methanolysis of *endo*-isofenchyl *p*-toluene sulphonate, which in addition to the products indicated above gives 24 % of *exo*-isofenchyl methyl ether, which is suggested to arise by a bimolecular substitution reaction. Formolysis of the above esters gives a mixture of formate esters which are not stable in formic acid, and slowly isomerize to a mixture of the formates of *exo*-fenchol and *exo*-isofenchol in which the latter predominates.

Deamination of the fenchylamines proceeds in the same general pattern as the solvolytic reactions of the esters, except that ring opening occurs in reaction of the *endo* isomer to give *p*-menthane derivatives (Hückel and Kern, 1969). This cannot occur from the ion (XXVIII), which, by analogy with the deamination of bornylamine, would be expected to be formed, but work on the pinanyl esters has shown that this ion can interconvert with the ion (XXIX), (Salmon and Whittaker, 1970), which is that formed from bornylamine and from pinane derivatives, and has been shown to ring open to give the observed *p*-menthanes, limonene, terpinolene and α-terpineol.

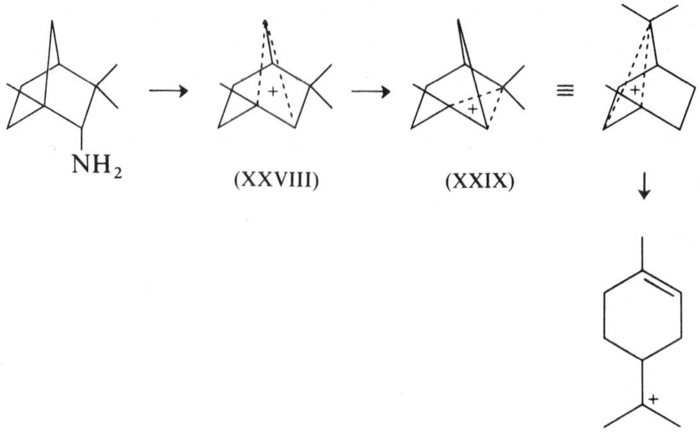

(XXVIII) (XXIX)

REFERENCES

Abraham, R. J., Bottom, F. H., Cooper, M. A., Salmon, J. R. and Whittaker, D. (1969). *Org. Mag. Res.* **I**, 51.

Acharya, S. P. and Brown, H. C. (1967). *J. Amer. Chem. Soc.* **89**, 1925.

Acharya, S. P., Brown, H. C., Suzuki, A., Nozawa, S. and Itoh, M. (1969). *J. Org. Chem.* **30**, 1.

Ansari, H. R. (1970). *Flavour Ind.* **1**, 252.

Arbuzov, B. A. and Isaeva, V. G. (1953). *Izvest. Akad. Nauk. SSSR Otd. Khim. Nauk.* 843.

Baeyer, A. (1894). *Chem. Ber.* **27**, 810.

Baeyer, A. (1898). *Chem. Ber.* **31**, 2067.

Baldwin, J. E. and Kreuger, S. M. (1969a). *J. Amer. Chem. Soc.* **91**, 2396.

Baldwin, J. E. and Kreuger, S. M. (1969b). *J. Amer. Chem. Soc.* **91**, 6444.

Baldwin, M. A., Banthorpe, D. V., Louden, A. and Waller, F. D. (1967). *J. Chem. Soc.* (*B*) 509.

Banthorpe, D. V. (1963). "Elimination Reactions". Elsevier, London.

Banthorpe, D. V. and Whittaker, D. (1966a). *Quart. Rev.* **20**, 373.

Banthorpe, D. V. and Whittaker, D. (1966b). *Chem. Rev.* **66**, 643.

Banthorpe, D. V. and Davies, H. ff. S. (1968a). *J. Chem. Soc.* (*B*) 1339.

Banthorpe, D. V. and Davies, H. ff. S. (1968b). *J. Chem. Soc.* (*B*) 1356.

Barber, L., Chapman, O. L. and Lassila, J. D. (1968). *J. Amer. Chem. Soc.* **90**, 5933.

Bardyshev, I. I., Shashkina, M. Ya. and Kulikov, V. I. (1966). *Zh. Org. Khim.* **2**, 1039.

Barthélémy, M., Monthéard, J-P. and Bessière-Chrétien, Y. (1968). *Bull. Soc. Chim. Fr.* 4881.

Bartlett, P. D. (1965). "Nonclassical Ions." W. A. Benjamin, New York.

Barton, D. H. R., Cookson, R. C., Klyne, W. and Shoppee, C. W. (1954). *Chem. Ind.* (*London*) 21.

Bates, R. B., Caldwell, E. S. and Klein, H. P. (1969). *J. Org. Chem.* **34**, 2615.

Bates, T. H., Best, J. V. F. and Ffrancon Williams, T. (1962a). *J. Chem. Soc.* 1521.

Bates, T. H., Best, J. V. F. and Ffrancon Williams, T. (1962b). *J. Chem. Soc.* 1531.

Beckmann, S. and Geiger, B. (1961). *Chem. Ber.* **94**, 1910.

Beckmann, S. and Geiger, B. (1962). *Chem. Ber.* **95**, 2101.
Bell, R. P. and Cox, B. G. (1970). *J. Chem. Soc.* (*B*) 194.
Bellamy, A. J. and Whitham, G. H. (1964). *J. Chem. Soc.* 4035.
Beltramé, P., Bunton, C. A. and Whittaker, D. (1960). *Chem. Ind.* (*London*) 557.
Beltramé, P., Bunton, C. A., Dunlop, A. and Whittaker, D. (1964). *J. Chem. Soc.* 658.
Berson, J. A. (1963). In "Molecular Rearrangements" (P. de Mayo, ed.), Vol. 1, pp. 111–231, Interscience, New York.
Berson, J. A., Bergmann, R. G., Hammons, J. H., McRowe, A. W., Remanic, A. and Houston, D. (1965). *J. Amer. Chem. Soc.* **87**, 3246.
Bertram, J. and Walbaum, H. (1894). *J. Pr. Chem.* (*II*) **49**, 1.
Bertram, J. and Helle, J. (1900). *J. Pr. Chem.* (*II*) **61**, 300.
Borowiecki, L., Zacharewicz, W. and Przystrupa, J. (1964). *Roczniki Chem.* **38**, 87.
Bredt, J. and Perkin, W. H., Jr. (1913). *J. Chem. Soc.* **103**, 2182.
Breslow, R., Groves, J. T. and Olin, S. S. (1966). *Tetrahedron Lett.* 4717.
Brown, H. C. (1962). "The Transition State." Chem. Soc. Special Publ. No. 16, pp. 140–158, London.
Brown H. C. (1966). *Chem. Brit.* 199.
Brown, H. C., Singh, K. P. and Garner, B. J. (1963). *J. Organometal. Chem.* **1**, 2.
Brown, H. C. and Suzuki, A. (1967). *J. Amer. Chem. Soc.* **89**, 1933.
Büchi, G. and Goldman, I. M. (1957). *J. Amer. Chem. Soc.* **79**, 4741.
Buechi, J. and Iconomou, N. (1965). *Pharm. Acta. Helv.* **40**, 421.
Bunton, C. A., Khaleeluddin, K. and Whittaker, D. (1961). *Nature, Lond.* **190**, 715.
Bunton, C. A., Khaleeluddin, K. and Whittaker, D. (1963). *Tetrahedron Lett.* 1825.
Bunton, C. A., O'Connor, Charmian and Whittaker, D. (1967). *J. Org. Chem.* **32**, 2812.
Burrows, W. D. and Eastman, R. H. (1959). *J. Amer. Chem. Soc.* **81**, 245.
Burns, W. D. P., Carson, M. S., Cocker, W. and Shannon, P. V. R. (1968). *J. Chem. Soc.* (*C*) 3073.
Burwell, R. L. (1951). *J. Amer. Chem. Soc.* **73**, 4461.
Carson, M. S., Cocker, W., Grayson, D. H., Pratt, A. C. and Shannon, P. V. R. (1968a). *J. Chem. Soc.* (*B*) 1136.
Carson, M. S., Cocker, W., Evans, S. M. and Shannon, P. V. R. (1968b). *Tetrahedron Lett.* 6153.
Carson, M. S., Cocker, W., Grayson, D. H. and Shannon, P. V. R. (1969). *J. Chem. Soc.* (*C*) 2220.
Carson, M. S., Cocker, W. and Kulkorni, P. B. (1970). *Tetrahedron Lett.* 669.
Clarke, P., Whiting, M. C., Papenmeier, G. and Reusch, W. (1962). *J. Org. Chem.* **27**, 3356.
Ciamician, G. and Silber, P. (1908). *Ber. Deut. Chem. Ges.* 1928.
Cocker, W., Shannon, P. V. R. and Staniland, P. A. (1966a). *J. Chem. Soc.* (*C*) 41.
Cocker, W., Shannon, P. V. R. and Staniland, P. A. (1966b). *J. Chem. Soc.* (*C*) 946.
Cocker, W., Shannon, P. V. R. and Staniland, P. A. (1967a). *J. Chem. Soc.* (*C*) 485.
Cocker, W., Shannon, P. V. R. and Staniland, P. A. (1967b). *J. Chem. Soc.* (*C*) 915.
Cocker, W., Pratt, A. C. and Shannon, P. V R. (1968a). *J. Chem. Soc.* (*C*) 484.
Cocker, W., Hanna, D. P. and Shannon, P. V. R. (1968b). *J. Chem. Soc.* (*C*) 489.
Cocker, W., Hanna, D. P., and Shannon, P. V. R. (1969). *J. Chem. Soc.* (*C*) 1302.
Corey, E. J. (1962). *J. Amer. Chem. Soc.* **84**, 2611.
Corey, E. J. and Burke, H. J. (1965). *J. Amer. Chem. Soc.* **78**, 174.
Corey, E. J., Burke, H. J. and Remers, W. A. (1956). *J. Amer. Chem. Soc.* **78**, 180.
Corey, E. J., Chow, S. W. and Scherrer, R. A. (1957). *J. Amer. Chem. Soc.* **79**, 5773.

Cookson, R. C., Hudec, J., Knight, S. A. and Whitear, B. R. D. (1963). *Tetrahedron* **19**, 1995.
Cooper, M. A., Salmon, J. R., Whittaker, D. and Scheidegger, U. (1967). *J. Chem. Soc.* (*B*) 1259.
Cooper, M. A. and Whittaker, D. (1968). Quoted by Banthorpe and Davies (1968).
Crowley, K. J. (1962). *Proc. Chem. Soc.* 245.
Crowley, K. J. (1968). *J. Org. Chem.* **33**, 3679.
Dauben, W. (1966). *J. Amer. Chem. Soc.* **88**, 2742.
Dauben, W. G., Lorber, M. and Fullerton, D. S. (1969). *J. Org. Chem.* **34**, 3587.
Dieffenbacher, A. and von Philipsborn, W. (1966). *Helv. Chim. Acta.* **49**, 57.
Doering, W. von E. and Wolfe, A. P. (1951). XIIth International Congress of Pure and Applied Chemistry, New York, p. 437 of Abstracts. *Perfum. and Essent. Oil Rec.* **42**, 414.
Doering, W. von E. and Lambert, J. B. (1963). *Tetrahedron* **19**, 1989.
Enklaar, C. J. (1907). *Rec. trav. chim.* **26**, 157.
Erman, W. F. (1964). *J. Amer. Chem. Soc.* **86**, 2887.
Erman, W. F. (1967). *J. Amer. Chem. Soc.* **89**, 3828.
Erman, W. F. and Gibson, T. W. (1969). *Tetrahedron* **25**, 2493.
Finch, A. M. T., Jr. and Vaughan, W. R. (1969). *J. Amer. Chem. Soc.* **91**, 1416.
Fordham, W. D. (1968). Reports on the progress of Applied Chemistry, pp. 264–273.
Forster, M. O. (1902). *J. Chem. Soc.* **81**, 267.
Frank, G. (1968). *J. Chem. Soc.* (*B*) 130.
Goering, H. L. and Schewene, C. B. (1965). *J. Amer. Chem. Soc.* **77**, 3516.
Goldblatt, L. H. and Palkin, S. (1941). *J. Amer. Chem. Soc.* **63**, 3517.
Gollnik, K., Schroeter, S., Ohloff, G., Schade, G. and Schenck, G. O. (1965). *Ann.* **687**, 14.
Gollnik, K. (1966a). *Tetrahedron Lett.* 327.
Gollnik, K. (1966b). *Tetrahedron Lett.* 5157.
Gollnik, K. and Schade, G. (1966a). *Tetrahedron* **22**, 123.
Gollnik, K. and Schade, G. (1966b). *Tetrahedron* **22**, 133.
Gollnik, K. and Schade, G. (1966c). *Tetrahedron Lett.* 2335.
Goryaev, M. I., Shabalina, V. I. and Dembitskii, A. D. (1964). *Zh. Obshch. Khim.* **34**, 3855.
Hanack, M. (1960). *Chem. Ber.* **844**, 93.
Hanack, M. and Hähnle, R. (1962). *Chem. Ber.* **95**, 191.
Hartshorn, M. P. and Wallis, A. F. A. (1964). *J. Chem. Soc.* 5254.
Hawkins, J. E. and Hunt, H. G. (1951). *J. Amer. Chem. Soc.* **73**, 5379.
Hawkins, J. E. and Burnis, W. A. (1959). *J. Org. Chem.* **24**, 1507.
Heckert, D. C. and Kropp, P. J. (1968). *J. Amer. Chem. Soc.* **90**, 4911.
Henderson, G. G. and Pollock, E. F. (1910). *J. Chem. Soc.* **97** 1620.
Hibbert, H. and Cannon, L. T. (1924). *J. Amer. Chem. Soc.* **46**, 119.
Hirstarvi, P. (1956). *Suomen Kemistilehti* **29B**, 138.
Hope, A. J. N. and Mitchell, S. (1954). *J. Chem. Soc.* 4217.
Houlihan, W. J., Levy, J. and Mayer, J. (1959). *J. Amer. Chem. Soc.* **81**, 4692.
Hückel, W. and Volkmann, D. (1963). *Ann.* **664**, 31.
Hückel, W. and Kern, H. J. (1965). *Ann.* **687**, 40.
Hückel, W. and Holzwarth, D. (1966) *Ann.* **697**, 69.
Hückel, W., Jennewein, C. M., Kern, H. J. and Vogt, O. (1968). *Ann.* **719**, 157.
Hückel, W. and Kern, H. J. (1969). *Ann.* **728**, 49.
Hughes, E. D., Ingold, C. K. and Rose, J. B. (1953). *J. Chem. Soc.* 3839.

Hunter, G. L. K. and Brogden, W. B., Jr. (1963). *J. Org. Chem.* **28**, 1679.
Hurst, J. J. and Whitham, G. H. (1960). *J. Chem. Soc.* 2864.
Hurst, J. J. and Whitham, G. H. (1963). *J. Chem. Soc.* 710.
Ipatieff, V. N., Pines, H., Dvorkovitz, V., Olberg, R. C. and Savoy, M. (1947). *J. Org. Chem.* **12**, 34.
Ipatieff, V. N., Germain, J. E., Thompson, W. W. and Pines, H. (1952). *J. Org. Chem.* **17**, 272.
Jennings, B. H. and Herschbach, G. B. (1965). *J. Org. Chem.* **30**, 3902.
Jerkunica, J. M., Borčić, S. and Sunko, D. E. (1965). *Tetrahedron Lett.* 4465.
Johnson, C. K., Dominy, B. and Reusch, W. (1963). *J. Amer. Chem. Soc.* **85**, 3894.
Kachler and Spitzer. (1863). *Monatsh.* **4**, 486.
Kergomard, A. (1953). *Ann. Chim. (France)* **8**, 153.
Klein, E. and Rojahn. W. (1965). *Chem. Ber.* **98**, 3045.
Komppa, G. and Hintikka, S. V. (1912). *Ann.* **387**, 294.
Komppa, G. and Hasselström, T. (1932). *Ann.* **496**, 164.
Kosima, H. (1961). *J. Chem. Soc. Japan* **82**, 115.
Kropp, P. J. (1966a). *J. Amer. Chem. Soc.* **88**, 4091.
Kropp, P. J. (1966b). *J. Amer. Chem. Soc.* **88**, 4926.
Kropp, P. J. Heckert, D. C. and Flautt, T. J. (1968). *Tetrahedron* **24**, 1385.
Kuwata, H. (1936). *J. Soc. Chem. Ind. Japan.* **39**, 392.
Lalande, R. and Ducasse, Y. (1961). *Compt. Rend.* **252**, 2114.
Libavius (1595). "Alchymia", J. Saurius, Ipensis P. Kopsii, Frankfurt, quoted by Finch and Vaughan (1969).
Lipp, A. (1921). *Chem. Ber.* **53**, 1815.
Liu, R. S. H. and Hammond, G. S. (1964). *J. Amer. Chem. Soc.* **86**, 1892.
Marsh, J. E. (1890). *J. Chem. Soc.* **57**, 828.
Matic, M. and Sutton, D. A. (1953). *J. Chem. Soc.* 349.
Mayer, R., Bochov, K. and Zieger, W. (1964). *Z. Chem.* **4**, 348.
Meerwein, H. and van Emster, K. (1920). *Chem. Ber.* **53**, 1815.
Meerwein, H. and van Emster, K. (1922). *Chem. Ber.* **55**, 2506.
Meerwein, H. and Joussen, J. (1922). *Chem. Ber.* **55**, 2529.
Meinwald, J., Eckell, A. and Erickson, K. L. (1965). *J. Amer. Chem. Soc.* **87**, 3532.
Meinwald, J. (1965). *J. Amer. Chem. Soc.* **87**, 5218.
Miki, T., Nishikawa, M. and Hagiwara, H. (1955). *Proc. Japan. Acad.* **31**, 718.
Mitzner, B. M. and Theimer, E. T. (1962). *J. Org. Chem.* **27**, 3359.
Mitzner, B. M., Theimer, E. T., Steinbach, L. and Wolt, J. (1965a). *J. Org. Chem.* **30**, 646.
Mitzner, B. M., Theimer, E. T. and Freeman, S. K. (1965b). *App. Spect.* **19**, 169.
Mitzner, B. M., Lemberg, S. and Theimer, E. T. (1966a). *Can. J. Chem.* **44**, 1090.
Mitzner, B. M., Mancini, V. J. and Lemberg, S. (1966d). *Can. J. Chem.* **44**, 2103.
Mitzner, B. M. and Lemberg, S. (1966b). *J. Org. Chem.* **31**, 2022.
Mitzner, B. M., Lemberg, S., Mancini, V. J. and Barth, P. (1966c). *J. Org. Chem.* **31**, 2419.
Mitzner, B. M., Mancini, V. J., Lemberg, S. and Theimer, E. T. (1968). *App. Spect.* **22**, 34.
Mitzner, B. M. and Mancini, V. J. (1969). *Am. Perfum. and Cosmet.* **84**, 37.
Nevell, T. P., de Salas, E. and Wilson, C. L. (1939). *J. Chem. Soc.* 1188.
Nickon, A., Lambert, J. L. and Oliver, J. E. (1966). *J. Amer. Chem. Soc.* **88**, 2787.
Norin, T. (1964). *Tetrahedron Lett.* **1**, 37.
Ohloff, G. and Klein, E. (1962). *Tetrahedron* **18**, 37.

Ohloff, G., Osiecki, J. and Djerassi, C. (1962). *Chem. Ber.* **95**, 1400.
Ohloff, G., Seibl, J. and Kovats, E. sz. (1964). *Ann.* **675**, 83.
Ohloff, G., Schulte-Elte, K. H. and Giersch, W. (1965). *Helv. Chim. Acta.* **48**, 1665.
Ohloff, G., Uhde, G., Thomas, A. F. and Kovats, E. sz. (1966). *Tetrahedron* **22**, 309.
Ourisson, G. and Rassat, A. (1960). *Tetrahedron Lett.* **21**, 16.
Parker, E. D. and Goldblatt, L. A. (1950). *J. Amer. Chem. Soc.* **72**, 2151.
Patton, E. L. (1950). *Am. Perfum. and Essent. Oil Rev.* **56**, 118.
Piatkowski, K., Kuczynski, H. and Kubik, A. (1966). *Rocz. Chem.* **40**, 213.
Pierson, G. O. and Runquist, O. A. (1969). *J. Org. Chem.* **34**, 3654.
Pines, H. and Eschinazi, H. (1955). *J. Amer. Chem. Soc.* **77**, 6314.
Pulkkinen, E. (1956). *Ann. Acad. Sci. Fenn. Ser. A. II.* No. 74.
Pulkkinen, E. (1957). *Suomen Kemistilehti* **30A**, 239.
Regan, A. F. (1969). *Tetrahedron* 3801.
Reusch, W., Anderson, D. F. and Johnson, C. K. (1968). *J. Amer. Chem. Soc.* **90**, 4988.
Richey, H. G., Jr., Grant, J. E., Garbacik, T. J. and Dull, D. L. (1965). *J. Org. Chem.* **30**, 3909.
Rudakov, G. A. and Kovatov, S. Ya. (1937). *J. Appl. Chem. USSR* **10**, 312.
Rudakov, G. A. and Artamonov, G. A. (1945). *J. Gen. Chem. USSR* **15**, 75.
Rudakov, G. A. and Marchevskii, A. T. (1953). *Sb. Statei Obshch. Khim.* **2**, 1432.
Rupe, H. (1900). *Chem. Ber.* **33**, 857.
Sabatier, P. and Senderens, J. B. (1901). *Compt. Rend.* **132**, 1256.
Salmon, J. R. and Whittaker, D. (1967). *J. Chem. Soc.* (*D*) 491.
Salmon, J. R. and Whittaker, D. (1971). *J. Chem. Soc.* (*B*), 1249.
Sasaki, T., Eguchi, S. and Yamada, H. (1971). *Tetrahedron Lett.* **99**.
Schenck, G. O. and Ziegler, K. (1944). *Naturwissenschaften* **32**, 157.
Schenck, G. O., Kinkel, K. G. and Mertens, H-J. (1953a). *Ann.* **584**, 125.
Schenck, G. O., Eggert, H. and Denk, W. (1953b). *Ann.* **584**, 177.
Schenck, G. O., Gollnik, K., Buchwald, G., Schroeter, S. and Ohloff, G. (1964). *Ann.* **674**, 93.
Schiff, R. (1880). *Chem. Ber.* **13**, 1402.
Schleyer, P. von R. (1964). *J. Amer. Chem. Soc.* **86**, 1856.
Schmidt, H. (1947). *Chem. Ber.* **80**, 520.
Schroeter, S. H. and Eliel, E. L. (1964). *J. Amer. Chem. Soc.* **86**, 2066.
Schroeter, S. H. and Eliel, E. L. (1965). *J. Org. Chem.* **30**, 1.
Schulte-Elte, K. H. and Ohloff, G. (1967). *Helv. Chim. Acta.* **50**, 153.
Semmler, F. W. (1894). *Chem. Ber.* **27**, 2520.
Semmler, F. W. (1901). *Chem. Ber.* **34**, 3126.
Simonsen, J. L. (1947). "The Terpenes", Vol. 1, 2nd Ed., Cambridge University Press, London.
Simonsen, J. L. (1949a). "The Terpenes", Vol. 2, 2nd Ed., Cambridge University Press, London.
Simonsen, J. L. (1949b). Previous reference, p. 63.
Smit, V. A., Semenovskii, A. V. and Kucherov, V. F. (1962). *Izv. Akad. Nauk. SSSR Otd. Khim. Nauk.* 470.
Spanninger, P. A. and von Rosenberg, J. L. (1969). *J. Org. Chem.* **34**, 3658.
Steinbach, L., Theimer, E. T. and Mitzner, B. M. (1964). *Can. J. Chem.* **42**, 959.
Steinbach, L. and Walt, J. (1965). *J. Org. Chem.* **30**, 646.
Strickler, H., Ohloff, G. and Kovats, E. sz. (1967). *Helv. Chim. Acta.* **50**, 759.
Sully, B. D. (1964). *Chem. Ind.* (*London*) 263.
Swann, G. (1948). *Ind. Chemist* **24**, 141.

Swann, G. and Cripwell, F. J. (1948). *Ind. Chem.* **24**, 5739.
Thomas, A. F., Schneider, R. A. and Meinwald, J. (1967). *J. Amer. Chem. Soc.* **89**, 68.
Tori, K. (1964). *Chem. Pharm. Bull.* (*Tokyo*) **12**, 1439.
Turro, N. J. (1967). "Molecular Photochemistry", W. A. Benjamin, New York.
Valenzuela, P. and Cori, O. (1967). *Tetrahedron Lett.* 3089.
Valkanas, G. and Iconomou, N. (1963). *Helv. Chim. Acta.* **46**, 1089.
van Bruggen, E. (1968). *Rec. Trav. chim.* **87**, 1134.
van Tamelen, E. E., McNarye, J. and Lornitzo, F. A. (1957). *J. Amer. Chem. Soc.* **79**, 1231.
van Tamelen, E. E. and Timmons, R. J. (1962). *J. Amer. Chem. Soc.* **84**, 1067.
Vaughan, W. R., Goetschel, C. T., Goodrow, M. H. and Warren, C. L. (1963). *J. Amer. Chem. Soc.* **85**, 2282.
Verghese, J. (1965). *Perfum. and Essent. Oil Rec.* **56**, 438.
Verghese, J. (1967). *Perfum. and Essent. Oil Rec.* **58**, 868.
Verghese, J. (1968). *Perfum. and Essent. Oil Rec.* **59**, 439, 876.
Verghese, J. (1969). *Perfum. and Essent. Oil Rec.* **60**, 25, 271.
von Rudloff, E. (1961). *Can. J. Chem.* **39**, 1.
Wallach, O. (1899). *Ann.* **305**, 242, 274.
Wallach, O. (1905). *Ann.* **339**, 94.
Watanabe, K., Pillai, C. N. and Pines, H. (1962). *J. Amer. Chem. Soc.* **84**, 3934.
Watson, G. and Gladden, G. W. (1964). *Perfum. and Essent. Oil Rec.* **55**, 793.
Whalley, E. (1958). *Can. J. Chem.* **36**, 228.
Whitham, G. H. (1961). *J. Chem. Soc.* 2232.
Williams, Claudia M. and Whittaker, D. (1970). *J. Chem. Soc. D.* 960.
Williams, Claudia M. and Whittaker, D. (1971a). *J. Chem. Soc.* (*B*) 668.
Williams, Claudia M. and Whittaker, D. (1971b). *J. Chem. Soc.* (*B*) 672.
Winstein, S. (1955). XIVth International Congress of Pure and Applied Chemistry, Zurich, p. 145 of Abstracts. Experimentia, Suppl. No. 2, 137.
Winstein, S. and Trifan, D. (1952a). *J. Amer. Chem. Soc.* **74**, 1147.
Winstein, S. and Trifan, D. (1952b). *J. Amer. Chem. Soc.* **74**, 1154.
Wolff, L. (1912). *Ann.* **394**, 97.
Wolinsky, J., Chollar, B. and Baird, M. D. (1962). *J. Amer. Chem. Soc.* **84**, 2775.
Wrolstad, R. E. and Jennings, W. G. (1965). *J. Chromatog.* **18**, 318.
Wystrach, V. P., Barnum, L. H. and Garber, H. (1957). *J. Amer. Chem. Soc.* **79**, 5786.
Zelinski, N. D. (1904). *J. Russ. Phys. Chem. Soc.* **36**, 770.
Zweifel, G. and Brown, H. C. (1964). *J. Amer. Chem. Soc.* **86**, 393.
Zweifel, G. and Whitney, C. C. (1966). *J. Org. Chem.* **31**, 4178.

3

THE SESQUITERPENES

J. S. ROBERTS

Lecturer, University of Stirling

I. Introduction

Sesquiterpenoids are defined as the group of C_{15} compounds derived by the assembly of three isoprenoid units and they are found in many forms of living systems of which higher plants are the principal member. Although the historical background to sesquiterpenoids can be traced back to the early nineteenth century the real impetus to the study of the varied and fascinating chemistry associated with these compounds stemmed from the pioneering efforts of Wallach, Semmler, and Ruzicka. In particular, Ruzicka, (Ruzicka, 1959; Ruzicka *et al.*, 1953) on a limited basis of fact but prophetic insight, proposed the Biogenetic Isoprene Rule which has proved to be the foundation-stone of terpenoid chemistry in general.

Research in the field of sesquiterpenoids has increased at an exponential rate closely paralleling modern methods of isolation, purification, structural elucidation and synthesis. At the time of writing there are some 600 known sesquiterpenoids which is an order of magnitude greater than 15 years ago. For this reason a comprehensive survey is well beyond the scope of this chapter and thus the main emphasis will be directed towards the structural types and their inter-relationships, with particular reference to the more recent advances made over the past decade.* Undoubtedly the major high-light of recent years has been the dramatic eruption of outstanding and elegant syntheses. In the solution of the inherently difficult problems of skeletal construction and stereochemical control associated with some of these syntheses, new methods have been introduced, which have endowed the synthetic organic chemist with a rich new array of armaments for future use. In particular, the work of Büchi, Corey, and Marshall and their respective research teams deserves special mention. Unfortunately space only permits the incorporation of brief outlines of some of these syntheses.†

At first sight, sesquiterpenoid structures present a bewildering collection of acyclic, monocyclic, bicyclic, tricyclic, and tetracyclic systems and it is only within the framework of the original Biogenetic Isoprene Rule (Ruzicka, 1959; Ruzicka *et al.*, 1953) and later extensions (Hendrickson, 1959; Parker *et al.*, 1967) that some semblance of relationship can be achieved. The focal point in this conceptual scheme is the utilization of *cis,trans*-farnesyl pyrophosphate (1, R = PP) and *trans,trans*-farnesyl pyrophosphate (2, R = PP) as starting points, with the proviso that nerolidyl pyrophosphate (3, R = PP) can also serve as a crucial building block. Suitable cyclizations of these precursors involving the removal of the pyrophosphate group by double bond participation (central and terminal) can be envisaged, as outlined in Scheme 1. Some of these "primary" structures (4)–(12) not only embody certain sesquiterpenoid structural types directly but they are also ideal candidates for further elaborations to the more exotic types via cyclizations, Wagner–Meerwein rearrangements, hydride shifts, etc.

* For full accounts in the progress of sesquiterpenoid chemistry, see Simonsen, J. L. and Barton, D. H. R. (1953). "The Terpenes", Vol. 3, Cambridge University Press, London. Barton, D. H. R. and de Mayo, P. (1957). *Quart. Rev.* **11**, 189. de Mayo, P. (1959). "Mono- and Sesquiterpenoids", Interscience, New York. Bryant, R. (1969). *In* "Rodd's Chemistry of Carbon Compounds" (S. Coffey, ed.), 1st Ed., Vol. II, Part C, pp. 256-368, Elsevier, Amsterdam. Ourisson, G., Munavalli, S. and Ehret, C. (1966). "International Tables of Selected Constants", No. 15, Data Relative to Sesquiterpenoids, Pergamon Press, Oxford. Mellor, J. M. and Munavalli, S. (1964). *Quart. Rev.* **18**, 270.

†Although the majority of syntheses described herein are of racemates, except when stated otherwise, one enantiomer is shown for convenience. This enantiomer is the naturally-occurring one.

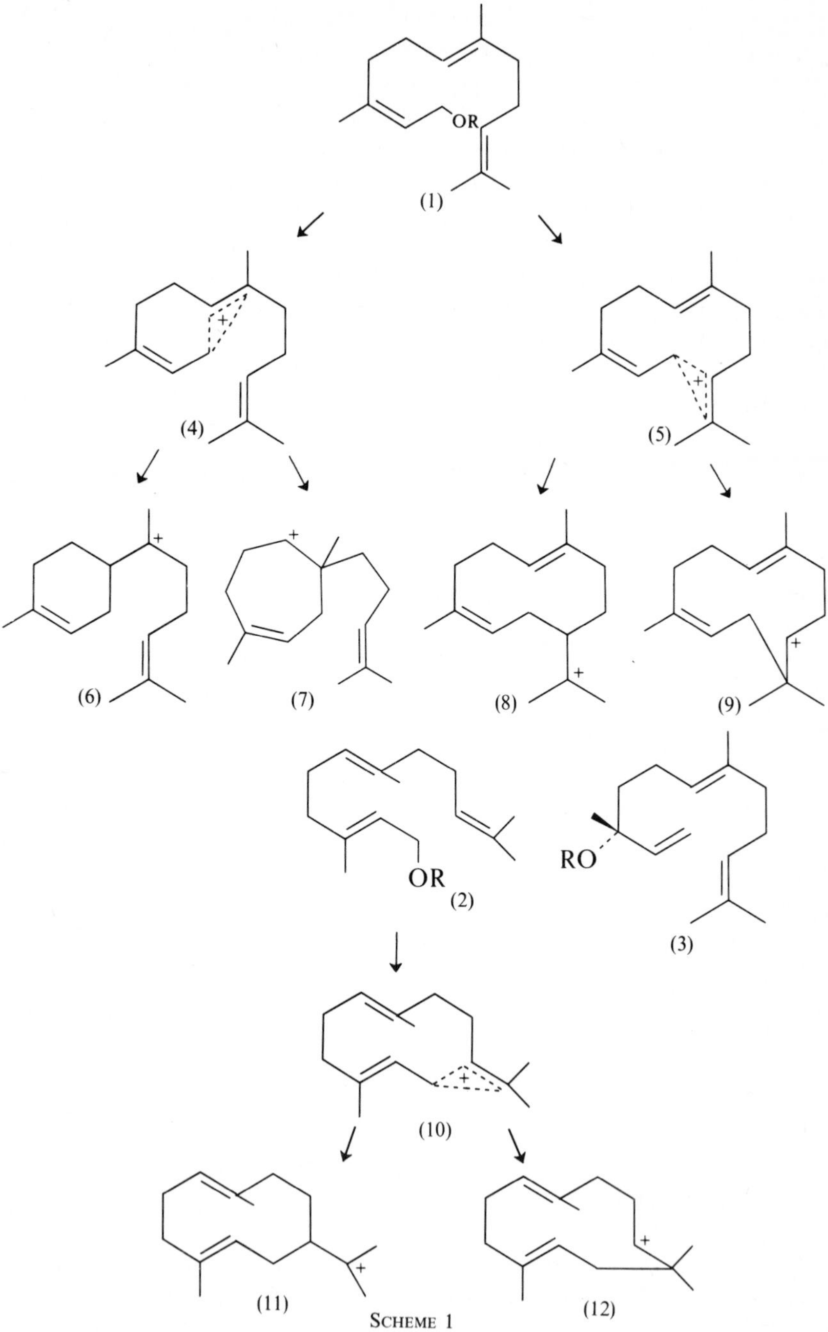

SCHEME 1

II. Farnesane

Both farnesol (2, R = H) and nerolidol (3, R = H), which have been isolated from many sources, are key precursors (as the pyrophosphate derivatives) in sesquiterpenoid biosynthesis. Although the gross structure of farnesol had been determined by Kerschbaum (1913), the all *trans* stereochemistry of the double bonds was only ascertained unambiguously by Burrell *et al.* (1966). At present there are only about fifteen farnesane-type sesquiterpenoids known including dendrolasin (13) (Quilico *et al.*, 1957; Sakai *et al.*, 1963), caparrapidiol (14) (Borges del Castillo, 1966), ngaione (15) (Birch *et al.*, 1953), davanone (16) (Sipma and van der Wal, 1968), *β*-sinensal (17) (Flath *et al.*, 1966; Teranishi *et al.*, 1968), and freelingyne (18) (Massy-Westropp *et al.*, 1966). Many of these compounds have been synthesized.

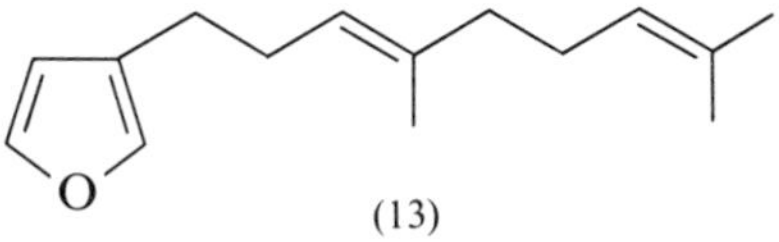

(13)

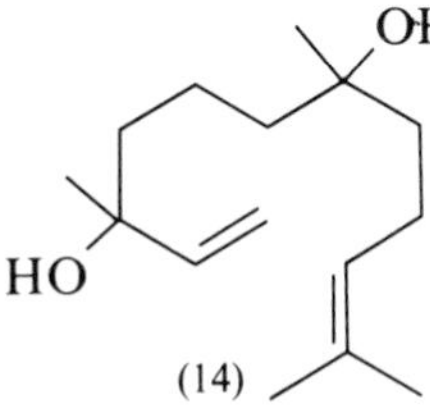

(14)

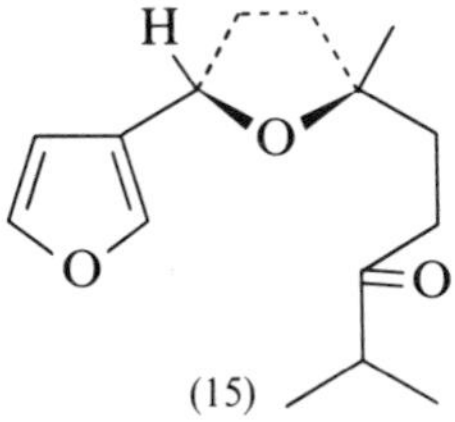

(15)

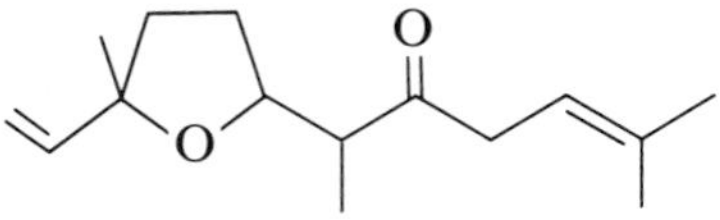

(16)

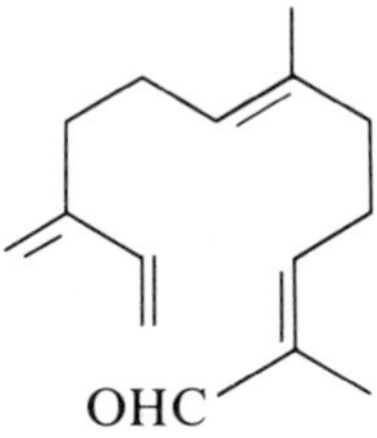

(17)

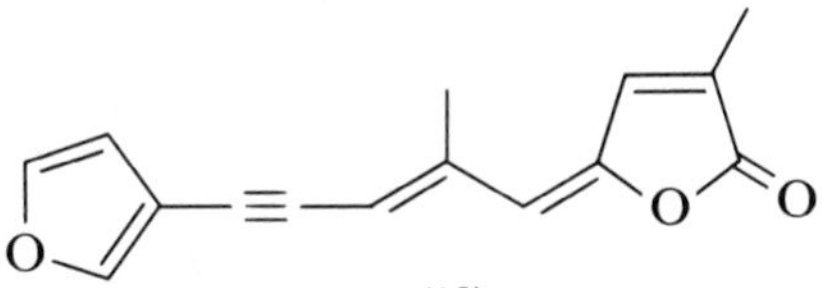

(18)

Recent interest in farnesane-type sesquiterpenoids has centred around the isolation and structural determination of the juvenile hormone (19, R = Et) from the giant silkworm moth *Hyalophora cecropia* (Trost, 1970). Although the complete physiological effects of juvenile hormone are not fully understood it is quite clear that it prevents the metamorphosis of immature insects. The powerful insecticidal potential of this effect has motivated no less than eleven syntheses (Cavill *et al.*, 1969; Corey *et al.*, 1968; Corey and Yamamoto, 1970; Dahm *et al.*, 1967; Findlay and MacKay, 1969; Johnson *et al.*, 1970; Johnson *et al.*, 1968; Mori *et al.*, 1969; Schulz and Sprung, 1969; van Tamelen and McCormick, 1970; Zurflüh *et al.*, 1968), some of which are either highly stereoselective or stereospecific. The plausible assumption that juvenile hormone is biosynthesized from farnesyl pyrophosphate by two successive methylations is partially supported by the co-occurrence of the mono-methylated analogue (19, R = Me) (Meyer *et al.*, 1968).

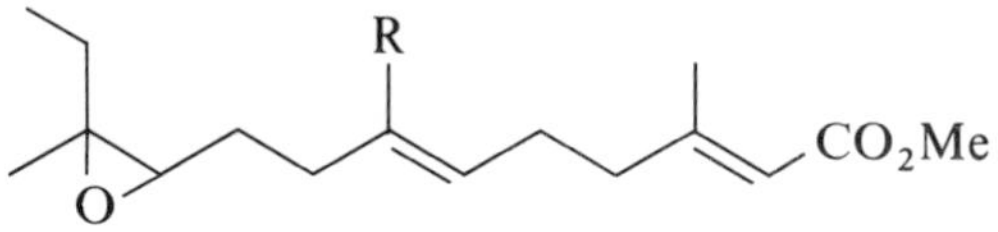

(19)

III. Mono- and Bicyclofarnesane

Although the majority of sesquiterpenoids can be derived either directly or indirectly by the processes outlined in Scheme 1, a relatively small group clearly originates from a *trans*-antiparallel cyclization of farnesyl pyrophosphate (20) resulting in the structural and stereochemical characteristics of rings A and B of higher terpenoids. The monocyclic group is exemplified by the important plant growth hormone, (+)-abscisic acid (21) (Cornforth *et al.*, 1965; Ohkuma *et al.*, 1965) and the unusual nor-sesquiterpenoid, Latia luciferin (22) (Shimomura and Johnson, 1968), the specific substrate in the bioluminescent system of the fresh-water limpet. A number of syntheses of abscisic acid have been reported but the most economical is that by Roberts *et al.* (1968). In three steps, α-ionone (23) was converted into (±)-abscisic acid by t-butyl chromate allylic oxidation to (24) followed by a Wittig reaction with carbethoxymethylenetriphenylphosphorane and subsequent hydrolysis. It should be noted, however, that the inclusion of the monocyclofarnesane group as true sesquiterpenoids is debatable, since some of them may very well be degradation products of, e.g., carotenoids.

The bicyclic class can be divided into two sub-groups which are antipodally related. Within the first group, characterized by the normal triterpenoid configuration, are such compounds as drimenol (25) (Appel *et al.*, 1959) and cinnamolide (26) (Canonica *et al.*, 1969). It is relevant to note that van Tamelen *et al.* (1963) and van Tamelen and Coates (1966) have achieved a

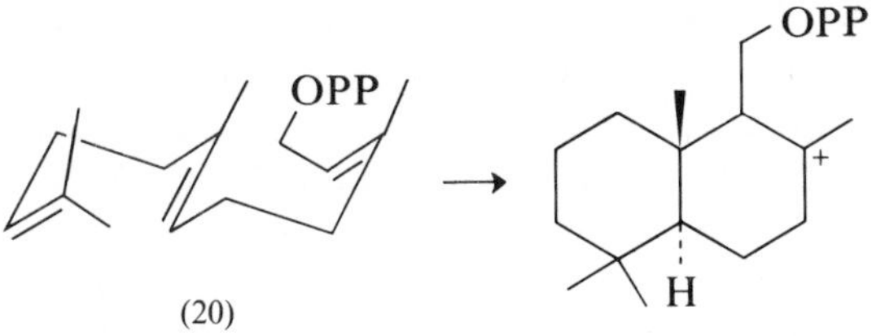

(20)

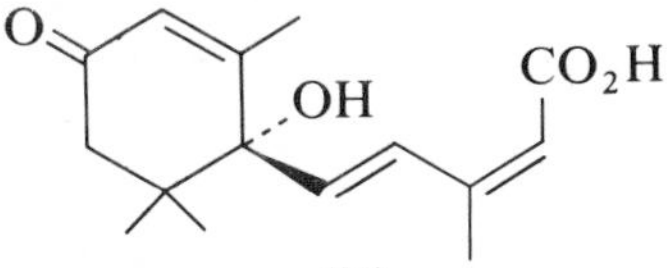

(21)

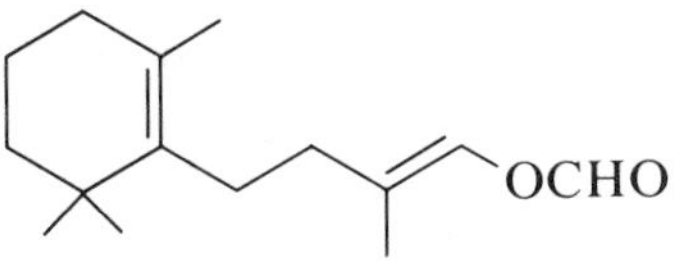

(22)

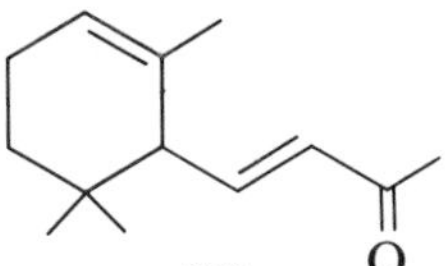

(23)

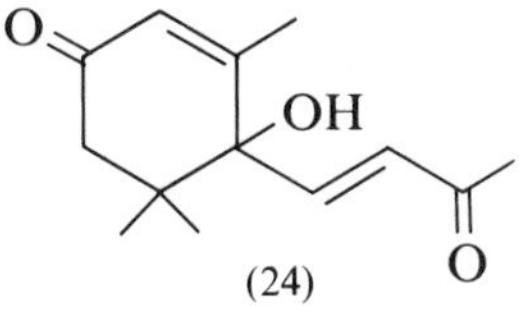

(24)

"biogenetic" synthesis of (±)-drimenol by acid-catalysed cyclization of the terminal monoepoxide of *trans,trans*-farnesyl acetate. One of the major products of this reaction was the bicyclic hydroxy-acetate (27) which was readily elaborated to drimenol. Iresin (28) (Djerassi and Burnstein, 1959; Djerassi and Rittel, 1957) is one of the very few examples belonging to the antipodal group. Pelletier and Prabhakar (1968) have synthesized (±)-iso-iresin (29) and related compounds.

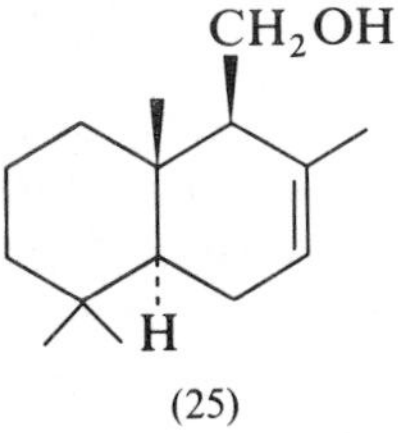

(25)

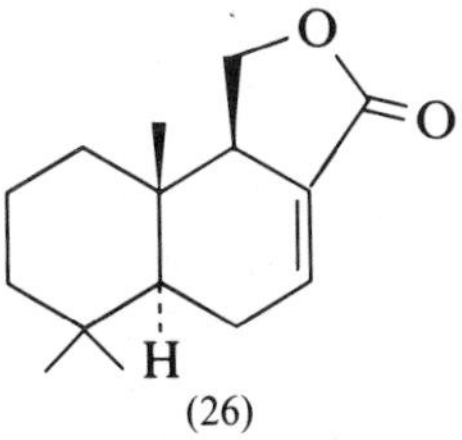

(26)

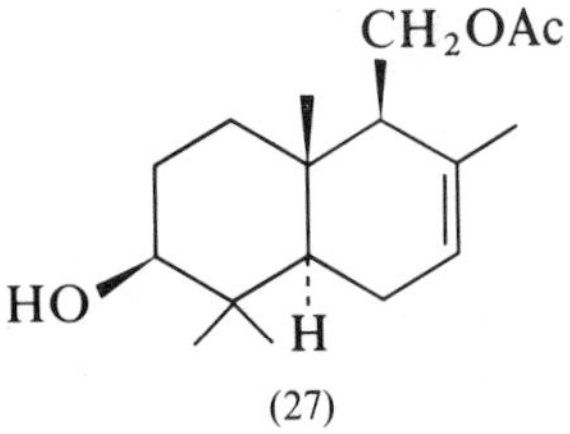

(27)

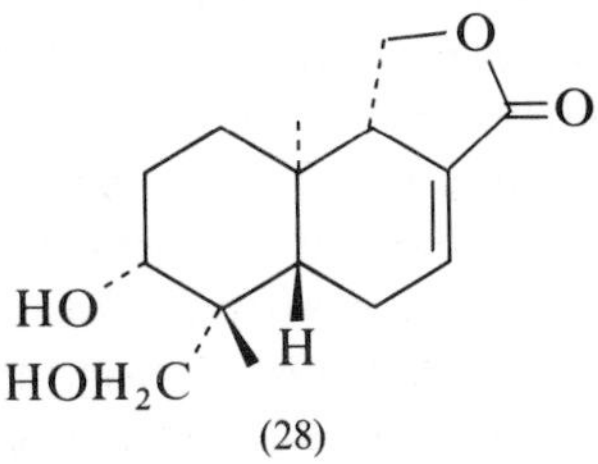

(28)

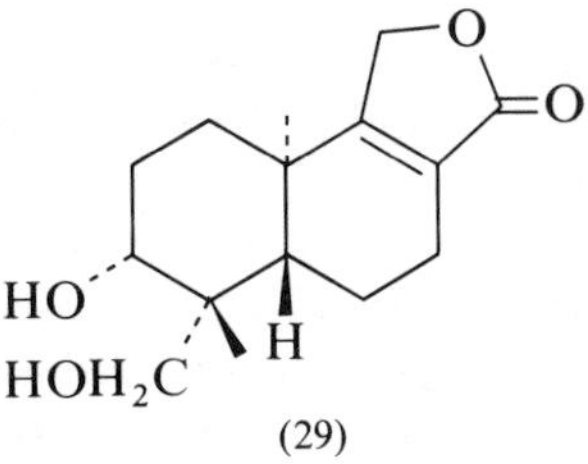

(29)

IV. Bisabolane, Curcumane, etc.

The sesquiterpenoids which come under this general heading are both numerous and widespread in Nature, particularly the hydrocarbons. Their biogenesis is usually considered in terms of the electronically-favoured cyclization of (1, R = PP) to cation (6). Typical examples are (+)-β-bisabolene (30), (−)-β-curcumene (31), (−)-α-zingiberene (32), nuciferal (33) (Sakai *et al.*, 1965), perezone (34) (Walls *et al.*, 1966), and bilabanone (35) (Irie *et al.*, 1967). A number of the hydrocarbons have been isolated in both optically active forms.

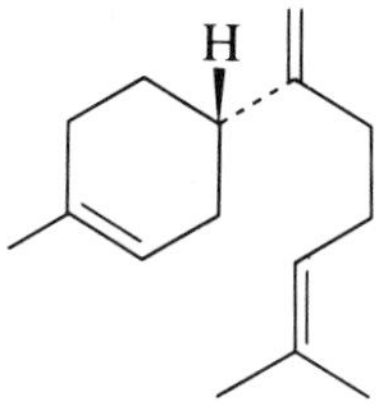

(30)

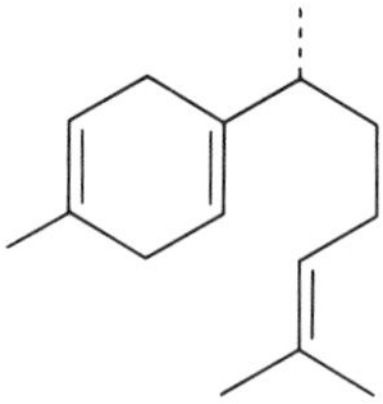

(31)

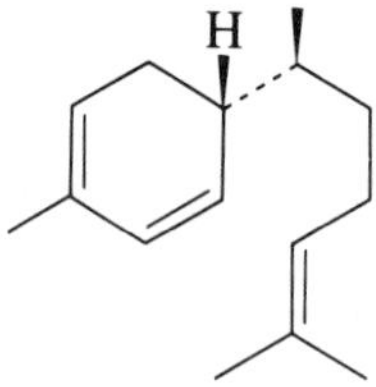

(32)

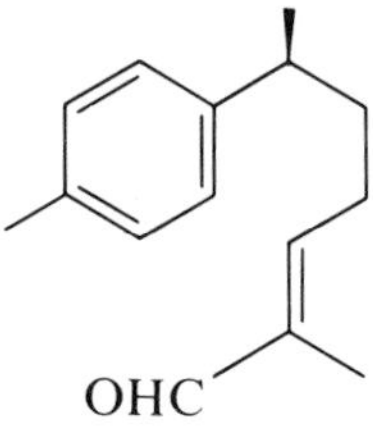

(33)

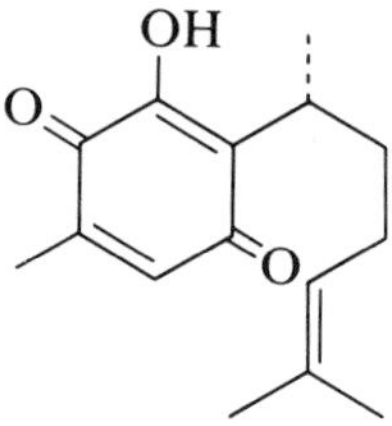

(34)

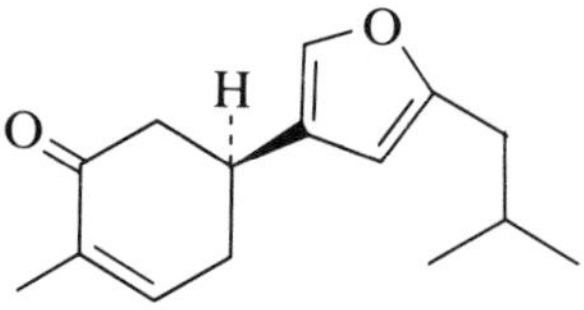

(35)

Sláma and Williams (1965) made the remarkable discovery that certain American paper products (e.g. *The New York Times*, *The Wall Street Journal*) contain a substance exhibiting juvenile hormone activity on the metamorphosis of the insect, *Pyrrhocoris apterus*. A year later Bowers and co-workers (Bowers *et al.*, 1966) isolated the active component from the balsam fir, *Abies balsamea*, and characterized it as (+)-juvabione (36), the methyl ester of the previously known todomatuic acid. Subsequently Černý *et al.* (1967) isolated both (+)-juvabione and the less active (+)-dehydrojuvabione (37) from the Czechoslovakian balsam fir wood. Syntheses of juvabione were soon reported (Ayyar and Krishna Rao, 1968; Mori and Matsui, 1968) culminating in two different and elegant routes by Pawson *et al.* (1970) and Birch *et al.* (1970). The former utilized R-(+)-limonene which was treated with disiamylborane and the adduct oxidized to yield the two diastereoisomeric alcohols (38, R = β-Me) and (38, R = α-Me). The alcohol (38, R = β-Me), on conversion to the corresponding nitrile, was treated with isobutyllithium and the resultant ketone (39) oxidatively modified to (+)-juvabione (36). Birch's synthesis hinged upon the acid-catalysed fission of the Diels–Alder adduct (40) which gave the 4-substituted cyclohexenone (41) which was then converted into (±)-juvabione by further elaboration.

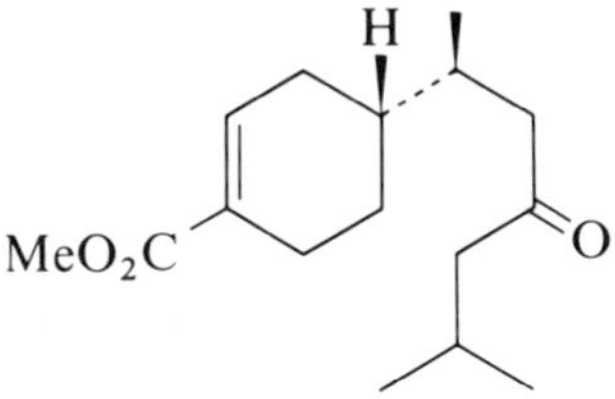

(36)

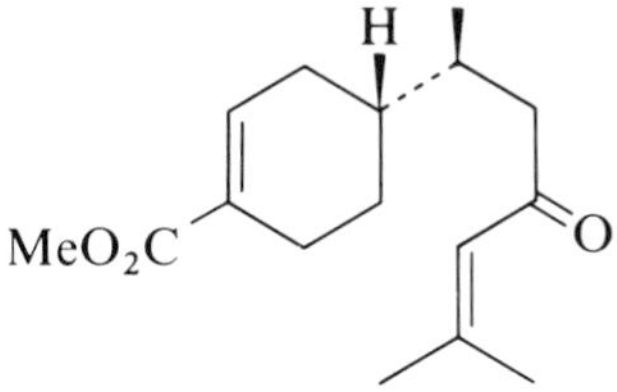

(37)

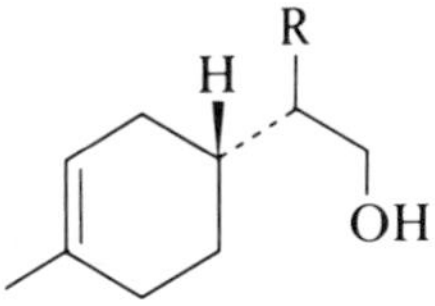

(38)

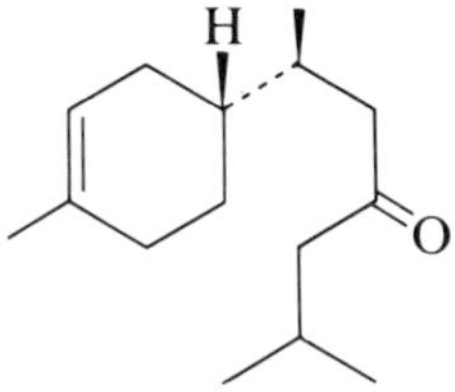

(39)

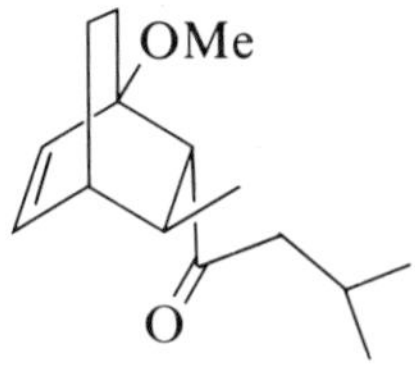

(40)

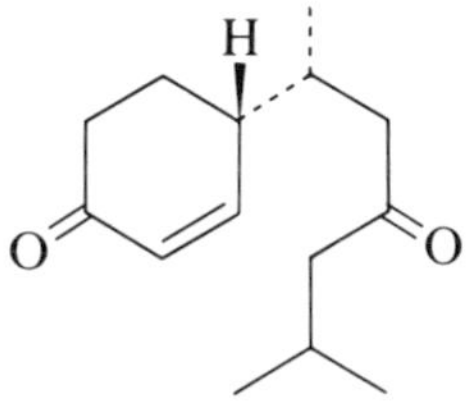

(41)

V. Sesquicarane, Bergamotane and Santalane

Deprotonation of (4) leads directly to the structure of sesquicarene (42) isolated by Ohta and Hirose (1968) from the essential oil of *Schisandra chinensis*. Five syntheses have now been reported for this sesquiterpene which is the isoprenoid homologue of 2-carene. All of these have depended upon an intramolecular carbene-olefin cyclization, viz. of (43, R = Me) (Coates and Freidinger, 1970; Mori and Matsui, 1970), (43, R = H) (Corey and Achiwa, 1969a) and (44) (Coates and Freidinger, 1970; Corey and Achiwa, 1969b; Nakatani and Yamanishi, 1969). More recently, Corey and Achiwa (1970) have utilized commercial farnesol which is predominantly the *trans, trans* isomer (*trans,trans*:*cis,trans*, 1·5:1) as the starting material. They have found that mercuric iodide not only catalysed the diazo decomposition of (44) but also isomerized the $\Delta^{2,3}$ double bond of the *trans,trans* analogue. Structurally related to sesquicarene is the sperm attractant, sirenin (45)

(Machlis *et al.*, 1968), produced by the female gametes of the water mould, *Allomyces*. This unique plant sex hormone has been ingeniously synthesized in a number of laboratories and again diazo decompositions of suitable substrates have played the key role (Corey and Achiwa, 1970; Corey *et al.*, 1969c; Grieco, 1969; Mori and Matsui, 1969; Mori and Matsui, 1970; Plattner *et al.*, 1970).

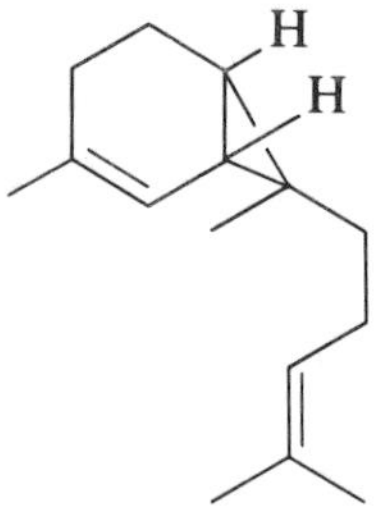

(42)

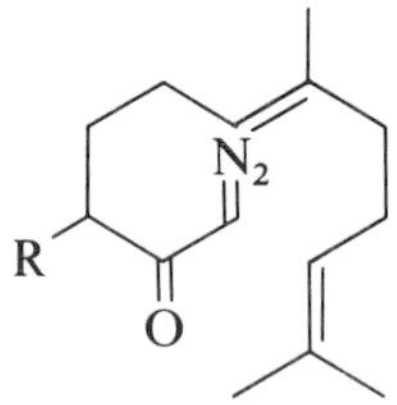

(43)

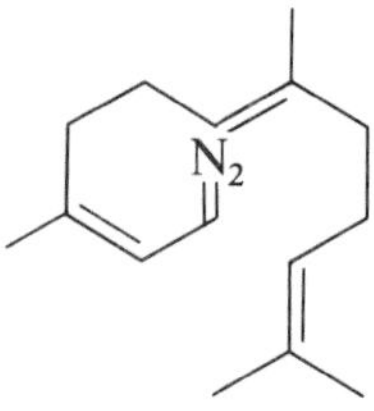

(44)

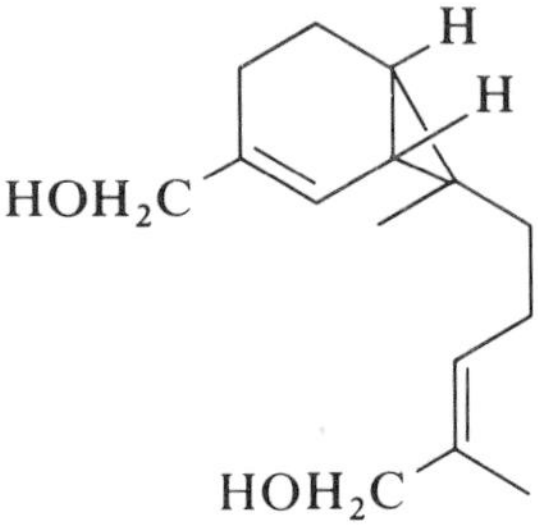

(45)

Cyclization of cation (6) in the electronically favoured sense with deprotonation of the resultant cation (46) leads directly to the gross structures of α- and β-bergamotene. The stereochemistries of these two sesquiterpenes have been in doubt for some time, but recently Gibson and Erman (1969), by synthesis of (−)-α-*cis*-bergamotene (47) and (+)-β-*cis*-bergamotene (48) from (−)-β-pinene, have conclusively demonstrated that the former is the naturally occurring isomer while the latter is not. Natural β-bergamotene should therefore be represented as (49) with the *trans* stereochemistry. α-*Trans*-bergamotene (50) is also known to be naturally-occurring.

The alternative cyclization of cation (6) leads to the cation (51) which, by a Wagner–Meerwein shift and deprotonation, produces α-santalene (52) and β-santalene (53). α-Santalene has a certain historical importance in that it was the first sesquiterpene for which the correct structure was proposed (Semmler, 1910) and indeed this structure was used by Ruzicka in his formulation of the Biogenetic Isoprene Rule. Alcohols related to these two sesqui-

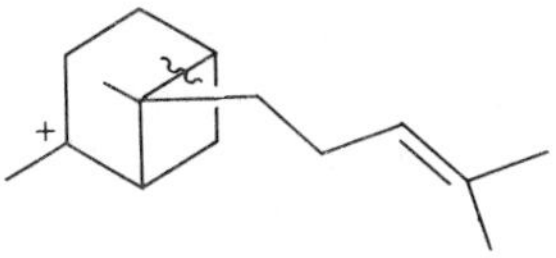

(46)

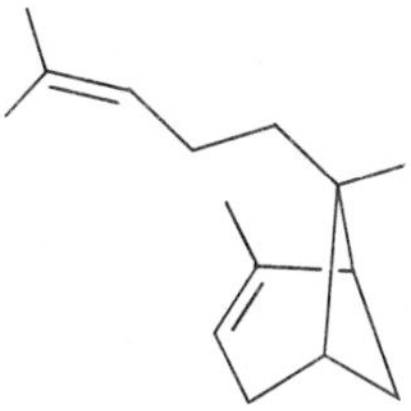

(47)

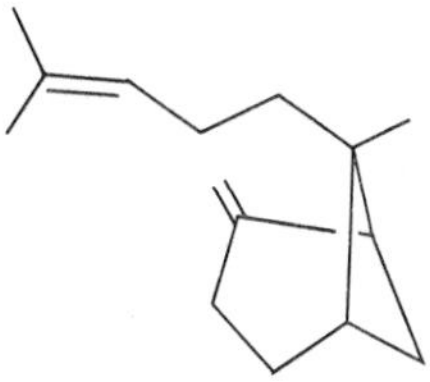

(48)

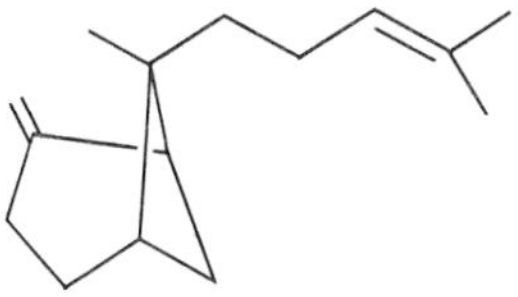

(49)

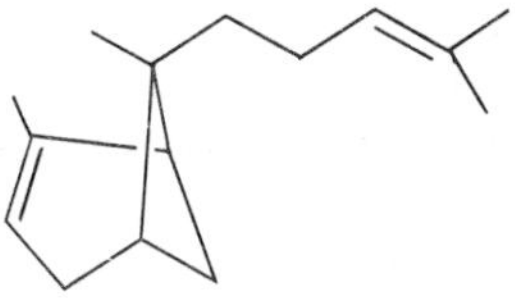

(50)

terpenes are well known and numerous syntheses of them have been recorded (Corey *et al.*, 1970). The antibiotic fumagillin (54) (Tarbell *et al.*, 1960) with its mixed terpenoid–polyketide origin can be related structurally to the bergamotane skeleton by oxidative cleavage of the bond shown in (46).

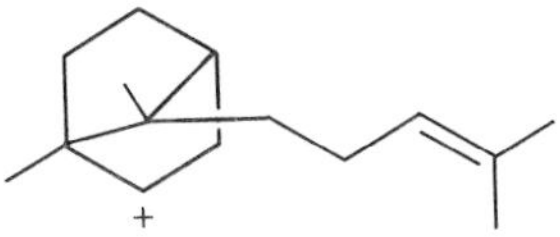

(51)

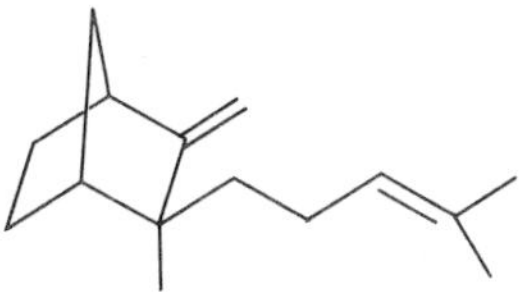

(52)

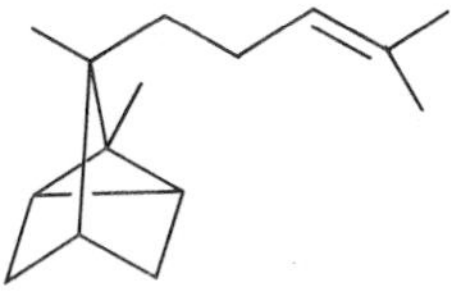

(53)

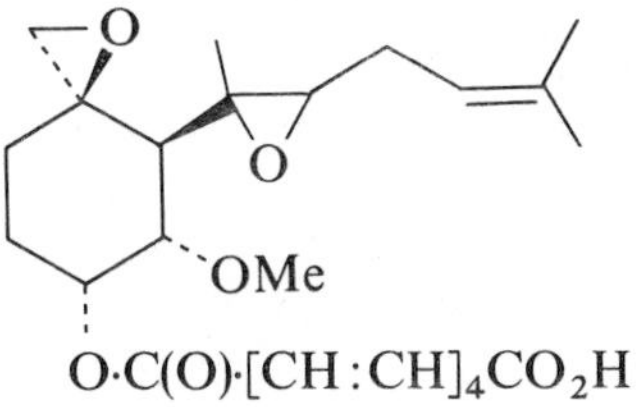

(54)

VI. Cadinane, Amorphane, Muurolane, Bulgarane, and Related Tricyclic Sesquiterpenoids

A number of sesquiterpenoids has been grouped together according to their ability to undergo dehydrogenation to cadalene (55). Originally much of the work on this group was carried out on impure samples with the result that many conflicting reports concerning stereochemistry and positions of double bonds were made. Gradually one stereochemically discrete class emerged, now referred to as the cadinanes (Koteswara Rao *et al.*, 1966) and exemplified by (+)-γ-cadinene (56). More recently, examples of the three other possible classes have been isolated and identified, e.g. (−)-α-muurolene (57) (Westfelt, 1966; Zabza *et al.*, 1966), (−)-ε-bulgarene (58) (Vlahov *et al.*, 1967a,b), and (−)-γ-amorphene (59) (Motl *et al.*, 1966). More unusual sesquiterpenoids belonging to the cadinane group are gossypol (60) (Adams *et al.*, 1938), the toxic yellow pigment of cotton seed, mansonone F (61) (Marini Bettólo *et al.*, 1965), a representative of a number of quinonoid compounds from the heartwood of *Mansonia Altissima* Chev., and the ring contracted hydroxy-ketone, oplopanone (62) (Takeda *et al.*, 1966). Before proceeding to the discussion of related tricyclic sesquiterpenoids, attention should be drawn to a distinct group of antipodal cadinane types which occur predominantly in the North Indian variety of vetiver oil. Examples of this group are (−)-γ_2-cadinene (63) (Kartha *et al.*, 1963) and the norsesquiterpenoid, khusitone (64) (Chakravatari, 1965).

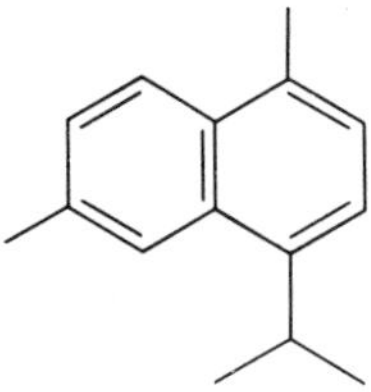

(55)

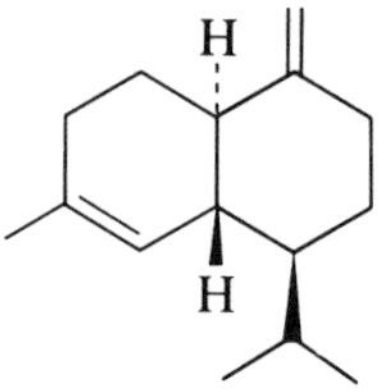

(56)

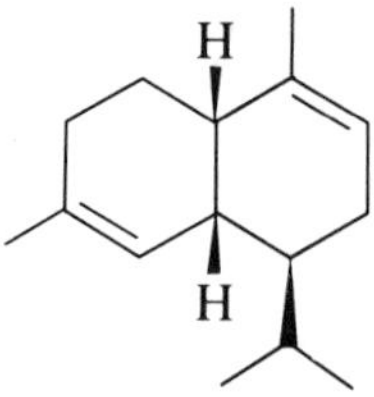

(57)

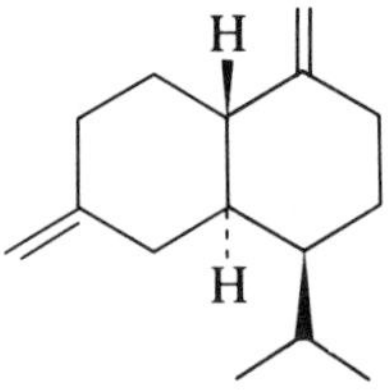

(58)

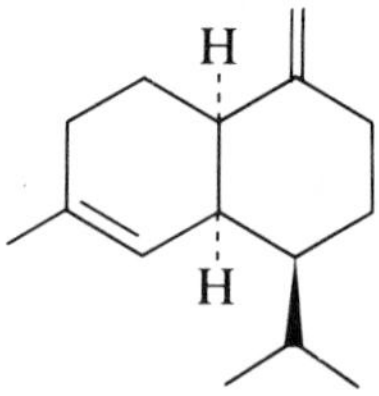

(59)

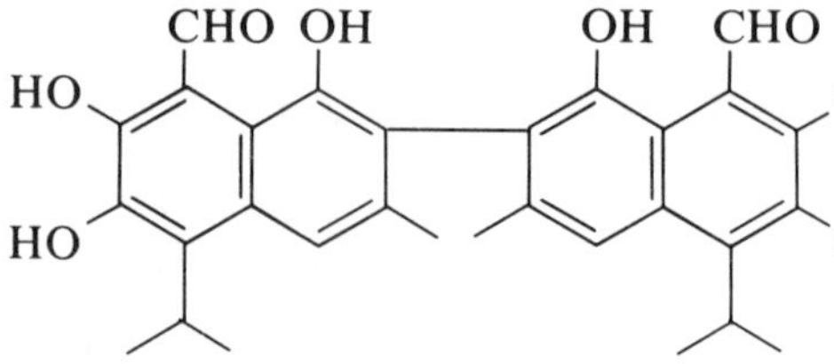

(60)

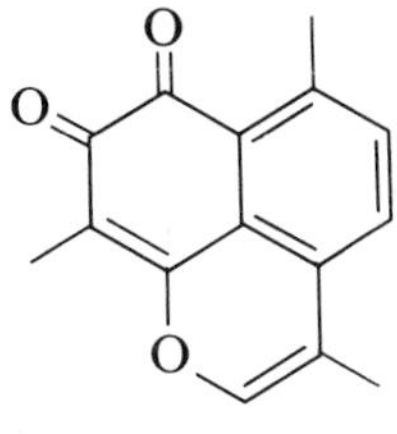

(61)

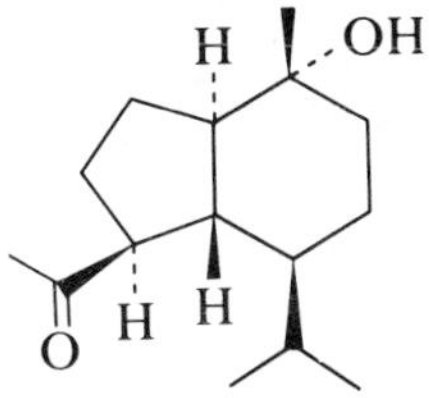

(62)

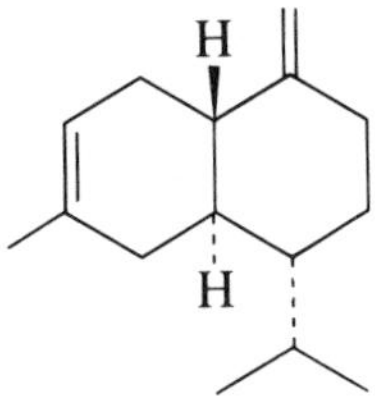

(63)

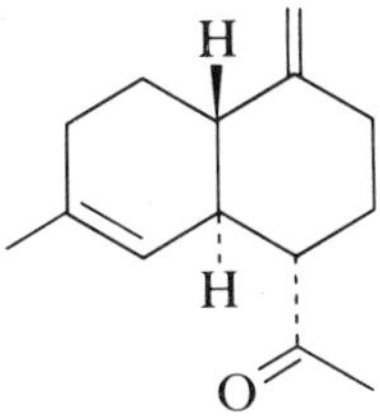

(64)

Closely related to and often co-occurring with these bicyclic hydrocarbons are the tricyclic sesquiterpenes, (−)-α-copaene (65) and (+)-α-ylangene (66). Proposals for the structure of α-copaene included many variants with a cyclopropane ring but eventually the correct structure was independently ascertained by two groups (de Mayo, 1965b; Kapadia *et al.*, 1965) and the absolute stereochemistry was derived by conversion of α-copaene into the well-known (−)-cadinene dihydrochloride (67). For some time the stereochemical relationship between α-copaene and α-ylangene was contested, but Ohta and Hirose (Ohta *et al.*, 1968; Ohta and Hirose, 1969), in a recent investigation of the acid-catalysed isomerization of these two sesquiterpenes, have demonstrated that (+)-α-ylangene is properly depicted as (66). The confusion appears to have arisen from the fact that (−)-α-ylangene is also a naturally-occurring sesquiterpene and this enantiomer is, in fact, the isopropyl epimer of (−)-α-copaene. Synthetic proof of the structure of α-copaene has been provided by Heathcock *et al.* (1967), who converted the readily available diketone (68) into the *cis*-decalin derivative (69) which, on base treatment, gave the tricyclic ketone (70) which was elaborated to α-copaene.

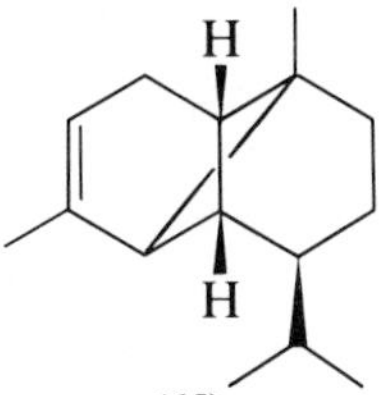

(65)

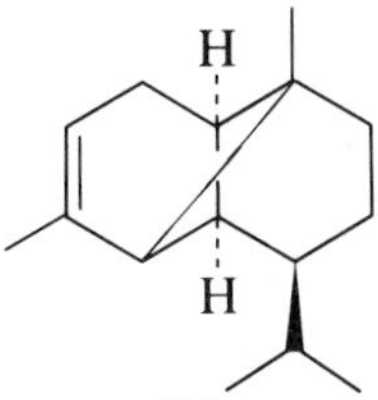

(66)

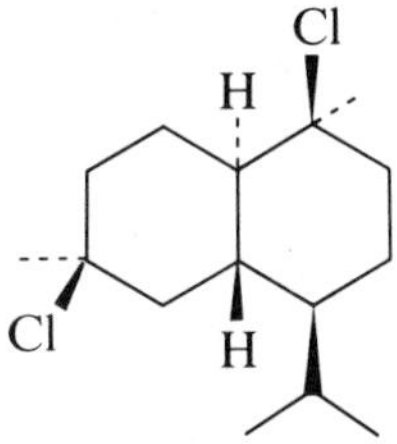

(67)

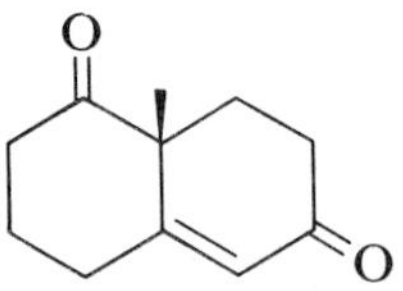

(68)

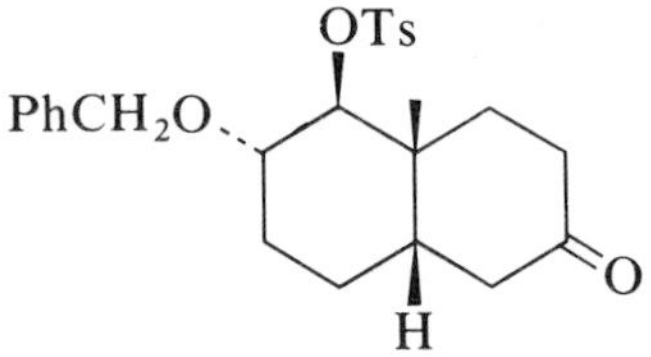

(69)

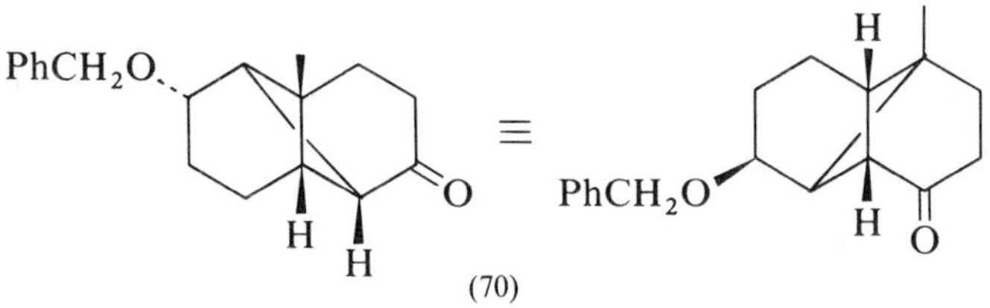

(70)

α- and β-cubebene (71) have been isolated from the essential oil of cubeb (Ohta *et al.*, 1966). Two independent syntheses of these sesquiterpenes have been reported (Piers *et al.*, 1969b; Tanaka *et al.*, 1969) in which formation of the tricyclic ring system was achieved by decomposition of the diazoketone (72).

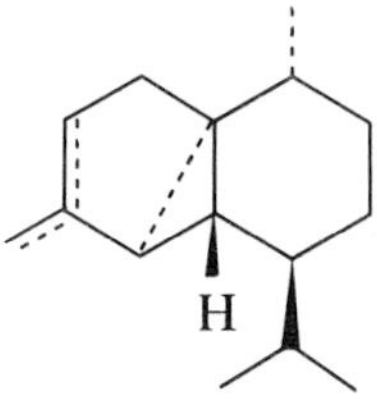

(71)

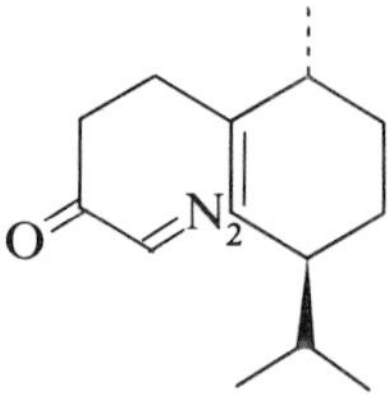

(72)

A biogenetic framework linking all these sesquiterpenoid types is summarized in Scheme 2. The formation of cation (73) is considered to arise by a 1,3-hydride shift from cation (8). It must be admitted, however, that other biogenetic postulates can be advanced.

By an elegant degradative study, de Mayo *et al.* (1963) deduced the structure and stereochemistry of helminthosporal (74), the toxin from *Helminthosporium sativum* responsible for serious cereal crop damage. Subsequently Corey and Nozoe (1965a) confirmed this structure by synthesis starting from (−)-carvomenthone, which also served to define the absolute stereochemistry. A biogenetic scheme was postulated by de Mayo *et al.* (1962) for the formation of helminthosporal which involved anti-Markownikoff cyclization of a cation such as (75) (i.e. the antipodal amorphane-type) to the cation (76) followed by a Wagner–Meerwein shift to (77) and oxidative

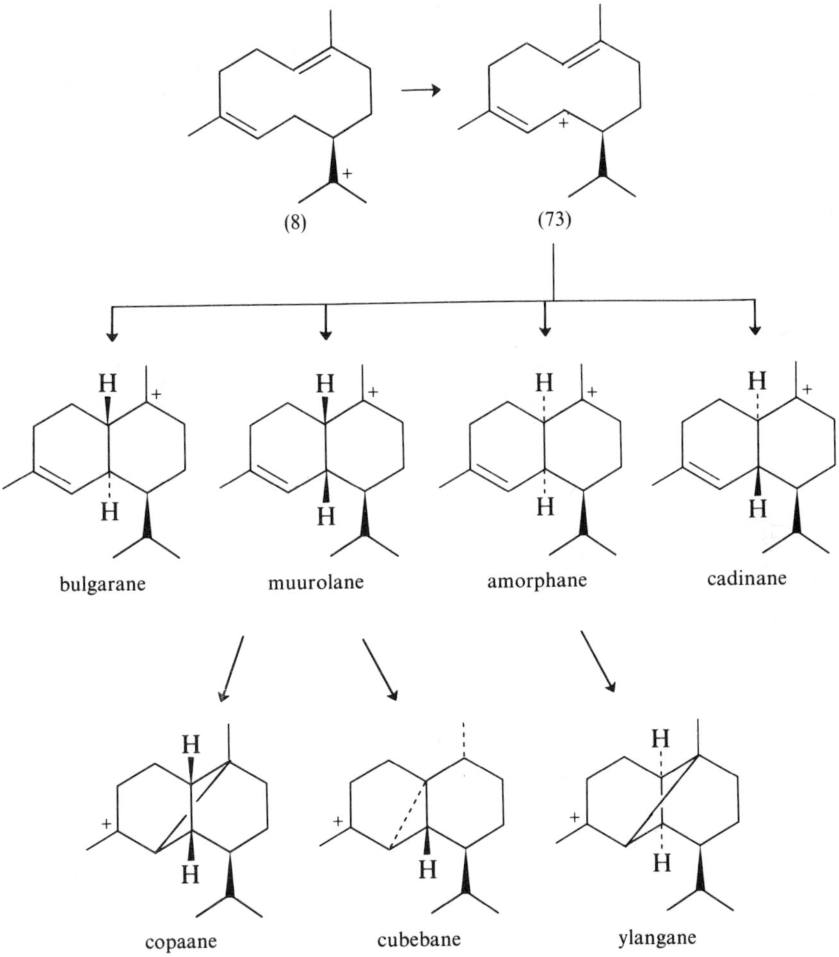

SCHEME 2

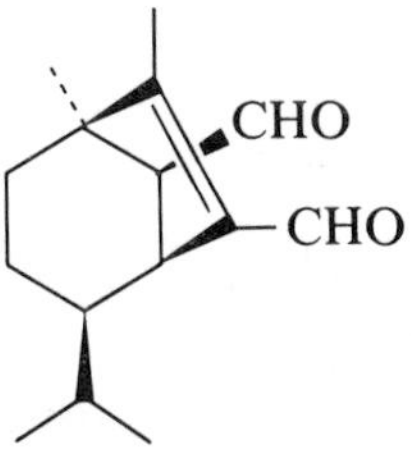

(74)

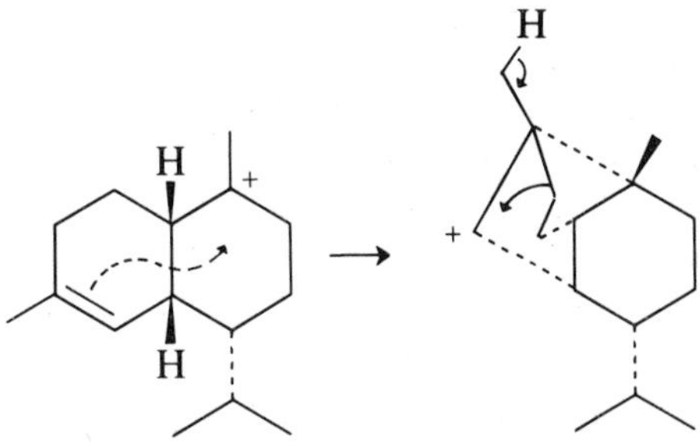

(75) (76)

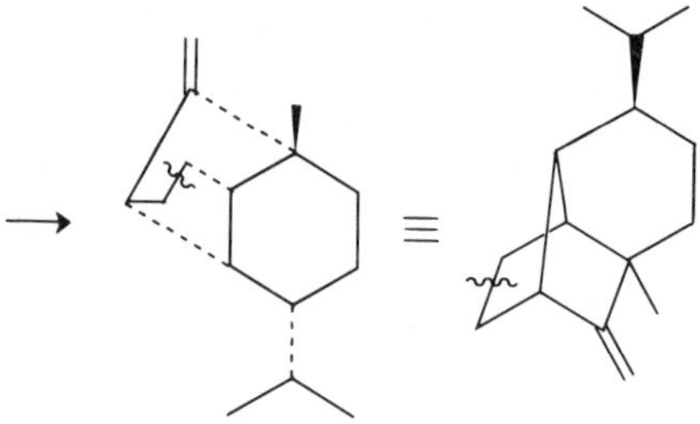

(77)

cleavage of the bond shown. This hypothesis not only withstood the test of labelling studies but a subsequent careful examination (de Mayo and Williams, 1965a) of the mould metabolites revealed the presence of the tricyclic hydrocarbon (77), now called (−)-sativene. Starting from the bromo-ester (78), derived from (+)-longifolene, de Mayo and Williams (1965a) achieved the synthesis of (+)-sativene (80) via the ring contracted olefin-ester (79). In outstanding fashion McMurry (1968) has also synthesized

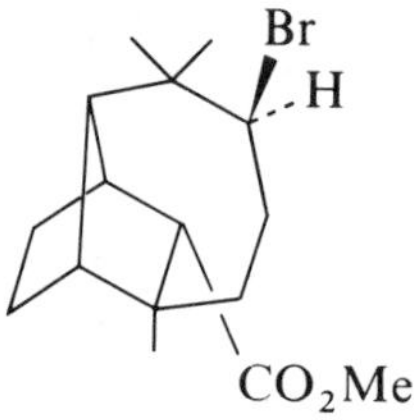

(78)

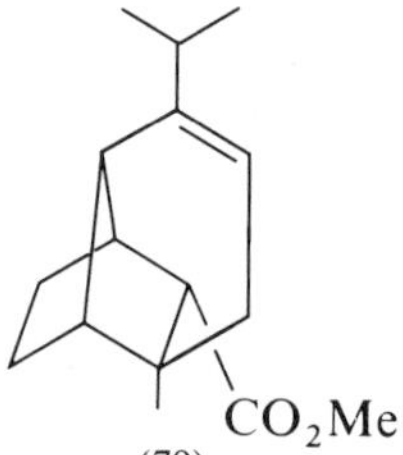

(79)

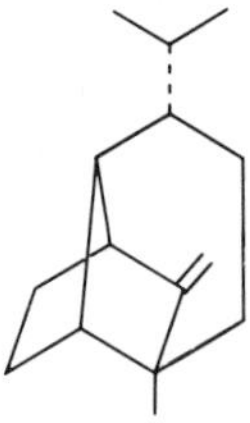

(80)

(±)-sativene by a method reminiscent of that used in the synthesis of copaene and ylangene. This synthesis hinged upon the successful construction of the *cis*-decalone compound (81) which, on base treatment, underwent an intramolecular alkylation to the tricyclic ketone (82) which was then converted

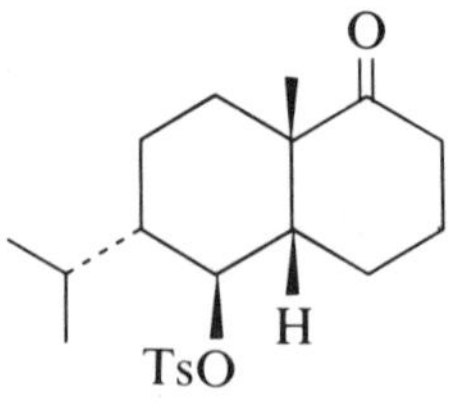

(81)

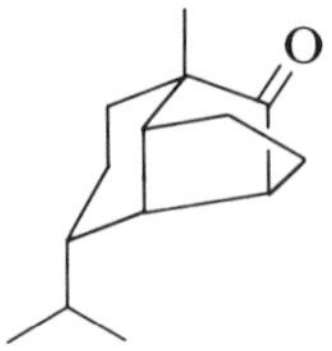

(82)

to (±)-sativene. Recently (+)-sativene (80) together with the tetracyclic analogue, (+)-cyclosativene (83), have been isolated (Smedman and Zavarin,

1968) from the cortical turpentine of California red fir. Kolbe-Haugwitz and Westfelt (1970) have determined the structure of copaborneol (84), mainly on the basis of its synthesis from a mixture of santalols. Biogenetically this structure can be derived from a muurolane-type precursor. McMurry (1969) and Smedman *et al.* (1969) have made some very interesting observations concerning the isomerizations of copacamphene (85), sativene, cyclosativene, and isosativene (86). Recently, two epimeric pairs of compounds (87, R = CH_2OH and R = CO_2H), derivatives of cyclocopacamphene, have been isolated (Andersen, 1970; Homma *et al.*, 1970; Kido *et al.*, 1969a) from vetiver oil.

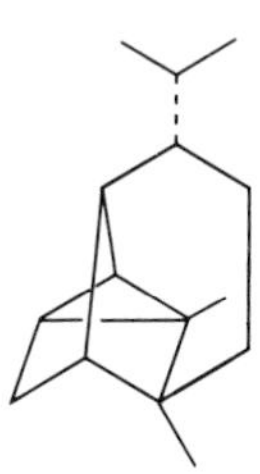

(83)

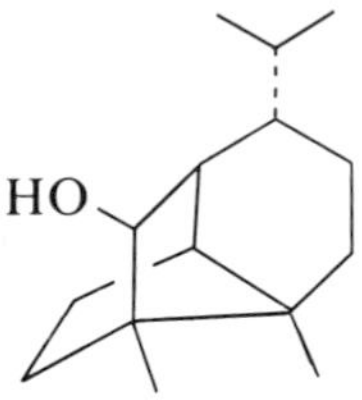

(84)

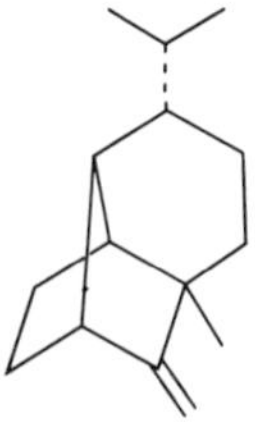

(85)

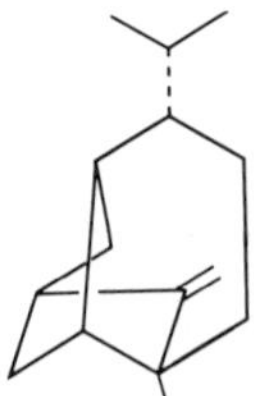

(86)

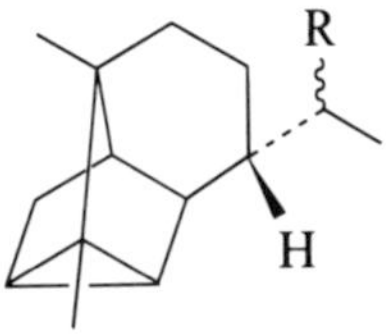

(87)

From the biogenetic standpoint the sesquiterpenoids of the picrotoxane type, exemplified by coriamyrtin (88, R = H) and tutin (88, R = OH) are classified under this general heading (Biollaz and Arigoni, 1969; Corbella *et al.*, 1969). The complex chemistry associated with this interesting group has been the subject of two recent reviews (Coscia, 1969; Porter, 1967) and thus will not be discussed here.

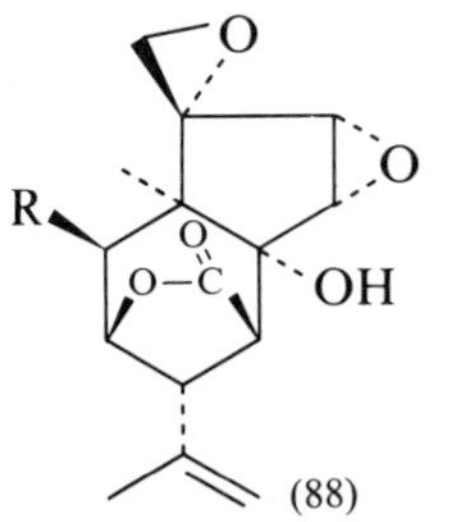

(88)

VII. Cuparane, Thujopsane, Widdrane, Chamigrane, Cedrane, Acorane, Laurane, etc.

The close structural and stereochemical relationships amongst various members of this group have elicited several biogenetic schemes most of which have involved the intermediacy of a bisabolane-type precursor (e.g. Scheme 3). The co-occurrence of several of these sesquiterpenes, e.g. α-cedrene (89), cuparene (90), thujopsene (91), widdrol (92) and β-chamigrene

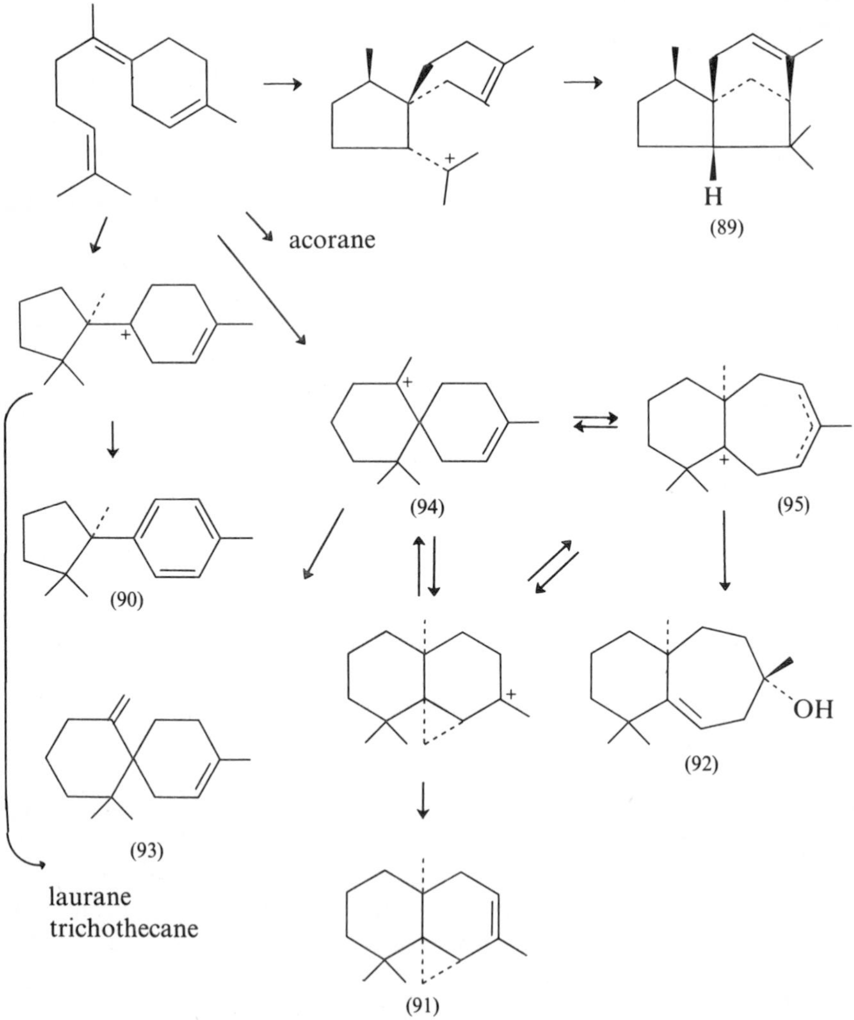

SCHEME 3

(93) has also been taken as indirect evidence of such a scheme (Yoshida *et al.*, 1967). In addition, *in vitro* studies have shown that β-chamigrene can be obtained from thujopsene and widdrol by acid-catalysed isomerization and dehydration respectively (Itô *et al.*, 1967). Dauben and Friedrich (1964) have also elegantly demonstrated the *in vitro* interconversion of thujopsene and widdrol. An alternative approach to the production of cations (94) and (95) involves cyclization of cation (96) derived from monocyclofarnesyl pyrophosphate. With this idea in mind, Kato *et al.* (1970a) have shown that both the *cis*-alcohol corresponding to cation (96) and the *trans* isomer (i.e. *trans* $\Delta^{2,3}$) can be dehydrated to α-chamigrene (97) in 25% yield.

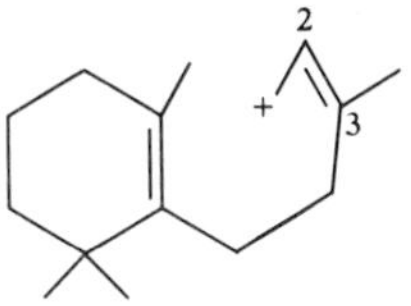

(96)

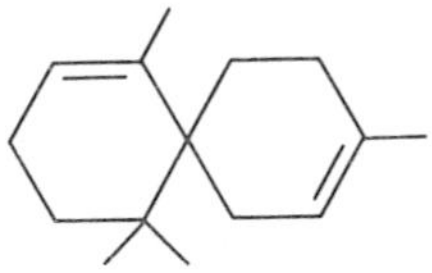

(97)

Each sub-group will now be considered separately. Several sesquiterpenoids of the cuparane skeleton have been isolated from the *Cupressaceae* family (Erdtman and Norin, 1966) and recently Lansbury and Hilfiker (1969) have reported a very neat synthesis of β-cuparenone (98) via acid-catalysed cyclization of the hydroxy-vinyl chloride (99) which was obtained from ethyl *p*-tolylacetate in three steps. An unusual quinonoid member of this group is the fungal pigment, helicobasidin (100) (Natori *et al.*, 1964).

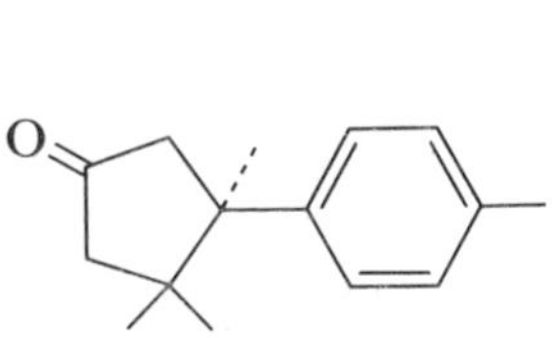

(98)

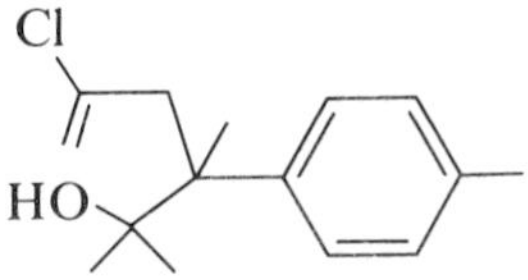

(99)

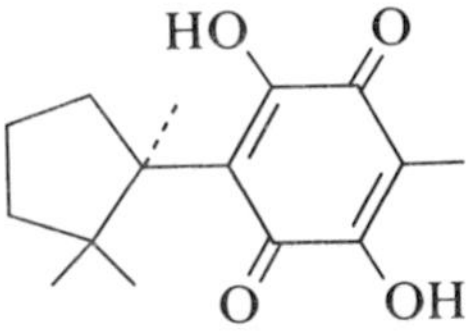

(100)

Thujopsene (91) and widdrol (92) are also common Cupressales constituents (e.g. *Widdringtonia*, *Juniperus*, *Thujopsis* species). The structural determinations of these compounds are due mainly to the degradative studies by Swedish (Enzell, 1962b; Norin, 1961) and Japanese (Itô *et al.*, 1963, 1964) chemists. In total, three syntheses of thujopsene (Büchi and White, 1964; Dauben and Ashcroft, 1963; Mori *et al.*, 1970) have been reported and Enzell (1962a) has completed the synthesis of widdrol.

In addition to the "biogenetic" synthesis of chamigrene, a rational synthesis has also been achieved (Tanaka *et al.*, 1967) in which the known α,β-unsaturated ketone (101) was converted into the spiro-dione (102) by a Diels–Alder reaction with 2-ethoxybuta-1,3-diene followed by acid hydrolysis of the resultant enol ether. A Grignard reaction with methylmagnesium iodide followed by dehydration yielded the ketone (103) which was converted into β-chamigrene by a Wittig reaction. Alternatively the ketone (103) could be prepared in 20% yield by using isoprene as the diene in the Diels–Alder reaction.

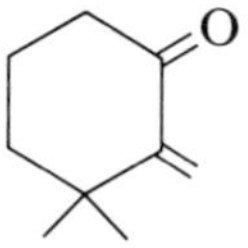

(101)

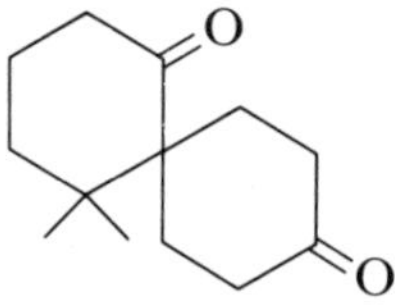

(102)

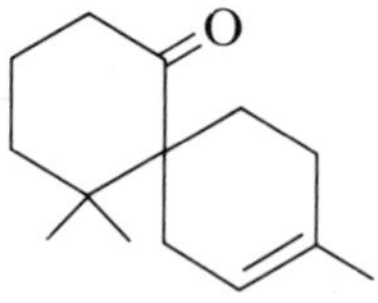

(103)

Although the alcohol, cedrol (104), had been isolated from cedar wood oil as early as 1841 by Walter, it was not until 1953 that the structure of α-cedrene (89) was finally ascertained by the rationalization of a sequence of degradative steps in which the dicarboxylic acid (105) and the tricarboxylic acid (106) played a crucial role (Plattner *et al.*, 1953; Stork and Breslow,

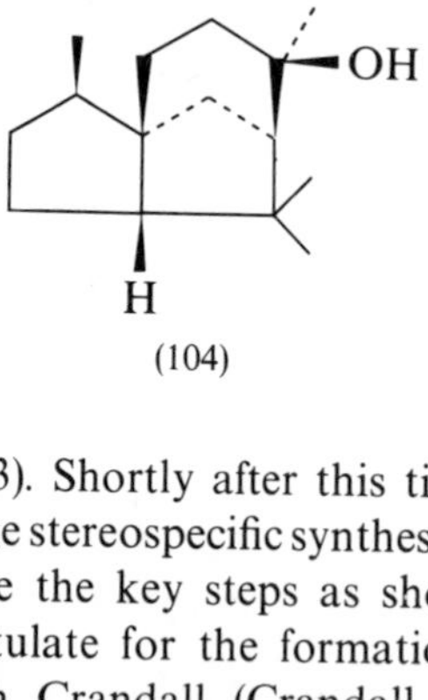

(104)

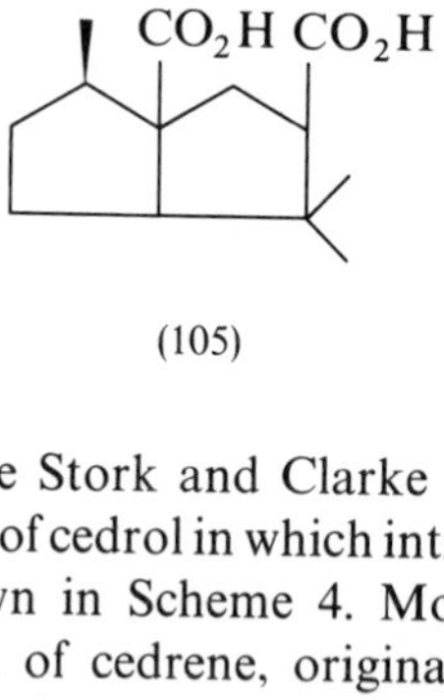

(105)

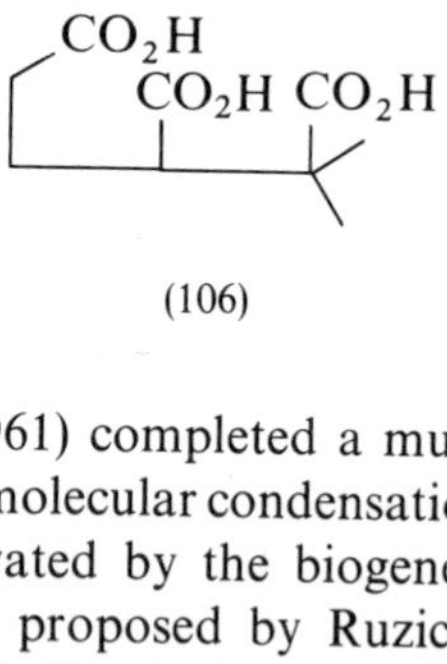

(106)

1953). Shortly after this time Stork and Clarke (1961) completed a multi-stage stereospecific synthesis of cedrol in which intramolecular condensations were the key steps as shown in Scheme 4. Motivated by the biogenetic postulate for the formation of cedrene, originally proposed by Ruzicka, both Crandall (Crandall and Lawton, 1969) and Corey (Corey *et al.*, 1969d) and their co-workers have synthesized α-cedrene and cedrol. Both syntheses were modelled along approximately parallel lines in the sense that the penultimate goal was the generation of the spiro-cation (107, R = +). The two pathways to this cation differed in several respects especially in the initial stages, but they practically coincided at the key spiro-dienone esters [108, R = Me (Corey *et al.*, 1969d) and R = Et (Crandall and Lawton, 1969)], which were prepared by similar $Ar_{1\text{-}5}$ participations from the appropriately substituted phenol precursors. Subsequent to the reports of these syntheses,

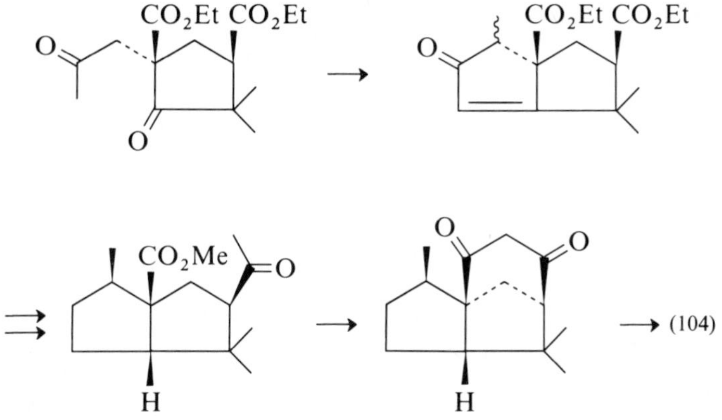

SCHEME 4

the alcohol (107, R = OH), named α-acorenol, was found to occur naturally in the wood of *Juniperus rigida*, and Tomita and Hirose (Tomita and Hirose, 1970a; Tomita *et al.*, 1970) demonstrated its facile acid-catalysed cyclization to (−)-α-cedrene in greater than 90% yield. From the same source γ- (109) and δ-acoradiene (110) have been isolated and these sesquiterpenes are identical with the α- and β-alaskenes isolated from *Chamaecyparis nootkatensis* by Andersen and Syrdal (1970).

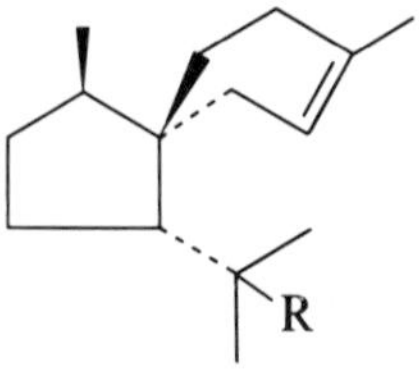

(107)

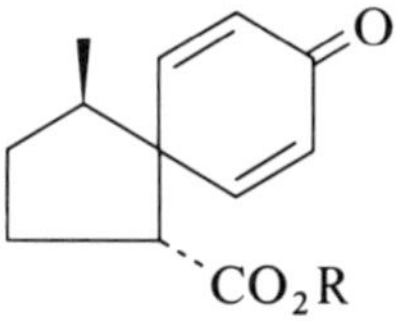

(108)

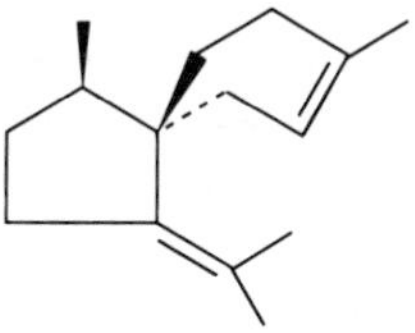

(109)

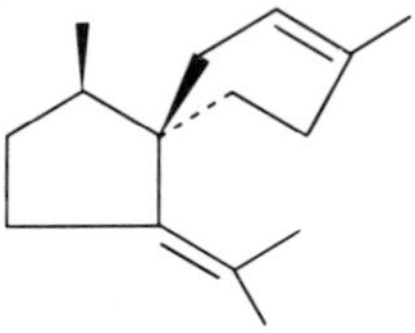

(110)

A number of highly oxygenated cedrane-type sesquiterpenoids, exemplified by jalaric acid (111), have been isolated (Singh *et al.*, 1969; Wadia *et al.*, 1969; Yates and Field, 1970; Yates *et al.*, 1970) from the basic hydrolysate of lac resin, the secretion of the insect, *Laccifer luca*.

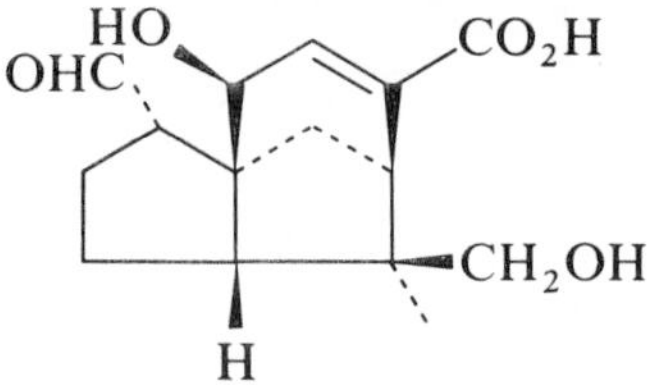

(111)

The diketone, acorone (112), isolated from the oil of sweet flag, *Acorus calamus*, was the first spiro sesquiterpenoid to be discovered. Although the gross skeleton was readily ascertained by degradation studies, the determination of the relative and absolute stereochemistries of acorone and its related co-occurring stereoisomers, isoacorone and cryptoacorone, proved to be a difficult task. Despite extensive chemical, dipole moment, and molecular rotation studies (Vrkoč *et al.*, 1964) and finally the X-ray analysis (McEachan *et al.*, 1966) of a derivative of acorone, the stereochemistries of these are still not certain.

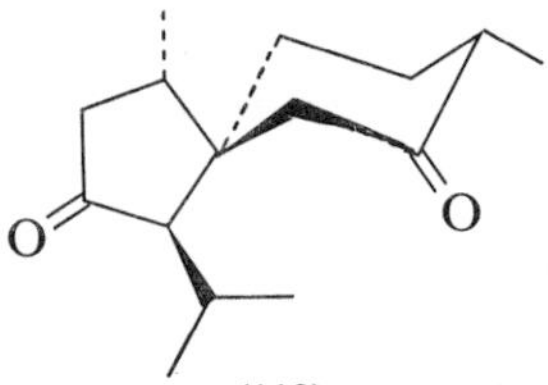

(112)

Closely related to the cuparane-type sesquiterpenoids are laurene (113) (Irie *et al.*, 1969b), laurinterol (114) (Cameron *et al.*, 1969; Irie *et al.*, 1970), isolaurinterol (115) (Irie *et al.*, 1970), and laurenisol (116) (Irie *et al.*, 1969a), all isolated from the *Laurencia* species of seaweed. The mollusc, *Aplysia kurodai*, which feeds on the above seaweed, produces aplysin (117) (Yamamura and Hirata, 1963) and it has been shown that laurinterol, on treatment with *p*-toluenesulphonic acid, is converted into aplysin.

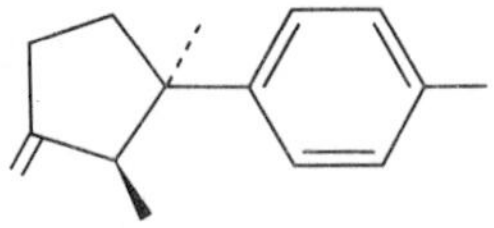

(113)

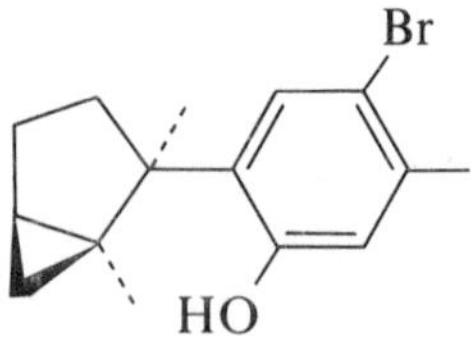

(114)

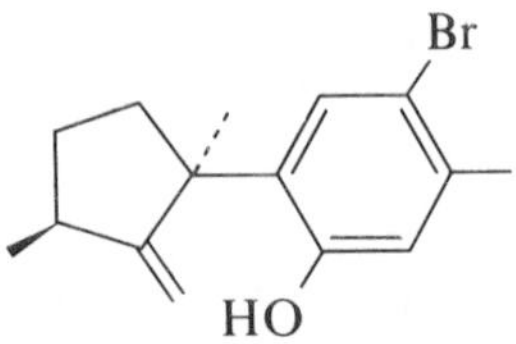

(115)

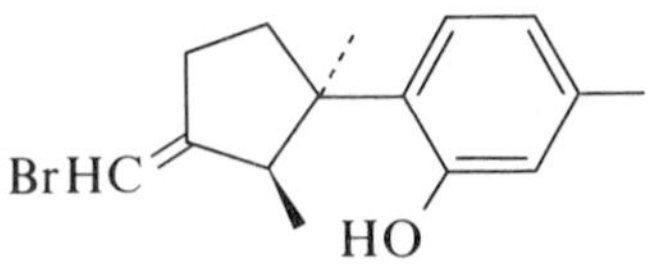

(116)

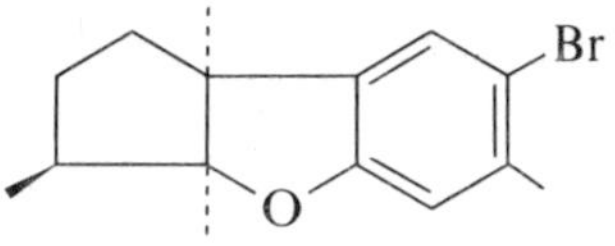

(117)

After extensive degradation studies, Fishman *et al.* (1960) concluded that the interesting fungal metabolite, trichothecin, had the structure (118). In 1964, however, a combination of the X-ray structural determination (Abrahamsson and Nilsson, 1964) of a derivative of trichodermin (119, $R^1 = H_2$, $R^2 = OAc$) and the conversion (Godfredsen and Vangedal, 1964, 1965) of trichodermin into trichothecolone acetate (119, $R^1 = O$, $R^2 = OAc$) necessitated a revision of the original structure to (119, $R^1 = O$, $R^2 =$ isocrotonyl). Since that time a number of related sesquiterpenoids has been isolated, e.g. trichodiene (120) (Nozoe and Machida, 1970) and verrucarin A (121) (Gutzwiller and Tamm, 1965; McPhail and Sim, 1965).

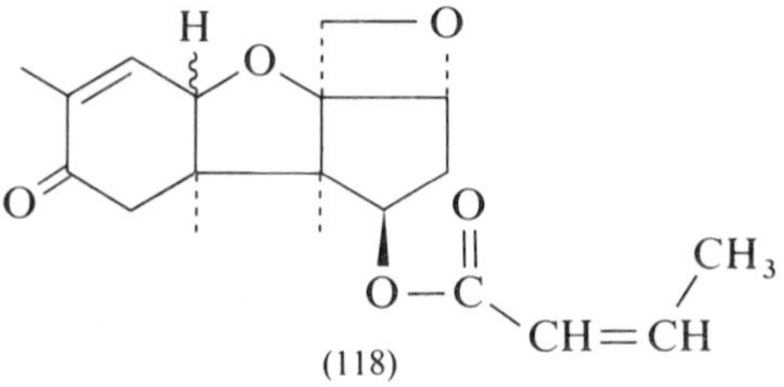

(118)

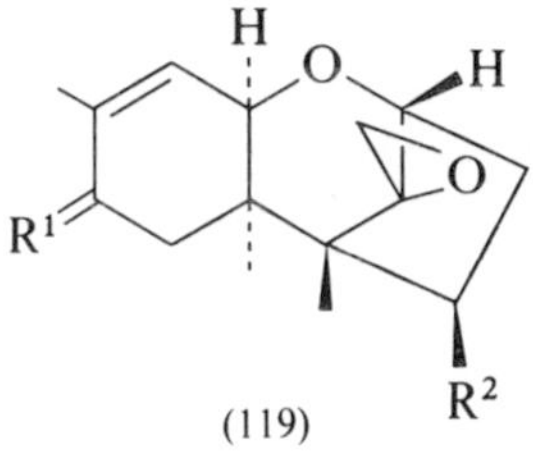

(119)

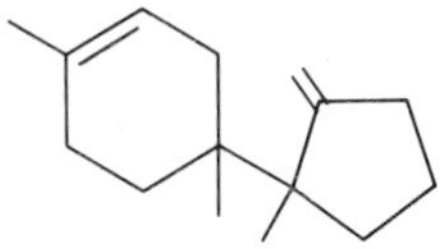

(120)

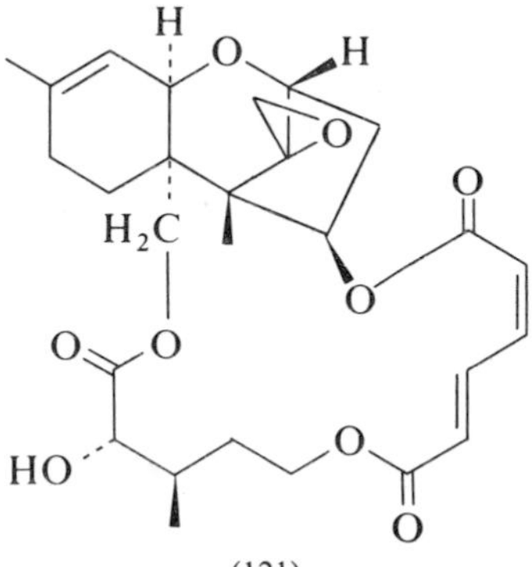

(121)

Finally, to conclude this section, we note that cyclization of a bisabolane-type precursor can lead to bazzanene (122), a new sesquiterpene isolated from *Bazzania pompeana* (Lac.) Mitt (Hayashi *et al.*, 1969).

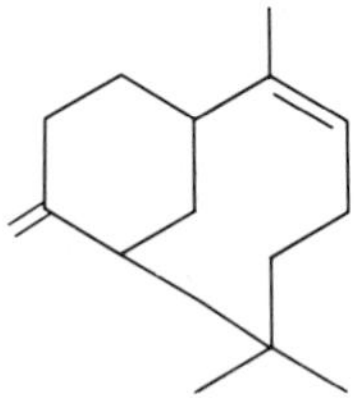

(122)

VIII. Daucane

The biogenesis of this small group of sesquiterpenoids, which includes carotol (123), daucol (124), and laserpitine (125) (Holub *et al.*, 1970), has been suggested and partially verified (Souček, 1962) in terms of cyclization of cation (7) with a concomitant 1,3-hydride shift. Both carotol and daucol, which are major components of carrot seed essential oil, have been extensively studied with the result that their stereochemistries have been firmly established by chemical and spectral methods (Levisalles and Rudler, 1967) as well as by X-ray analysis (Bates *et al.*, 1969).

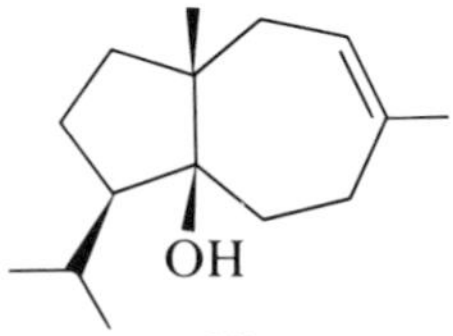

(123)

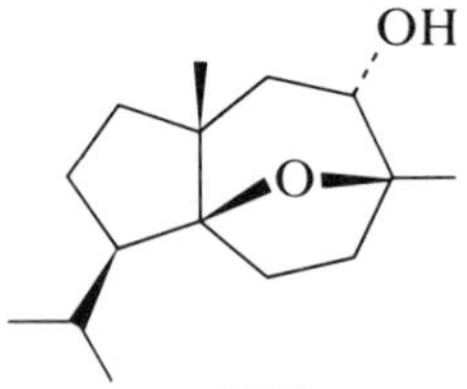

(124)

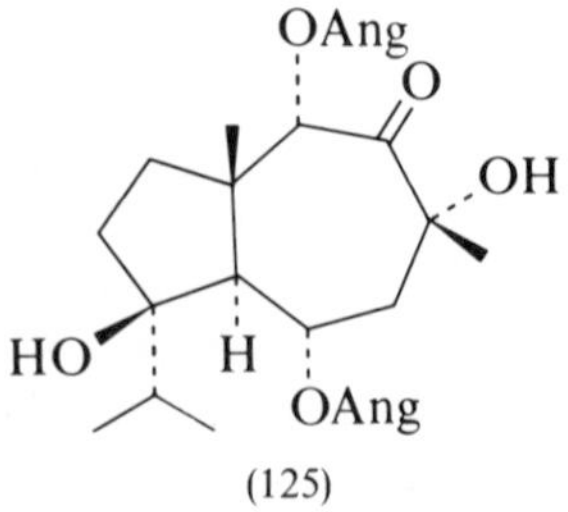

(125)

IX. Caryophyllane, Humulane, and Related Sesquiterpenoids

One of the most interesting and widely-occurring sesquiterpenes is caryophyllene (126). Through the years many valiant attempts were made to elucidate the structure of this unusual hydrocarbon. Knowing now the ease with which caryophyllene and some of its derivatives can undergo both transannular and deep-seated rearrangements one can sympathize with the earlier workers in this field. In 1954, the word acrobatic was aptly used (Nickon, 1954) to describe the molecular agility of caryophyllene and today this word is all the more appropriate in the light of recent investigations. The structural elucidation of caryophyllene was eventually completed in the early 1950s by the piecing together of an intricate jigsaw whose varied components comprised rational degradative studies, diagnosis of the nine-membered ring by infra-red studies, structural determination of some of the rearrangement products with a rationalization of their geneses, and finally X-ray studies. Acid-catalysed rearrangement of caryophyllene leads to two major products, caryolan-1-ol (127) and clovene (128), the structural assignments of which were instrumental in the unravelling of the caryophyllene structure. The structure of clovene has been confirmed by an elegant synthesis (Doyle *et al.*, 1965). In addition to clovene, another hydrocarbon rearrangement product, neoclovene (129), has been identified (Parker *et al.*, 1969) and its synthesis, partially modelled upon its proposed mode of formation from caryophyllene, has been achieved (McKillop *et al.*, 1967). Caryolan-1-ol, on

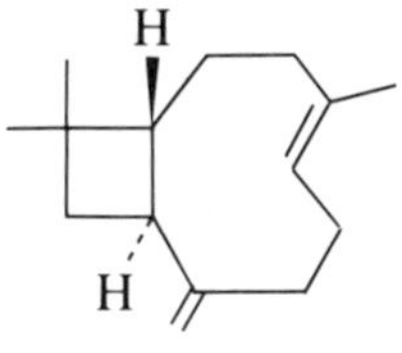

(126)

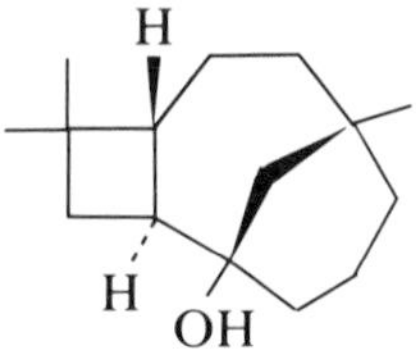

(127)

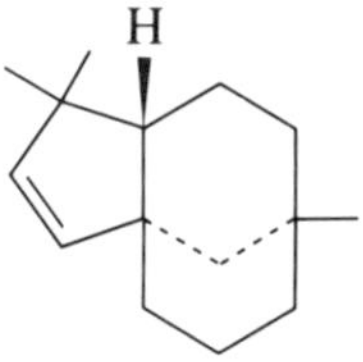

(128)

(129)

dehydration with polyphosphoric acid, gives rise to a plethora of products of which isoclovene (130) (Clunie and Robertson, 1961), pseudoclovene A (131) (Ferguson *et al.*, 1967), and pseudoclovene B (132) (Parker, personal communication) have been positively identified. Isocaryophyllene (133), the *cis* double bond isomer of caryophyllene, yields neoclovene and the tricyclic olefin (134) on acid-catalysed rearrangement (Gollnick *et al.*, 1970). These rearrangement products related to caryophyllene are by no means exhaustive but only serve to illustrate the point made at the beginning of this section.

One of the major achievements in sesquiterpene synthesis is that of caryophyllene and isocaryophyllene. Essentially there are three problems associated

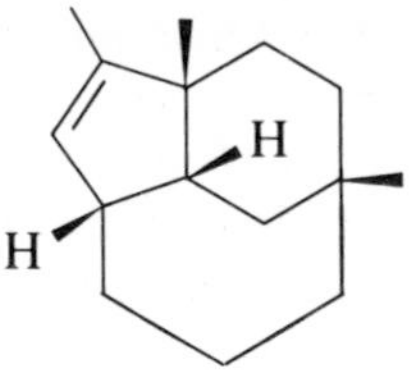

(130)

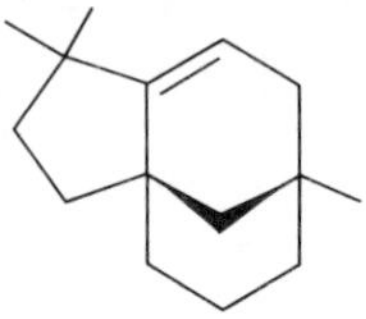

(131)

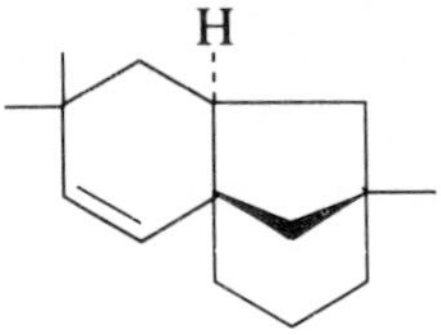

(132)

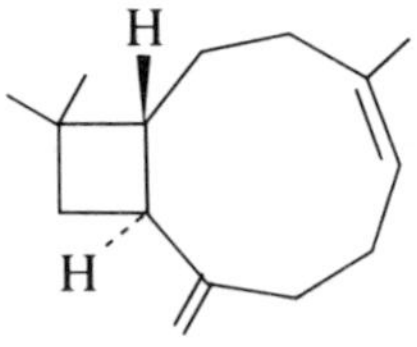

(133)

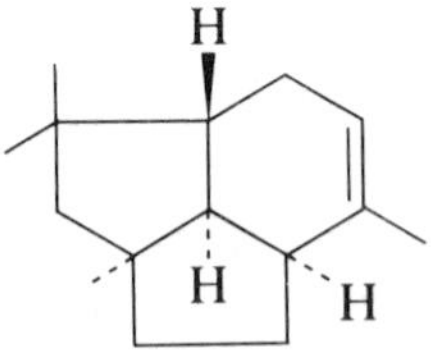

(134)

with the synthesis of these hydrocarbons, viz. the *trans* fusion of a four-membered ring, the construction of an appropriately substituted nine-membered ring and the control of the stereochemistry of the endocyclic double bond. These three aspects were brilliantly handled by Corey *et al.* (1964a), who employed the following route (Scheme 5):

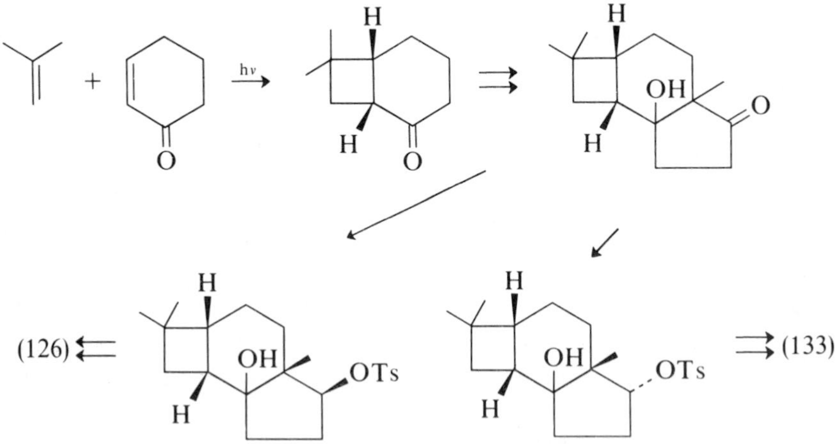

SCHEME 5

Humulene, like caryophyllene, eluded structural assignment for many years until it was recognized by Šorm and others (Šorm *et al.*, 1954) that humulene was a tetramethyl-cycloundecatriene. Confusion arose, however, as to the positions and stereochemistries of the three double bonds. These aspects were finally resolved when it was demonstrated (Dev, 1960; Sutherland and Waters, 1961) that naturally-occurring humulene is a mixture of at least two isomers, α- (135) and β-humulene (136). The α-isomer has the all *trans* stereochemistry as deduced (Hartsuck and Paul, 1964; McPhail *et al.*, 1964) from the X-ray analysis of the bis-silver nitrate complex. Pertinent

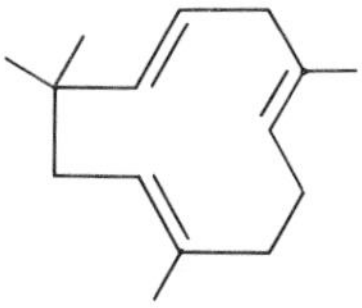

(135)

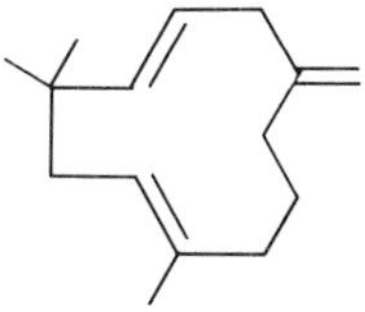

(136)

to the almost universal co-occurrence of caryophyllene and humulene is the fact that Greenwood *et al.* (1968) have converted humulene into caryophyllene by an *in vitro* sequence. Thus, treatment of humulene with N-bromosuccinimide in aqueous acetone resulted in the formation of the tricyclic bromohydrin (137) which was dehydrated to the tricyclic bromide (138). Reduction of (138) with lithium aluminium hydride gave caryophyllene (126) and humulene (135) in 30% and 10% yields respectively.

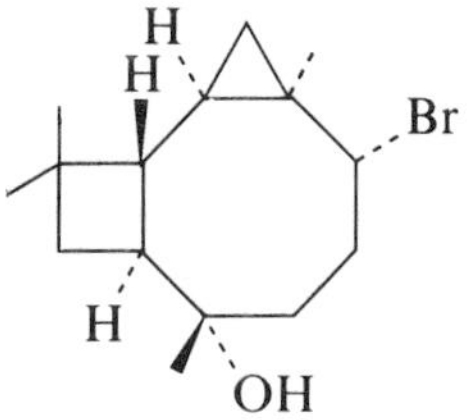

(137)

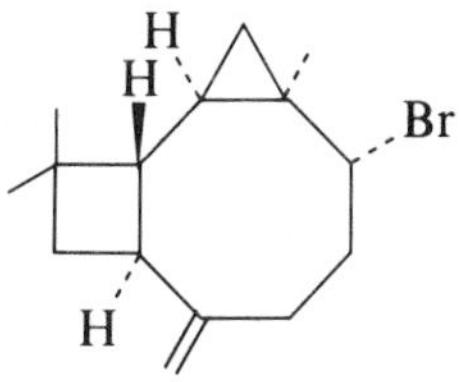

(138)

Corey and Hamanaka (1967) have synthesized humulene by the following route. A Wittig reaction between the aldehyde (139) and the ylide (140), followed by removal of the two protecting groups and treatment of the diol with phosphorus tribromide, yielded the *trans,cis,trans*-dibromide (141).

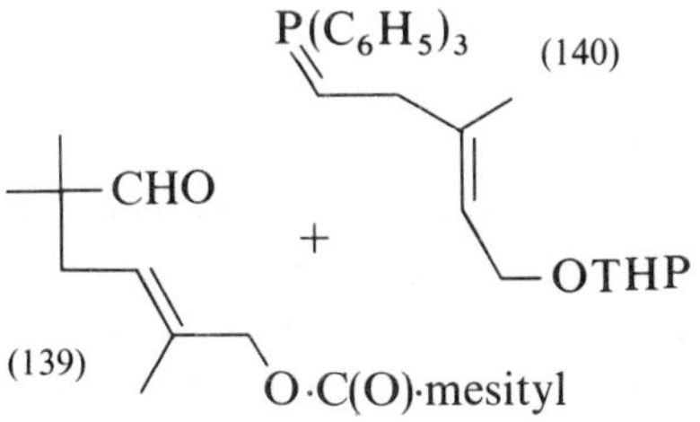

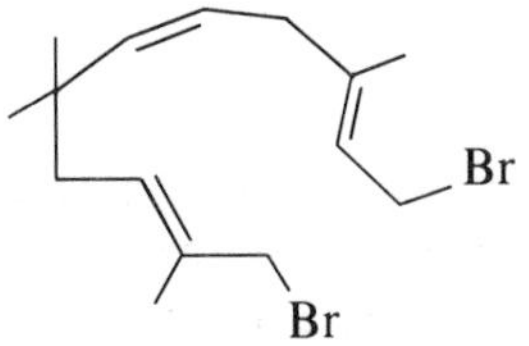

(141)

Nickel carbonyl-induced cyclization yielded the *cis* isomer of humulene (142) which was photochemically isomerized to humulene in the presence of diphenyl disulphide.

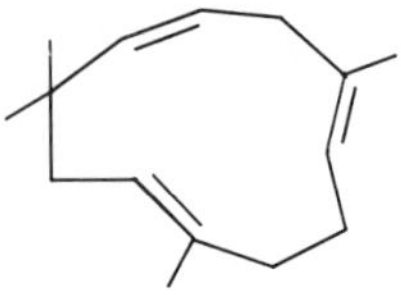

(142)

α-Caryophyllene alcohol (143), now known as apollan-11-ol (Nickon *et al.*, 1970), was considered for many years to be an acid-catalysed rearrangement product of caryophyllene. This, however, was disproven by two independent studies (Gemmell *et al.*, 1970; Nickon *et al.*, 1970) both of which, in addition to elucidating the structure of the alcohol, clearly demonstrated its origin from humulene. Corey and Nozoe (1965b) reported an elegant and economical synthesis of this alcohol as shown in Scheme 6.

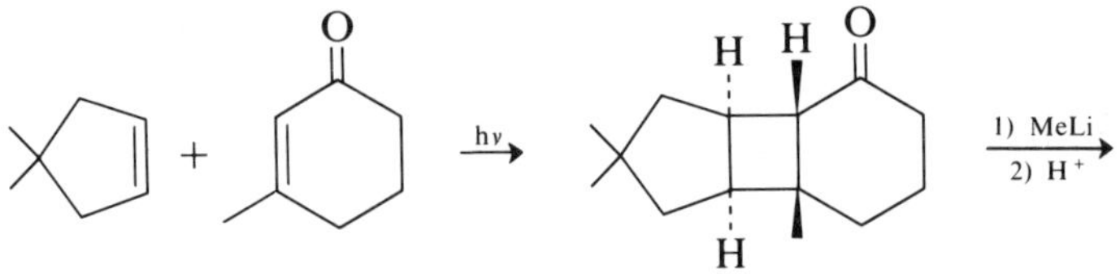

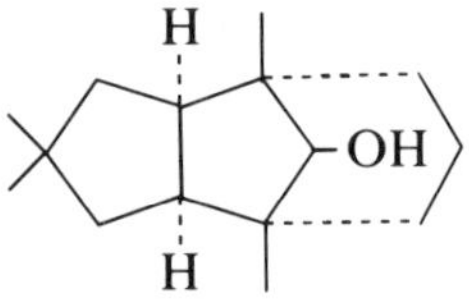

(143)

SCHEME 6

A number of very interesting sesquiterpenoid fungal metabolites have been isolated and these are included under this heading in view of their probable biogenetic origins from a humulene-type precursor (Scheme 7).

Illudin M (144, R = H) and illudin S (144, R = OH) have been isolated (Matsumoto *et al.*, 1965; McMorris and Anchel, 1965; Nakanishi *et al.*, 1965) from the bioluminescent mushroom, *Clitocybe illudens*, as has the related compound, illudol (145) (McMorris *et al.*, 1967). Matsumoto *et al.* (1968) have achieved a synthesis of illudin M in which the expert manipulation of

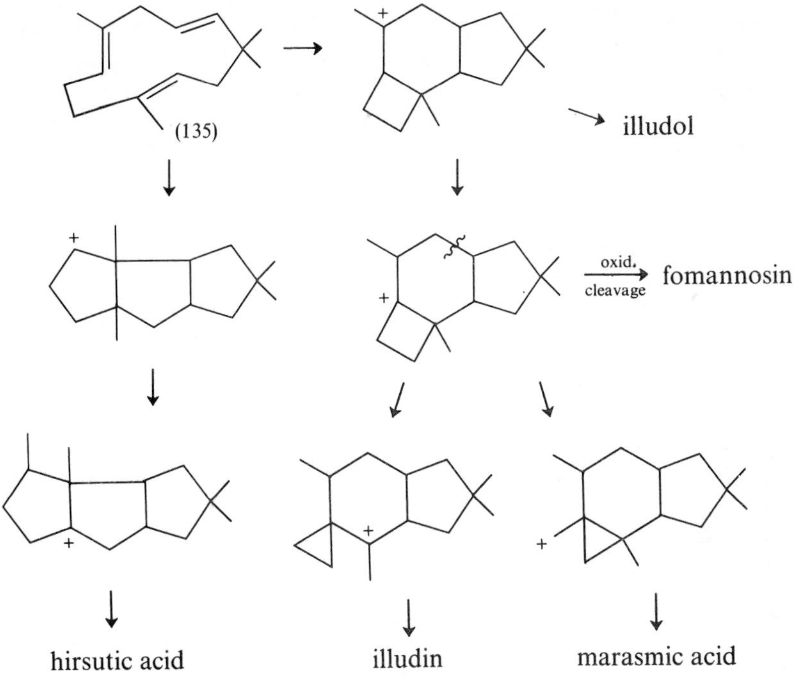

SCHEME 7

various functional groups is a major highlight. In essence, the synthetic sequence involved the addition of the β-keto-sulphoxide ketal (146) to the substituted cyclopentenone (147). This Michael product (148) underwent a Pummerer rearrangement to (149) which, on heating in ethanol, yielded the diketone (150). Intramolecular aldolization of (150) gave (151) which, on acetylation, reaction with methylmagnesium iodide, borohydride reduction and deketalization (also elimination of acetic acid), yielded (±)-illudin M (144, R = H).

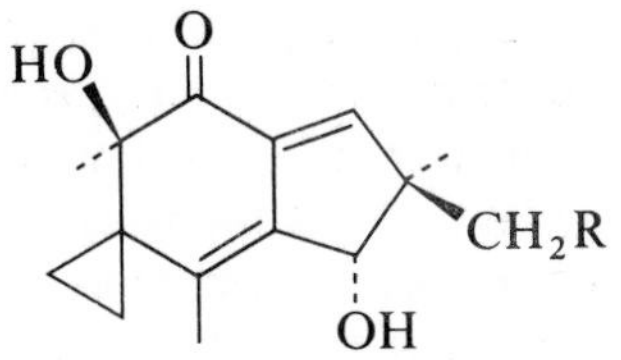

(144)

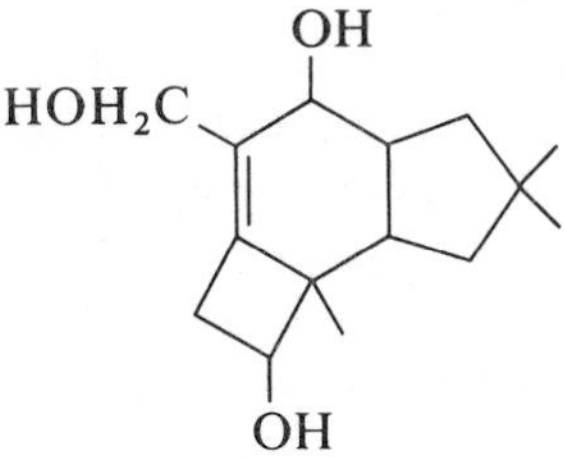

(145)

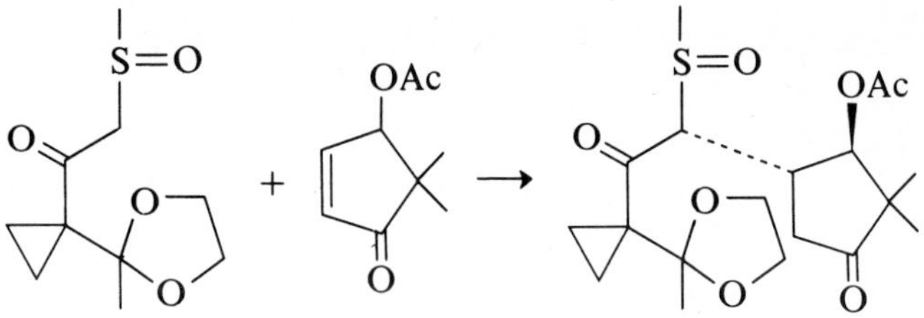

(146) (147) (148)

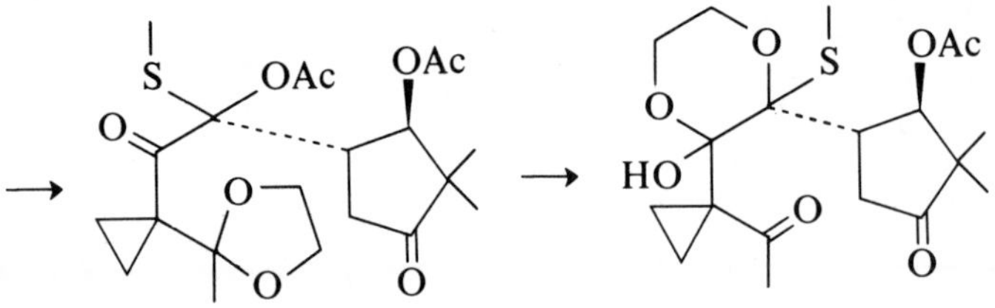

(149) (150)

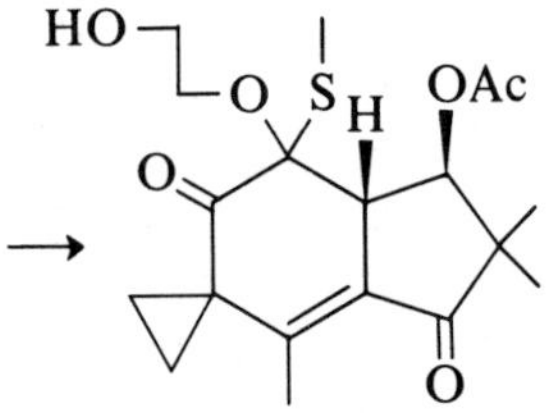

(151)

In an outstanding degradative study, Dugan *et al.* (1966) have elucidated the structure of marasmic acid (152), the antibiotic metabolite from *Marasmius conigenus.* A remarkable attempt to synthesize this complex molecule has been made by de Mayo and co-workers (Helmlinger *et al.*, 1970) using no less than four individual photochemical reactions at various stages. The tetracyclic ketone (153) was procured from the photochemical combination of cyclopentane-1,2-dione enol acetate with spiro[2,4]hept-5-ene. This was converted into the enone (154) which was then rearranged to the tricyclic ketol (155). Subsequent lead tetra-acetate oxidation, esterification, and elimination of methanol yielded the enone-ester (156, R = H). Singlet oxygen addition to this compound, followed by reduction of the hydroperoxide, gave the hydroxy-enone-ester (156, R = OH) which was photolysed in the presence of vinylene carbonate to give the unstable tricyclic carbonate (157). Reduction, esterification with pivaloyl chloride, and dehydration gave the diester (158). Addition of diazomethane and photolysis of the pyrazoline derivative gave the pentacyclic compound (159). Periodate cleavage of the glycol formed by hydrolysis and subsequent decarboxylation gave (160), which is an isomer of methyl marasmate.

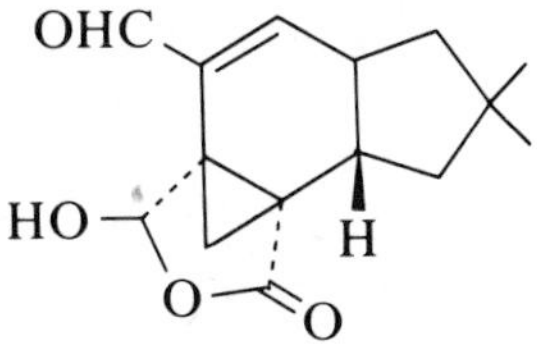
(152)

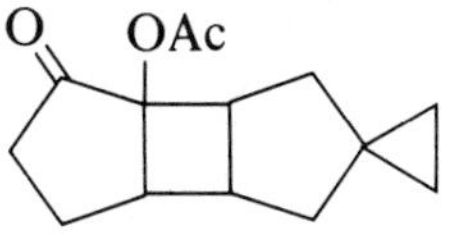
(153)

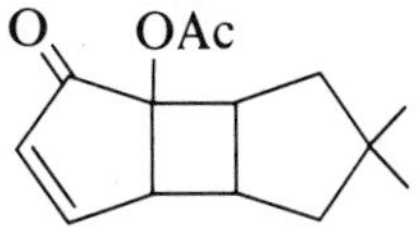
(154)

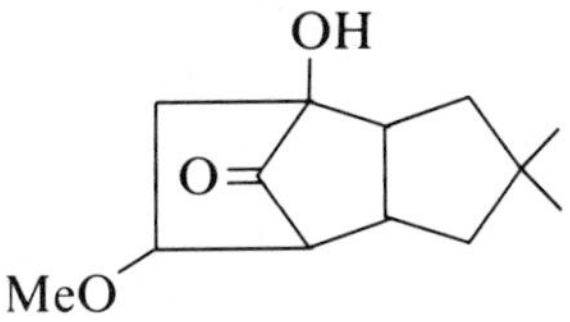
(155)

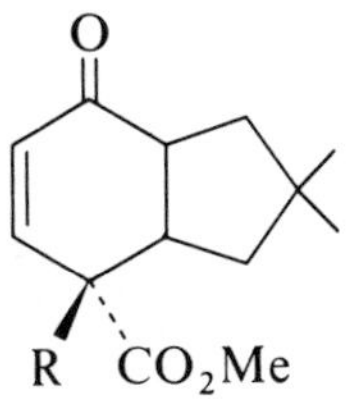
(156)

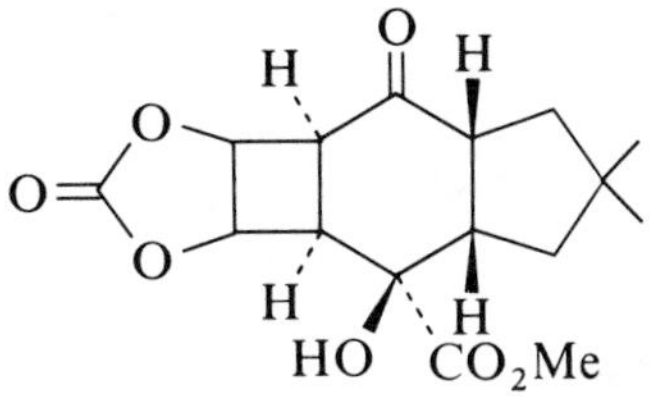
(157)

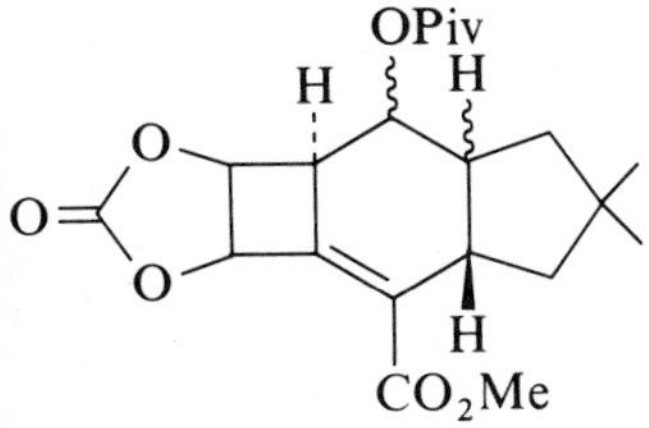
(158)

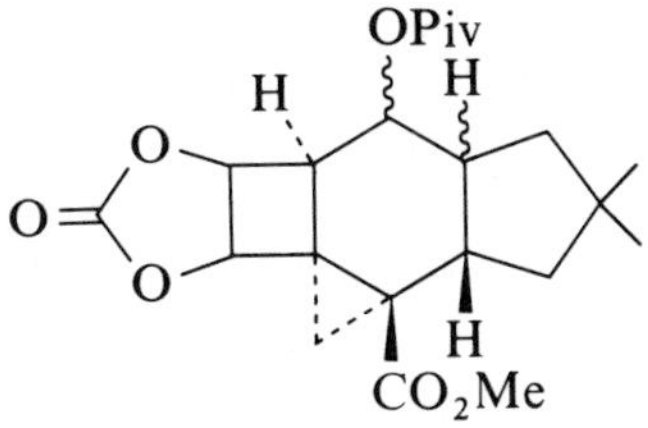
(159)

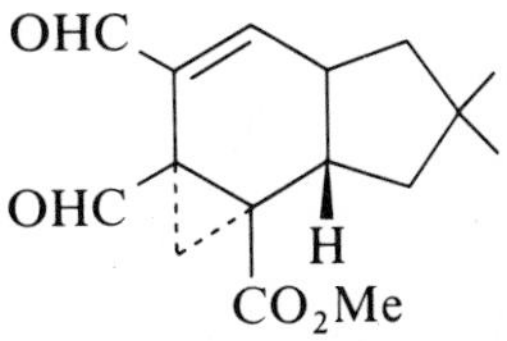
(160)

The unique structural features of fomannosin (161), a metabolite from the wood-rotting fungus, *Fomes annosus* (Fr.) Kart, have been deduced (Kepler *et al.*, 1967) from the X-ray analysis of a heavy atom derivative. Hirsutic acid C (162) is elaborated from the mould, *Stereum hirsutum*, and its structure was ascertained (Comer *et al.*, 1967) largely on the basis of X-ray studies.

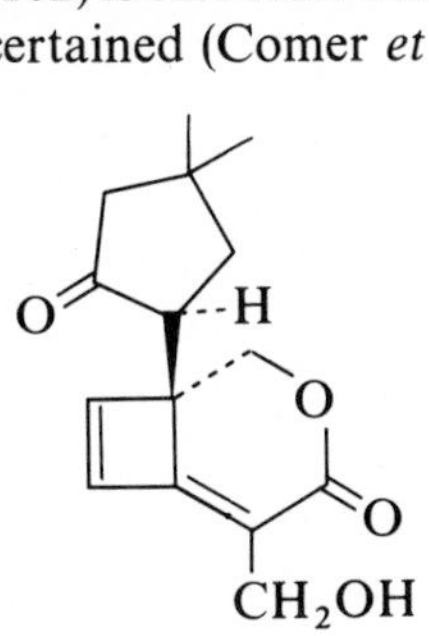
(161)

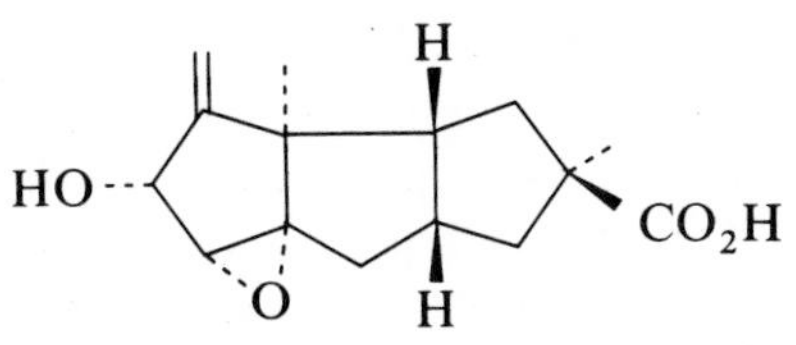
(162)

X. Himachalane, Longifolane, Longibornane and Longipinane

By analogy with the biogenetic scheme outlined for the formation of the cadinane-type sesquiterpenoids and the related tricyclic compounds, a similar scheme can be invoked for the structural types under this heading. Thus, cyclization of cation (163), derived from cation (9), leads to cation (164) which is the ideal precursor of the himachalenes which have been studied extensively by Joseph and Dev (1968). Further cyclization of (164) leads to (165) and (166) which are the obvious progenitors of the closely related tricyclic compounds. The inter-relationships of this group are exactly analogous to the more familiar monoterpene compounds, camphene, borneol, pinene and tricyclene.

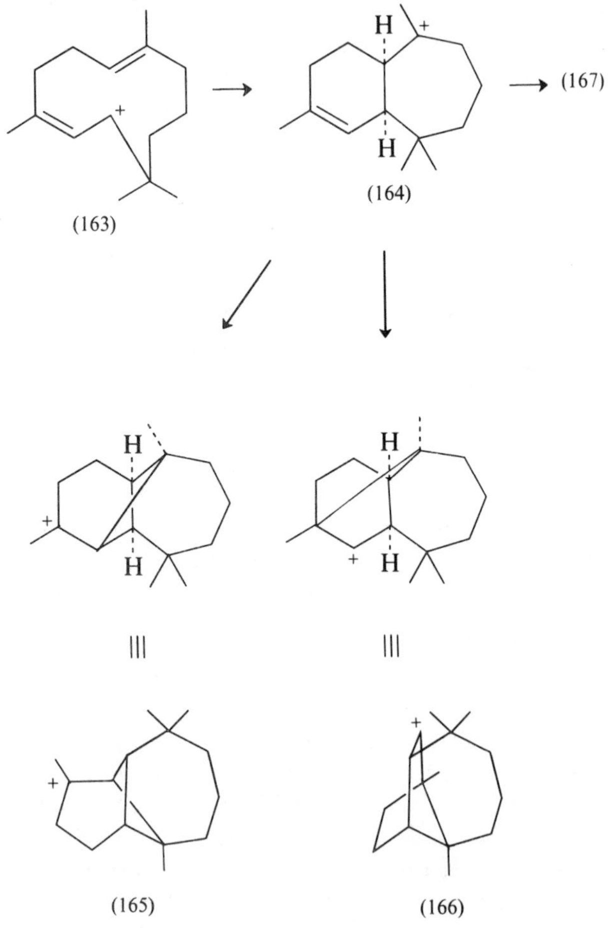

de Mayo and co-workers (Challand *et al.*, 1969) have carried out a neat synthesis of (±)-β-himachalene (167) based upon the photochemical production of the tricyclic ketone (170) from (168) and (169). Reduction and mesylation of (170) followed by base treatment gave the ring-opened ketone (171) which was converted in six steps into β-himachalene.

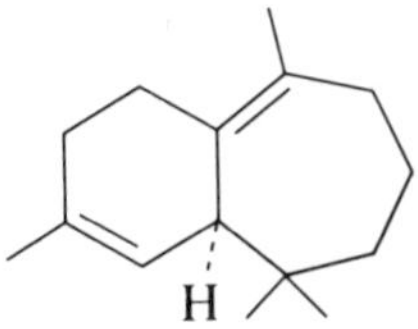

(167)

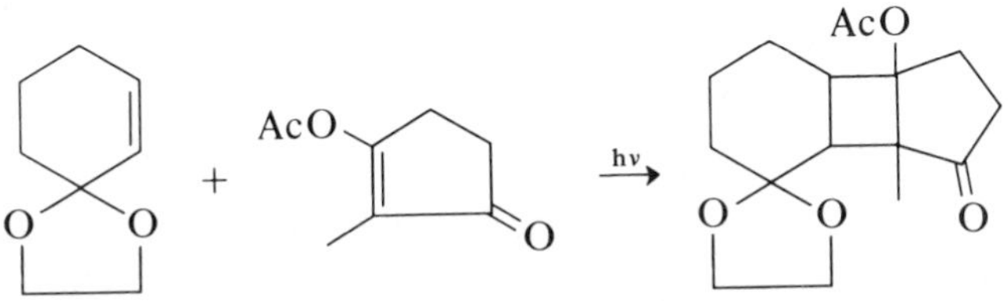

(168) (169) (170)

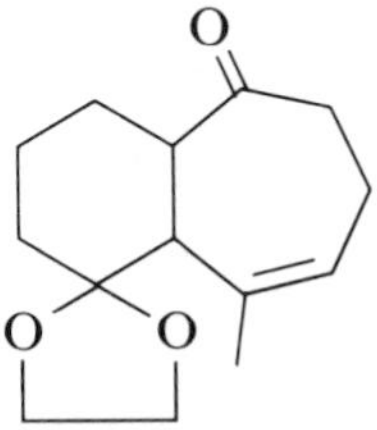

(171)

Longifolene (172), a constituent of the essential oil of many *Pinus* species, was originally studied by Simonsen and co-workers. Its structural elucidation proved difficult but finally it was determined from the X-ray analysis of its hydrochloride (Moffett and Rogers, 1953). Ourisson (1964) and Dev (Prahad *et al.*, 1970) have investigated many of the interesting transannular and isomerization reactions associated with longifolene. Yet another of Corey's ingenious syntheses includes that of longifolene (Corey *et al.*, 1964b). To summarize, this synthesis involved the successful five-step ring expansion of the Wieland–Miescher ketone (173) to (174) which underwent an intramolecular Michael cyclization to the tricyclic diketone (175) which was then

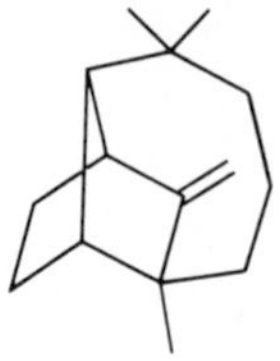

(172)

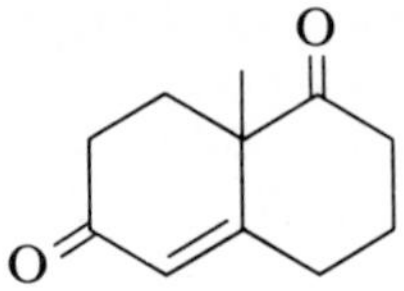

(173)

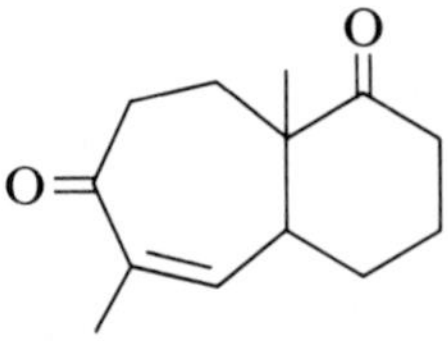

(174)

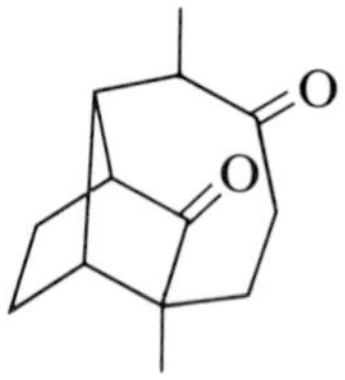

(175)

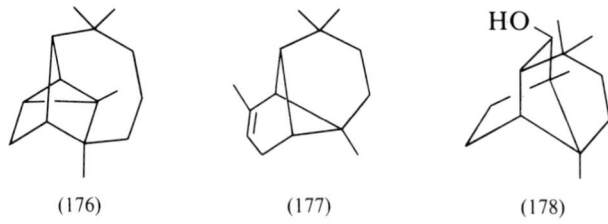

suitably elaborated to longifolene. The related compounds, longicyclene (176) (Nayak and Dev, 1963), α-longipinene (177) (Erdtman and Westfelt, 1963), and longiborneol (178) (Naffa and Ourisson, 1954) have also been isolated from various *Pinus* species.

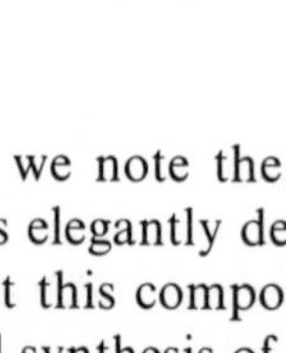

(176)

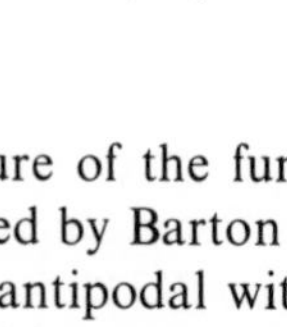

(177)

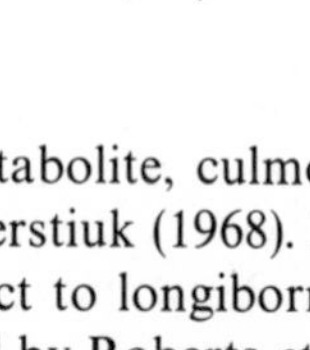

(178)

Finally we note the structure of the fungal metabolite, culmorin (179), which was elegantly determined by Barton and Werstiuk (1968). It is interesting that this compound is antipodal with respect to longiborneol (178). A rational synthesis of culmorin has been described by Roberts *et al.* (1969). who employed the bicyclic diketone (180), derived from tetrahydroeucarvone, as the basic building block. In six steps this diketone was converted into the keto-diester (181) which was cyclized and decarboxylated to the tricyclic diketone (182). Barton and Werstiuk had previously converted the (−)-diketone (182) to culmorin by reduction with sodium in *n*-propanol.

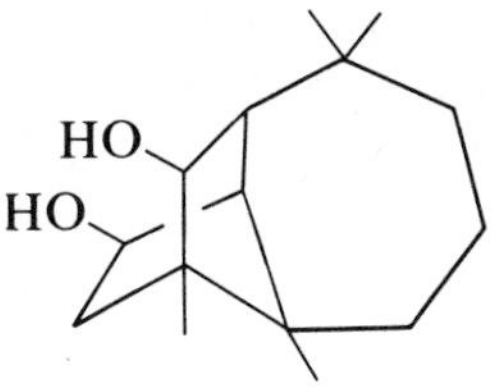

(179)

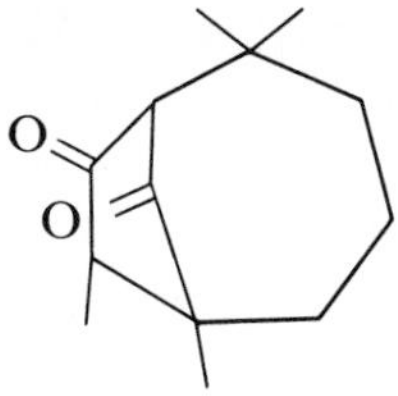

(180)

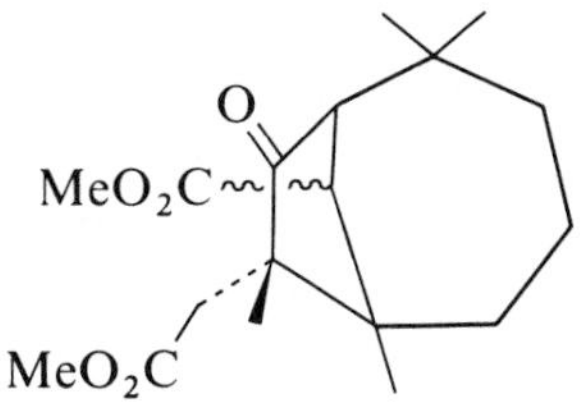

(181)

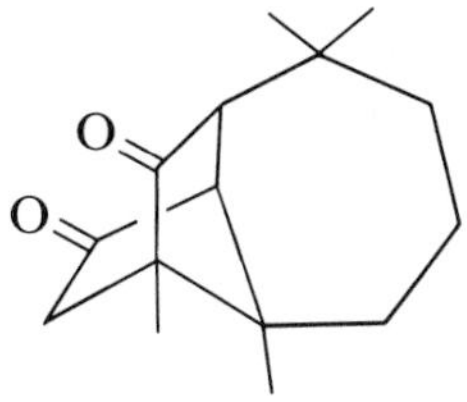

(182)

XI. Germacrane

Although no ten-membered ring sesquiterpenoids had been recognized as such at the time, Ruzicka *et al.* (1953) suggested that the cation (11) could play an important role in the biogenesis of the elemane-, eudesmane-, and guaiane-type sesquiterpenoids. Since that time the evidence supporting this postulate has been largely circumstantial with the one exception that labelled 11,12-dihydrocostunolide (183) has been incorporated into α-santonin (184) during biosynthesis (Barton *et al.*, 1968). The first sesquiterpenoid with the germacrane skeleton, pyrethrosin (185), was identified by Barton and de Mayo in 1957 (Barton and de Mayo, 1957; Barton *et al.*, 1960). Subsequently,

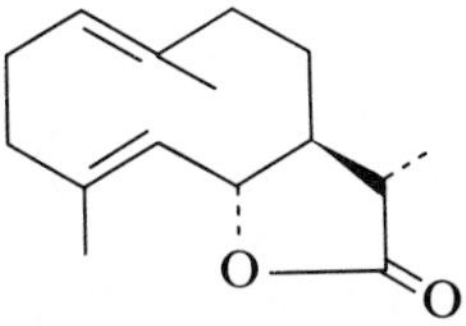

(183)

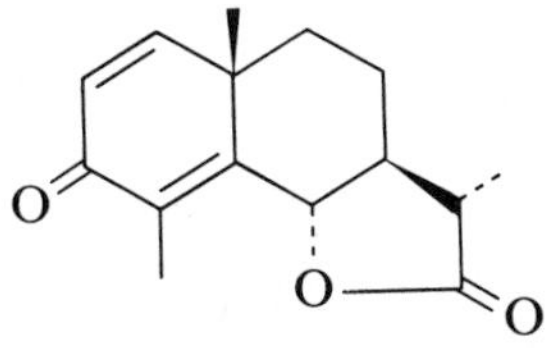

(184)

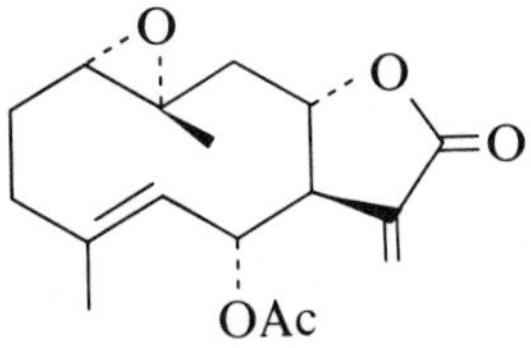

(185)

as a result of extensive studies by Bhattacharyya, Šorm, Takeda and their co-workers, a large number of germacranolides have been isolated and identified (Šorm and Dolejš, 1965). Essentially these compounds fall into three distinct classes, viz. the 6α,12-olides, the 8α,12-olides and the furanogermacranes as exemplified by cnicin (186) (Samek *et al.*, 1969; Yoshioka *et al.*, 1970b), chamissonin (187) (L'Homme *et al.*, 1969), and linderalactone (188) (Takeda *et al.*, 1964, 1969). It is extremely interesting that Kupchan *et al.* (1969a) have recently isolated a number of anti-tumour germacranolides, e.g. elephantin (189).

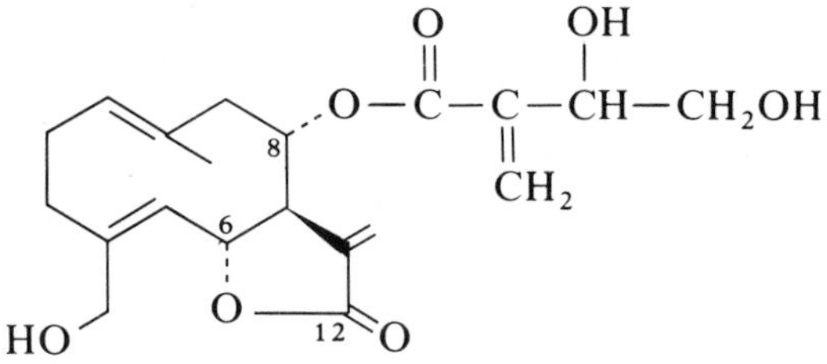

(186)

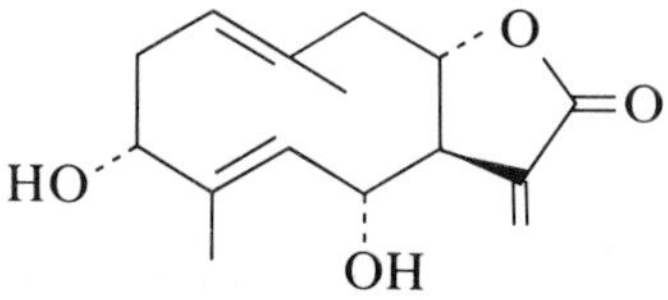

(187)

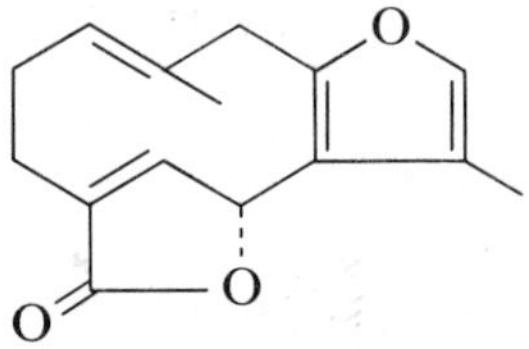

(188)

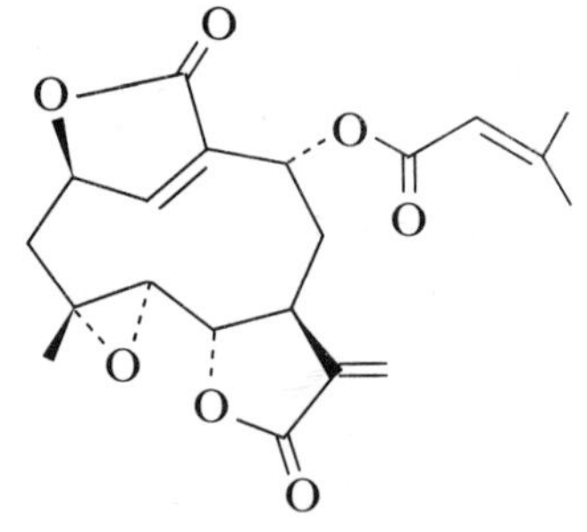

(189)

The indirect evidence for the biosynthetic intermediacy of the germacrane-type sesquiterpenes is two-fold. The first concerns the extreme ease with which some of the compounds can undergo *in vitro* acid-catalysed cyclizations. This fact has been witnessed on many occasions and to illustrate the point the following examples are quoted. Pyrethrosin (185), on treatment with acetic anhydride and toluene-*p*-sulphonic acid, gives rise to β-cyclopyrethrosin (190), which has, in fact, been found to be naturally-occurring

(Doskotch and El-Feraly, 1969). Cyclization of costunolide (191), originally studied by Bhattacharyya *et al.*, is smoothly effected by Amberlite cation exchange resin giving rise to α- (192) and β-cyclocostunolide (193) (Jain and McCloskey, 1969). Recently Morikawa and Hirose (1969) have isolated both

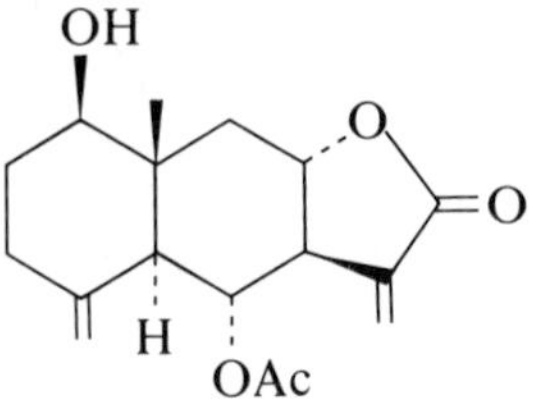
(190)

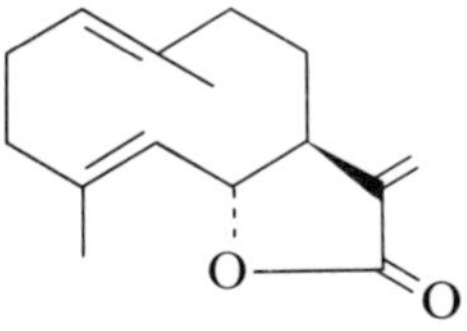
(191)

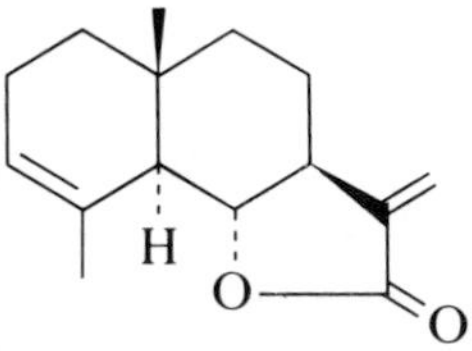
(192)

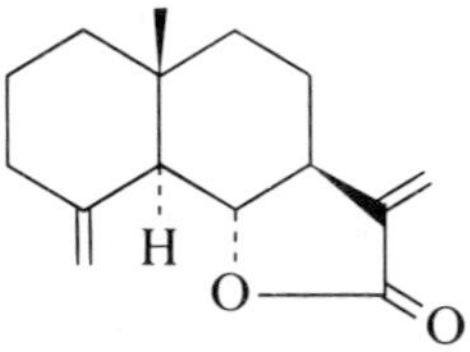
(193)

germacrene-C (194) and germacrene-D (195) (Yoshihara *et al.*, 1969) from different parts of the plant, *Kadsura japonica*. In contact with silica gel, germacrene-C gives rise to selina-4,6-diene (196) and selina-4(14),6-diene (197) while germacrene-D affords (+)-γ-muurolene, (−)-α-amorphene, (+)-δ-cadinene and (+)-γ-cadinene. The biogenetic significance of germacrene-D is perhaps even more fundamental in the light of the fact that it can be photoisomerized to α- (198) and β-bourbonene (199) together with β-copaene. The structural elucidation of the bourbonenes, isolated from

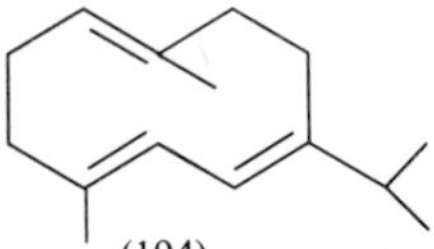
(194)

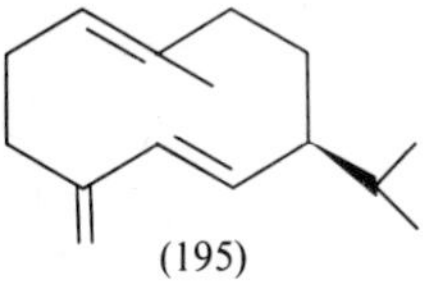
(195)

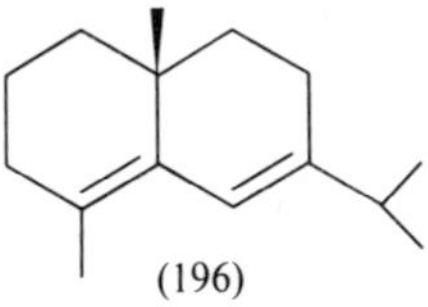
(196)

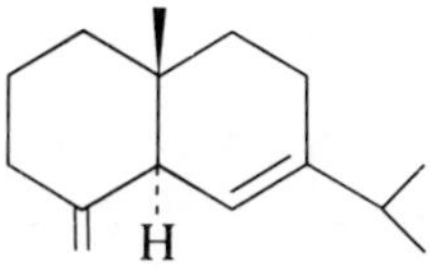
(197)

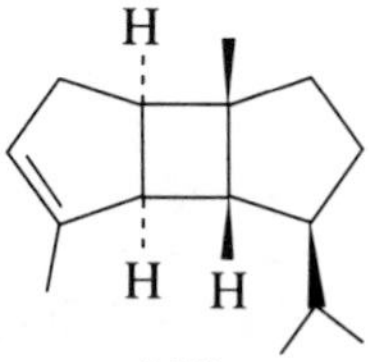
(198)

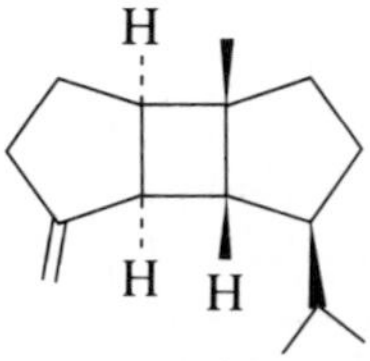
(199)

Geranium bourbon oil, was carried out by Křepinský *et al.* (1966a,b) and elegantly confirmed by two photochemically-based syntheses (Brown, 1968; White and Gupta, 1968). Heathcock and Badger (1968) have also noted the interesting photoisomerization of the cyclodecadienone (200) to the tricyclic ketones (201) and (202), related to the copaane/ylangane and bourbonane skeletons.

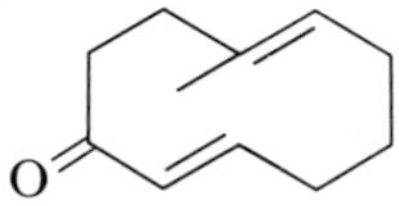

(200)

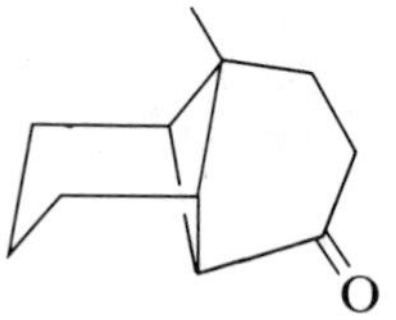

(201)

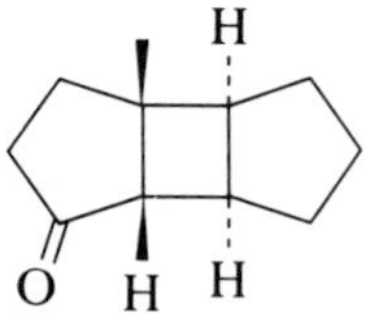

(202)

The *in vitro* studies by Brown *et al.* (1967) also illustrate the probable intermediacy of germacrane-type sesquiterpenoids. These authors have found that the germacratriene (203, R = H_2), chemically derived from the naturally-occurring germacrone (203, R = O), can be converted into eudesmane-type compounds by treatment with electrophilic reagents. For example, the bromohydrin (204, R = Br) was obtained on reaction with N-bromosuccinimide in aqueous acetone and the two dienes (205, R = H) and (206) by percolation through Grade I alumina. Subsequently, Brown

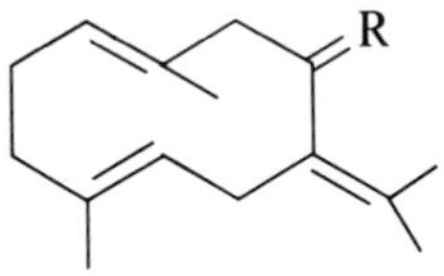

(203)

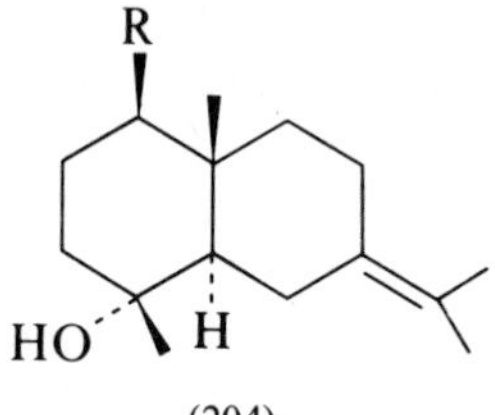

(204)

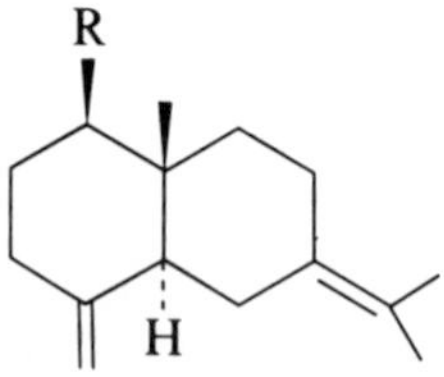

(205)

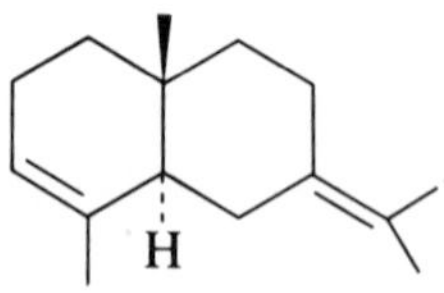

(206)

and Sutherland (1968) made the important observation that whereas the epoxide (207) cyclized to (204, R = OH) and (205, R = OH) on aqueous acid treatment, the epoxide (208) yielded the guaiane-type compounds (209) and (210) on similar treatment.

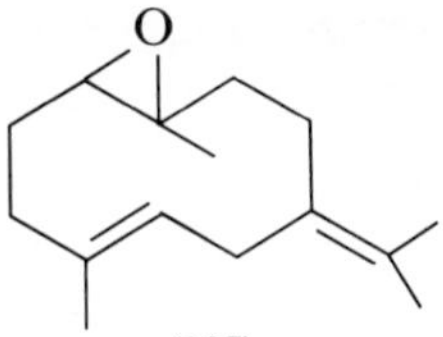

(207)

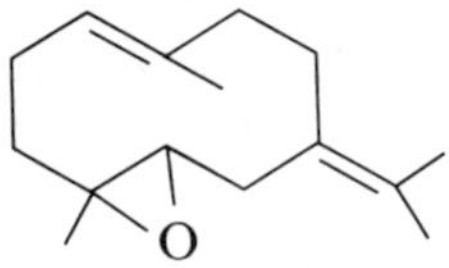

(208)

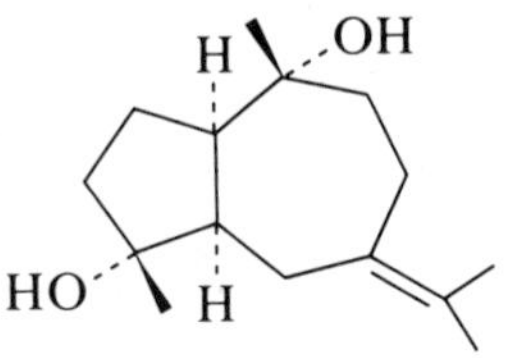

(209)

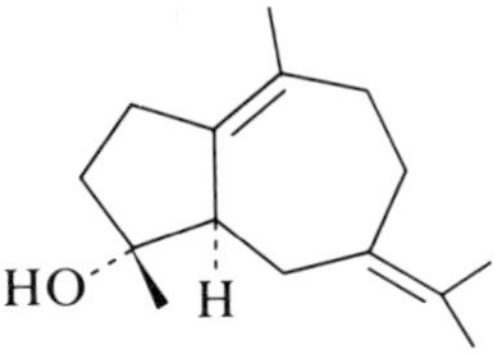

(210)

The second major piece of evidence to implicate germacrane sesquiterpenoids as important biogenetic precursors is derived from detailed studies of their co-occurrence with closely related bicyclic eudesmane and guaiane types. Again this evidence is gleaned from many observations and it will suffice to note a few examples. Germacrene-B, i.e. (203, R = H_2), selina-4(14),7(11)-diene (205, R = H) and selina-3,7(11)-diene (206) co-occur in oil of hops (Hartley and Fawcett, 1969). Various *Ambrosia* species [e.g. *A. psilostachya* DC. (Yoshioka and Mabry, 1969) and *A. confertiflora* DC. (Yoshioka *et al.*, 1970a)] elaborate germacranolides together with either eudesmanolides or pseudoguaianolides.

Not only do a number of the germacrane-type sesquiterpenoids exhibit interesting transannular reactions but the effect of the *trans,trans*-cyclodeca-1,5-diene moiety is also manifest by an anomalous ultra-violet absorption at *c.* 210 nm (i.e. the result of transannular conjugation). These effects have motivated two independent lines of study regarding the conformation of the ten-membered ring. The first of these is the examination of suitable derivatives by X-ray methods and there are now several examples of this application. For instance, Allen and Rogers (1967) have demonstrated that germacrene-B, as a 1:1 complex with silver nitrate, exists in the conformation (211) and this assignment is reinforced by the X-ray analyses of the silver nitrate adducts of costunolide (212) (Šorm *et al.*, 1970) and pregeijerene (213) (Coggon *et al.*, 1970). In a different approach, viz. the use of the Nuclear Overhauser Effect (NOE), Bhacca and Fischer (1969) have concluded that the observed signal enhancements in an n.m.r. study of dihydrotamaulipin-A acetate are best accommodated by the conformation (214). Takeda (1970) has also used this technique successfully to ascertain the preferred conformations of a number of furanogermacranes and Yoshioka and Mabry (1969) have demonstrated by their n.m.r. studies that isabelin exists in the two conformations (215) and (216) in the ratio 10:7 at room temperature.

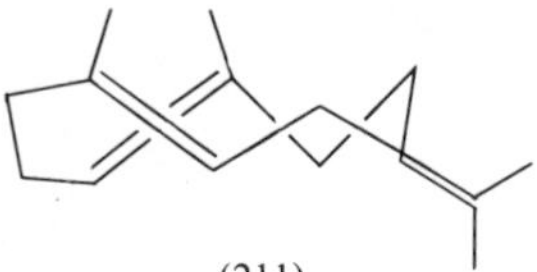
(211)

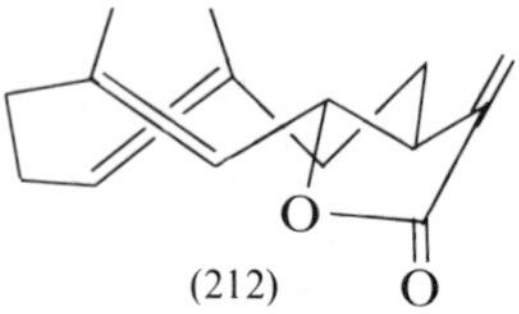
(212)

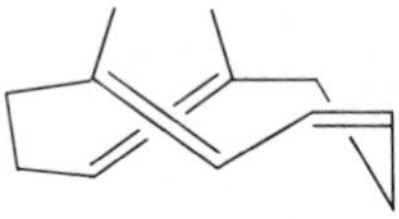
(213)

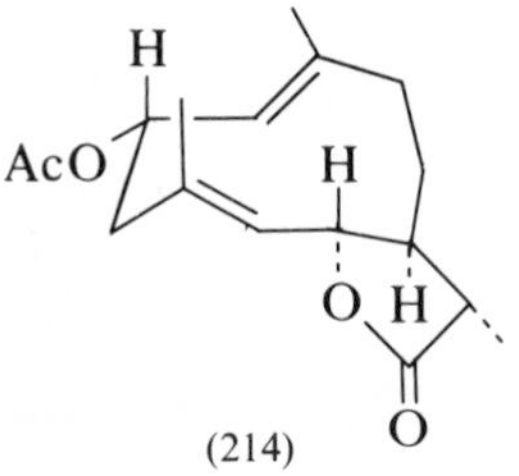
(214)

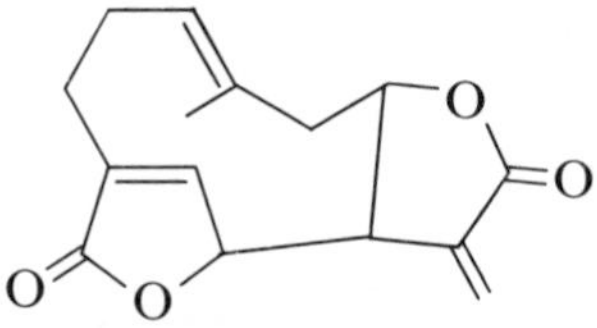
(215)

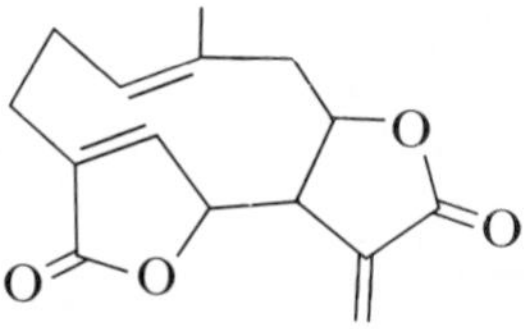
(216)

To date the only recorded synthesis of a germacranolide is that of dihydrocostunolide by Corey and Hortmann (1963). This was achieved by conversion of the keto-lactone (217), derived from santonin, into the diene (218). This diene, on photolysis, underwent a conrotatory cycloreversion to the triene (219) which was hydrogenated to dihydrocostunolide (183).

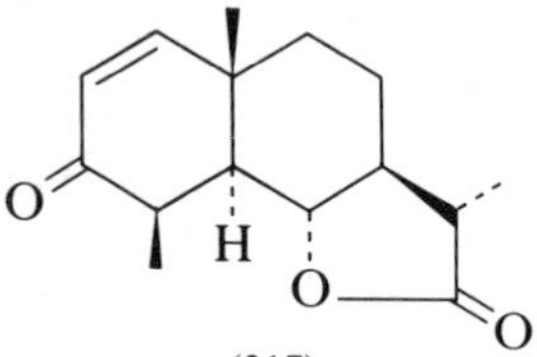
(217)

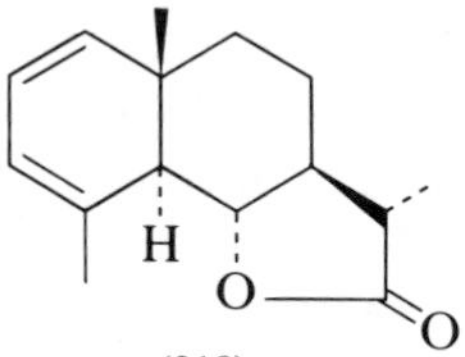
(218)

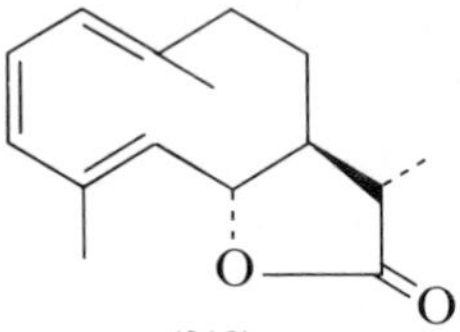
(219)

XII. Elemane

The formation of the elemane-type sesquiterpenoids has been explained in terms of a Cope rearrangement of a germacra-1(10),4-diene precursor. Thus, compounds such as elemol (220), saussurea lactone (221) and isolinderalactone (222) have been isolated from natural sources. However, the claim that these compounds are naturally-occurring has to be tempered with the fact that careful isolation procedures often reveal the presence of the corresponding germacrane-type precursor. This has been strikingly demonstrated (Jones and Sutherland, 1968) in the case of the plant, *Hedycarya angustifolia.* Although its essential oil is described as a rich source of elemol, careful extraction of the leaves yielded primarily hedycaryol (223). Kupchan *et al.* (1969b), however, in their search for anti-tumour agents, have isolated a number of oxidatively modified elemane dilactones, e.g. vernolepin (224), the formation of which, via a germacradiene-14,15-olide precursor in the course of isolation, seems unlikely, although not impossible.

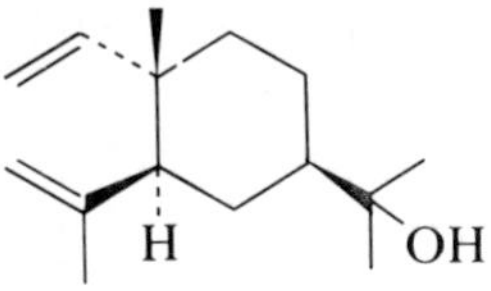

(220)

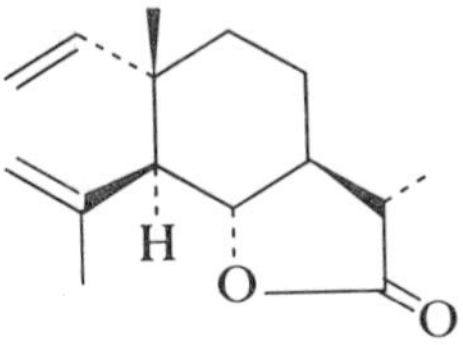

(221)

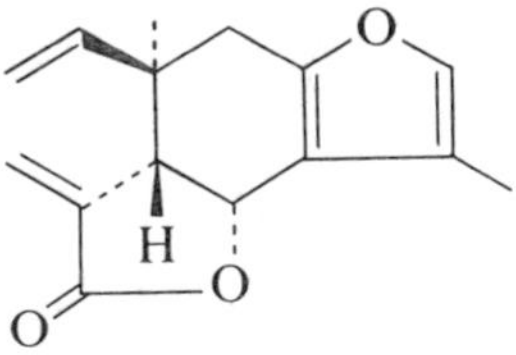

(222)

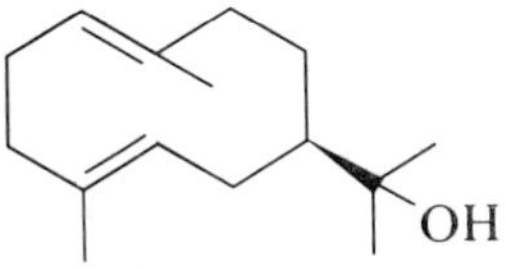

(223)

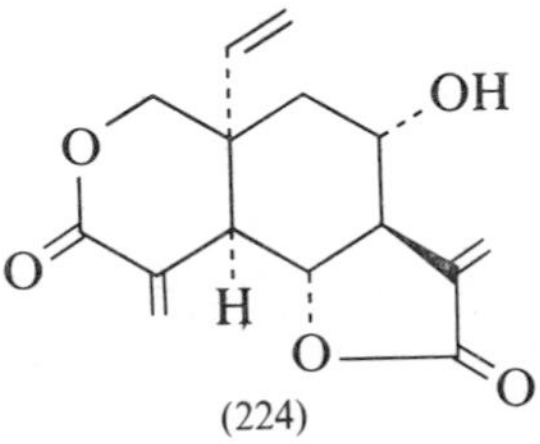

(224)

Recently, a number of interesting studies have been carried out on the reversibility of the Cope rearrangement (Takeda, 1970; Takeda *et al.*, 1970; Tori and Horibe, 1970). It seems that the results obtained so far are consistent with the requisite geometries (i.e. chair-like transition state) of both the elemadiene and germacradiene counterparts. NOE and X-ray studies also accord with these results.

Corey and Broger (1969) have accomplished a very elegant synthesis of elemol (220) involving, as the key step, the nickel carbonyl-induced cyclization of the ester-dibromide (225). The major product of this reaction was not the germacrane-type ester (227) but the elemane-type (226) which was converted into elemol.

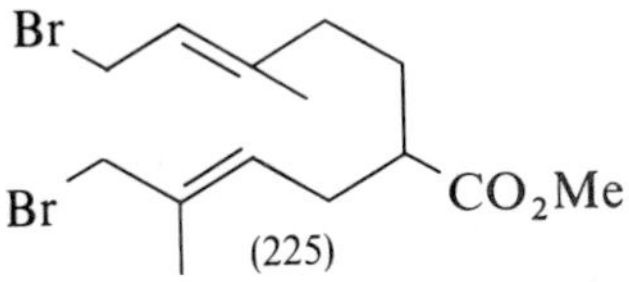

(225)

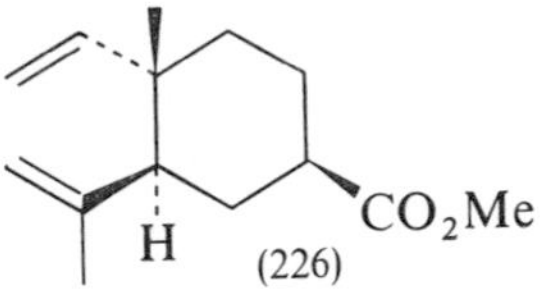

(226)

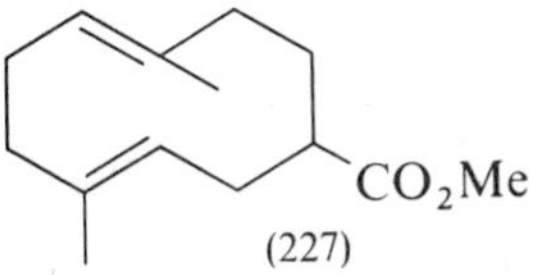

(227)

XIII. Eudesmane (Selinane)

The logical precursor of this group (Cocker and McMurry, 1960) is the cation (11) which can undergo a favourable cyclization to the *trans* decalin system, exemplified by cryptomeridiol (228) (Sumimoto *et al.*, 1963). Originally this group was characterized by the formation of eudalene (229) on dehydrogenation. There are now many examples of this group including those with furan and lactone rings (usually 6α,12-olides or 8β,12-olides). From the synthetic standpoint, the work of Marshall and his collaborators

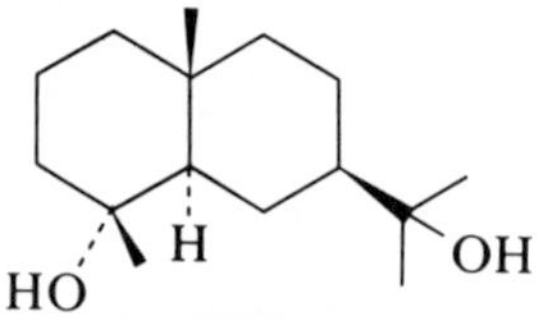

(228)

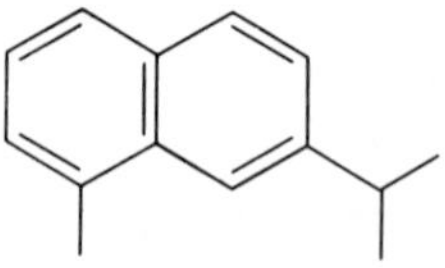

(229)

is noteworthy. They (Marshall *et al.*, 1966b) have synthesized β-eudesmol (230), β-selinene (231) and costol (232) by elaboration of the key *trans*-decal-2-one (234) which was derived from the well-known octalone (233) in six steps. Thus, conversion of the ketone (234) into the ester (235, R = β-CO_2Me) via the nitrile (235, R = α-CN) and treatment with methyl-lithium afforded β-eudesmol. Treatment of the nitrile (235, R = α-CN) with methyl-lithium (and subsequent epimerization) followed by a Wittig reaction afforded β-selinene. Finally, costol was obtained by lithium aluminium hydride reduction of the sodio derivative of (236) which was also derived from the ketone (234).

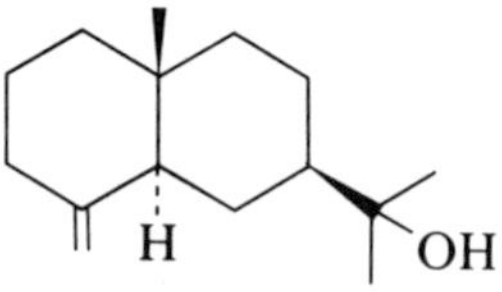

(230)

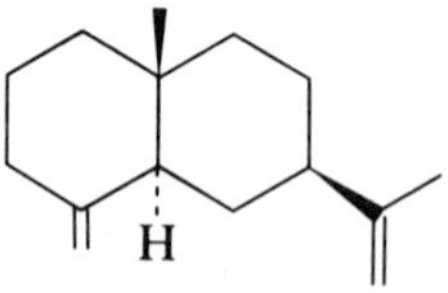

(231)

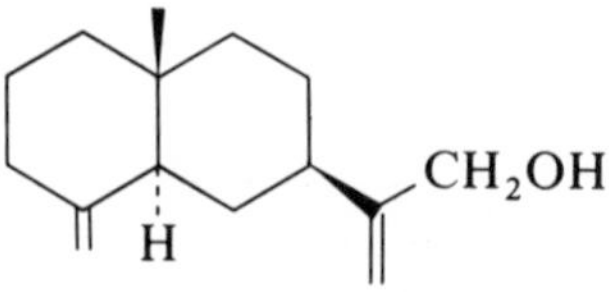

(232)

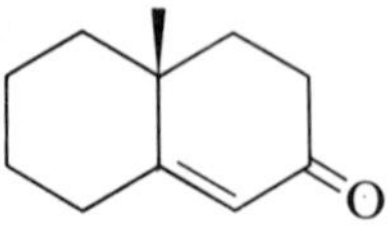

(233)

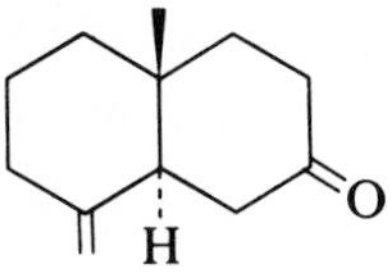

(234)

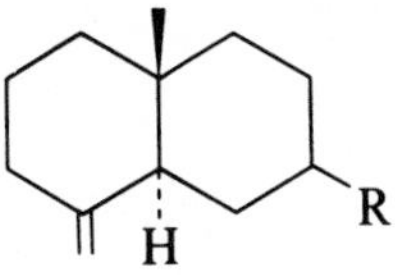

(235)

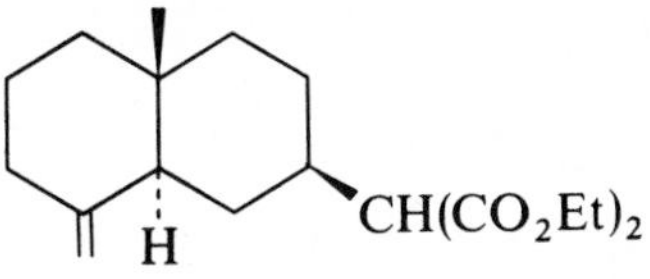

(236)

Undoubtedly the most important member of this group is α-santonin (184) which occurs extensively in various species of the genus *Artemisia*. This sesquiterpenoid has commanded the attention of organic chemists for well over a century principally because of the vast array of interesting chemical and photochemical transformations associated with it. All of these aspects have been thoroughly reviewed and will not be discussed in this chapter. Marshall *et al.* (1966a) have also contributed to this area by their syntheses of telekin (237) and alantolactone (238). Thus, the ketone (239), derived from Hagemann's ester in an economical number of steps, was converted into the lactone (240). This lactone, on oxidation with singlet oxygen and carbomethoxylation, gave the hydroxy-lactone (241) which was elaborated to telekin by lithium aluminium hydride reduction of the sodio enolate of (241) and allylic oxidation of the intermediate triol (242). The hydroxy-lactone (243), on hydrogenation and dehydration, gave the lactone (244) which was converted into alantolactone by a sequence analogous to that described for telekin.

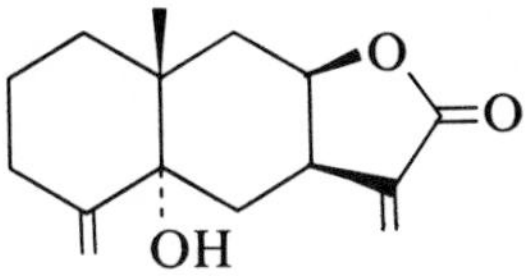

(237)

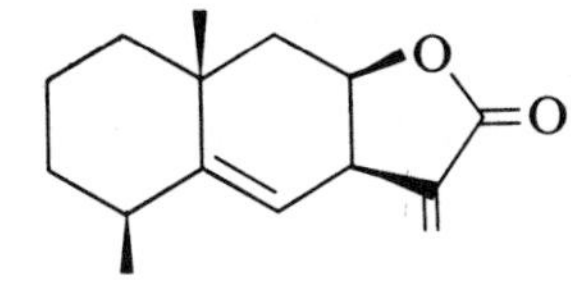

(238)

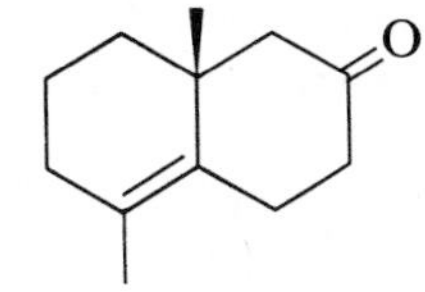

(239)

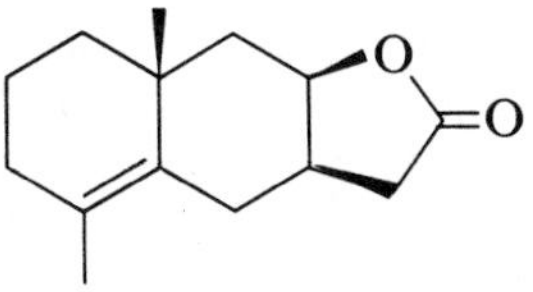

(240)

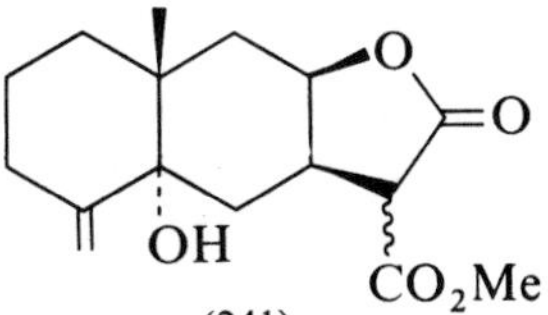

(241)

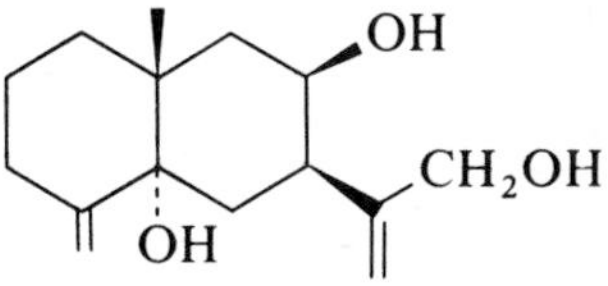

(242)

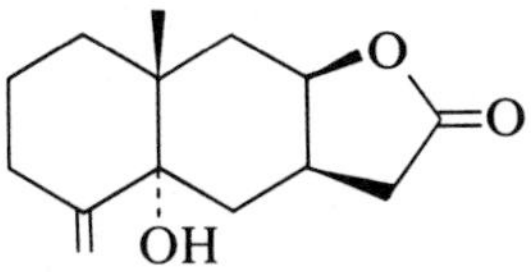

(243)

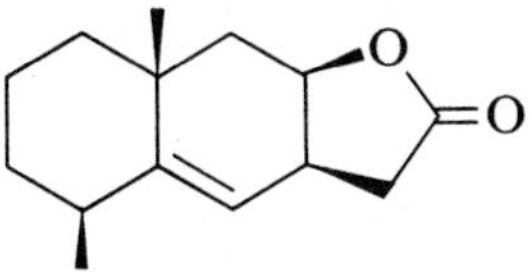

(244)

There are relatively few examples of *cis*-fused eudesmane sesquiterpenoids. Hortmann (1968) has suggested that these types may be derived by a thermally-allowed disrotatory closure of the hypothetical precursor (245) leading to (246) and/or (247). The structure of the sesquiterpenoid, occidentalol, has, in fact, been revised by Hortmann and De Roos (1969) in favour of (246). Recently, several nor-sesquiterpenoids e.g. chamaecynone (248) (Nozoe *et al.*, 1966) and dehydrochamaecynenol (249) (Asao *et al.*, 1968) have been isolated from *Chamaecyparis formosensis* and a precursor such as (247) (or nor-derivative) seems reasonable. Co-occurring with

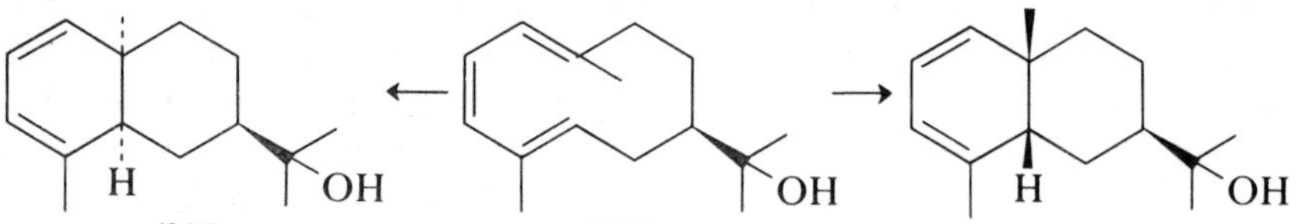

(246) (245) (247)

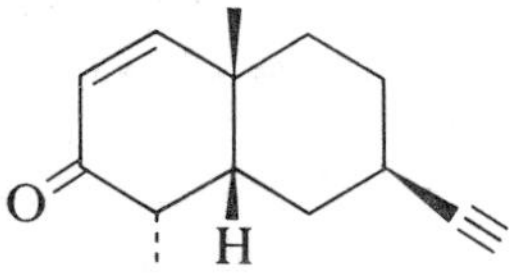

(248)

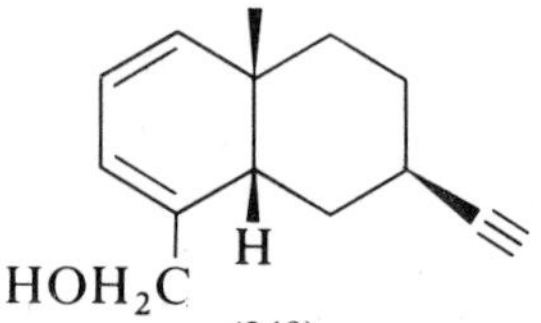

(249)

occidentalol in *Thuja occidentalis* is the aromatic compound, occidol (250) (Nakazaki, 1962). Recently Tomita and Hirose (1970b) have identified the unusual sesquiterpenoid, occidenol (251), isolated from *Thuja koraiensis*. Its genesis from the 2,3-epoxide of (245) by a divinyloxiran–dihydrooxepin rearrangement is likely. Nozoe *et al.* (1967) have reported a rational synthesis of chamaecynone (248) starting from α-santonin.

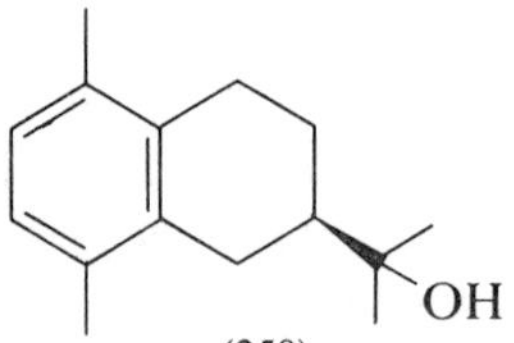

(250)

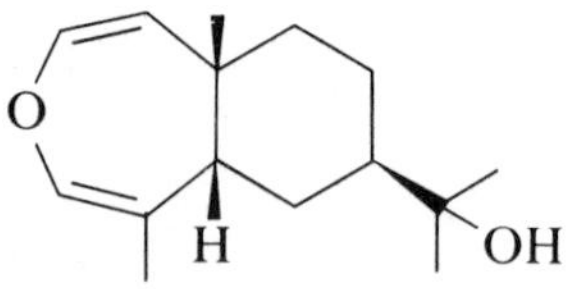

(251)

XIV. Eremophilane, Valerane, Valencane, Vetispirane and Tricyclovetivane

Various biogenetic schemes have been postulated to account for those sesquiterpenoids which can be formally related to a eudesmane-type precursor via a 1,2-alkyl migration. One such postulate (McSweeney *et al.*,

1970) is summarized in Scheme 8 in which cyclization of the two cyclodeca-1,6-diene conformers (252) and (253), derived from double bond isomerization of (11), followed by the appropriate 1,2-alkyl shifts is envisaged. Each individual sub-group will now be considered separately.

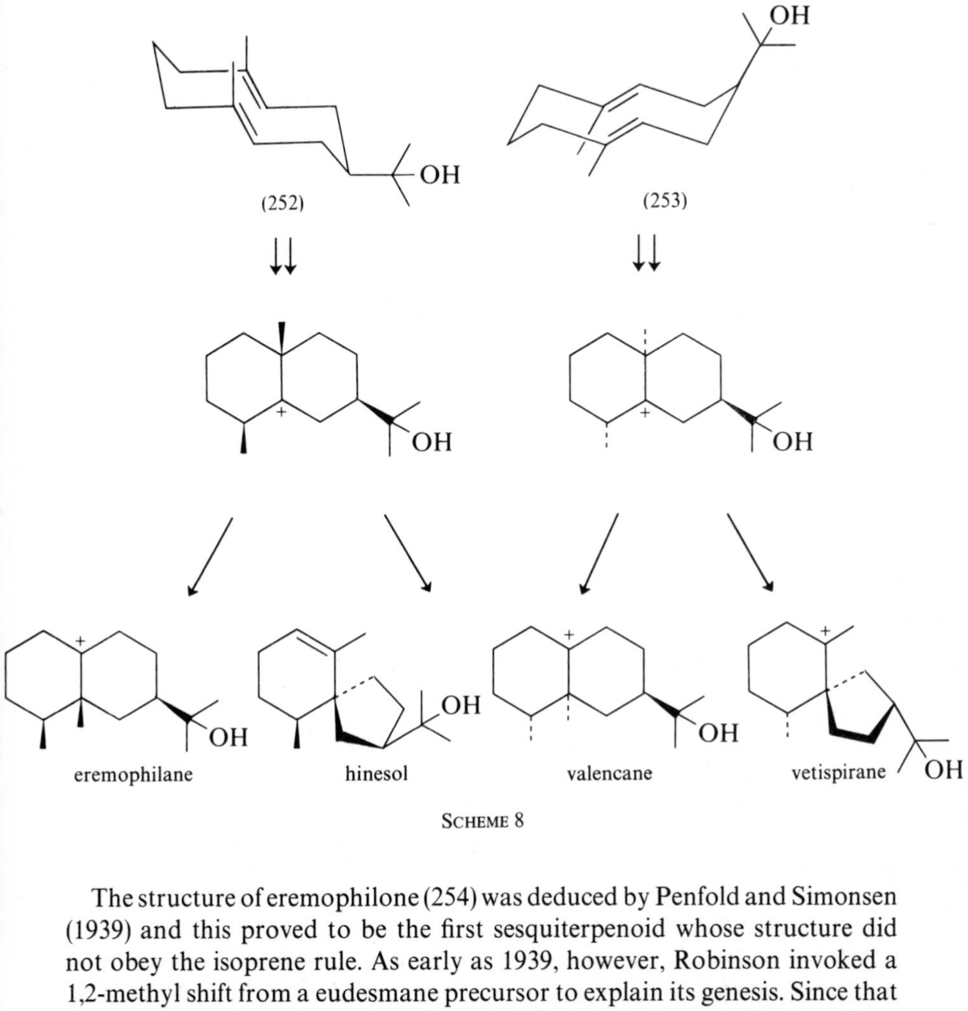

SCHEME 8

The structure of eremophilone (254) was deduced by Penfold and Simonsen (1939) and this proved to be the first sesquiterpenoid whose structure did not obey the isoprene rule. As early as 1939, however, Robinson invoked a 1,2-methyl shift from a eudesmane precursor to explain its genesis. Since that time a relatively large number of eremophilane sesquiterpenoids have been isolated and identified (Pinder, 1968), principally by Novotný and his co-workers. The *Petasites* species is a particularly rich source of this group, producing such compounds as eremophilenolide (255) and petasalbin (256) (Novotný *et al.*, 1964). Biogenetically related to the eremophilane type are

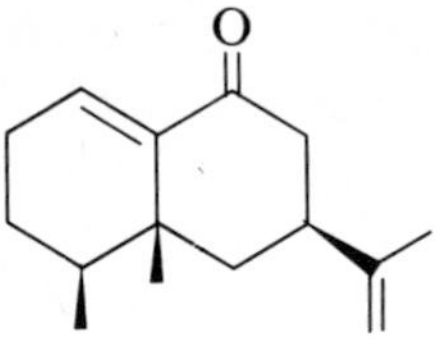
(254)

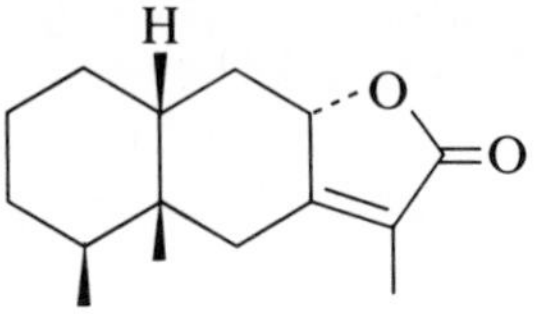
(255)

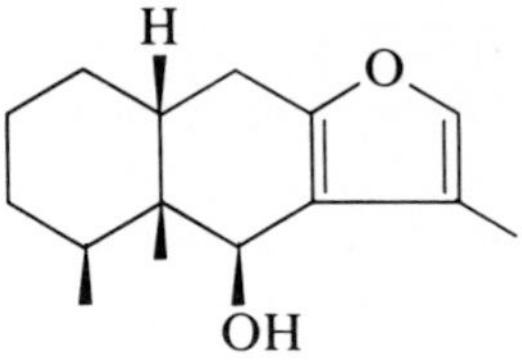
(256)

the modified sesquiterpenoids, bakkenolide A (257) (Shirahata *et al.*, 1969), maturinone (258) and cacalol (259), the latter co-occurring with decompostin (260) (Brown and Thomson, 1969; Harmatha *et al.*, 1969; Joseph-Nathan *et al.*, 1970).

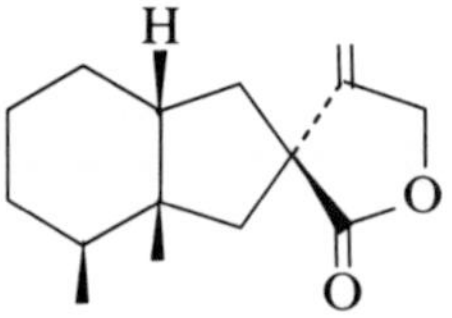
(257)

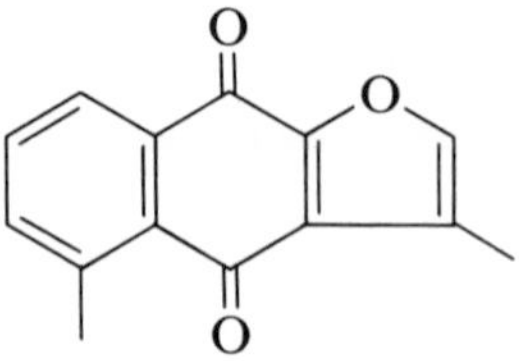
(258)

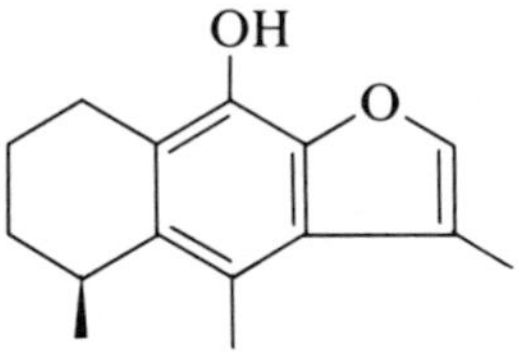
(259)

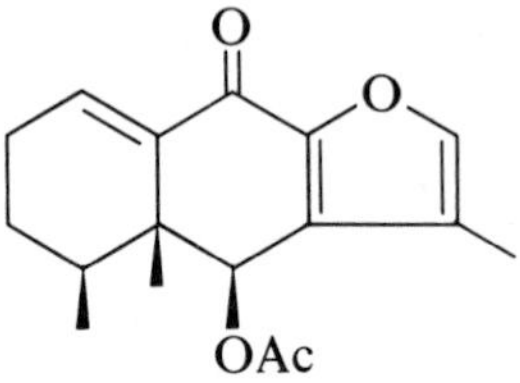
(260)

The structure (261) initially assigned to eremophilene has been revised to (262) by two groups. In the first place, Piers and Keziere (1969) have described a stereoselective synthesis of the diene (261) and demonstrated its non-identity with natural eremophilene and furthermore they showed that eremoligenol (263) could be dehydrated to eremophilene (262). This structural revision was reinforced by the work of Křepinský *et al.* (1968). Coates and Shaw (1970a) have also confirmed this assignment by the stereoselective synthesis of both (±)-eremophilene and eremoligenol. These syntheses centred around a novel procedure for the removal of the keto group in the keto-ester (264). This process involved conversion to the corresponding methoxymethyl enol ether and subsequent reduction with lithium in liquid ammonia to afford the less stable axial ester (265). Treatment of this compound with methyl-lithium yielded eremoligenol (263) which, in turn, yielded eremophilene (262) on dehydration.

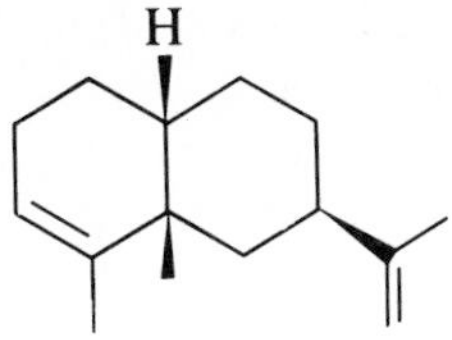
(261)

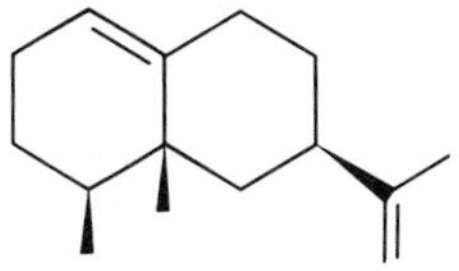
(262)

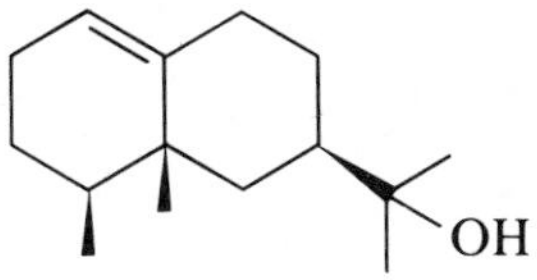
(263)

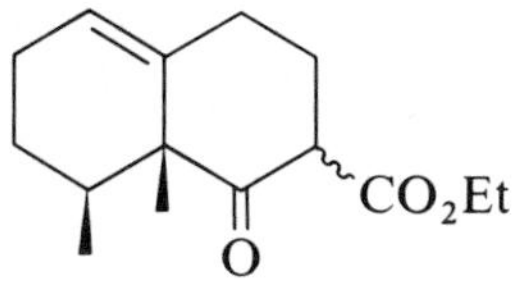
(264)

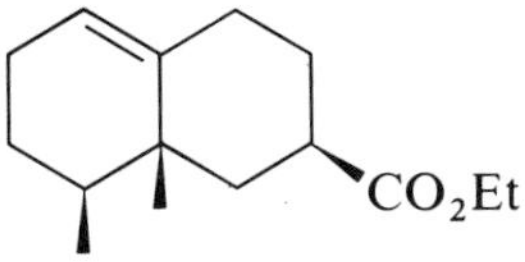
(265)

The valerane-type sesquiterpenoids are rather a rare group of which valeranone (266, R = H), fauronyl acetate (266, R = OAc) and cryto-fauronol (267) (Hikino *et al.*, 1965) are examples. The structure and stereo-chemistry of valeranone was the subject of extensive international debate but finally it was settled in 1963 by Hikino *et al.* (1965a). As final proof of this assignment Marshall *et al.* (1968) have synthesized the antipode of valeranone from (−)-dihydrocarvone by two different routes.

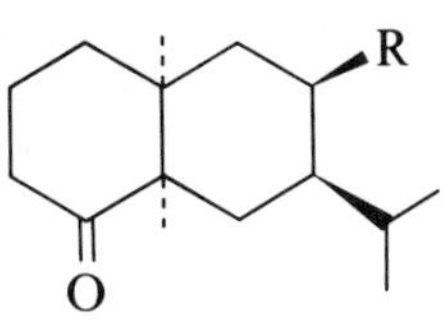
(266)

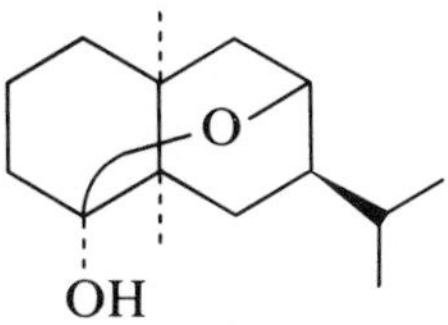
(267)

MacLeod (1965) reported the isolation and characterization of nootkatone (268, R = O), nootkatene (269), and valencene (268, R = H_2). The absolute stereochemistry of these three important sesquiterpenoids, all isolated from the essential oil of grapefruit, was determined by comparison with an anti-podal eremophilane derivative. Two syntheses of (±)-nootkatone have been reported (Marshall and Ruden, 1970; Pesaro *et al.*, 1968). Coates and Shaw (1970a) have also confirmed the structures of valencene (268, R = H_2) and valerianol (270) (Jommi *et al.*, 1967) by base equilibration of (265) to the more stable equatorial isomer (271) which, on treatment with methyl-lithium, gave valerianol and subsequent dehydration, valencene. The isolation and structural elucidation of the novel tetracyclic sesquiterpenes, ishwarane (272, R = H_2) (Govindachari *et al.*, 1970) and ishwarone (272, R = O) (Fuhrer *et al.*, 1970; Ganguly *et al.*, 1969; Govindachari *et al.*, 1969), obtained from the roots of *Aristolochia indica*, have been reported by Govindachari *et al.* From a biogenetic point of view, these compounds belong to the valencane group.

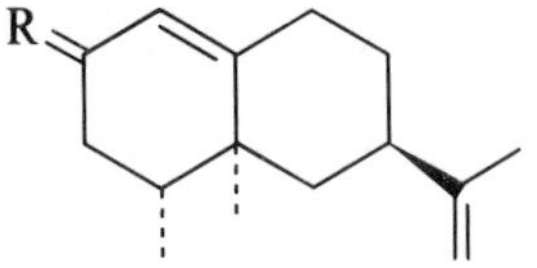
(268)

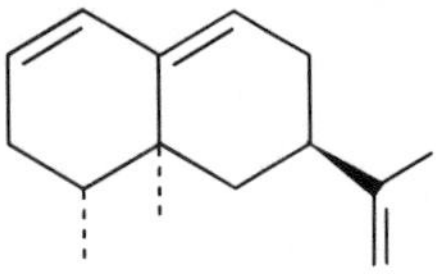
(269)

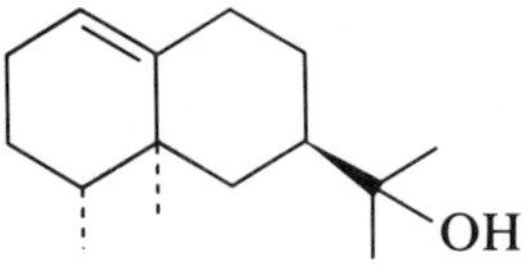
(270)

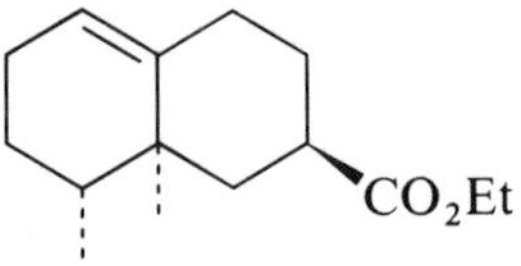
(271)

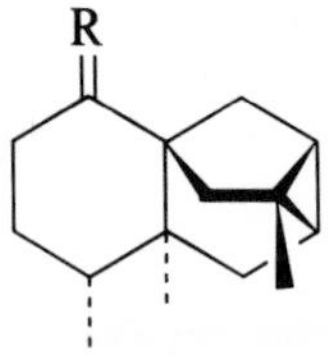
(272)

For many years the structures of α- and β-vetivone, two constituents of the essential oil of vetiver, had been considered in terms of the hydroazulenone (273). The reasoning for this assignment was based on extensive chemical studies (especially for β-vetivone) and in particular the obtention of vetivazulene (274) by dehydrogenation was deemed persuasive. In 1967, however, both de Mayo (Endo and de Mayo, 1967, 1969) and Marshall (Marshall and Andersen, 1967) and their co-workers, by two different lines of investigation, demonstrated that α-vetivone is defined by structure (275) and, as such, is sometimes referred to as isonootkatone. Marshall *et al.* (1967) followed up this assignment by an unambiguous synthesis which involved the annelation of 4-isopropylidene-2-carbomethoxycyclohexanone with *trans*-3-penten-2-one to give (276). Elaboration of the carbomethoxy group to a methyl group completed the synthesis of α-vetivone.

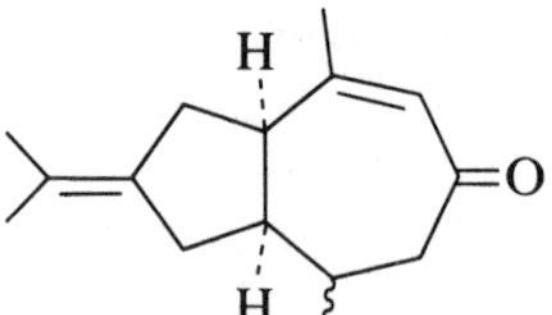
(273)

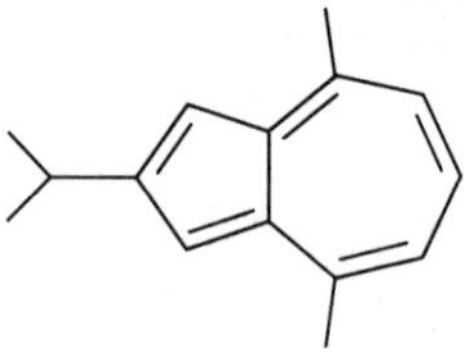
(274)

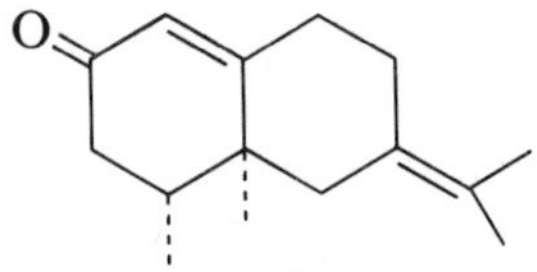
(275)

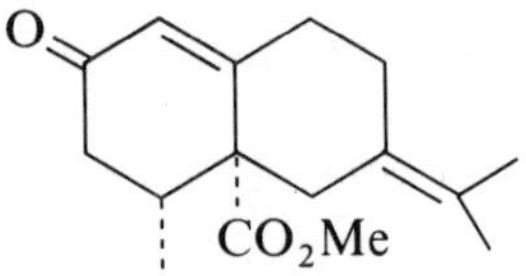
(276)

In the same year Marshall and Johnson (1970) showed that the structure of β-vetivone also required drastic revision to the spiro[4,5] system (277). This re-assignment was necessitated by the non-identity of the three synthetic ketones (278, R = Me) (two *cis* and one *trans*) with the supposed *meso*-ketone (278, R = β-Me) derived from β-vetivone by chemical means. A consideration of the reported chemical reactions of β-vetivone, together with the stepwise degradation of β-vetivone to the spiro[4,5]decane (279) which was independently synthesized from the dienone (280), led Marshall *et al.* to the structural elucidation of β-vetivone. This was beautifully proved by synthetic means which involved photoisomerization of (280) to (281) followed by acid-catalysed rearrangement to (282). This enone was then elaborated to a mixture of methyl ketones (283, R^1 = H, R^2 = $COCH_3$) and (283, R^1 = $COCH_3$, R^2 = H), which were converted to β-vetivone. Hinesol (284) (see Scheme 8), which can be converted (Yosioka and Kimura, 1965, 1969) to the antipode of naturally-occurring β-vetivone, was also obtained in racemic form from the methyl ketone (283, R^1 = H, R^2 = $COCH_3$). An alternative synthesis of hinesol has also been reported by Marshall and Brady (1969).

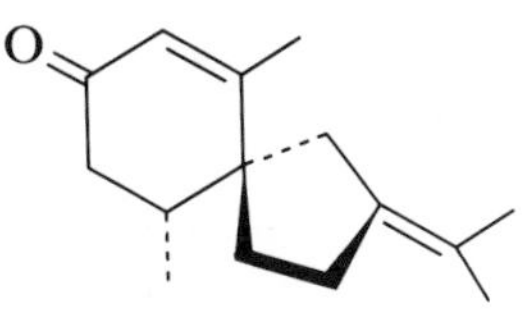

(277)

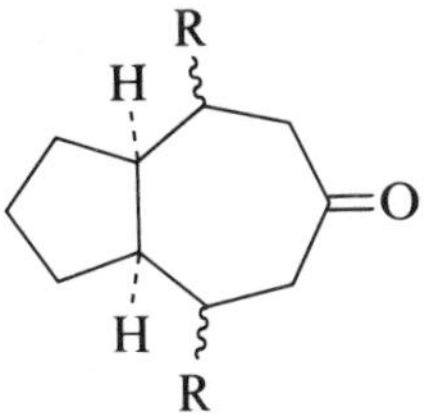

(278)

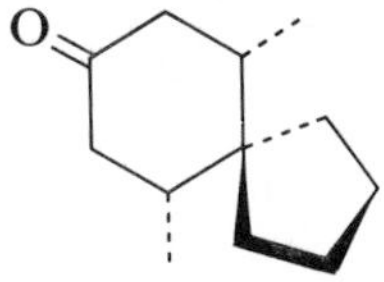

(279)

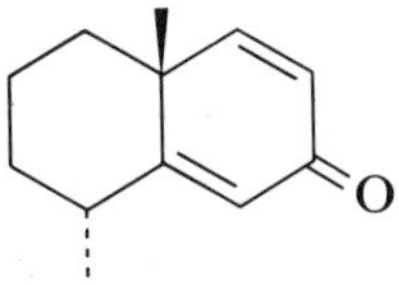

(280)

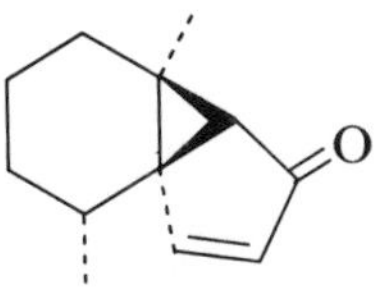

(281)

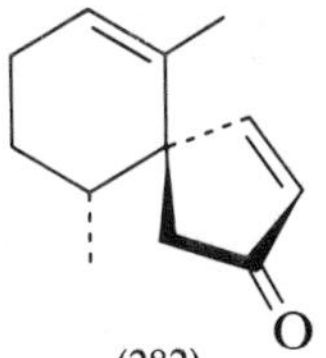

(282)

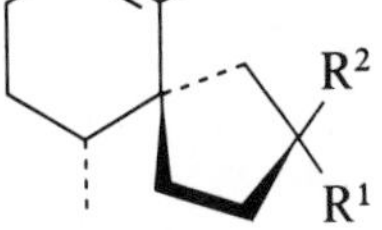

(283)

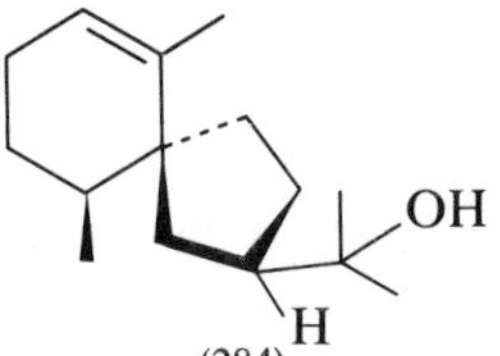

(284)

Recently a number of tricyclic sesquiterpenoids have been isolated from various vetiver oils; these include such compounds as zizanoic acid (285, R = CO_2H) (Kido *et al.*, 1967, 1968), tricyclovetivene (285, R = Me) (Sakuma and Yoshikoshi, 1968), and khusimol (285, R = CH_2OH) (Coates *et al.*, 1969). Biogenetic schemes (Kido *et al.*, 1967; MacSweeney *et al.*, 1970; Nigam *et al.*, 1968) have been offered to account for the gross structure and stereochemistry of this group. Kido *et al.* (1969b) have reported a synthesis of epizizanoic acid (286) starting from (+)-methyl camphene carboxylate (287). Conversion to the corresponding aldehyde and aldolization with acetone gave the enone (288), which was converted to the cyano-diketone (289) and hence to the tricyclic ester (290). Hydroxylation to the glycol and rearrangement of the derived mono-mesylate gave two epimeric (at C-4) ketones (291) on prolonged base equilibration. The synthesis was completed by a Wittig reaction on the 4β-H epimer.

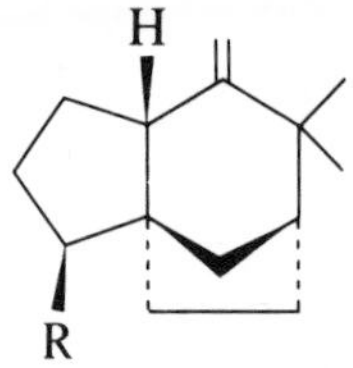

(285)

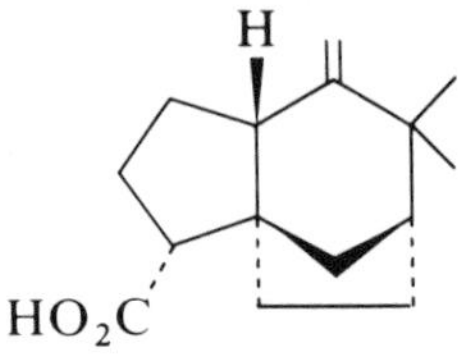

(286)

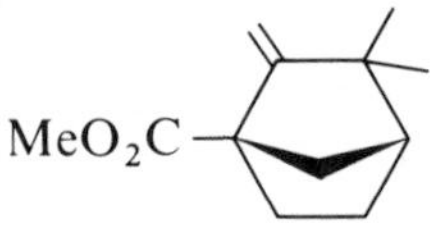

(287)

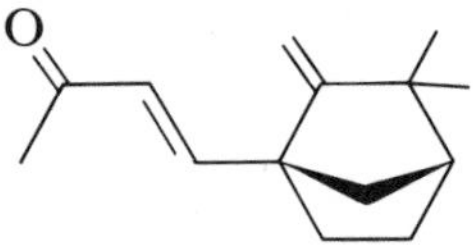

(288)

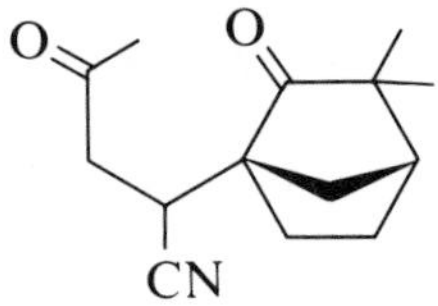

(289)

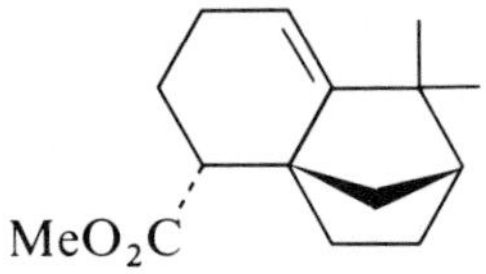

(290)

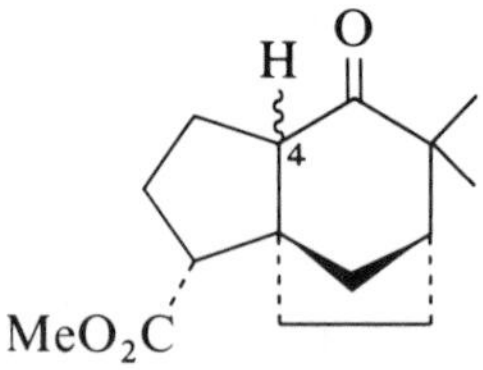

(291)

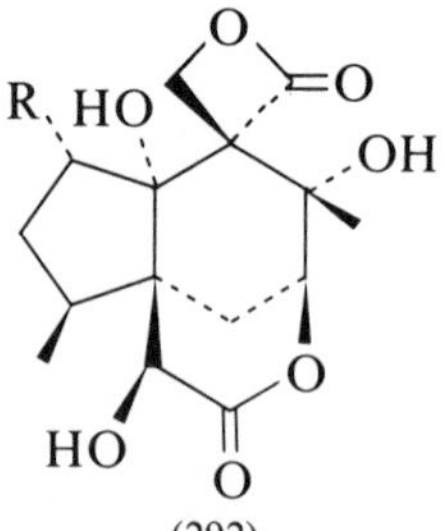

(292)

Structurally related to the tricyclovetivane group are the two toxic sesquiterpenoids, anisatin (292, R = OH) and neoanisatin (292, R = H), isolated from the Japanese star anise (Yamada *et al.*, 1968).

XV. Guaiane

The anti-Markownikoff cyclization of cation (11) to cations (293) and/or (294) has long been accepted as the mode of formation of the guaiane-type sesquiterpenoids, which, as a result of extensive work on the lactone derivatives by Šorm, Herz, Geissman, Romo and others, constitutes the largest single group of sesquiterpenoids.

The simpler members of this group are rather few in number and include such compounds as guaiol (295) and bulnesol (296). Two total syntheses of ($\pm$)-bulnesol have now been reported. In their synthesis, Marshall and Partridge (1969) used an appropriately functionalized bicyclo[4,3,1]decane precursor (297) which, on solvolysis, gave the olefin–acetate (298). This compound was converted into bulnesol *via* the corresponding (epimerized) 7β-carbomethoxy derivative. Kato *et al.* (1970b) on the other hand, used a suitably substituted *cis*-fused decalin system (299) which, on solvolysis, gave the same ester (300) as Marshall and Partridge had prepared. Piers and

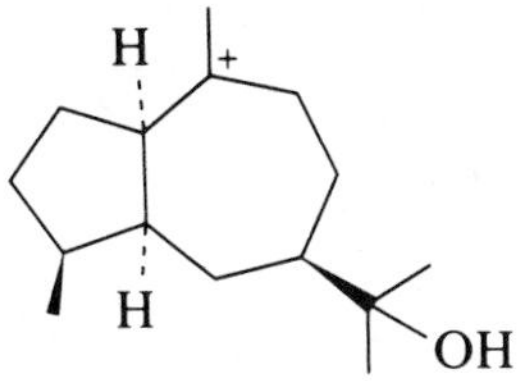

(293)

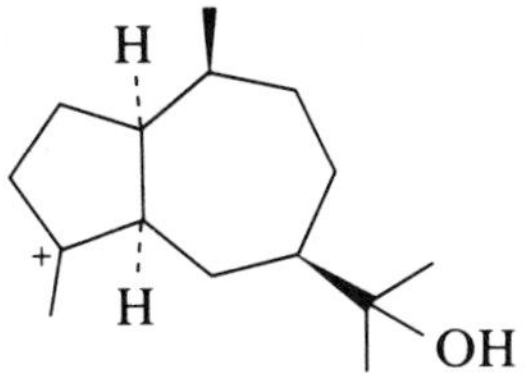

(294)

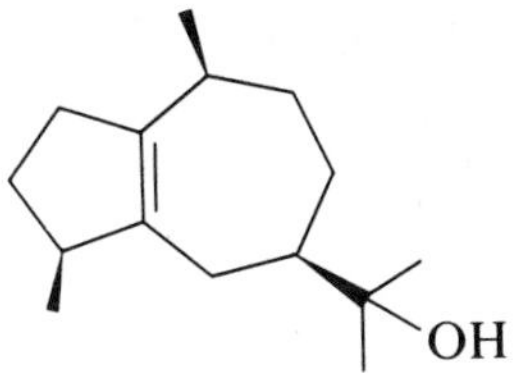

(295)

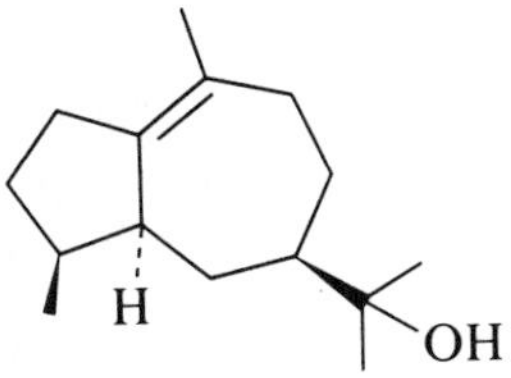

(296)

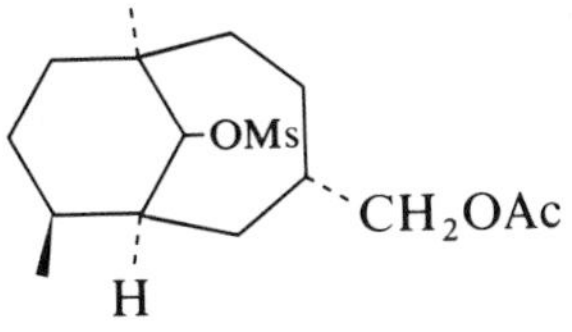

(297)

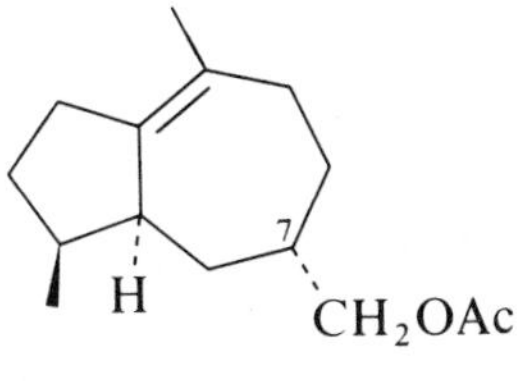

(298)

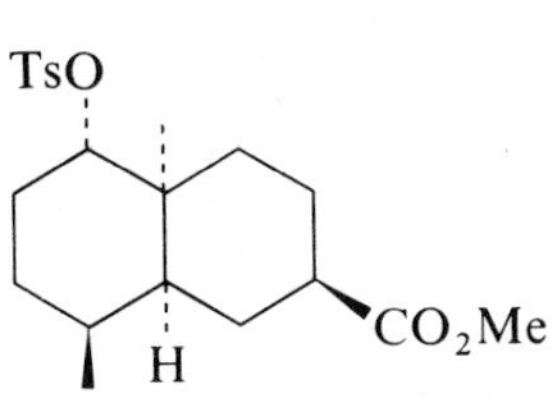

(299)

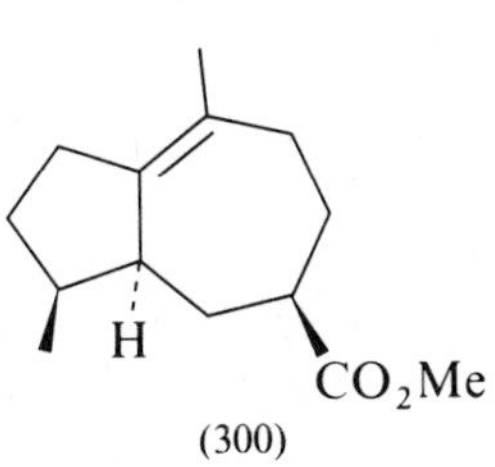

(300)

Cheng (1969), in an alternative approach, made use of the well-documented photochemical transposition of the dienone (301) to the substituted hydroazulenone (302) which was readily converted into α-bulnesene (303), the dehydration product of bulnesol.

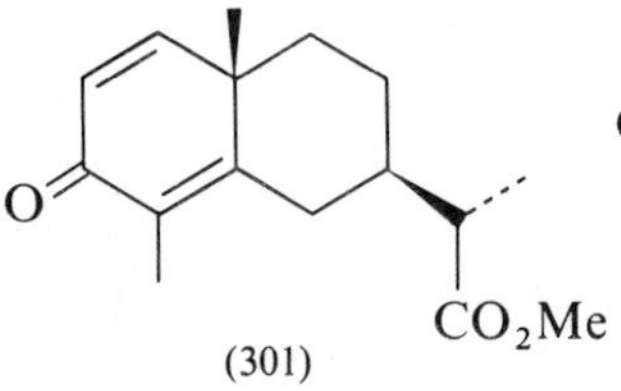

(301)

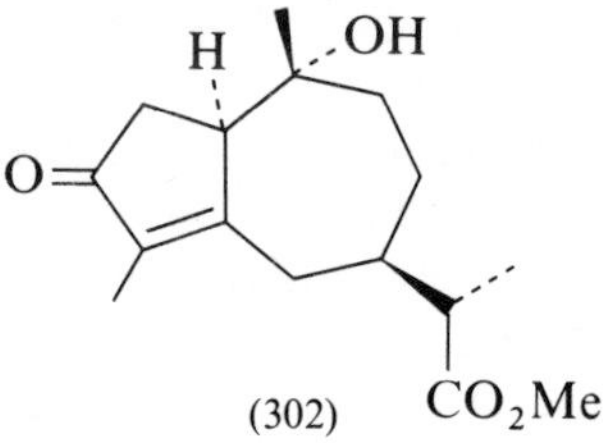

(302)

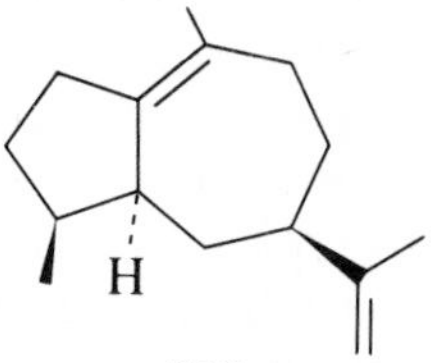

(303)

From the basic guaiane skeleton a number of interesting tricyclic sesquiterpenoids are derivable. These include kessane (304) (Hikino *et al.*, 1963, 1967), cyperene (305) (Trivedi *et al.*, 1964), patchouli alcohol (306), and seychellene (307). Until 1963 the structure of patchouli alcohol was considered (Büchi *et al.*, 1961) to be (308) and to verify this Büchi *et al.* (1964) set out to

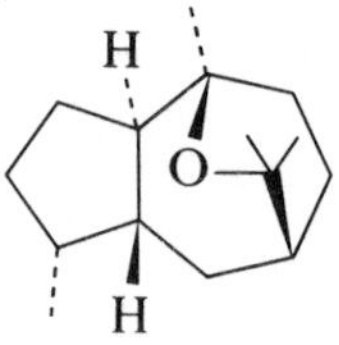

(304)

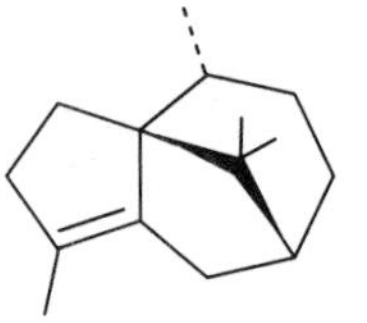
(305)

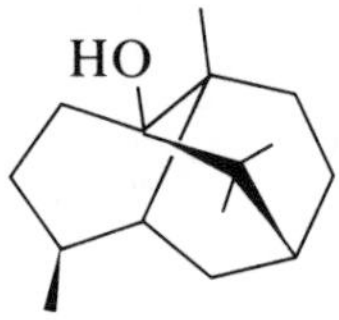

(306)

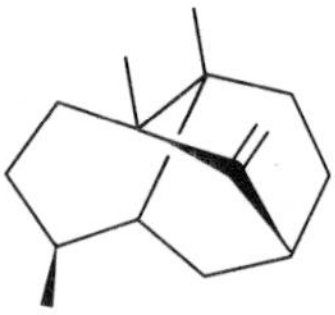
(307)

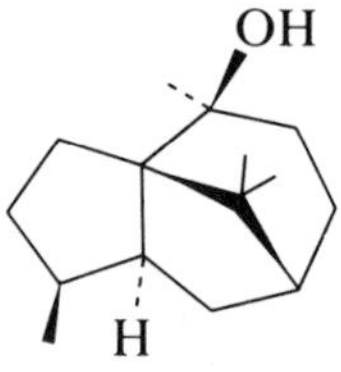

(308)

synthesize the "related" hydrocarbon, α-patchoulene (309). This they accomplished by converting (−)-homocamphor (310) into β-patchoulene (311). The α-epoxide of β-patchoulene was then isomerized to the hydroxy-olefin (312) which was transformed to α-patchoulene in four steps. Since the acetate of patchouli alcohol had already been converted into α-patchoulene by pyrolysis and α-patchoulene epoxide could be converted into patchouli alcohol, the above synthetic sequence appeared to constitute a structural proof. However, the X-ray analysis of the patchouli alcohol diester of chromic acid revealed structure (306) for patchouli alcohol (Dobler *et al.*, 1963). Thus pyrolysis of patchouli alcohol acetate to give α-patchoulene is attended by a skeletal rearrangement, and, remarkably, the conversion of α-patchoulene epoxide to patchouli alcohol proceeds with precisely the reverse skeletal change.

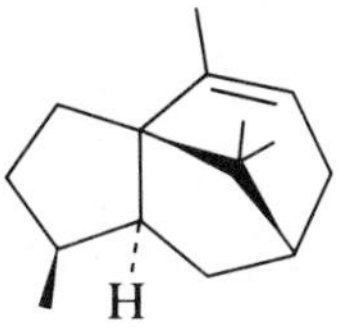

(309)

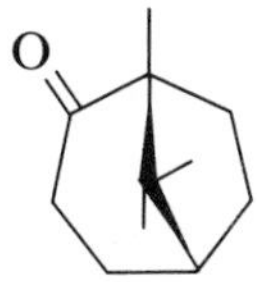

(310)

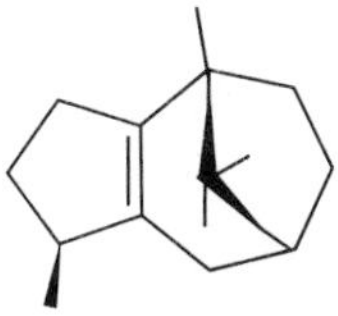

(311)

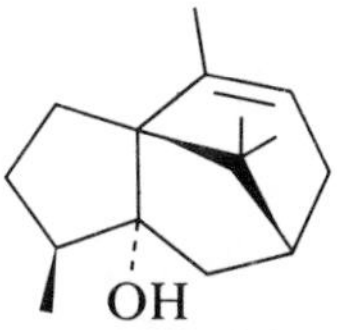

(312)

Danishefsky and Dumas (1968) have also reported a very elegant synthesis of (±)-patchouli alcohol. This involved the Diels–Alder reaction of methyl vinyl ketone with 2,6,6-trimethyl-cyclohexa-2,4-dienone to give (313). Reduction and equilibration afforded the *exo*-dihydro derivative which was converted into the alcohol (314) on treatment with vinyl-lithium. This, in turn, was converted into the allylic alcohol (315) which gave two epimeric alcohols on catalytic hydrogenation. One of these, on treatment with phosphorus tribromide and reductive cyclization with sodium in tetrahydrofuran, yielded patchouli alcohol (306). Kato *et al.* (1970b) have synthesized kessane (304) by solvolysis of the mesylate (316) *cf.* solvolysis of (299).

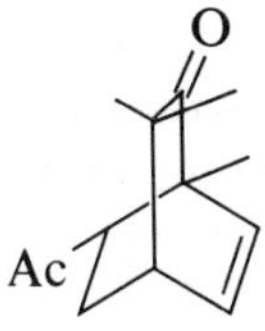

(313)

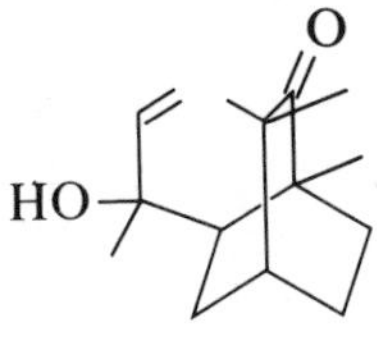

(314)

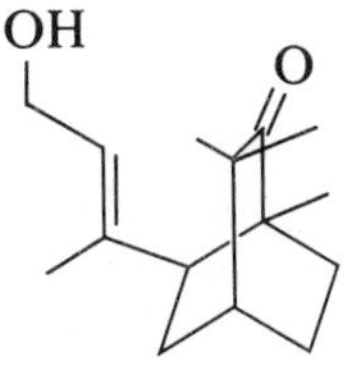

(315)

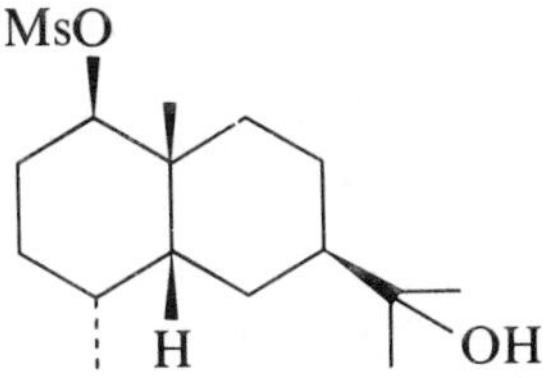

(316)

Seychellene (307) co-occurs with a number of simpler guaiane sesquiterpenes in the leaves of *Pogostemon patchouli* and its structural elucidation is the result of a systematic degradation study carried out by Wolff and Ourisson (1969). Piers *et al.* (1969c) have accomplished a beautifully-conceived synthesis of seychellene which involved the stereoselective conversion of the known enone (317) into the *cis*-fused decalone (318) in nine steps. Treatment of (318) with base yielded the tricyclic ketone (319) which was then converted into seychellene.

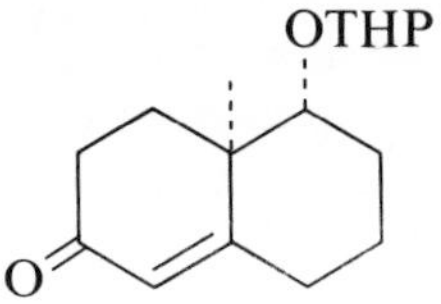

(317)

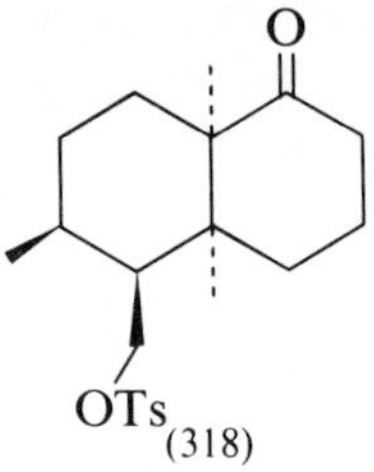

(318)

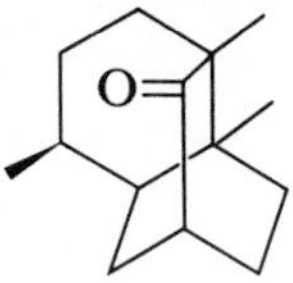

(319)

The family Compositae, of which there are thirteen tribes and which comprise some 20,000 species, is a rich source of sesquiterpenoid lactones (Geissman and Irwin, 1970). Although only a small fraction of these species (about 1 %) has been investigated to date a large number of sesquiterpenoid lactones (*ca.* 200) have been isolated and identified. Of this number the majority are guaianolides or near relatives and thus it is impossible within the scope of this chapter even to scratch the surface of this important area of sesquiterpenoid research. A number of factors have contributed towards the lowering of the activation barrier to structural elucidation in this field; these include sophisticated use of n.m.r., c.d. and o.r.d. spectroscopic techniques and the ultimate weapon—X-ray analysis. Even with the limited number of species investigated, there are some indications that advances in the field of chemotaxonomy have been made. Future results will undoubtedly provide further evidence for some of the chemotaxonomic theories and may ultimately lead to an understanding of the enzymatic processes which govern sesquiterpenoid lactone production.

We shall simply note that guaianolides have been identified which have three of the four possible lactone fusions, *viz.*, 6α,12-olide, 8α, and 8β,12-olide (320). Similarly, pseudoguaianolides (Herz, 1968; Romo and Romo de Vivar, 1967) are known with 6β,12-olide, 8α, and 8β,12-olide structures (321). Normally the guaianolides have the 1α,5α *cis* ring fusion while the pseudoguaianolides have the 1α,5β-Me stereochemistry. The pseudoguaian-6β,12-olides are also normally characterized by a 10β-methyl group and the

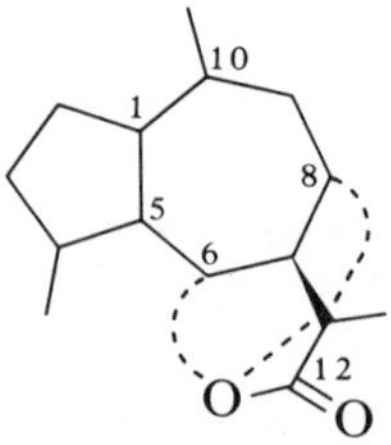

(320)

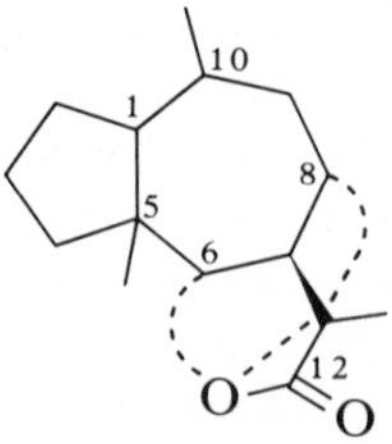

(321)

8α(β),12-olides by a 10α-methyl group. A number of modified guaianolides and pseudoguaianolides are also known. These include such compounds as psilostachyin (322) (Mabry *et al.*, 1966), psilotropin (323) (de Silva and Geissman, 1970), xanthanol (324) (Winters *et al.*, 1969), carabrone (325) (Minato and Horibe, 1968; Minato *et al.*, 1964), and mexicanin H (326) (Romo *et al.*, 1966).

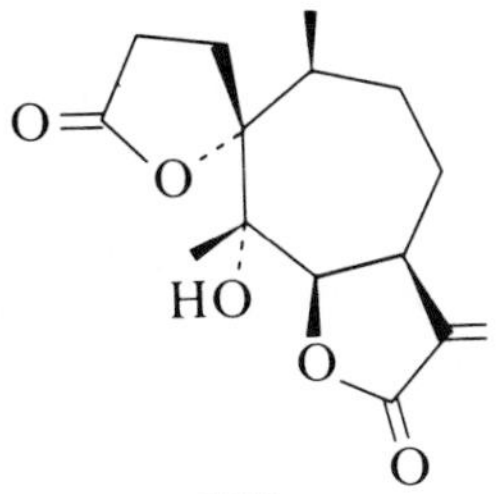

(322)

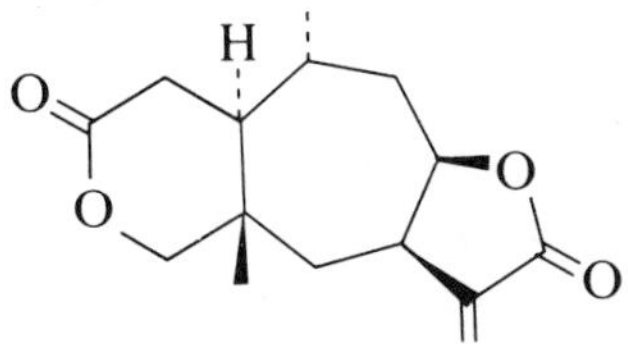

(323)

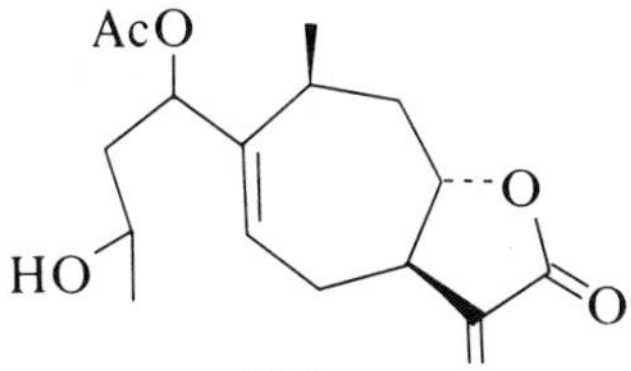

(324)

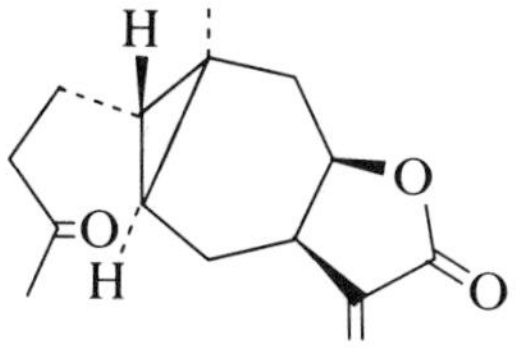

(325)

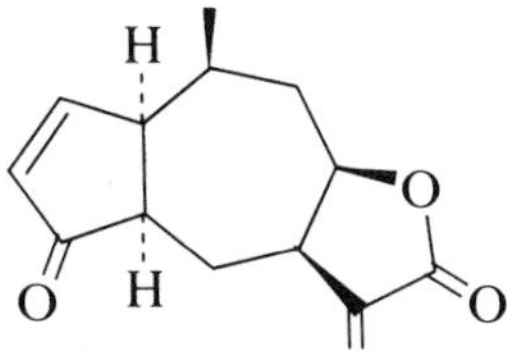

(326)

Synthetic routes to the guaianolides and pseudoguaianolides present many problems in terms of functional group manipulation and stereochemical control. It is therefore not surprising that the successes in this area are few in number. White *et al.* (1969) made use of dihydro-0-acetylisophotosantonic acid lactone (327) as starting material which was converted into naturally-occurring desacetoxymatricarin (328, R = α-Me) in five steps. Utilizing an intermediate in this sequence, Marx and White (1969) also synthesized the C-11 epimeric compound, achillin (328, R = β-Me). Hendrickson *et al.* (1968), on the other hand, isomerized the monoepoxide of santonin (329) to the diketone (330), the 3-diazo derivative of which was

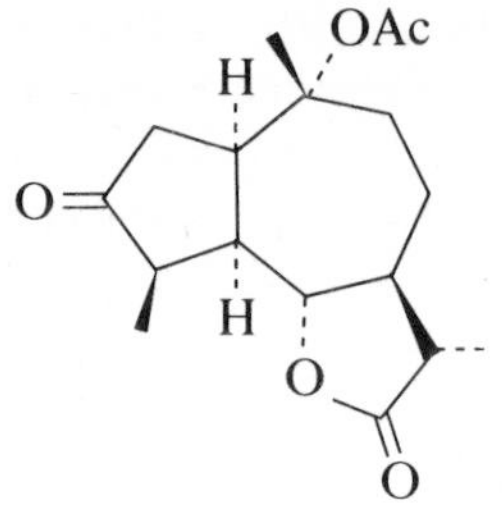

(327)

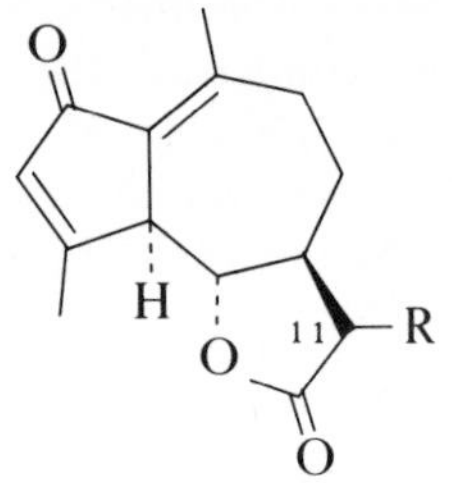

(328)

reduced to the enone (331). The 1β-bromo derivative of (331) was converted into (332) by acidic methanolysis and this, on treatment with trifluoroacetic acid and thionyl chloride, yielded the *cis*-lactone (333) (both epimers) which were rearranged to (334). Unfortunately, however, this compound has the wrong stereochemistry at C-1 in relation to the naturally-occurring pseudoguaianolides.

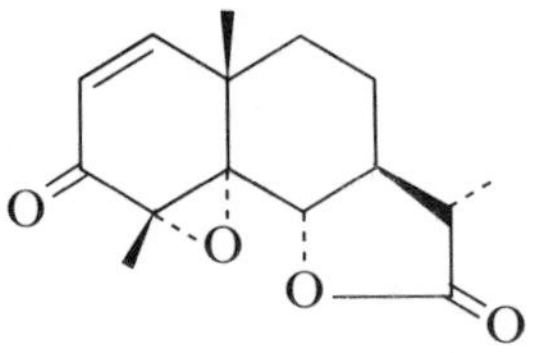

(329)

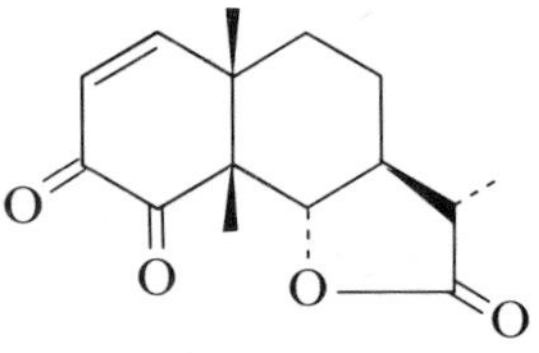

(330)

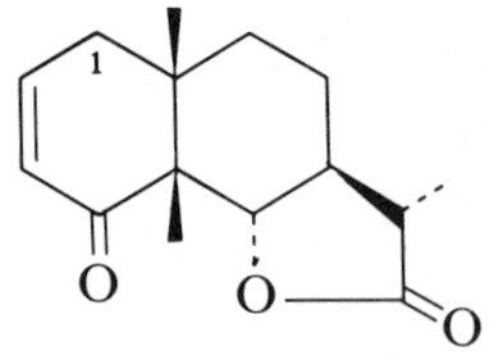

(331)

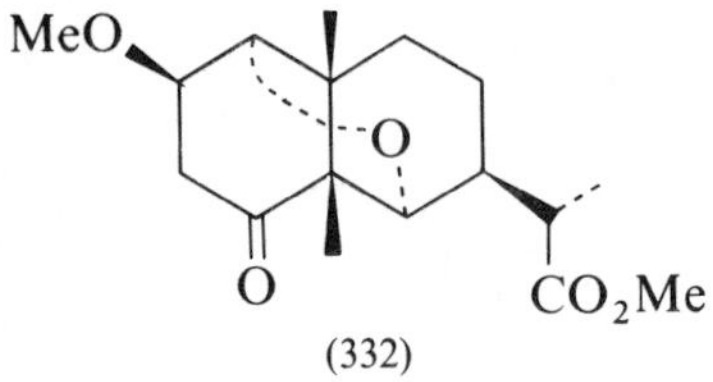

(332)

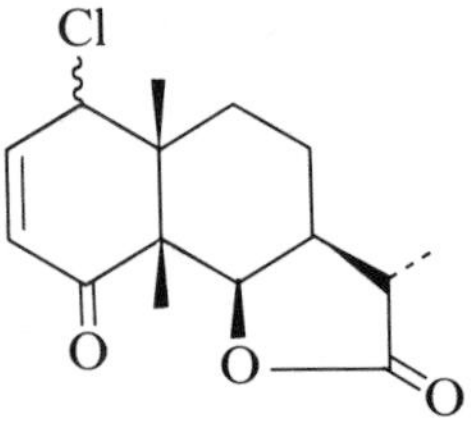

(333)

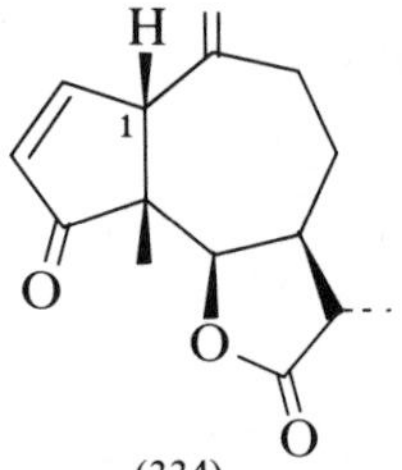

(334)

XVI. Maaliane, Aristolane and Aromadendrane

In 1960, Bates *et al.* (1960) suggested that a logical precursor of those sesquiterpenoids bearing a *gem*-dimethyl cyclopropane ring would be (335), derived from cation (10) by deprotonation. This postulate is now all the more attractive in view of the recent isolation of this very sesquiterpene, bicyclogermacrene, from the peel oil of *Citrus junos* (Nishimura *et al.*, 1969). It is interesting to note that the related compounds, germacrene-B and -D, spathulenol (336) and globulol (337) co-occur with bicyclogermacrene.

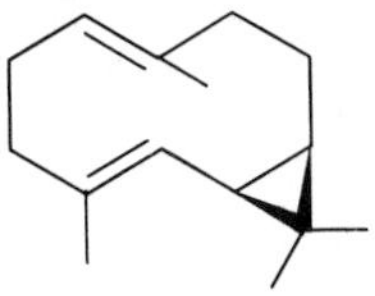
(335)

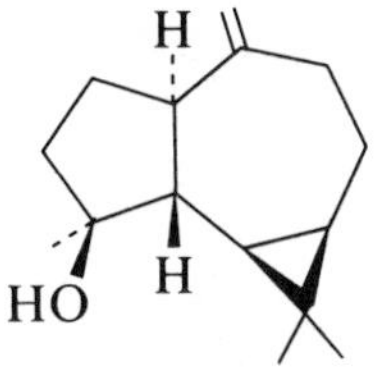

(336)

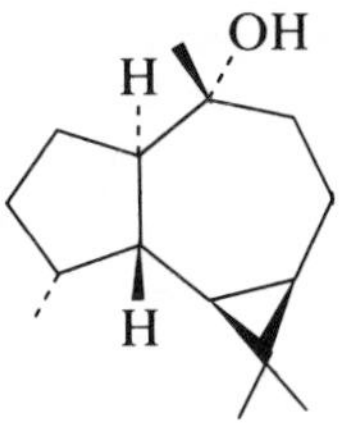

(337)

To date there are only two compounds, viz. β-maaliene (338) and maaliol (339), which can be derived from (335) by a straightforward Markownikoff cyclization. The structural elucidation and syntheses of these compounds have been reported by Bates *et al.* (1960); Büchi *et al.* (1959). There are, however, a number of sesquiterpenoids which are related to the maaliane skeleton by a 1,2-methyl shift. These include (+)-1(10)-aristolene (340) (Büchi *et al.*, 1962), 9-aristolene (341, R = H_2) (Vrkoč *et al.*, 1963), aristolone (341, R = O) (Büchi *et al.*, 1962; Furukawa, 1961), and debilone (342) (Křepinský *et al.*, 1970). In 1965, Carboni *et al.* (1965) reported the isolation

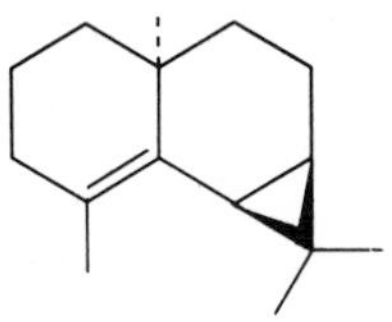
(338)

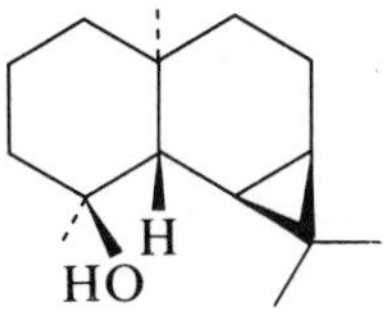

(339)

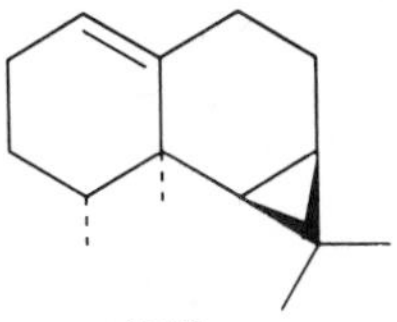
(340)

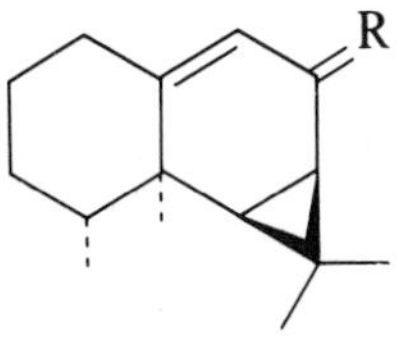

(341)

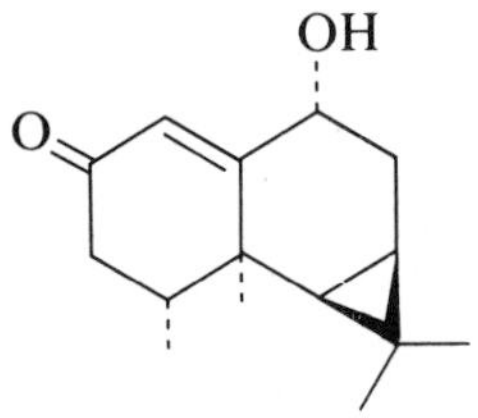

(342)

of α-ferulene (i.e. the antipode of 9-aristolene) from the latex of *Feruli communis* L. This sesquiterpene has also been isolated from the gorgonian *Pseudopterogorgia americana* together with (−)-1(10)-aristolene, (+)-γ-maaliene (343) and the unique "sylvestrene-like" hydrocarbon, (+)-β-gorgonene (344) (Weinheimer *et al.*, 1968). Two syntheses of aristolone have been reported. That by Berger *et al.* (1968) involved the formation of (345) by condensation of 2,3-dimethylcyclohexanone with methyl vinyl ketone. This enone was then transformed into the isomeric enone (346) which, on reaction with 2-diazopropane, photolysis of the resultant pyrazoline and bromination/dehydrobromination, yielded aristolone. Piers *et al.* (1969a) on the other hand, accomplished the synthesis *via* intramolecular cyclization of the olefinic diazoketone (347) also procured from 2,3-dimethylcyclohexanone.

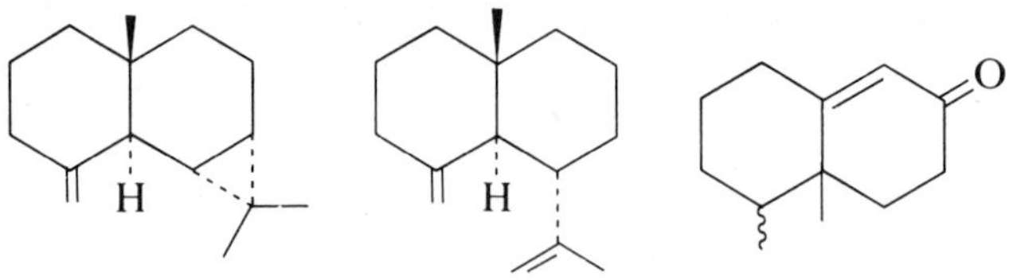

(343) (344) (345)

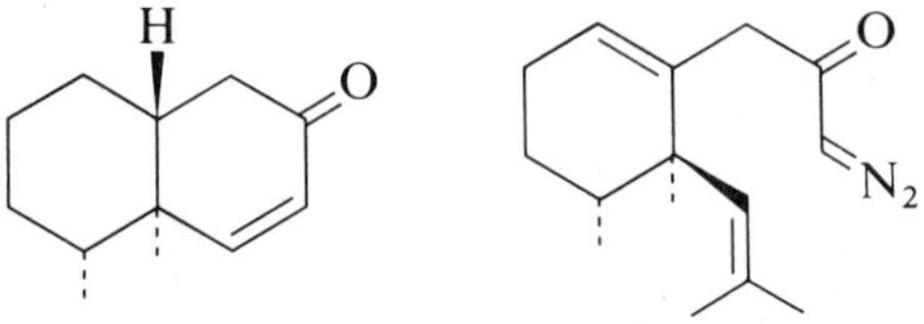

(346) (347)

Coates and Shaw (1970b) have recently reported the synthesis of (±)-1(10)-aristolene. This involved the formation of (348) by condensation of the monopyrrolidine enamine of 2-methylcyclohexane-1,3-dione with pent-3-en-2-one. Selective removal of the conjugated carbonyl group and carbomethoxylation yielded (349) which was converted into the enone (350) by treatment with methyl-lithium and 1% concentrated hydrochloric acid. Conversion to the corresponding 2-pyrazoline with hydrazine and subsequent thermolysis yielded (±)-1(10)-aristolene.

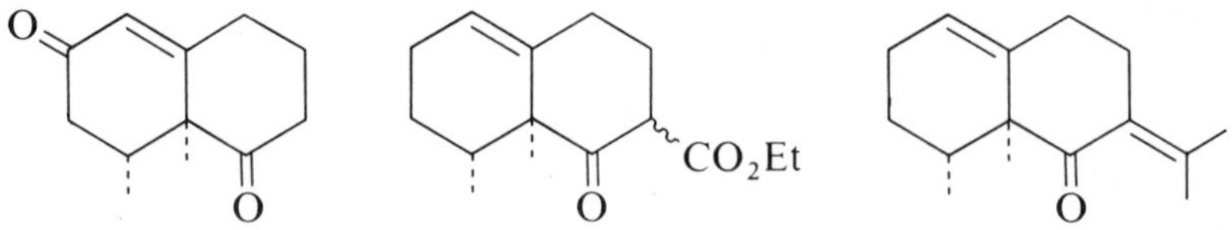

(348) (349) (350)

Anti-Markownikoff cyclization of (335) leads to a perhydroazulene skeleton which is exemplified in the sesquiterpenoids, cyclocolorenone (351, R = O), α-gurjunene (351, R = H_2) (Palmade *et al.*, 1963), and aromadendrene (352). Büchi *et al.* (1966) ascertained the stereochemical features of cyclocolorenone by the synthesis of its C-1 epimer starting from O-acetyliso-photosantonic acid lactone. In an outstanding synthetic study Büchi *et al.* (1969) have also defined the stereochemistry of aromadendrene (352) and alloaromadendrene (353) and related alcohols. This synthetic sequence started from (−)-perillaldehyde (354) which was converted into the diene (355). A Diels–Alder reaction of (355) with acrolein gave the tricyclic aldehyde (356) which was transformed into the diol-monotosylate (357). This was smoothly converted into the tricyclic ketone (358) and hence to the enantiomer of naturally-occurring aromadendrene by a Wittig reaction.

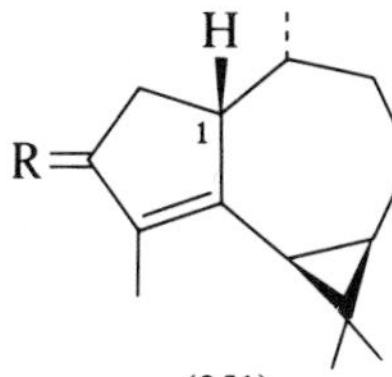

(351)

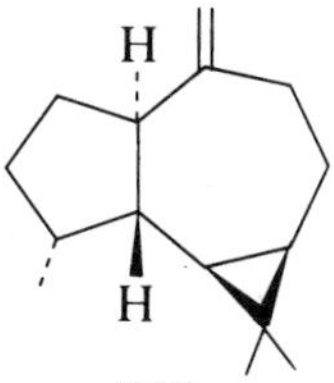

(352)

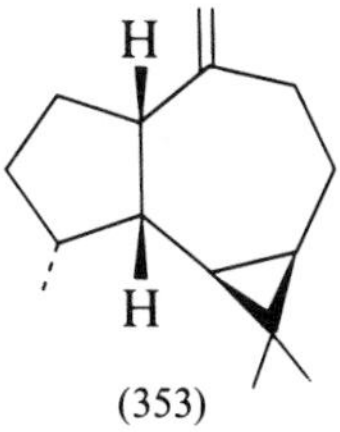

(353)

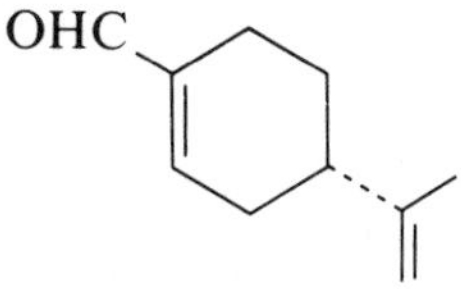

(354)

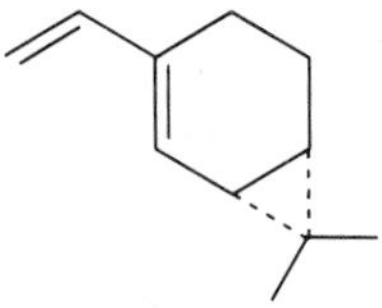
(355)

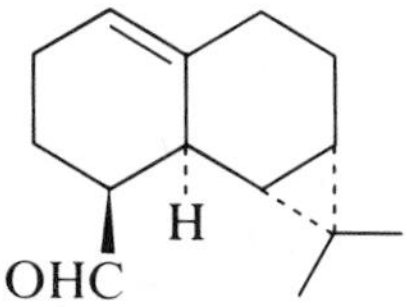

(356)

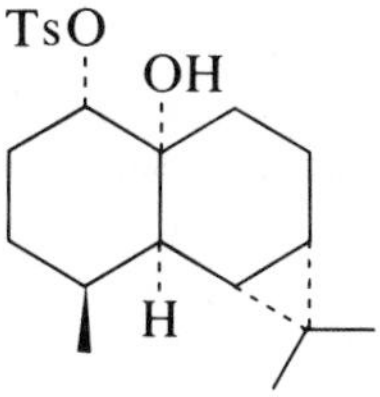

(357)

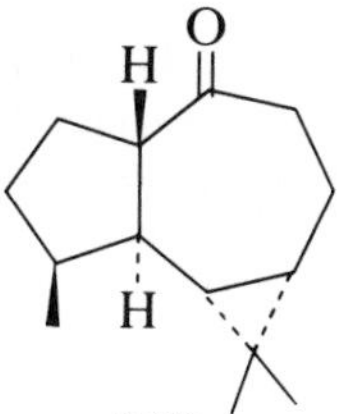

(358)

XVII. Non-farnesyl Sesquiterpenoids

There are relatively few sesquiterpenoids which cannot be derived from a farnesyl precursor either directly or indirectly. Some recent examples of this group include furoventalene (359) (Weinheimer and Washecheck, 1969), the keto-lactone (360) (Bohlmann *et al.*, 1969), and humbertiol (361) (Raulais and Billet, 1970). These types of compounds are perhaps derived from the combination of a C_{10} monoterpene unit—cyclic or acyclic—with a C_5 unit such as dimethylallyl pyrophosphate or isopentenyl pyrophosphate.

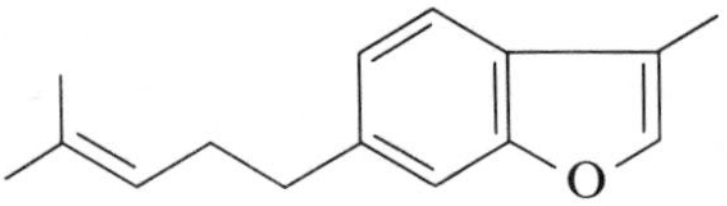

(359)

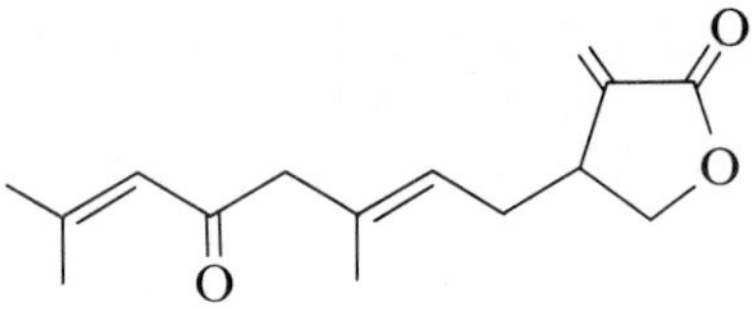

(360)

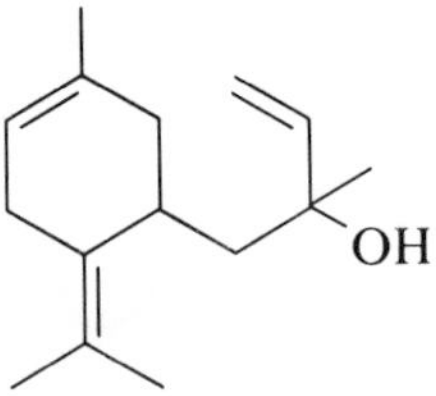

(361)

REFERENCES

Abrahamsson, S. and Nilsson, B. (1964). *Proc. Chem. Soc.* 188.

Adams, R., Morris, R. C., Geissman, T. A., Butterbaugh, D. J. and Kirkpatrick, E. C. (1938). *J. Amer. Chem. Soc.* **60**, 2193.

Allen, F. H. and Rogers, D. (1967). *Chem. Commun.* 588.

Andersen, N. H. (1970). *Tetrahedron Lett.* 1755.

Andersen, N. H. and Syrdal, D. D. (1970). *Tetrahedron Lett.* 2277.

Appel, H. H., Brooks, C. J. W. and Overton, K. H. (1959). *J. Chem. Soc.* 3322.

Asao, T., Ibe, S., Takase, K., Cheng, Y. S. and Nozoe, T. (1968). *Tetrahedron Lett.* 3639.

Ayyar, K. S. and Krishna Rao, G. S. (1968). *Can. J. Chem.* **46**, 1467.

Barton, D. H. R., Böckmann, O. C. and de Mayo, P. (1960). *J. Chem. Soc.* 2263.

Barton, D. H. R. and de Mayo, P. (1957). *J. Chem. Soc.* 150.

Barton, D. H. R., Moss, G. P. and Whittle, J. A. (1968). *J. Chem. Soc.* (*C*) 1813.

Barton, D. H. R. and Werstiuk, N. H. (1968). *J. Chem. Soc.* (*C*) 148.

Bates, R. B., Büchi, G., Matsuura, T. and Schaffer, B. R. (1960). *J. Amer. Chem. Soc.* **82**, 2327.

Bates, R. B., Green, C. D. and Sneath, T. C. (1969). *Tetrahedron Lett.* 3461.

Berger, C., Franck-Neumann, M. and Ourisson, G. (1968). *Tetrahedron Lett.*, 3451.

Bhacca, N. S. and Fischer, N. H. (1969). *Chem. Commun.* 68.

Biollaz, M. and Arigoni, D. (1969). *Chem. Commun.* 633.

Birch, A. J., Macdonald, P. L. and Powell, V. H. (1970). *J. Chem. Soc.* (*C*) 1469.

Birch, A. J., Massy-Westropp, R. A. and Wright, S. E. (1953). *Austral. J. Chem.* **6**, 385.

Bohlmann, F., Zdero, C. and Grenz, M. (1969). *Tetrahedron Lett.* 2417.

Borges del Castillo, J., Brooks, C. J. W. and Campbell, M. M. (1966). *Tetrahedron Lett.* 3731.

Bowers, W. S., Fales, H. M., Thompson, M. J. and Uebel, E. C. (1966). *Science, N.Y.* **154**, 1020.
Brown, E. D., Solomon, M. D., Sutherland, J. K. and Torre, A. (1967). *Chem. Commun.* 111.
Brown, E. D. and Sutherland, J. K. (1968). *Chem. Commun.* 1060.
Brown, M. (1968). *J. Org. Chem.* **33**, 162.
Brown, P. M. and Thomson, R. H. (1969). *J. Chem. Soc.* (*C*) 1184.
Büchi, G., Erickson, R. E. and Wakabayashi, N. (1961). *J. Amer. Chem. Soc.* **83**, 927.
Büchi, G., Greuter, F. and Tokoroyama, T. (1962). *Tetrahedron Lett.* 827.
Büchi, G., Hofheinz, W. and Paukstelis, J. V. (1969). *J. Amer. Chem. Soc.* **91**, 6473.
Büchi, G., Kauffman, J. M. and Loewenthal, H. J. E. (1966). *J. Amer. Chem. Soc.* **88**, 3403.
Büchi, G., MacLeod, W. D., Jr. and Padilla O, J. (1964). *J. Amer. Chem. Soc.* **86**, 4438.
Büchi, G., Schach v. Wittenau, M. and White, D. M. (1959). *J. Amer. Chem. Soc.* **81**, 1968.
Büchi, G. and White, J. D. (1964). *J. Amer. Chem. Soc.* **86**, 2884.
Burrell, J. W. K., Garwood, R. F., Jackman, L. M., Oskay, E. and Weedon, B. C. L. (1966). *J. Chem. Soc.* (*C*) 2144.
Cameron, A. F., Ferguson, G. and Robertson, J. M. (1969). *J. Chem. Soc.* (*B*) 692.
Canonica, L., Corbella, A., Gariboldi, P., Jommi, G., Křepinský, J., Ferrari, G. and Casagrande, C. (1969). *Tetrahedron* **25**, 3895.
Carboni, S., Da Settimo, A., Malaguzzi, V., Marsili, A. and Pacini, P. L. (1965). *Tetrahedron Lett.* 3017.
Cavill, G. W. K., Laing, D. G. and Williams, P. J. (1969). *Austral. J. Chem.* **22**, 2145.
Černý, V., Dolejš, L., Labler, L., Šorm, F. and Sláma, K. (1967). *Tetrahedron Lett.* 1053; (1967). *Collect. Czech. Chem. Commun.* **32**, 3926.
Chakravarti, K. K. (1965). *Indian J. Chem.* **3**, 324.
Challand, B. D., Hikino, H., Kornis, G., Lange, G. and de Mayo, P. (1969). *J. Org. Chem.* **34**, 794.
Clunie, J. S. and Robertson, J. M. (1961). *J. Chem. Soc.* 4382.
Coates, R. M., Farney, R. F., Johnson, S. M. and Paul, I. C. (1969). *Chem. Commun.* 999.
Coates, R. M. and Freidinger, R. M. (1970). *Tetrahedron* **26**, 3487.
Coates, R. M. and Shaw, J. E. (1970a). *J. Org. Chem.* **35**, 2597.
Coates, R. M. and Shaw, J. E. (1970b). *J. Amer. Chem. Soc.* **92**, 5657.
Cocker, W. and McMurry, T. B. H. (1960). *Tetrahedron* **8**, 181.
Coggon, P., McPhail, A. T. and Sim, G. A. (1970). *J. Chem. Soc.* (*B*) 1024.
Comer, F. W., McCapra, F., Qureshi, I. H. and Scott, A. I. (1967). *Tetrahedron* **23**, 4761.
Corbella, A., Gariboldi, P., Jommi, G. and Scolastico, C. (1969). *Chem. Commun.* 634.
Corey, E. J. and Achiwa, K. (1969a). *Tetrahedron Lett.* 1837.
Corey, E. J. and Achiwa, K. (1969b). *Tetrahedron Lett.* 3257.
Corey, E. J. and Achiwa, K. (1970). *Tetrahedron Lett.* 2245.
Corey, E. J., Achiwa, K. and Katzenellenbogen, J. A. (1969c). *J. Amer. Chem. Soc.* **91**, 4318.
Corey, E. J. and Broger, E. A. (1969). *Tetrahedron Lett.* 1779.
Corey, E. J., Girotra, N. N. and Mathew, C. T. (1969c). *J. Amer. Chem. Soc.* **91**, 1557.
Corey, E. J. and Hamanaka, E. (1967). *J. Amer. Chem. Soc.* **89**, 2758.
Corey, E. J. and Hortmann, A. G. (1963). *J. Amer. Chem. Soc.* **85**, 4033.
Corey, E. J., Katzenellenbogen, J. A., Gilman, N. W., Roman, S. A. and Erickson, B. W. (1968). *J. Amer. Chem. Soc.* **90**, 5618.

Corey, E. J., Kirst, H. A. and Katzenellenbogen, J. A. (1970). *J. Amer. Chem. Soc.* **92**, 6314.
Corey, E. J., Mitra, R. B. and Uda, H. (1964a). *J. Amer. Chem. Soc.* **86**, 485.
Corey, E. J. and Nozoe, S. (1965a). *J. Amer. Chem. Soc.* **87**, 5728.
Corey, E. J. and Nozoe, S. (1965b). *J. Amer. Chem. Soc.* **87**, 5733.
Corey, E. J., Ohno, M., Mitra, R. B. and Vatakencherry, P. A. (1964b). *J. Amer. Chem. Soc.* **86**, 478.
Corey, E. J. and Yamamoto, H. (1970). *J. Amer. Chem. Soc.* **92**, 6636.
Cornforth, J. W., Milborrow, B. V., Ryback, G. and Wareing, P. F. (1965). *Nature, Lond.* **205**, 1269.
Coscia, C. J. (1969). *In* "Cyclopentanoid Terpene Derivatives", (W. I. Taylor and A. R. Battersby, eds.) Marcel Dekker, New York.
Crandall, T. G. and Lawton, R. G. (1969). *J. Amer. Chem. Soc.* **91**, 2127.
Dahm, K. H., Trost, B. M. and Roeller, H. (1967). *J. Amer. Chem. Soc.* **89**, 5292.
Danishefsky, S. and Dumas, D. (1968). *Chem. Commun.* 1287.
Dauben, W. G. and Ashcroft, A. C. (1963). *J. Amer. Chem. Soc.* **85**, 3673.
Dauben, W. G. and Friedrich, L. E. (1964). *Tetrahedron Lett.* 2675.
de Mayo, P., Robinson, J. R., Spencer, E. Y. and White, R. W. (1962). *Experientia* **18**, 359.
de Mayo, P., Spencer, E. Y. and White, R. W. (1962). *J. Amer. Chem. Soc.* **84**, 494; *idem* (1963). *Can. J. Chem.* **41**, 2996.
de Mayo, P. and Williams, R. E. (1965a). *J. Amer. Chem. Soc.* **87**, 3275.
de Mayo, P., Williams, R. E., Büchi, G. and Feairheller, S. H. (1965b). *Tetrahedron* **21**, 619.
de Silva, L. B. and Geissman, T. A. (1970). *Phytochem.* **9**, 59.
Dev, S. (1960). *Tetrahedron* **9**, 1.
Djerassi, C. and Burnstein, S. (1959). *Tetrahedron* **7**, 37.
Djerassi, C. and Rittel, W. (1957). *J. Amer. Chem. Soc.* **79**, 3528.
Dobler, M., Dunitz, J. D., Gubler, B., Weber, H. P., Büchi, G. and Padilla O, J. (1963). *Proc. Chem. Soc.* 383.
Doskotch, R. W. and El-Feraly, F. S. (1969). *Canad. J. Chem.* **47**, 1139.
Doyle, P., Maclean, I. R., Murray, R. D. H., Parker, W. and Raphael, R. A. (1965). *J. Chem. Soc.* 1344.
Dugan, J. J., de Mayo, P., Nisbet, M., Robinson, J. R., and Anchel, M. (1966). *J. Amer. Chem. Soc.* **88**, 2838.
Endo, K. and de Mayo, P. (1967). *Chem. Commun.* **89**; *idem* (1969). *Chem. Pharm. Bull.* **17**, 1324.
Enzell, C. (1962a). *Tetrahedron Lett.* 185.
Enzell, C. (1962b). *Acta Chem. Scand.* **16**, 1553.
Erdtman, H. and Norin, T. (1966). *Fortschr. Chem. Organ. Naturstoffe* **24**, 206.
Erdtman, H. and Westfelt, L. (1963). *Acta Chem. Scand.* **17**, 2351.
Ferguson, G., Hawley, D. M., McKillop, T. F. W., Martin, J., Parker, W. and Doyle, P. (1967). *Chem. Commun.* 1123.
Findlay, J. A. and MacKay, W. D. (1969). *Chem. Commun.* 733.
Fishman, J., Jones, E. R. H., Lowe, G. and Whiting, M. C. (1960). *J. Chem. Soc.* 3948.
Flath, R. A., Lundin, R. E. and Teranishi, R. (1966). *Tetrahedron Lett.* 295.
Fuhrer, H., Ganguly, A. K., Gopinath, K. W., Govindachari, T. R., Nagarajan, K., Pai, B. R. and Parthasarathy, P. C. (1970). *Tetrahedron* **26**, 2371.
Furukawa, S. (1961). *J. Pharm. Soc. Japan* **81**, 570.

Ganguly, A. K., Gopinath, K. W., Govindachari, T. R., Nagarajan, K., Pai, B. R. and Parthasarathy, P. C. (1969). *Tetrahedron Lett.* 133.

Geissman, T. A. and Irwin, M. A. (1970). *Pure and Applied Chem.* **21**, 167.

Gemmell, K. W., Parker, W., Roberts, J. S., Robertson, J. M. and Sim, G. A. (1970). *J. Chem. Soc* (*B*) 947.

Gibson, T. W. and Erman, W. F. (1969). *J. Amer. Chem. Soc.* **91**, 4771.

Godtfredsen, W. O. and Vangedal, S. (1964). *Proc. Chem. Soc.* 188; *idem* (1965). *Acta Chem. Scand.* **19**, 1088.

Gollnick, K., Schade, G., Cameron, A. F., Hannaway, C., Roberts, J. S. and Robertson, J. M. (1970). *Chem. Commun.* 248.

Govindachari, T. R., Mohamed, P. A. and Parthasarathy, P. C. (1970). *Tetrahedron* **26**, 615.

Govindachari, T. R., Nagarajan, K. and Parthasarathy, P. C. (1969). *Chem. Commun.* 823.

Greenwood, J. M., Solomon, M. D., Sutherland, J. K. and Torre, A. (1968). *J. Chem. Soc.* (*C*) 3004.

Grieco, P. A. (1969). *J. Amer. Chem. Soc.* **91**, 5660.

Gutzwiller J. and Tamm, Ch. (1965). *Helv. Chim. Acta* **48**, 157.

Harmatha, J., Samek, Z., Novotný, L., Herout, V. and Šorm, F. (1969). *Collect. Czech. Chem. Commun.* **34**, 2792.

Hartley, R. D. and Fawcett, C. H. (1969). *Phytochem.*, **8**, 637, 1793.

Hartsuck, J. A. and Paul, I. C. (1964). *Chem. Ind.* (*London*) 977.

Hayashi, S., Matsuo, A. and Matsuura, T. (1969). *Experientia* **25**, 1139.

Heathcock, C. H. and Badger, R. A. (1968). *Chem. Commun.* 1510.

Heathcock, C. H., Badger, R. A. and Patterson, J. W., Jr. (1967). *J. Amer. Chem. Soc.* **89**, 4133.

Helmlinger, D., de Mayo, P., Nye, M., Westfelt, L. and Yeats, R. B. (1970). *Tetrahedron Lett.* 349.

Hendrickson, J. B. (1959). *Tetrahedron* **7**, 82.

Hendrickson, J. B., Ganter, C., Dorman, D. and Link, H. (1968). *Tetrahedron Lett.* 2235.

Herz, W. (1968). *In* "Recent Advances in Phytochemistry", (T. J. Mabry, ed.), Vol. 1, p. 229, North-Holland Publishing Co., Amsterdam.

Hikino, H., Hikino, Y., Takeshita, Y., Meguro, K. and Takemoto, T. (1965a). *Chem. Pharm. Bull.* **13**, 1408.

Hikino, H., Hikino, Y., Takeshita, Y., Shirata, K. and Takemoto, T. (1963). *Chem. Pharm. Bull.* **11**, 547; (1967). **15**, 321.

Hikino, H., Takeshita, Y., Hikino, Y. and Takemoto, T. (1965b). *Chem. Pharm. Bull.* **13**, 631.

Holub, M., Tax, J., Sedmera, P. and Šorm, F. (1970). *Collect. Czech. Chem. Commun.* **35**, 3597.

Homma, A., Kato, M., Wu, M.-D., and Yoshikoshi, A. (1970). *Tetrahedron Lett.* 231.

Hortmann, A. G. (1968). *Tetrahedron Lett.* 5785.

Hortmann, A. G. and De Roos, J. B. (1969). *J. Org. Chem.* **34**, 736.

Irie, T., Fukuzawa, A., Izawa, M. and Kurosawa, E. (1969a). *Tetrahedron Lett.* 1343.

Irie, H., Kimura, H., Otani, N., Ueda, K. and Uyeo, S. (1967). *Chem. Commun.* 678.

Irie, T., Suzuki, M., Kurosawa, E. and Masamune, T. (1970). *Tetrahedron* **26**, 3271.

Irie, T., Suzuki, T., Yasunari, Y., Kurosawa, E. and Masamune, T. (1969b). *Tetrahedron* **25**, 459.

Itô, S., Endo, K. and Nozoe, T. (1963). *Chem. Pharm. Bull.* **11**, 132; *idem* (1964). *Tetrahedron Lett.* 3375.
Itô, S., Endo, K., Yoshida, T., Yatagai, M. and Kodama, M. (1967). *Chem. Commun.* 186.
Jain, T. C. and McCloskey, J. E. (1969). *Tetrahedron Lett.* 2917.
Johnson, W. S., Brocksom, T. J., Loew, P., Rich, D. H., Werthemann, L., Arnold, R. A., Li, T. and Faulkner, J. (1970). *J. Amer. Chem. Soc.* **92**, 4463.
Johnson, W. S., Li, T., Faulkner, D. J. and Campbell, S. F. (1968). *J. Amer. Chem. Soc.* **90**, 6225.
Jommi, G., Křepinský, J., Herout, V. and Šorm, F. (1967). *Tetrahedron Lett.* 677.
Jones, R. V. H. and Sutherland, M. D. (1968). *Chem. Commun.* 1229.
Joseph, T. C. and Dev, S. (1968). *Tetrahedron* **24**, 3809, 3841.
Joseph-Nathan, P., Negrete, C. and Gonzalez, P. (1970). *Phytochem.*, **9**, 1623.
Kapadia, V. H., Nagasampagi, B. A., Naik, V. G. and Dev, S. (1965). *Tetrahedron* **21**, 607.
Kartha, C. C., Kalsi, P. S., Shaligram, A. M., Chakravarti, K. K. and Bhattacharyya, S. C. (1963). *Tetrahedron* **19**, 241.
Kato, T., Kanno, S. and Kitahara, Y. (1970a). *Tetrahedron* **26**, 4287.
Kato, M., Kosugi, H. and Yoshikoshi, A. (1970b). *Chem. Comm.* 185.
Kato, M., Kosugi, H. and Yoshikoshi, A. (1970c). *Chem. Commun.* 934.
Kepler, J. A., Wall, M. E., Mason, J. E., Basset, C., McPhail, A. T. and Sim, G. A. (1967). *J. Amer. Chem. Soc.* **89**, 1260.
Kerschbaum, M. (1913). *Chem. Ber.* **46**, 1732.
Kido, F., Sakuma, R., Uda, H. and Yoshikoshi, A. (1969a). *Tetrahedron Lett.* 3169.
Kido, F., Uda, H. and Yoshikoshi, A. (1967). *Tetrahedron Lett.* 2815; (1968). *idem, ibid,* 1247.
Kido, F., Uda, H. and Yoshikoshi, A. (1969b). *Chem. Commun.* 1335.
Kolbe-Haugwitz, M. and Westfelt, L. (1970). *Acta Chem. Scand.* **24**, 1623.
Křepinský, J., Jommi, G., Samek, Z. and Šorm, F. (1970). *Collect. Czech. Chem. Commun.* **35**, 745.
Křepinský, J., Motl, O., Dolejš, L., Novotný, L., Herout, V. and Bates, R. B. (1968). *Tetrahedron Lett.* 3315.
Křepinský, J., Samek, Z. and Šorm, F. (1966a). *Tetrahedron Lett.* 3209.
Křepinský, J., Samek, Z., Šorm, F. and Lamparsky, D. (1966b). *Tetrahedron Lett.* 359.
Kupchan, S. M., Aynehchi, Y., Cassady, J. M., Schnoes, H. K. and Burlingame, A. L. (1969a). *J. Org. Chem.* **34**, 3867.
Kupchan, S. M., Hemingway, R. J., Werner, D. and Karim, A. (1969b). *J. Org. Chem.* **34**, 3903.
Lansbury, P. T. and Hilfiker, F. R. (1969). *Chem. Commun.* 619.
Levisalles, J. and Rudler, H. (1967). *Bull. Soc. chim. France* 2059.
L'Homme, M. F., Geissman, T. A., Yoshioka, H., Porter, T. H., Renold, W. and Mabry, T. J. (1969). *Tetrahedron Lett.* 3161.
Mabry, T. J., Miller, H. E., Kagan, H. B. and Renold, W. (1966). *Tetrahedron* **22**, 1139.
Machlis, L., Nutting, W. H. and Rapoport, H. (1968). *J. Amer. Chem. Soc.* **90**, 1674.
MacLeod, W. D., Jr. (1965). *Tetrahedron Lett.* 4779.
MacSweeney, D. F., Ramage, R. and Sattar, A. (1970). *Tetrahedron Lett.* 557.
Marini Bettòlo, G. B., Casinovi, C. G. and Galeffi, C. (1965). *Tetrahedron Lett.* 4857.
Marshall, J. A. and Andersen, N. H. (1967). *Tetrahedron Lett.* 1611.
Marshall, J. A. and Brady, S. F. (1969). *Tetrahedron Lett.* 1387.
Marshall, J. A., Bundy, G. L. and Fanta, W. I. (1968). *J. Org. Chem.* **33**, 3913.

Marshall, J. A., Cohen, N. and Hochstetler, A. R. (1966a). *J. Amer. Chem. Soc.* **88**, 3408.
Marshall, J. A., Faubl, H. and Warne, T. M., Jr. (1967). *Chem. Commun.* 753.
Marshall, J. A. and Johnson, P. C. (1970). *J. Org. Chem.* **35**, 192.
Marshall, J. A. and Partridge, J. J. (1969). *Tetrahedron* **25**, 2159.
Marshall, J. A., Pike, M. T. and Carroll, R. D. (1966b). *J. Org. Chem.* **31**, 2933.
Marshall, J. A. and Ruden, R. A. (1970). *Tetrahedron Lett.* 1239.
Marx, J. N. and White, E. H. (1969). *Tetrahedron* **25**, 2117.
Massy-Westropp, R. A., Reynolds, G. D. and Spotswood, T. M. (1966). *Tetrahedron Lett.* 1939.
Matsumoto, T., Shirahama, H., Ichihara, A., Fukuoka, Y., Takahashi, Y., Mori, Y. and Watanabe, M. (1965). *Tetrahedron* **21**, 2671.
Matsumoto, T., Shirahama, H., Ichihara, A., Shin, H., Kagawa, S., Sakan, F., Matsumoto, S. and Nishida, S. (1968). *J. Amer. Chem. Soc.* **90**, 3280.
McEachan, C. E., McPhail, A. T. and Sim, G. A. (1966). *J. Chem. Soc.* (*C*) 579.
McKillop, T. F. W., Martin, J., Parker, W. and Roberts, J. S. (1967). *Chem. Commun.* 162.
McMorris, T. C. and Anchel, M. (1965). *J. Amer. Chem. Soc.* **87**, 1594.
McMorris, T. C., Nair, M. S. R. and Anchel, M. (1967). *J. Amer. Chem. Soc.* **89**, 4562.
McMurry, J. E. (1968). *J. Amer. Chem. Soc.* **90**, 6821.
McMurry, J. E. (1969). *Tetrahedron Lett.* 55; *ibid* (1970). 3731, 3735.
McPhail, A. T., Reed, R. I. and Sim, G. A. (1964). *Chem. Ind.* (*London*) 976.
McPhail, A. T. and Sim, G. A. (1965). *Chem. Commun.* 350.
Meyer, A. S., Schneidermann, H. A., Hanzmann, E. and Ko, J. H. (1968). *Proc. Nat. Acad. Sci. U.S.* **60**, 853.
Minato, H. and Horibe, I. (1968). *J. Chem. Soc.* (*C*) 2131.
Minato, H., Nosaka, S. and Horibe, I. (1964). *J. Chem. Soc.* 5503.
Moffett, R. H. and Rogers, D. (1953). *Chem. Ind.* (*London*) 916.
Mori, K. and Matsui, M. (1968). *Tetrahedron* **24**, 3127.
Mori, K. and Matsui, M. (1969). *Tetrahedron Lett.* 4435.
Mori, K. and Matsui, M. (1970). *Tetrahedron* **26**, 2801.
Mori, K., Ohki, M., Kobayashi, A. and Matsui, M. (1970). *Tetrahedron* **26**, 2815.
Mori, K., Stalla-Bourdillon, B., Ohki, M., Matsui, M. and Bowers, W. S. (1969). *Tetrahedron* **25**, 1667.
Morikawa, K. and Hirose, Y. (1969). *Tetrahedron Lett.* 1799.
Motl, O., Romañuk, M. and Herout, V. (1966). *Collect. Czech. Chem. Commun.* **31**, 2025.
Naffa, P. and Ourisson, G. (1954). *Bull. Soc. chim. France* 1410.
Nakanishi, K., Ohashi, M., Tada, M. and Yamada, Y. (1965). *Tetrahedron* **21**, 1231.
Nakatani, Y. and Yamanishi, T. (1969). *Agric. Biol. Chem.* **33**, 1805.
Nakazaki, M. (1962). *Bull. Chem. Soc. Japan* **35**, 1387.
Natori, S., Nishikawa, H. and Ogawa, H. (1964). *Chem. Pharm. Bull.* **12**, 236.
Nayak, U. R. and Dev, S. (1963). *Tetrahedron Lett.* 243.
Nickon, A. (1954). *Perfum. Essent. Oil Rec.* **45**, 149.
Nickon, A., Iwadare, T., McGuire, F. J., Mahajan, J. R., Narang, S. A. and Umezawa, B. (1970). *J. Amer. Chem. Soc.* **92**, 1688.
Nigam, I. C., Komae, H., Neville, G. A., Radecka, C. and Paknikar, S. K. (1968). *Tetrahedron Lett.* 2497.
Nishimura, K., Shinoda, N. and Hirose, Y. (1969). *Tetrahedron Lett.* 3097.
Norin, T. (1961). *Acta Chem. Scand.* **15**, 1676.

Novotný, L., Herout, V. and Šorm, F. (1964). *Collect. Czech. Chem. Commun.* **29**, 2189.
Nozoe, T., Asao, T., Ando, M. and Takase, K. (1967). *Tetrahedron Lett.* 2821.
Nozoe, T., Cheng, Y. S. and Toda, T. (1966). *Tetrahedron Lett.* 3663.
Nozoe, S. and Machida, Y. (1970). *Tetrahedron Lett.* 2671.
Ohkuma, K., Addicott, F. T., Smith, O. E. and Thiessen, W. E. (1965). *Tetrahedron Lett.* 2529.
Ohta, Y. and Hirose, Y. (1968). *Tetrahedron Lett.* 1251.
Ohta, Y. and Hirose, Y. (1969). *Tetrahedron Lett.* 1601.
Ohta, Y., Ohara, K. and Hirose, Y. (1968). *Tetrahedron Lett.* 4181.
Ohta, Y., Sakai, T. and Hirose, Y. (1966). *Tetrahedron Lett.* 6365.
Ourisson, G. (1964). *Proc. Chem. Soc.* 274.
Palmade, M., Pesnelle, P., Streith, J. and Ourisson, G. (1963). *Bull. Soc. chim. France* 1950.
Parker, W. Personal communication.
Parker, W., Raphael, R. A. and Roberts, J. S. (1969). *J. Chem. Soc.* (*C*) 2634.
Parker, W., Roberts, J. S. and Ramage, R. (1967). *Quart. Rev.* **21**, 331.
Pawson, B. A., Cheung, H.-C., Gurbaxani, S. and Saucy, G. (1970). *J. Amer. Chem. Soc.* **92**, 336.
Pelletier, S. W. and Prabhakar, S. (1968). *J. Amer. Chem. Soc.* **90**, 5318.
Penfold, A. R. and Simonsen, J. L. (1939). *J. Chem. Soc.* 87.
Pesaro, M., Bozzato, G. and Schudel, P. (1968). *Chem. Commun.* 1152.
Piers, E., Britton, R. W. and de Waal, W. (1969a). *Can. J. Chem.* **47**, 831.
Piers, E., Britton, R. W. and de Waal, W. (1969b). *Tetrahedron Lett.* 1251.
Piers, E., Britton, R. W. and de Waal, W. (1969c). *Chem. Commun.* 1069.
Piers, E. and Cheng, K. F. (1969). *Chem. Commun.* 562.
Piers, E. and Keziere, R. J. (1969). *Can. J. Chem.* **47**, 137.
Pinder, A. R. (1968). *Perfum. Essent. Oil Rec.* **59**, 280, 645.
Plattner, J. J., Bhalerao, U. T. and Rapoport, H. (1970). *J. Amer. Chem. Soc.* **92**, 3429.
Plattner, P. A., Fürst, A., Eschenmoser, A., Keller, W., Kläui, H., Meyer, St. and Rosner, M. (1953). *Helv. Chim. Acta* **36**, 1845.
Porter, L. A. (1967). *Chem. Rev.* **67**, 441.
Prahlad, J. R., Nayak, U. R. and Dev, S. (1970). *Tetrahedron* **26**, 663.
Quilico, A., Piozzi, F. and Pavan, M. (1957). *Tetrahedron Lett.* **1**, 177.
Koteswara Rao, M. V. R., Krishna Rao, G. S. and Dev, S. (1966). *Tetrahedron* **22**, 1977.
Raulais, D. and Billet, D. (1970). *Bull. Soc. chim. France* 2401.
Roberts, D. L., Heckman, R. A., Hege, B. P. and Bellin, S. A. (1968). *J. Org. Chem.* **33**, 3566.
Roberts, B. W., Poonian, M. S. and Welch, S. C. (1969). *J. Amer. Chem. Soc.* **91**, 3400.
Romo, J. and Romo de Vivar, A. (1967). *Fortschr. Chem. Org. Naturstoffe* **25**, 90.
Romo, J., Romo de Vivar, A. and Joseph-Nathan, P. (1966). *Tetrahedron Lett.* 1029.
Ruzicka, L. (1959). *Proc. Chem. Soc.* 341.
Ruzicka, L., Eschenmoser, A. and Heusser, H. (1953). *Experientia* **9**, 357.
Sakai, T., Nishimura, K. and Hirose, Y. (1963). *Tetrahedron Lett.* 1171.
Sakai, T., Nishimura, K. and Hirose, Y. (1965). *Bull. Chem. Soc. Japan* **38**, 381.
Sakuma, R. and Yoshikoshi, A. (1968). *Chem. Commun.* 41.
Samek, Z., Holub, M., Herout, V. and Šorm, F. (1969). *Tetrahedron Lett.* 2931.
Schulz, H. and Sprung, I. (1969). *Angew. Chem. Int. Ed. Eng.* **8**, 271.
Semmler, F. W. (1910). *Chem. Ber.* **43**, 1893.
Shimomura, O. and Johnson, F. H. (1968). *Biochemistry* **7**, 1734.

Shirahata, K., Kato, T., Kitahara, Y. and Abe, N. (1969). *Tetrahedron* **25**, 3179.
Singh, A. N., Upadhye, A. B., Wadia, M. S., Mhaskar, V. V. and Dev, S. (1969). *Tetrahedron* **25**, 3855.
Sipma, G. and van der Wal, B. (1968). *Rec. Trav. chim.* **87**, 715.
Sláma, K. and Williams, C. M. (1965). *Proc. Nat. Acad. Sci. U.S.* **54**, 411.
Smedman, L. and Zavarin, E. (1968). *Tetrahedron Lett.* 3833.
Smedman, L. A., Zavarin, E. and Teranishi, R. (1969). *Phytochem.* **8**, 1457.
Šorm, F. and Dolejš, L. (1965). "Guaianolides and Germacranolides", Hermann, Paris.
Šorm, F., Streibl, M., Jarolím, V., Novotný, L., Dolejš, L. and Herout, V. (1954). *Collect. Czech. Chem. Commun.* **19**, 570.
Šorm, F., Suchý, M., Holub, M., Línek, A., Hadinec, I. and Novák, C. (1970). *Tetrahedron Lett.* 1893.
Souček, M. (1962). *Collect. Czech. Chem. Commun.* **27**, 2929.
Stork, G. and Breslow, R. (1953). *J. Amer. Chem. Soc.* **75**, 3291.
Stork, G. and Clarke, F. H. Jr. (1961). *J. Amer. Chem. Soc.* **83**, 3114.
Sumimoto, M., Ito, H., Hirai, H. and Wada, K. (1963). *Chem. Ind.* (*London*) 780.
Sutherland, M. D. and Waters, O. J. (1961). *Austral. J. Chem.* **14**, 596.
Tanaka, A., Uda, H. and Yoshikoshi, A. (1967). *Chem. Commun.* 188.
Tanaka, A., Uda, H. and Yoshikoshi, A. (1969). *Chem. Commun.* 308.
Takeda, K. (1970). *Pure and Applied Chem.* **21**, 181.
Takeda, K., Horibe, I. and Minato, H. (1970). *J. Chem. Soc.* (*C*) 1142,
Takeda, K., Minato, H. and Ishikawa, M. (1964). *J. Chem. Soc.* 4578; *idem* (1969). *J. Chem. Soc.* (*C*) 1491.
Takeda, K., Minato, H. and Ishikawa, M. (1966). *Tetrahedron Supplement* **7**, 219.
Tarbell, D. S., Carman, R. M., Chapman, D. D., Huffman, K. R. and McCorkindale, N. J. (1960). *J. Amer. Chem. Soc.* **82**, 1005.
Teranishi, R., Thomas, A. F., Schudel, P. and Büchi, G. (1968). *Chem. Commun.* 928.
Tomita, B. and Hirose, Y. (1970a). *Tetrahedron Lett.* 143.
Tomita, B. and Hirose, Y. (1970b). *Tetrahedron Lett.* 235.
Tomita, B., Isona, T. and Hirose, Y. (1970). *Tetrahedron Lett.* 1371.
Tori, K. and Horibe, I. (1970). *Tetrahedron Lett.* 2811.
Trivedi, B., Motl, O., Smolíková, J. and Šorm, F. (1964). *Tetrahedron Lett.* 1197.
Trost, B. M. (1970). *Accounts Chem. Res.* **3**, 120.
van Tamelen, E. E. and Coates, R. M. (1966). *Chem. Commun.* 413.
van Tamelen, E. E. and McCormick, J. P. (1970). *J. Amer. Chem. Soc.* **92**, 737.
van Tamelen, E. E., Storni, A., Hessler, E. J. and Schwartz, M. (1963). *J. Amer. Chem. Soc.* **85**, 3295.
Vlahov, R., Holub, M. and Herout, V. (1967a). *Collect. Czech. Chem. Commun.* **32**, 822.
Vlahov, R., Holub, M., Ognjanov, I. and Herout, V. (1967b). *Collect. Czech. Chem. Commun.* **32**, 808.
Vrkoč, J., Jonáš, J., Herout, V. and Šorm, F. (1964). *Collect. Czech. Chem. Commun.* **29**, 539.
Vrkoč, J., Křepinský, J., Herout, V. and Šorm, F. (1963). *Tetrahedron Lett.* 225, 735.
Wadia, M. S., Khurana, R. G., Mhaskar, V. V. and Dev, S. (1969). *Tetrahedron* **25**, 3841.
Walls, F., Padilla, J., Joseph-Nathan, P., Giral, F., Escobar, M. and Romo, J. (1966). *Tetrahedron* **22**, 2387.
Weinheimer, A. J. and Washecheck, P. H. (1969). *Tetrahedron Lett.* 3315.

Weinheimer, A. J., Washecheck, P. H., van der Helm, D. and Hossain, M. B. (1968). *Chem. Commun.* 1070.
Westfelt, L. (1966). *Acta Chem. Scand.* **20**, 2829, 2841, 2852.
White, E. H., Eguchi, S. and Marx, J. N. (1969). *Tetrahedron* **25**, 2099.
White, J. D. and Gupta, D. N. (1968). *J. Amer. Chem. Soc.* **90**, 6171.
Winters, T. E., Geissman, T. A. and Safir, D. (1969). *J. Org. Chem.* **34**, 153.
Wolff, G. and Ourisson, G. (1969). *Tetrahedron* **25**, 4903.
Yamada, K., Takada, S., Nakamura, S. and Hirata, Y. (1968). *Tetrahedron* **24**, 199.
Yamamura, S. and Hirata, Y. (1963). *Tetrahedron* **19**, 1485.
Yates, P. and Field, G. F. (1970). *Tetrahedron* **26**, 3135.
Yates, P., Field, G. F. and Burke, P. M. (1970). *Tetrahedron* **26**, 3159.
Yoshida, T., Endo, K., Itô, S. and Nozoe, T. (1967). *Yakugaku Zasshi* **87**, 434.
Yoshihara, K., Ohta, Y., Sakai, T. and Hirose, Y. (1969). *Tetrahedron Letters* 2263.
Yoshioka, H. and Mabry, T. J. (1969). *Tetrahedron* **25**, 4767.
Yoshioka, H., Renold, W., Fischer, N. H., Higo, A. and Mabry, T. J. (1970a). *Phytochem.* **9**, 823.
Yoshioka, H., Renold, W. and Mabry, T. J. (1970b). *Chem. Commun.* 148.
Yosioka, I. and Kimura, T. (1965). *Chem. Pharm. Bull.* **13**, 1430; (1969). *idem, ibid,* **17**, 856.
Zabza, A., Romaňuk, M. and Herout, V. (1966). *Collect. Czech. Chem. Commun.* **31**, 3373.
Zurflüh, R., Wall, E. N., Siddall, J. B. and Edwards, J. (1968). *J. Amer. Chem. Soc.* **90**, 6224.

4

THE DI- AND SESTERTERPENES PART 1: THE DITERPENES

J. R. HANSON

Lecturer in Chemistry, University of Sussex

I. Introduction

The diterpenoids form a large group of C-20 substances derived from geranylgeraniol pyrophosphate. They are mainly of fungal or plant origin and include the resin acids and the gibberellin plant growth hormones. The chemistry of these substances has attracted organic chemists for over 150 years. Unfortunately, the examples originally selected for study came from pine wood resin. This is a mixture of closely related diterpene acids which is difficult to purify. Hence a considerable portion of the earlier literature is confused by authors unwittingly working with mixtures. However, over the last fifteen years the application of chromatographic techniques and of physical methods, particularly nuclear magnetic resonance spectroscopy, has revolutionized this field and now barely a month passes without the description of a new diterpene.

Despite their structural variety, it is possible to classify the diterpenoid substances on clear biogenetic principles (Ruzicka, 1953). In this chapter we intend to present this classification and then to outline the typical methods employed for structural determination. Following this we shall describe the evidence for the stereochemistry of various centres. Much of the chemistry of the diterpenoids is common to any alicyclic series and finds ready analogy in the steroid series. However, there are some features which are peculiar to the diterpenoids. For example, there has been effort directed at biogenetically patterned experiments. Unlike the triterpenes and steroids, oxygenation is much more frequently found on the pendant groups and, thus, there are more examples of ether and lactone formation reactions. Furthermore, the non-bonded interactions between these pendant groups provide the driving force for skeletal deformation and a number of interesting rearrangements. Hence, in discussing typical reactions of the series, we shall emphasize these aspects. Without the stimulus of pharmaceutical interest, synthetic activity in the diterpenoid field was slower than in the steroid series and the first total synthesis was not reported until 1954 although compounds of similar structure had been described in 1939.

II. The Classification of the Diterpenes

The clearest classification of terpenoid substances is based on biogenetic considerations. There are a few open chain diterpenes, including the parent geranylgeraniol and its isomer geranyl linalöol, whilst phytol forms a part of the chlorophyll molecule and the side chain of vitamins E and K_1. However, the majority of compounds are cyclic. In the lower terpenes the characteristic mode of cyclization is based on the terminal pyrophosphate leaving group generating a carbonium ion which then may attack the starter isopropylidene unit leading, for example, to the monoterpenes based on the limonene skeleton, and to the sesquiterpenes containing the germacrane skeleton. Compounds based on the further cyclization of these rings (e.g. the pinane and eudesmane series) are well-known. In the diterpene series there are only a few examples of this type of cyclization of geranylgeranyl pyrophosphate (I). The hydrocarbon, cembrene (III) from pine resins contains this fourteen membered ring whilst a group of diols such as the α- and β-duvatriene-1,3- and -1,5-diols have been found in tobacco. Several of the compounds in this group possess an ether bridge across the macrocyclic ring as in incensole (from frankincense) and in the lactone eunicin isolated from the marine organism *Eunicea mammosa*. Cyclization of the macrocyclic ring may lead (Erdtman *et al.*, 1967) to the carbon skeleton of verticillol (II) and to the taxicin series [e.g. taxisin-1 (IV)]. These are an interesting group of substances found in the yew tree (*Taxus baccata*) (Della Casa de Marcano and Halsall, 1969).

The major mode of cyclization involves the formation of perhydronaphthalene and perhydrophenanthrene derivatives (Ruzicka, 1953). Biogenesis is initiated by protonation of the double bond of the starter isopropylidene unit. A number of definite stages are found in this mode of

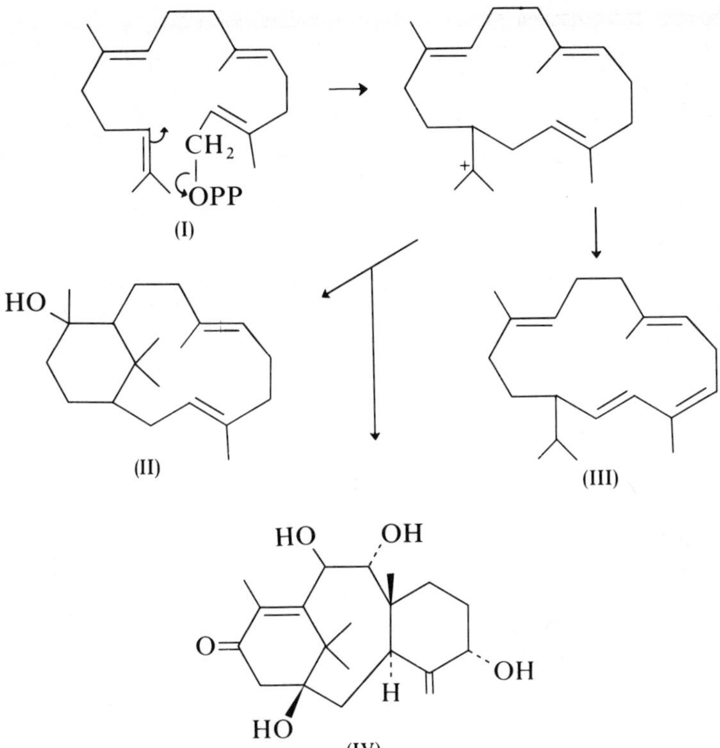

FIG. 1. The macrocyclic diterpenes and their cyclization products.

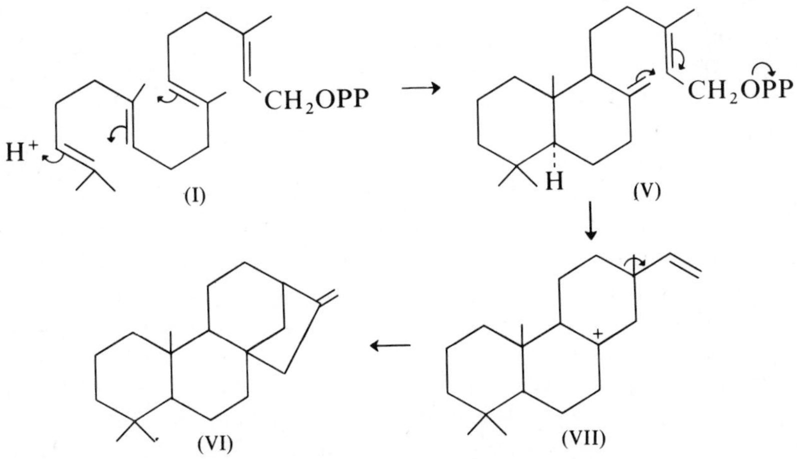

FIG. 2. Stages in the cyclization of geranylgeranyl pyrophosphate.

cyclization. Initially, geranylgeranyl pyrophosphate (I) cyclizes to form a bicyclic labdane intermediate (V). Immediately, one of the first characteristic features of the diterpenoid substances becomes apparent, the formation of normal and antipodal A/B ring junctions. The former belong to the same optical series as the steroids whilst the latter belong to the opposite series. The nature of this ring junction is a function of the relative orientation of the cyclizing double bonds and may be visualized as arising through different modes of coiling the open chain precursor on an enzyme surface. Examples of both series are quite widespread, and there is even a report of both the ring fusions occurring alongside each other in the same plant species. Subsequent modification of the labdadienol may lead, on the one hand, to compounds related to manöol (VIII) (a geraniol linalöol isomerism), and on the other, to labdanolic acid (IX) and its relatives. A fairly common modification of the side chain involves the formation of a furan ring as in the diterpenoid bitter principle marrubiin (X). Indeed, in some plant species each of the oxidative stages leading to the furan ring have been found to co-occur. Skeletal variations that are sometimes found involve the migration of a methyl group from C-10 to C-9, possibly initiated by protonation of the C-8 double bond and a hydride shift from C-9 to C-8. The methyl migration may form part of a "back-bone" or "friedo" rearrangement in which there is a hydride shift from C-5 to C-10 and a further methyl group migration from C-4 to C-5, completed by discharge of the carbonium ion on ring A [e.g. clerodin (XI)]. Characteristically, all the migrating groups possess a *trans* diaxial relationship to each other. Oxygenation patterns that are frequently found involve C-15 as an alcohol, acid, and often part of a furan ring, C-13 as an alcohol and sometimes part of an ether with C-8 and C-18, or C-19 as part of an alcohol, acid (or C-19 as a lactone function). However, examples are known of oxygenation at nearly every other centre. Since the cyclizing agent is the biological equivalent of H^+ attacking an olefin rather than an epoxide, such as squalene epoxide, the 3-hydroxyl group characteristic of the triterpenes and steroids is not such a common feature of the diterpenes.

The next cyclization involves the pyrophosphate acting as a leaving group, initiating cyclization to form the tricyclic pimaradienes. The existence of epimers at the vinyl:methyl substitution, already apparent at the bicyclic level (cf. manöol and 13-epimanöol) is carried through into this group (cf. pimaric acid and sandaracopimaric acid) and would appear to determine the orientation of ring D in the tetracyclic series. However, as yet a correlation does not appear to exist between the co-occurrence of di-, tri- and tetra-cyclic diterpenes with a consistent C-13 configuration. The tricyclic pimaradiene (XII) may readily act as a progenitor of the abietadiene (XIII) and cassaine (XIV) diterpenes. An alternative mode of cyclization, again involving a "friedo" or "backbone" type of rearrangement, may lead to the rosane

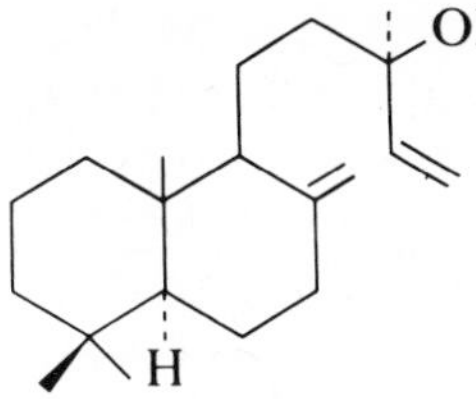

(VIII)

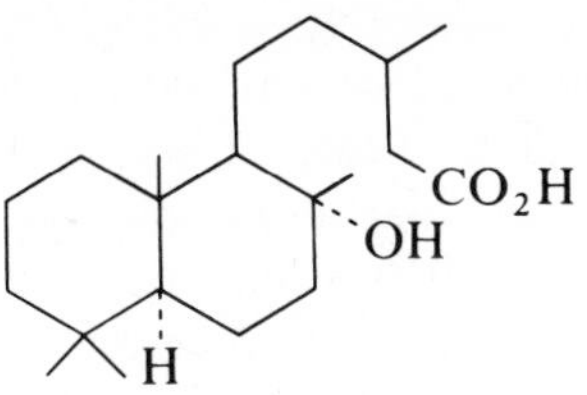

(IX)

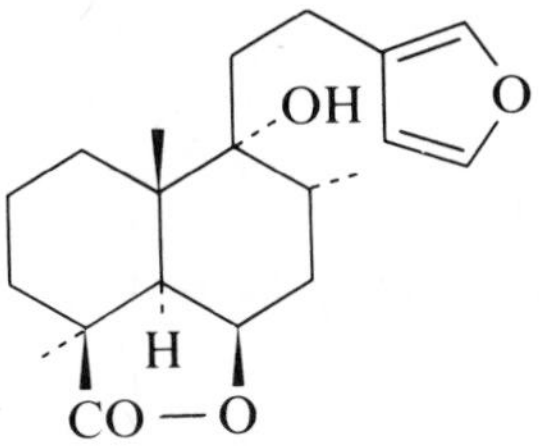

(X)

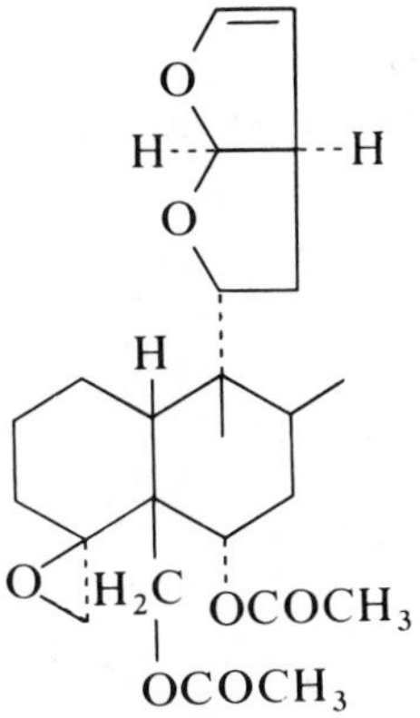

(XI)

FIG. 3. Typical bicyclic diterpenoids.

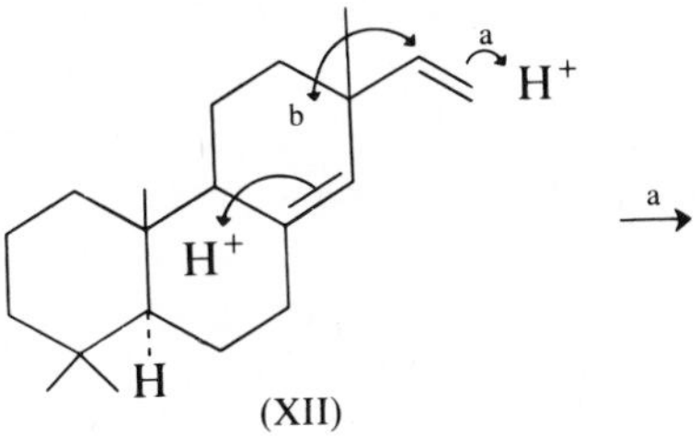

(XII)

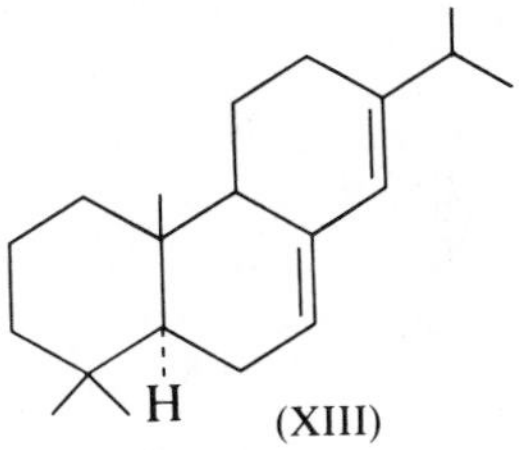

(XIII)

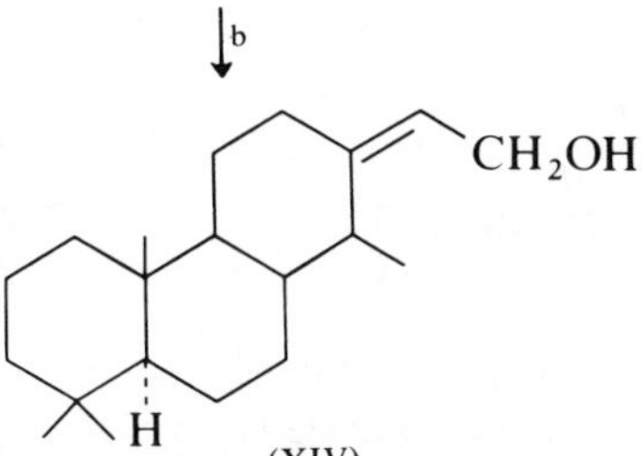

(XIV)

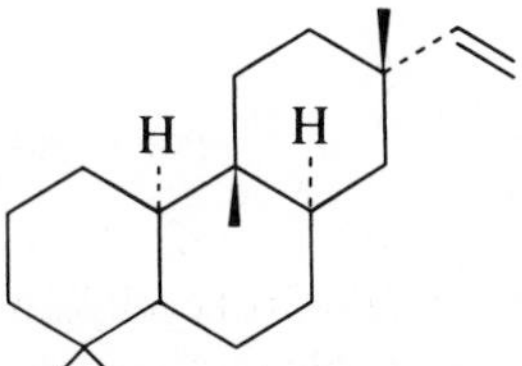

(XV)

FIG. 4. The formation of tricyclic diterpenoids.

diterpenes and to compounds such as rimuene (XV). Skeletal variations found in the tricyclic series not only involve this backbone rearrangement but also the aromatization of ring C. Oxygenation frequently occurs at C-12, C-18 and C-19 although again examples are known at many other centres.

The tetracyclic diterpenes are a closely related series which may be fitted into this biogenetic scheme. A decade and a half ago Wenkert (1955) suggested that the tetracyclic diterpenes might arise through cyclization of suitably oriented pimaradienes (XVI). This theory, which has subsequently been elaborated, involves a non-classical carbonium ion (XVII) which could then collapse in a variety of ways leading to compounds of the kaurene (XVIII) (and phyllocladene) atisine (XIX) and beyerene (XX) skeleta. The isolation

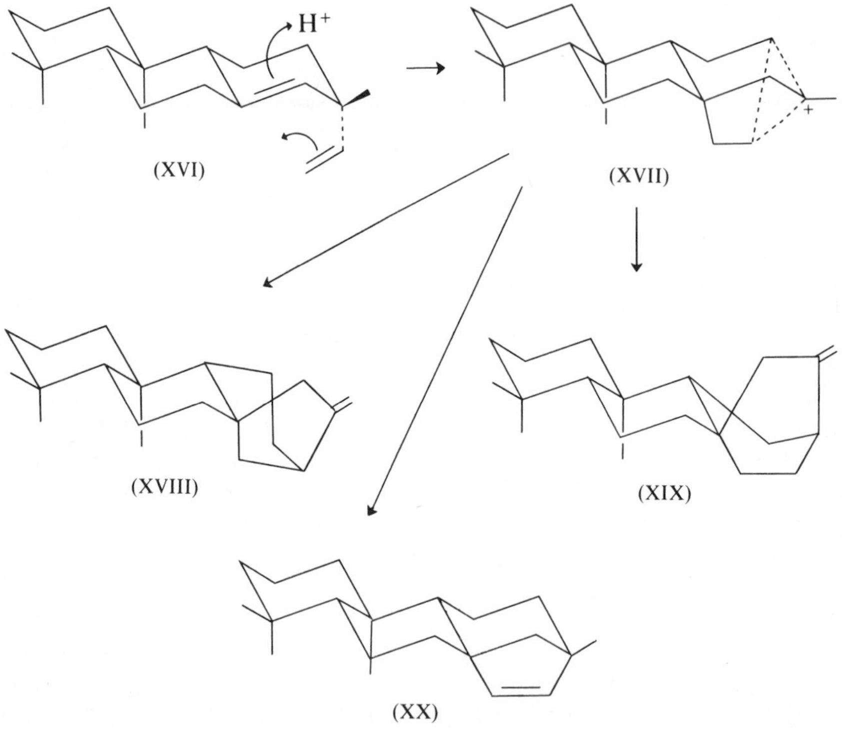

FIG. 5. The cyclization of (−)-pimaradiene.

(Hugel *et al.*, 1965) of cyclokaurane diterpenes from *Trachylobium verrucosum* is of considerable interest since they possess a carbon skeleton based on this intermediate carbonium ion. The gibberellins belong to the tetracyclic series and their carbon skeleton could arise by migration of the 7–8 bond

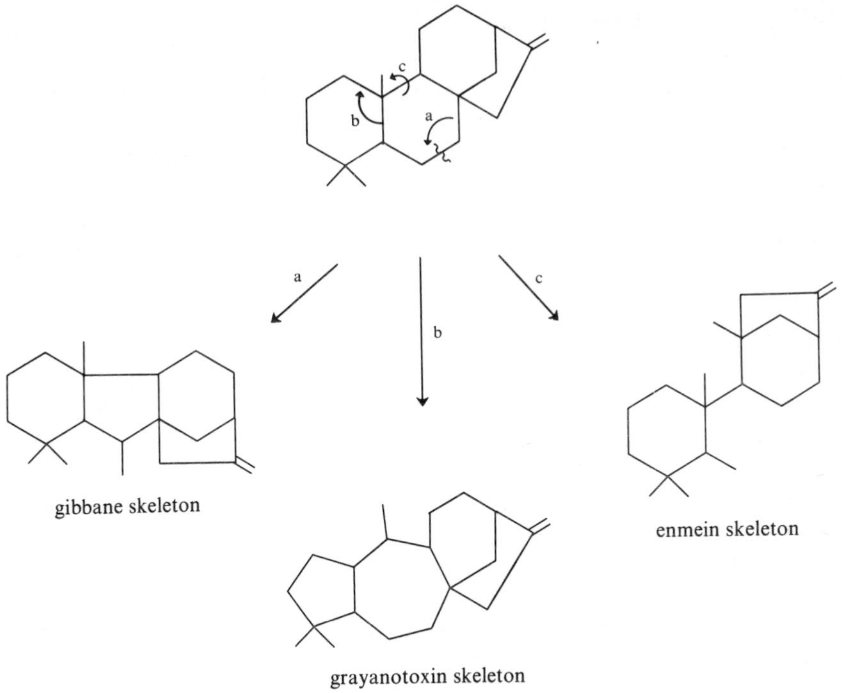

FIG. 6. The structural relationship of some tetracyclic diterpenes.

to C-6 and the extrusion of C-7 (pathway a) (Birch *et al.*, 1955). On the other hand, migration of the 5–10 bond to C-1 (pathway b) may lead to the grayanotoxins, which are a group of plant poisons. The furanoid skeleton of cafestol might be formed by a simple Wagner–Meerwein rearrangement at C-18. Cleavage of ring B of (−)-kaurene between C-6 and C-7 and rotation about the 9–10 bond leads to the enmein series (pathway c). Oxygenation is most commonly found at C-3, C-19, C-7 and on ring D. There are a group of alkaloids which possess an additional oxazolidine ring across ring A. Some retain the C-20 carbon skeleton of kaurene but a more complex and highly oxygenated group form the aconite alkaloids occurring in Delphinium and Aconitum species. In these the C-17 carbon atom is lost and the C-9 → C-8 bond has migrated to C-9–C-14 in the atisine skeleton thus generating a seven-membered ring B.

The classification which has been outlined above is based on a biogenetic analysis of the structures which are found in the diterpene series.

Experimental biosynthetic evidence to confirm these relationships is slowly accumulating and is described in another chapter.

III. Structure Determination

There are four phases to the determination of the structure of a natural product. These involve the initial characterization and definition of the functional groups. This is followed by the determination of the underlying carbon skeleton. The third phase involves placing the functional groups on this carbon skeleton and the final phase involves the determination of both the relative and absolute stereochemistry of the complete molecule. Each stage, and the methods used in terpene chemistry, can be readily seen in a study of diterpenoid structural determination. In this section we shall give a general description of the methods used.

Present day methods of characterization are much more rigorous than those of earlier times and involve a mixture of physical and chemical techniques and thus the identity of compounds is less often a subject of dispute. In Fig. 7 we list typical functional groups found in this series and the physical methods used to characterize them. From this information and evidence such as microhydrogenation and per-acid titration, the number of double bonds may be found, and hence, from the analysis, the number of rings present. At this stage, it is often clear that the compound may be a bicyclic labdane or a tetracyclic, possibly kauranoid, diterpene and structure determination may then rest on a judicious set of interrelationships. These may involve selective removal of the oxygen functions to define the parent hydrocarbon or alternatively, a partial synthesis of a common degradation product. The more drastic procedure of dehydrogenation common in earlier studies is less often undertaken these days despite advances in vapour-phase chromatography capable of separating aromatic hydrocarbons. An example of the first type of skeletal interrelationship is shown in the accompanying degradation and interrelationship of 7-hydroxy-(XXII) and 7,18-dihydroxy-kaurenolide (XXI) with (−)-kaurane (XXIII) (Cross *et al.*, 1963). This work illustrates methods of removal of hydroxyl, carbonyl and acid functions under conditions which avoid the epimerization of ring-junctions. An example (Gopinath *et al.*, 1961) of the second type of degradation to a common intermediate is found in the ozonolysis and hydrogen peroxide oxidation of polyalthic acid (XXIV) to (XXV), a compound which was also prepared in the opposite enantiomeric series from neoabietic acid (XXVI).

The spectral characteristics of a diterpene can define the number of terminal pendant groups, such as $C{-}CH_3$, primary alcohol, carboxylic acid and $C{=}CH_2$. Assuming that the diterpene obeys the isoprene rule and possesses a common skeleton, this information can define the series to which a compound belongs. Thus, the bicyclic labdane diterpenes possess six pendant groups, the tricyclic pimarane–abietane series five, the tetracyclic

Group	I.R.	U.V.	N.M.R.
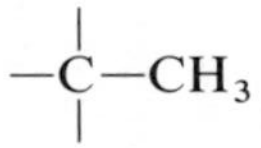	—	—	Singlet $\tau \sim 9{\cdot}0$
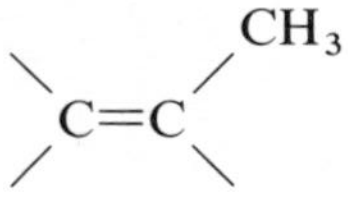	—	—	$\tau \sim 8{\cdot}2$–$8{\cdot}4$
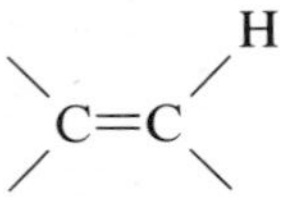	1650 cm^{-1} below 1000 cm^{-1} Characterizes type	If conj. then apply Woodward's rules to λ max.	$\tau \sim 3$–5 Integral and multiplicity characterizes type
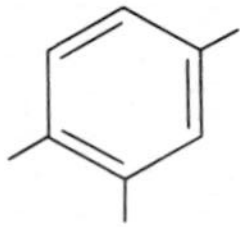	1600 cm^{-1}	260–280 nm	$\tau \sim 1{\cdot}5$–3 Integral and multiplicity characterizes type
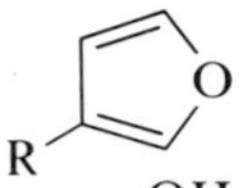	—	220 nm	$\tau \sim 2{\cdot}6$–$3{\cdot}3$
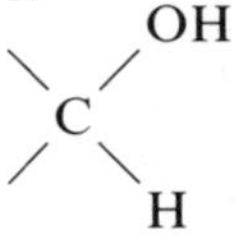	~3500 cm^{-1}	— (unless phenol)	$\tau \sim 5{\cdot}5$–$6{\cdot}5$ Integral distinguishes p and s
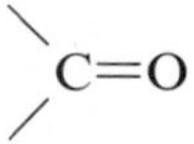	1680–1800 cm^{-1} Characterizes ring size and type (e.g. aldehyde, cyclopentanone to substn. or lactone)	If conj. then apply Woodward's rules to λ max.	Can use for $—CH_2{\cdot}CO$ $[CH_3CO{\cdot}O \sim 7{\cdot}9]$
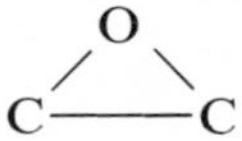	Poor	—	$\tau \sim 6{\cdot}9$–$7{\cdot}4$

FIG. 7. Spectral characterization of diterpenoid substances.

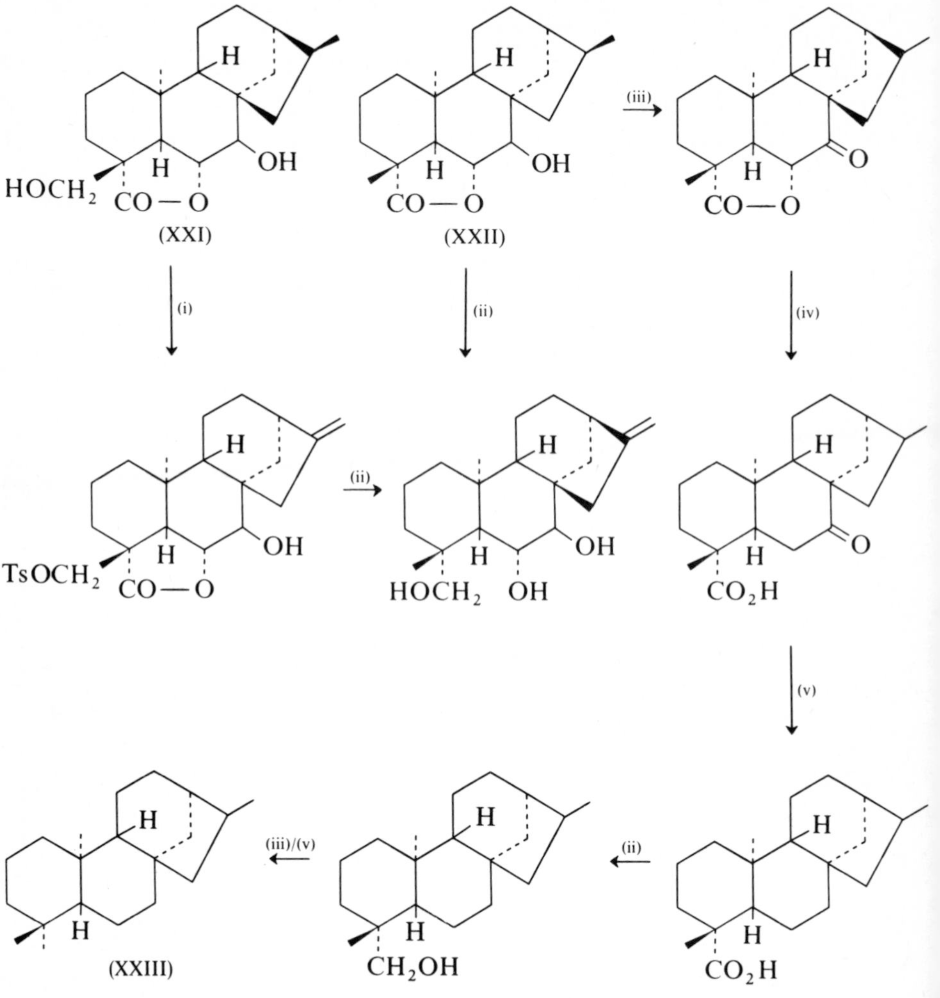

(i) TsCl/pyr.; (ii) $LiAlH_4$; (iii) CrO_3/H_2SO_4; (iv) Zn/AcOH; (v) N_2H_4/KOH.

FIG. 8. The degradation of the kaurenolides.

kauranoid diterpenes four, whilst the C-20 gibberellins possess five and the C_{19} gibberellins four.

Selenium dehydrogenation of the bicyclic labdane diterpenes gives mainly 1,2,5-trimethylnaphthalene and 1,7-dimethylphenanthrene, the latter arising through cyclization. 1,1,4,7-Tetramethylperinaphthene has also been isolated from a group of these diterpenes. An analysis of this method is

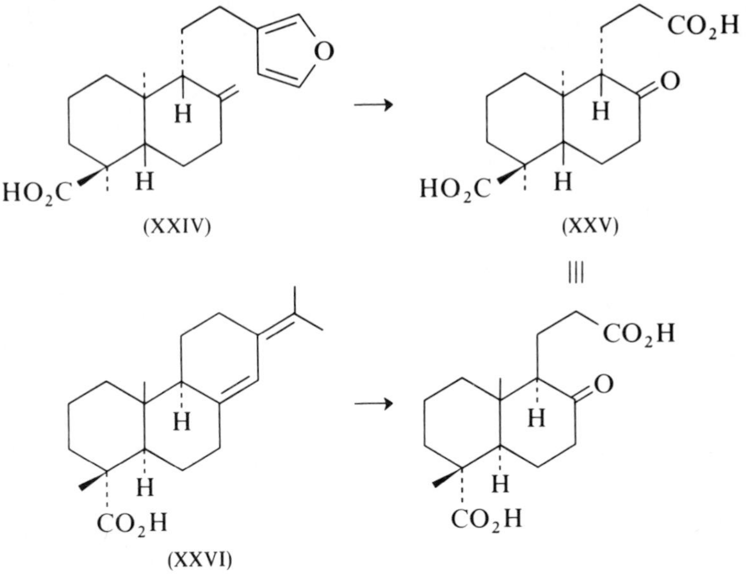

FIG. 9. The degradation of polyalthic acid.

found in the degradation of biformene (Carmen *et al.*, 1961). In the tricyclic series retene (1-methyl-7-isopropylphenanthrene) and 1,7-dimethylphenanthrene are typical dehydrogenation products. The tetracyclic series although they can give these phenanthrenes, do so far less readily as ring D acts as a

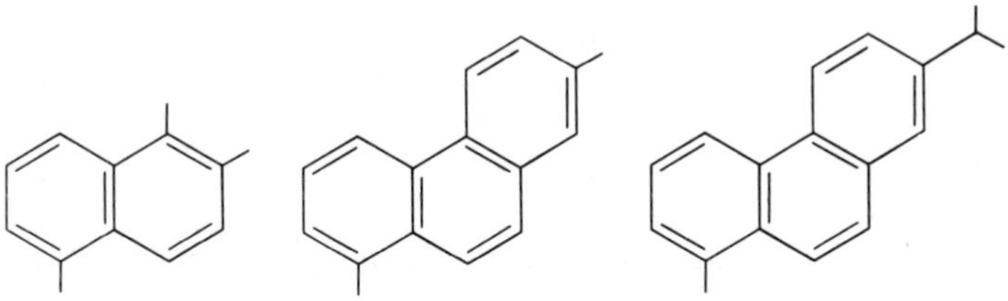

FIG. 10. Diterpenoid dehydrogenation products.

meta bridge blocking aromatization. However a very careful analysis of the dehydrogenation method was used in the determination of the structure of the diterpene alkaloid anhydroignavinol (XXVII). Three of the alkyl phenanthrenes that were obtained are drawn to show their origin from the parent alkaloid.

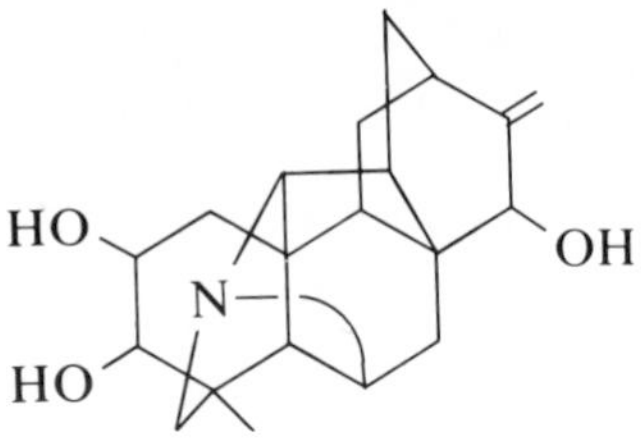

(XXVII)

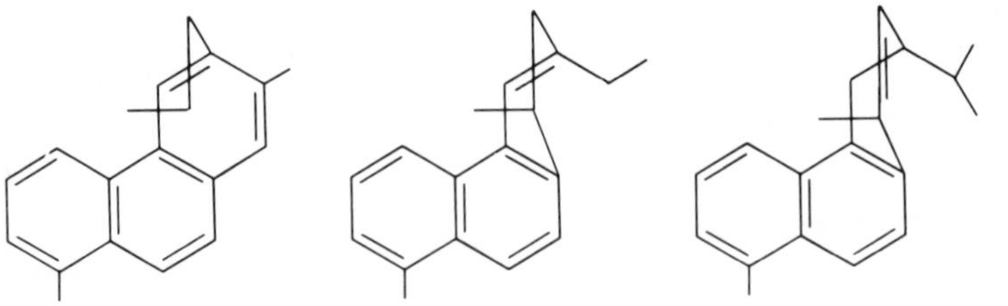

1,7-dimethyl-6n-propyl- 1,8-dimethyl-3-ethyl- 1,8-dimethyl-3-isopropyl-
-phenanthrenes

FIG. 11. Some dehydrogenation products of anhydroignarinol.

In a number of cases dehydrogenation has given evidence for the location of other functional groups on the carbon skeleton. For example carbonyl groups appear in the final product as phenols. Dehydrogenation of an alcohol may generate a carbonium ion in such a manner as to produce a methyl migration and this can lead to structural information. An example of this can be seen in degradation of columbin (XXVIII) (Barton *et al.*, 1956). On heating, columbin readily decarboxylated to give decarboxycolumbin, from which decarboxyoctahydrocolumbinic acid, and hence, its corresponding alcohol may be obtained by hydrogenolysis and hydrogenation. Dehydrogenation of this alcohol gave 1,5-dimethyl-2-naphthoic acid. However, Wolff–Kishner reduction of the ketone and subsequent dehydrogenation gave only 1-methol-2-naphthoic acid. Hence a tertiary methyl group was placed adjacent to the carbonyl group to account for the Wagner–Meerwein rearrangement during dehydrogenation.

Another example of this is found in the chemistry of abietic acid. Dehydrogenation of abietic acid afforded 1-methyl-7-isopropylphenanthrene, whilst dehydrogenation of the corresponding alcohol (abietinol) gave 1-ethyl-7-isopropylphenanthrene. The use of Grignard reagents to insert extra methyl groups at sites of oxygenation has received some attention. The structural work on cassaic acid, the diterpenoid moiety of the *Erythrophleum* alkaloids, provides several examples of this (Humber and Taylor, 1955). Cassaic acid (XXIX) was converted to its diketone and the latter subjected

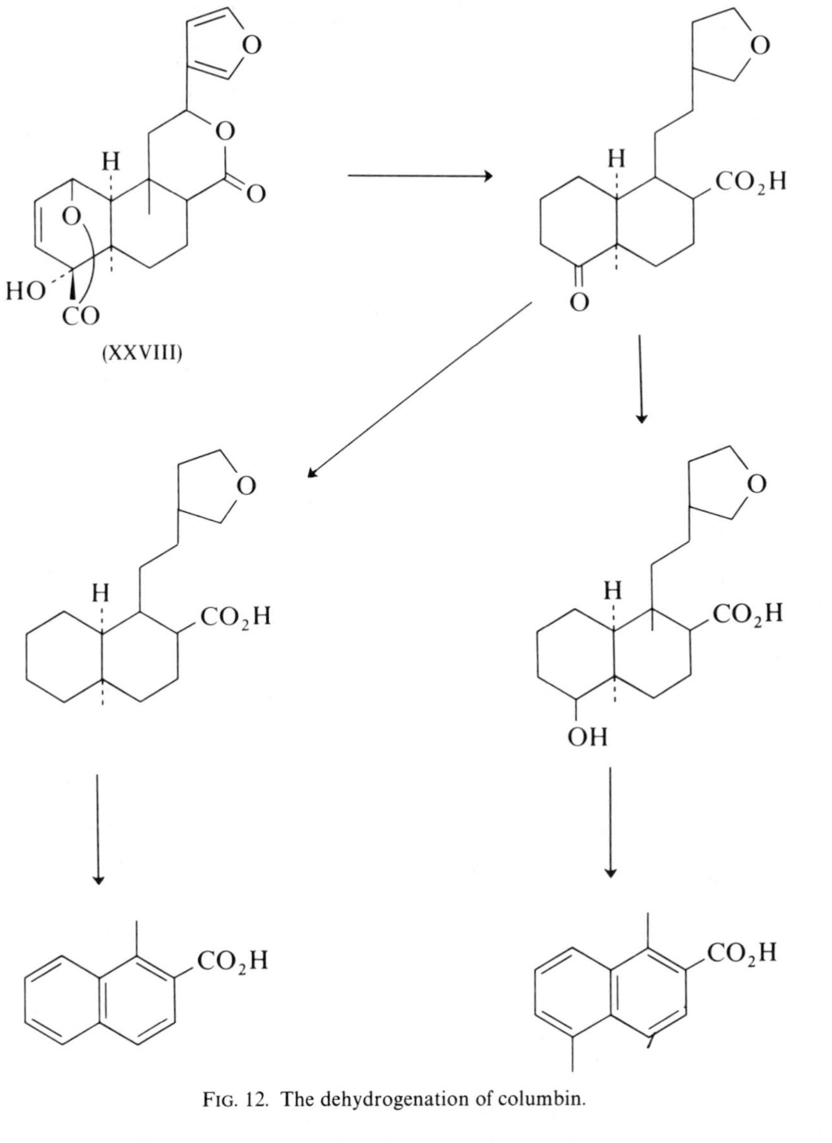

FIG. 12. The dehydrogenation of columbin.

to a Wolff–Kishner reduction to afford cassanic acid (XXX). This on selenium dehydrogenation gave 1,2,8-trimethylphenanthrene. The position of the carboxyl group in cassanic acid was established by converting the ester to isopropyl group with methyl magnesium iodide and then dehydrogenating this to a hydrocarbon shown by synthesis to be 1,8-dimethyl-7-isobutyl-phenanthrene. The position of the ring B oxygen function was determined

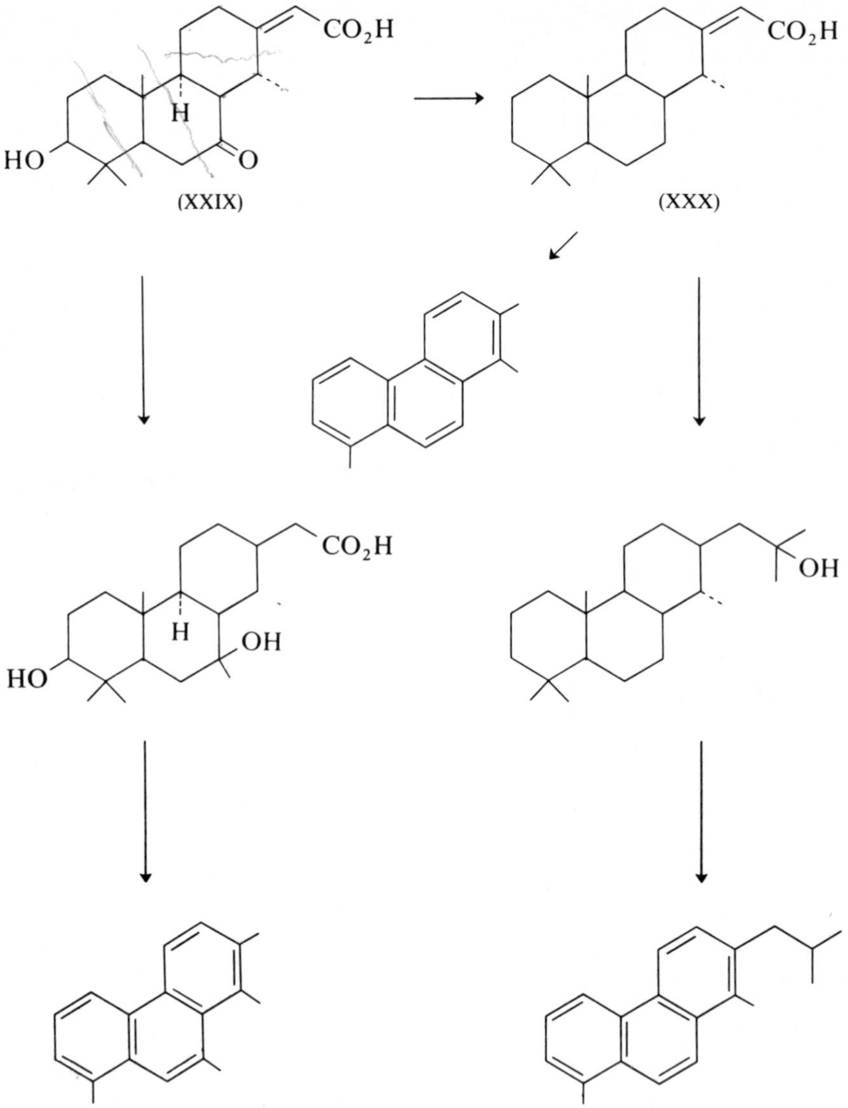

FIG. 13. The degradation of cassaic acid.

by labelling the ketone position with methyl magnesium iodide and then dehydrogenating the product to the corresponding methylphenanthrene. A method of locating the position of double bonds uses the reaction of the corresponding epoxide with methyl magnesium iodide. Although this was successful in the dihydropimaric acid series (Harris and Sanderson, 1948), it gave incorrect information in the case of the diterpene hydrocarbon rimuene.

Oxidative degradation has played an important part in determining the underlying carbon skeleton of diterpenoid substances. As with dehydrogenation it can lack specificity and is often wasteful of material. Ozonolysis has formed a valuable method. Thus, ozonolysis of the terminal methylene characteristic of the kaurene–phyllocladene series has given formaldehyde on the one hand and a norketone shown (in the majority of cases by its infra-red spectrum) to be a cyclopentanone. Ozonolysis or potassium permanganate oxidation of the manöol side chain gave a diketone which readily underwent an internal aldol condensation to form a tricyclic unsaturated ketone which has played an important part in structural and synthetic work in this series (see p. 186). On the other hand, the stepwise opening of rings and the application of Blanc's rule with ketone or anhydride formation, a sequence which formed an important part of steroid degradations, has played a less significant role in the diterpene series.

In determining the position of functional groups spectroscopic methods have played an increasingly valuable role. For example, infrared absorption may reveal the presence of a hydroxyl group whilst the number, position and multiplicity of the C$\underline{H}$·OH protons in the n.m.r. spectrum can give information on whether it is a primary, secondary or tertiary alcohol. Oxidation, giving either acidic or ketonic material, may then confirm this whilst the carbonyl absorption in the infra-red can characterize the ring-size of a ketone. The number of methylene groups adjacent to the ketone may be obtained in several ways. For example, deuteration studies were used to locate the carbonyl group of ketomanoyl oxide at the unusual 2-position, mass spectroscopy being used to define the number of protons exchanged with deuterium. Alternatively oxidation with selenium dioxide can afford a diketone as in the degradation of gibberic acid to gibberdionic acid (Cross, 1954).

In defining the position and stereochemistry of hydroxyl groups, and to a lesser extent carbonyl functions, differential shifts of various proton resonances, usually $C{-}CH_3$, between aromatic and non-aromatic solvents have been employed. If an aromatic solvent molecule such as benzene or pyridine selectively solvates a hydroxyl group it may deshield an adjacent group. This has been used in distinguishing between substituents on the axial or equatorial groups at C-4. In the axial case (C-19) a substituent may have a definite effect on the C-20 methyl resonance. This forms an alternative to pK_a measurements. The position of hydroxymethyl and acetoxymethyl proton resonances have also been used to define the stereochemistry at this centre (Gaudemer *et al.*, 1964).

An interesting example of the combined use of spectroscopic and chemical techniques came in the degradation of cafestol (XXXI) (Djerassi *et al.*, 1959).

Hydrogenolysis of the furan system leads to the norcafestanone (XXXII). Location of the carbonyl group on ring A was established by stepwise bromination and dehydrobromination, introducing at each step a double bond in conjugation with the ketone and thus forming a chromophore in the ultra-violet spectrum.

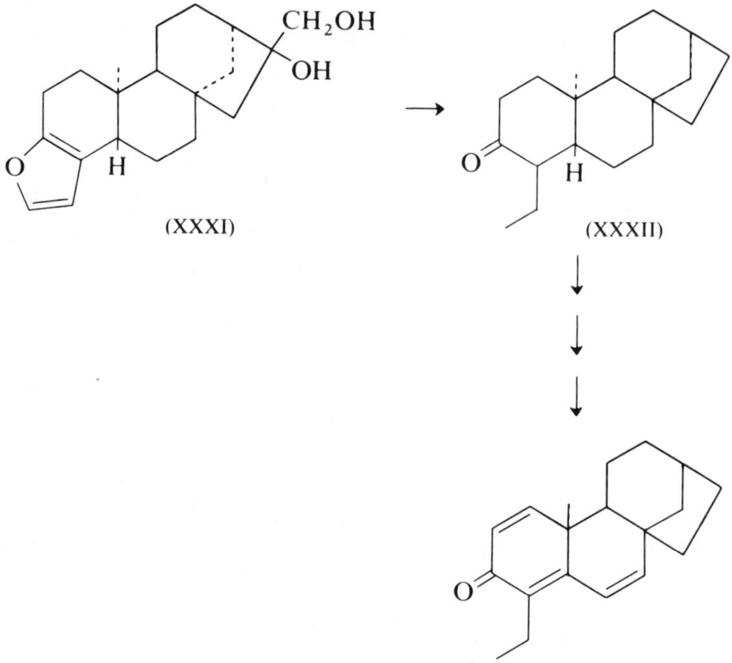

FIG. 14. The bromination and dehydrobromination of norcafestanone.

IV. The Stereochemistry of the Diterpenes

Physical and chemical methods have played an important part in determining the stereochemistry of each class of the diterpenoids. In the bicyclic series the key compounds have been manöol and agathene dicarboxylic acid. The stereochemistry of manöol (XXXIII) rests upon a number of pieces of evidence. The normal *trans* A/B fusion was demonstrated by interrelationship with dehydroabietic acid (XXXVI) (see later). The stereochemistry at C-9 also follows the normal pattern with an α-hydrogen atom. There are four main pieces of evidence for this. Firstly manöol was degraded to a lactone (XXXV) related to ambreinolide. This is an oxidation product of the

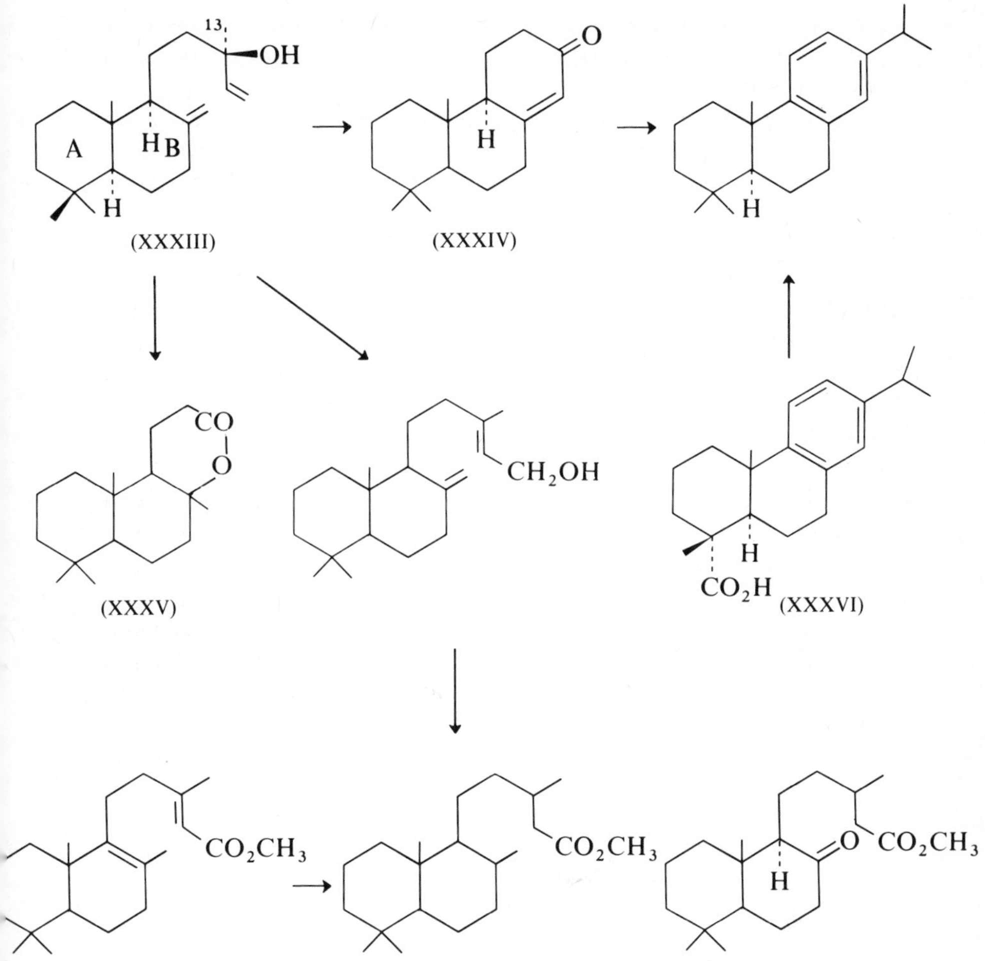

(XXXIII) (XXXIV) (XXXV) (XXXVI) (XXXVII) (XXXVIII) (XXXIX)

FIG. 15. The stereochemistry of manöol.

triterpene ambrein of known stereochemistry. Secondly, the tricyclic $\alpha\beta$-unsaturated ketone (XXIV) obtained by oxidation of manöol (Hosking, 1936), had an optical rotatory dispersion comparable to that of cholest-4-en-3-one (Grant and Hodges, 1960). This supplemented an earlier extensive series of molecular rotation arguments (Klyne, 1953). Thirdly, catalytic hydrogenation (Halsall and Cocker, 1956) of methyl $\Delta^{8,9}$-labdanolate (XXXVII) gave methyl dihydrocativate (XXXVIII) which had in turn been related to manöol. Here it was assumed that reduction took place from the less-hindered α-face of the molecule to give an equatorial side-chain. Finally, ozonolysis (Halsall and Cocker, 1957) of methyl labdanolate under mild conditions gave a norketone (XXXIX) which was stable to alkali, a proof which was strengthened by the subsequent isolation of the less-stable isomer under similar conditions during a synthetic sequence. The configuration at C-13 of the bicyclic diterpenes has been the subject of some dispute (Barltrop *et al.*, 1960). Molecular rotation arguments related manöol at this centre to R-(−)-linalöol. In the case of sclareol (XL) which has been related to manöol, hydrogen bonding invalidated a number of arguments for the stereochemistry of this centre. However, a direct correlation has been established by the isolation from this fragment of the molecule of the γ-lactone of γ-hydroxy-γ-methylhexanoic acid (XLI) of known absolute configuration (Soucek and Vlad, 1964). Thus, sclareol possessed the "R" configuration at this centre. In sclareol there is an additional asymmetric centre at C-8 which was defined as follows. Firstly, oxidation of sclareol gave an enol-ether (XLII). Further oxidation of this led to the formulation of a *trans* fused-γ-lactone (XLIII) which then underwent acid-catalysed isomerization to a more stable (*cis*)-γ-lactone (Ruzicka *et al.*, 1942; Schenk *et al.*, 1952). Secondly, manöol underwent epoxidation predominantly from the "α" face of the molecule to give an epoxide which was reduced with lithium aluminium hydride to sclareol. Thirdly, pyrolysis of sclareol diacetate regenerated the exocyclic methylene of manöol, suggesting that the C-8 acetoxyl group possessed an equatorial conformation (Ohloff, 1958).

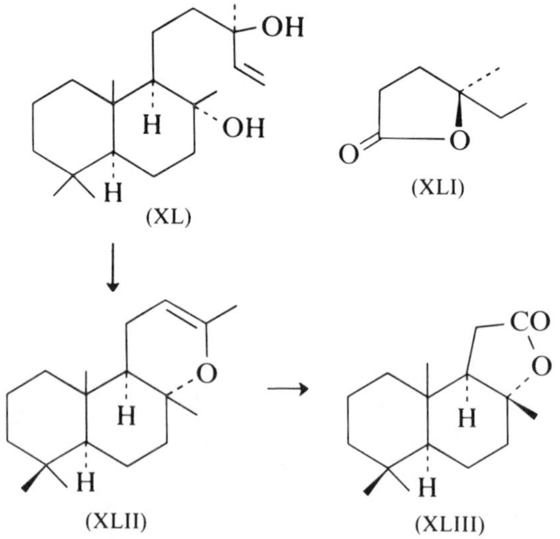

FIG. 16. Some aspects of the stereochemistry of sclareol.

In the case of the C-13 epimeric ethers manoyl oxide and 13-epimanoyl oxide, the stereochemistry at C-13 was assigned on three pieces of evidence. Each rests on the assumption that ring C remains in a chair form in which the vinyl and methyl substituents take up axial or equatorial conformations. Thus, in a study of mass spectral intensities (Hodges and Reed, 1960), it was assumed that an axial group would be expelled more readily than an equatorial. Secondly (Wenkert *et al.*, 1961), in the different epimers the methyl and vinyl proton resonances appear to be deshielded to different extents by the oxygen ring. Thirdly, on more secure grounds, the vinyl group was cleaved to afford the corresponding carboxylic acid. The relative rates of hydrolysis of the methyl esters were then studied in order to distinguish the more hindered axial epimer. In each case an axial (β) vinyl group was assigned to 13-epimanoyl oxide (Giles *et al.*, 1962).

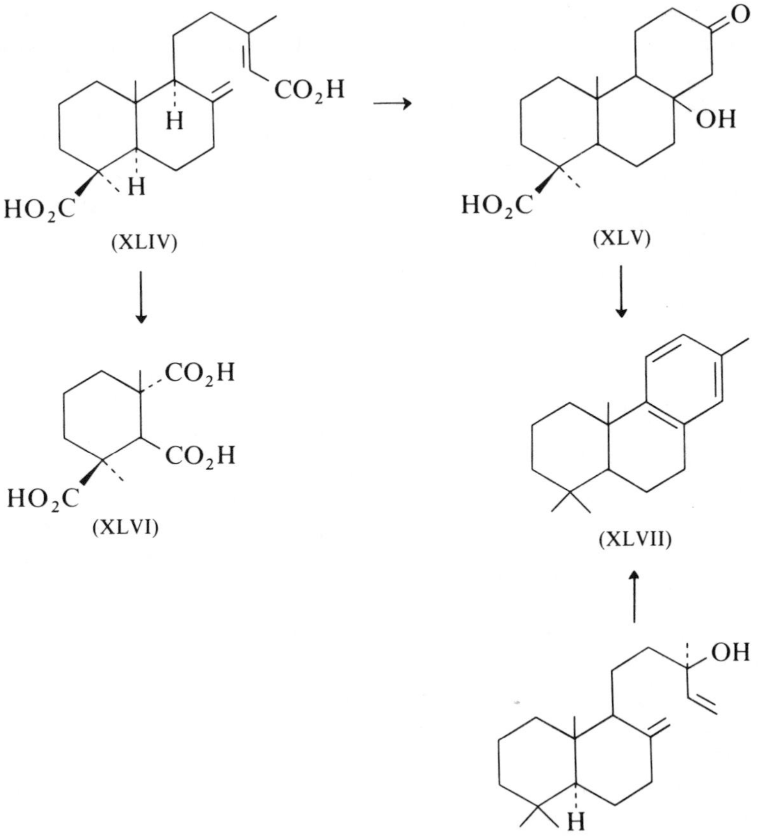

FIG. 17. Some aspects of the stereochemistry of agathic acid.

The bicyclic diterpenes of the labdanolic acid subgroup fall into two classes, those possessing a $\Delta^{13,14}$ olefin and those in which C-13 is a saturated centre. In the latter series an additional centre of asymmetry is created about which stereoisomers are known. To simplify the discussion we will treat the unsaturated compounds first. Agathic acid (XLIV) is the best known of these. The stereochemistry of the A/B fusion of agathic acid was established (Ruzicka *et al.*, 1948) by correlation with manöol. A β-hydroxyketone (XLV) was obtained by ozonolysis of agathic acid. Ring C was converted to an aromatic ring and the ester grouping reduced to a methyl to give a hydrocarbon (XLVII) which was also produced from manöol. The stereochemistry of the carboxyl group at C-4 was established (Ruzicka and Bernold, 1941) by prolonged oxidation of agathic acid with alkaline potassium permanganate. An optically active acid was isolated possessing the structure (XLVI) and derived from ring A. Hence the carboxyl group at C-4 was an axial substituent. Molecular rotation arguments led to the demonstration that the side chain was β-oriented at C-9, a result supported by studies with the tricyclic iso-

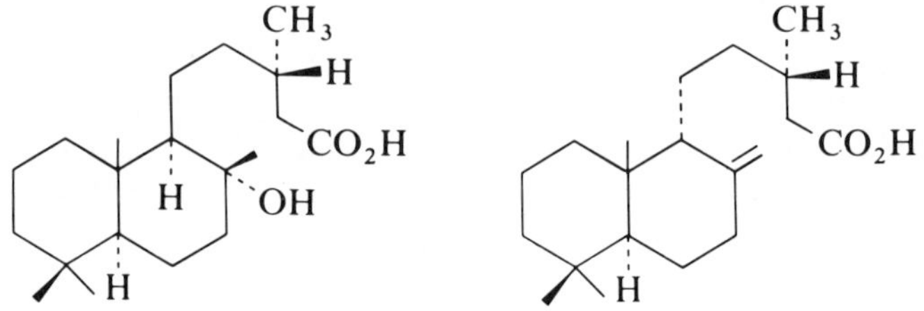

(XLVIII) (XLIX)

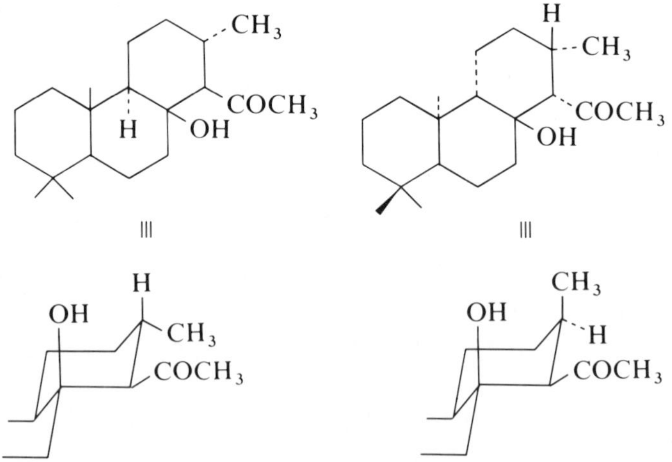

FIG. 18. The stereochemistry at C-13 of labdanolic and eperuic acids.

agathic acid. A study of the position of the side chain methyl resonances in the n.m.r. spectrum indicated that the methyl carboxyl groups were *cis* to one another.

Labdanolic acid possesses a normal A/B fusion since it may be derived from sclareol. Eperuic acid is related to labdanolic acid in the sense that it is antipodal to it, except at C-13 at which it possesses the same absolute stereochemistry. This was established by converting the acids to tricyclic compounds. Labdanolic acid reacted with methyl lithium to form a methyl ketone. Dehydration and cleavage of the C-8 olefin gave a C-8 ketone. The dione from labdanolic acid readily cyclized under basic conditions whilst that from eperuic acid did not cyclize. Under acidic conditions both cyclized giving an unsaturated ketone. The difference in ease of cyclization can be rationalized if the compound derived from eperuic acid possessed an axial methyl at C-13 introducing strong diaxial interactions in the cyclic product. Thus labdanolic acid was assigned the stereochemistry (XLVIII) and eperuic acid (XLIX) (Overton and Renfrew, 1967).

The stereochemistry of the diterpenoid bitter principle marrubiin (L) has been the subject of some controversy (Appleton *et al.*, 1967; Wheeler *et al.*, 1967). A relationship with a degradation product (LI) of ambrein enabled the A/B ring junction to be defined. However the presence of an oxygen function at C-6 vitiated the stereochemical interpretation of physical data such as pK_a and n.m.r. measurements on the functional groups at C-4. Once

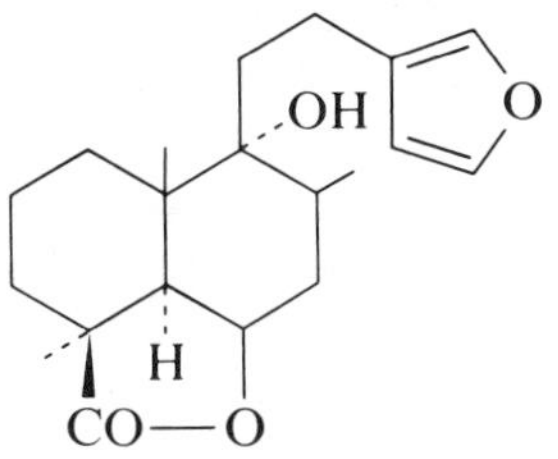

(L)

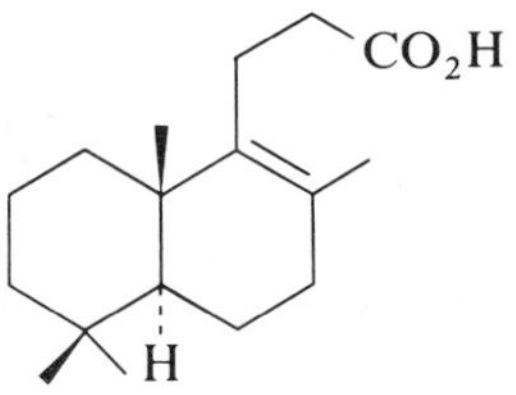

(LI)

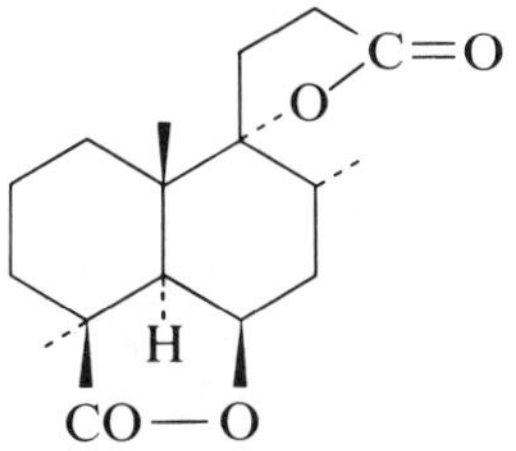

(LII)

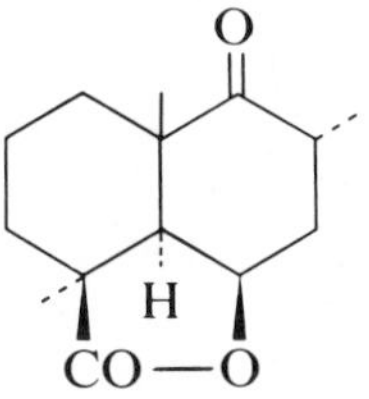

(LIII)

FIG. 19. Some degradation products of marrubiin.

the C-6 functional group was removed n.m.r. data were clearly interpreted in terms of an axial function. This type of interaction has been the cause of a number of misinterpretations of data. At C-6, n.m.r. data clearly indicated an equatorial proton and hence the lactone ring was *cis* and β-oriented. The reduction of a carbonyl at position 6 is influenced by the substituent at C-4. In the presence of a 4-carboxyl, reduction with sodium borohydride and with lithium in liquid ammonia furnished the axial alcohol. However, in the presence of an ester whilst lithium aluminium hydride reduction gave an axial alcohol, lithium in liquid ammonia gave an equatorial product. Ready ether and lactone formation between C-19 and C-6 is in accord with a 6β-axial substituent in the natural series. Solvent shift data in the n.m.r. spectra led to the conclusion that the methyl group at C-8 was equatorial and α, whilst the formation of a $\Delta^{8(9)}$ endocyclic olefin on dehydration of the C-9 alcohol with phosphoryl chloride in pyridine was an indication of the *trans* diaxial relationship between the C-9 hydroxyl group and the C-8 proton. In the absence of pyridine, the $\Delta^{9(11)}$ olefin is formed possibly by internal attack of phosphoryl chloride. Alternative evidence for the C-9 stereochemistry of marrubiin came from the partial synthesis of the degradation product (LII) and its C-9 epimer from (LIII) under stereochemically defined conditions.

The tricyclic diterpenes fall into four main classes: those possessing the pimarane, the abietane, the rosane, and the cassaine skeleton. These show many stereochemical features in common with the bicyclic diterpenes. The existence of C-13 epimers, already noted in the bicyclic series, is found amongst the tricyclic diterpenes. Three closely related pimaradiene acids are well-known. These are pimaric acid (LIV) (Harris and Sanderson, 1948), sandaracopimaric acid (LV) and isopimaric acid (LVI). Pimaric acid and isopimaric acid were shown (Edwards and Howe, 1959) to have identical stereochemistry at C-4, C-5 and C-10 by degradation to the acid (LVII). Furthermore, cleavage of the vinyl double bond of pimaric and sandaracopimaric acids gave a nor-aldehyde from which the same nor-acid could be obtained by Wolff–Kishner reduction. They, therefore, differ only in stereochemistry at C-13. On partial hydrogenation of pimaric and isopimaric acids it was possible to produce the dihydro acids in which the vinyl group had been reduced. By isomerizing the nuclear double bond to the $\Delta^{8(9)}$ position it was possible to eliminate the asymmetry at C-9. Since two different compounds were produced these acids must differ at C-13. With concentrated sulphuric acid these dihydro-acids undergo a characteristic lactonization giving lactones of the type (LVIII). Under more vigorous conditions the dihydro-γ-lactones are isomerized to δ-lactones (LIX). It was suggested that the mixture containing the smaller amount of δ-lactone was that in which a bulky axial C-13 substituent was generated on formation of the δ-lactone.

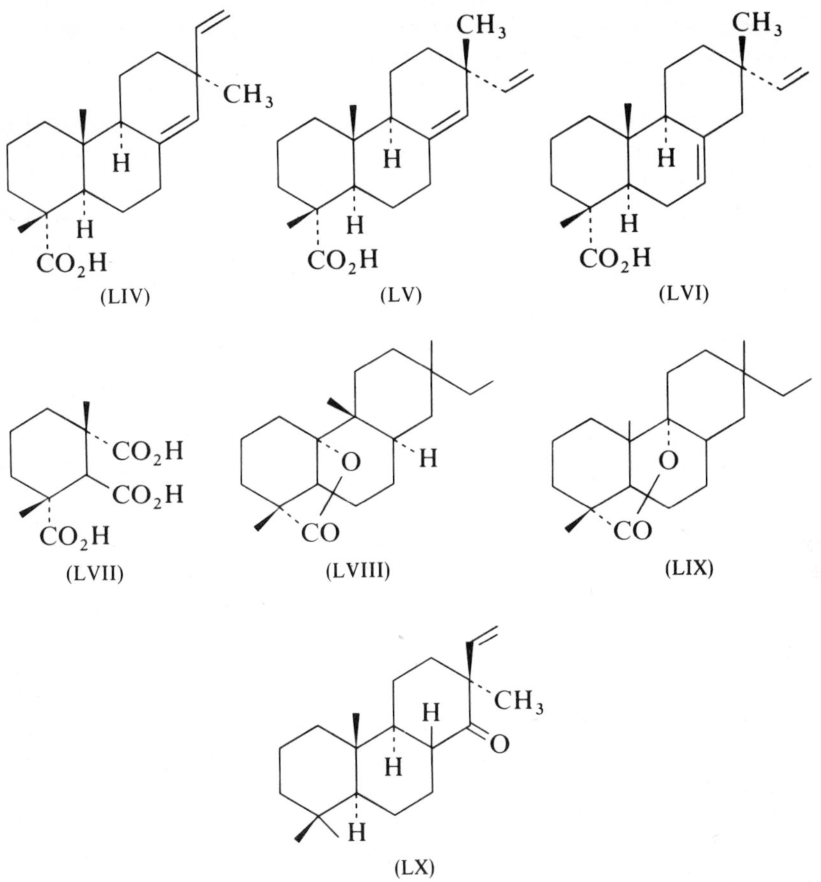

FIG. 20. Aspects of the stereochemistry of the tricyclic diterpenes.

This led to the conclusion that the vinyl group was β-oriented in pimaric acid. Although the point is of conformational interest, in view of the small differences actually observed some caution was suggested in the interpretation of these results. It was subsequently shown that not only were pimaric acid and isopimaric acid isomeric at C-13, but also that the nuclear double bond occupied the 8(14) position in pimaric acid and 7(8) in isopimaric acid. Evidence for the absolute configuration at C-9 came from a study (Bose, 1960) of the plain optical rotatory dispersion curves. Dihydropimaric acid gave a positive curve whilst dihydroisopimaric acid gave a negative curve. Using cholest-4-ene and cholest-5-ene as reference compounds the acids were assigned an α-hydrogen atom at C-9. A direct chemical relationship with the steroids also served to establish the orientation of the C-9 proton. Whilst sandaracopimaradiene has been partially synthesized from compounds in the androstane series, the most conclusive evidence for the stereochemistry of the pimarane diterpenes stems from the total synthesis of pimaradiene and sandaracopimaradiene (Ireland and Schiess, 1963). The intermediate ketone (LX) which was transformed into sandaracopimaradiene, showed the homoconjugation in its ultraviolet spectrum to be expected from an axially oriented β-vinyl ketone. This was absent from its 13-epimer with an equatorial vinyl group.

The stereochemistry of abietic acid has played a key part in establishing stereochemical relationships in this series. Vigorous oxidation of both abietic and pimaric acid gives a tricarboxylic acid (LVII) derived from ring A. This acid is optically inactive and hence the 1,3 carbonyl groups must be *cis* to each other. Dissociation constant data indicated (Barton and Schmeidler, 1948) that the other carboxyl group was *trans* to these two carboxyl groups. Hence the A/B fusion must be *trans*. The C-9 hydrogen atom was assigned an α-configuration on the basis of molecular rotation arguments whilst abietic acid shows a plain optical rotatory dispersion curve similar to that of cholest-3,5-diene. Again the stereochemistry has been placed beyond doubt by total synthesis.

Rosenonolactone (LXI) is the best known of some diterpenoid metabolites of *Tricothecium roseum* (Ellestad *et al.*, 1965). Its carbon skeleton represents a rearrangement of the pimarane skeleton. Rosenonolactone is epimerized by base to a compound possessing a more hindered carbonyl group. This epimerization involves the conversion of *trans* B/C ring junction to a *cis* B/C junction. Oxidation, hydrolysis and retro-aldol cleavage of rosenonolactone affords two fragments [(LXII) and (LXIII)] representing rings A and C. The absolute configuration of each of these was determined. Finally, X-ray analysis of dibromorosololactone confirmed the stereochemistry. A group of alcohols from the wood of *Erythroxylon monogynum* has been shown (Connolly *et al.*, 1966a) by chemical interrelationship with rosadiene to possess the enantiomeric stereochemistry. However, the structures of ring A of the erythroxydiols [(LXIV) and (LXV)] represent alternative modes of discharge of the friedo-carbonium ion. A similar series is also found with the isopimarane stereochemistry in the hydrocarbons dolabradiene and rimuene (Connolly *et al.*, 1966b).

The stereochemistry of the tetracyclic diterpenes was the subject of considerable speculation (Scott *et al.*, 1964), confused at one stage by reports on "mirene" which subsequently turned out to be a mixture. Although phyllocladene was the first tetracyclic diterpene to be isolated along with the closely related alcohol phyllocladanol, it represents at present the sole example of the stereochemistry (LXVI) in this series (Briggs *et al.*, 1962). Acid-catalysed isomerization of phyllocladene gave isophyllocladene which was degraded

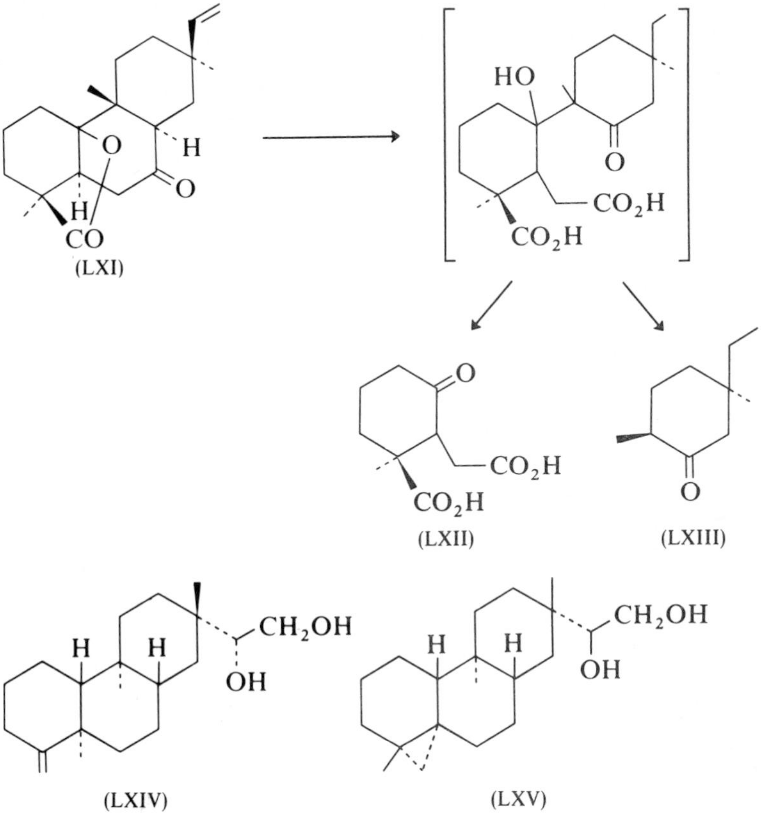

FIG. 21. Diterpenes of the friedo-pimarane stereochemistry.

(Grant and Hodges, 1960a) to podocarp-8(14)-en-13-one (LXVII). This compound is a degradation product of manöol (Hosking, 1936) and it is of known stereochemistry (Grant and Hodges, 1960b) whilst its racemate has been synthesized (Barltrop and Rogers, 1958). This established the normal nature of the A/B ring fusion. Since the optical rotatory dispersion curves of the nor-ketone (LXVII) and the keto-ester (LXIX) showed positive Cotton effects, ring D was assigned a β-absolute configuration.

Kaurene occurs in both enantiomeric forms and possesses a skeleton of considerable importance amongst the diterpenes. Most of the known kauranoid diterpenes possess the antipodal A/B fusion. The conversion of 7-hydroxykaurenolide (Cross *et al.*, 1963), steviol, and the diterpene alkaloid garryfoline (Mosettig *et al.*, 1961) into the 16-epimeric (–)-kauranes together with degradations (Briggs *et al.*, 1963; Hanson, 1963) of (–)-kaurene

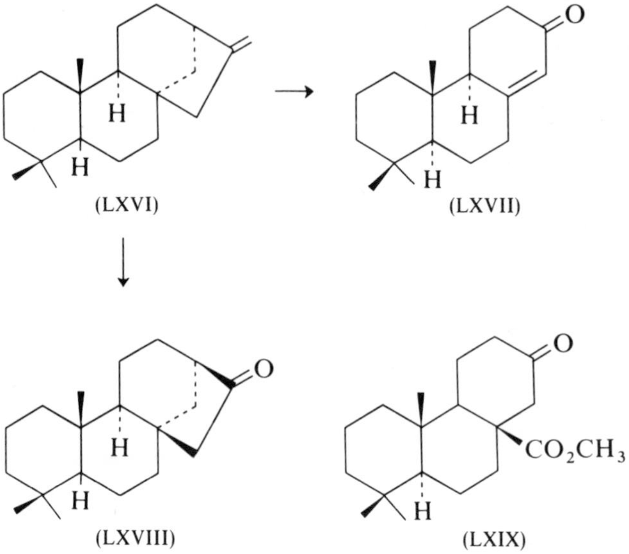

FIG. 22. Some degradation products of phyllocladene.

provided evidence for the stereochemistry (LXX) of (−)-kaurene. The value of these compounds particularly the kaurenolides, in providing stereochemical evidence lies in the presence of oxygen functions close to the asymmetric centres. The stereochemistry of the kaurenolides [(LXXI) and (LXXII)] involved several pieces of evidence (Cross *et al.*, 1962). At C(4) it was possible to distinguish between the axial lactone function and the equatorial hydroxymethyl substituent of 7,18-dihydroxykaurenolide. Thus, an equatorial 18-toluene-*p*-sulphonate underwent hydrogenolysis whilst an axial 19-toluene-*p*-sulphonate underwent hydrolysis. Secondly, the equatorial acids were stronger than the axial acids. The *trans* diaxial relationship of the oxygen atoms on ring B was demonstrated in a number of ways. Reduction of the 7-ketones with sodium borohydride led to a 7-alcohol epimeric at this centre with the natural alcohols. Comparison of the relative rates of elimination of the 7-toluene-*p*-sulphonates showed little difference between the epimers and thus there was no *trans* diaxial relationship between the β-hydrogen atom and the leaving group. Furthermore whereas hydrolysis of the natural kaurenolides gave the corresponding 6,7-diols in high yield, hydrolysis of their 7-epimers gave a 25–30 % yield of the 6-deoxy-7-ketones by the diaxial elimination of water. The natural diols were inert to oxidation with sodium periodate whilst their 7-epimers reacted rapidly. Hence in the kaurenolides there is a *trans* relationship between the 6- and 7-substituents. The presence of a 19 → 6 diaxial lactone ring precluded a *cis* A/B ring

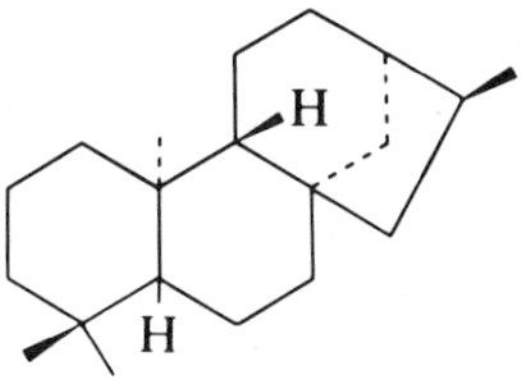

(LXX)

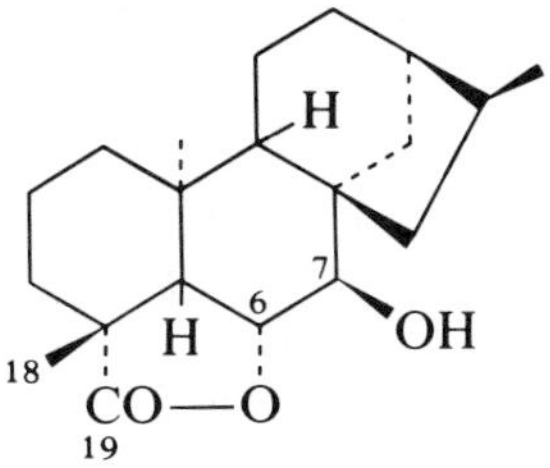

(LXXI)

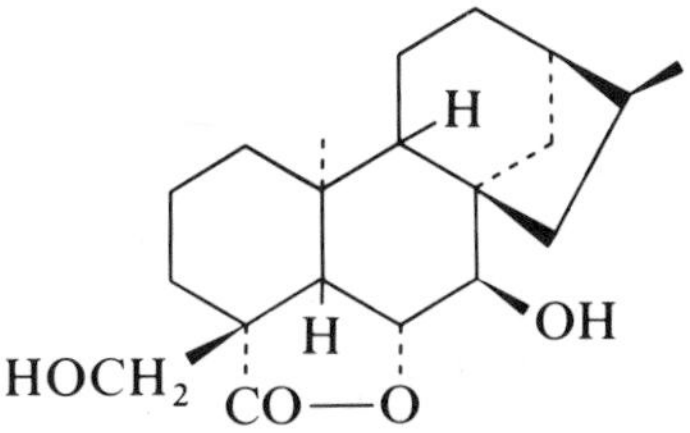

(LXXII)

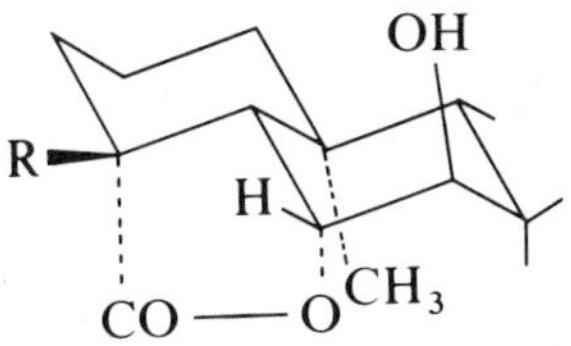

(LXXIII)

FIG. 23. The stereochemistry of the kaurenolides.

junction. Furthermore, oxidation of the 7-epihydroxy 19 → 6 lactone with lead tetraacetate gave a 7 → 20 ether, thus demonstrating a *cis* relationship between the 7-epihydroxyl group and the angular methyl group. Therefore, the relative stereochemistry of the oxygen functions and the angular methyl group required the presence of the fragment (LXXIII) [R = CH_3 and CH_2OH respectively] in the kaurenolides. Since methyl 7-oxo-(−)-kauran-19-oate showed a large positive Cotton effect in its optical rotatory dispersion curve, whilst the corresponding lactone showed a negative effect superimposed on a positive background, the lactone ring on formation must close axially into a negative quadrant and, therefore, (LXXIII) represents the absolute stereochemistry of rings A and B.

The 16-ketones showed positive Cotton effects and hence ring D possessed a β-absolute configuration. Thus, (LXXI) represents the overall stereochemistry of the kaurenolides. 7-Hydroxykaurenolide was reduced to (−)-kaurane through a reaction sequence which precluded epimerization at any of the ring junctions. Hence (LXX) represents the stereochemistry of (−)-kaurene. Other pieces of evidence came from the degradation of steviol and kaurene itself, particularly from experiments in which ring D was opened and a carbonyl group placed on ring C at C-13.

The beyerane skeleton has a number of representatives amongst the tetracyclic diterpenes. The stereochemistry of this group has been established through several interrelationships. One of the simplest derivatives and amongst the earliest to be isolated (Baarschers *et al.*, 1962), was stachenone possessing a 3-ketone. The optical rotatory dispersion curve of this 3-ketone implied an antipodal ring junction. Interrelationship with beyerol (Jefferies *et al.*, 1962), monogynol and isosteviol (Hanson, 1967) together with a study (Jefferies, 1967) of the n.m.r. spectra of the series particularly the influence of ring D substituents on the C-10 angular methyl group, showed that ring D was α-oriented. Carbonyl functions were introduced at C-15 and C-16 and studies of their O.R.D. curves confirmed this assignment. Finally the complete series of hydrocarbons have been synthesized by stereochemically defined processes.

V. Aspects of the Chemistry of the Diterpenes

Many of the general features of alicyclic chemistry may be exemplified from the chemistry of diterpenoid substances. Thus there are examples of addition from the less-hindered face of the molecule, of *trans* diaxial elimination and of the selective hydrolysis of hindered and non-hindered groups. Steric hindrance by the pendant groups particularly the angular methyl group at C-10, plays an important part in determining the stereochemical outcome in many of these reactions. However, in this section we shall select reactions of particular interest in diterpene chemistry.

Ruzicka's proposal (Ruzicka, 1953) that the pimaradienes originate from cyclization of labdane or manöol precursor has stimulated the study of the *in vitro* acid catalysed cyclizations of this series (McCreadie and Overton, 1968; Wenkert *et al.*, 1968).

Manöol (LXXIV), 13-epimanöol and the *cis* and *trans* labdadienols give the same mixture of products, that is asymmetry in the side-chain is destroyed prior to cyclization. Treatment with acetic acid–sulphuric acid mixtures for 3 hours affords a C-13 epimeric mixture of pimaradienes (LXXV). After 150 hours this contains about 50% of a rosadiene (LXXVI). This cyclization has been extended to a synthesis of rosenonolactone. Using formic acid as the cyclizing agent, the tricyclic pimaradiene sandaracopimaradiene and rosadiene are formed together with a 14α-hibyl formate (LXXVII). Labelling experiments (Edwards and Mootoo, 1969) have established that the latter (Bose, 1960) formed through an eight-membered system rather than through a cyclization of a pimaradiene.

The isomerization of the nuclear double bond of the pimaradienes is also encountered. Thus the treatment of the sandaracopimaradiene series with hydrogen chloride in chloroform gives an equilibrium 6:2:1 mixture of the $\Delta^{8(9)}$, $\Delta^{8(14)}$, and $\Delta^{7(8)}$ isomers. Similarly, treatment of the rosadiene (LXXVII) with hydrogen chloride in acetic acid yields rimuene.

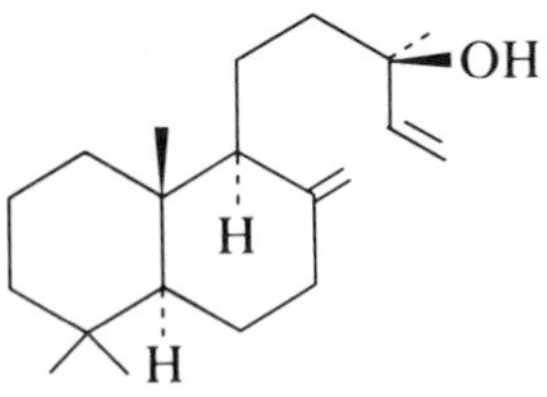

(LXXIV)

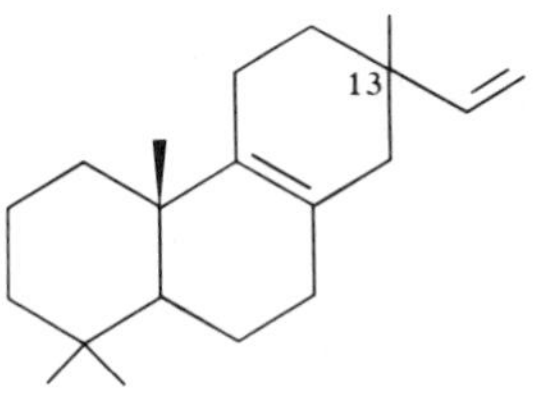

(LXXV)

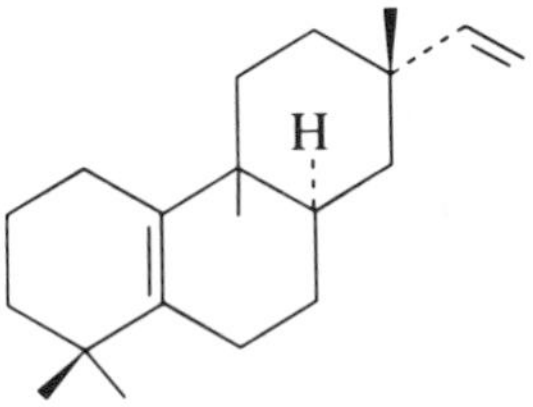

(LXXVI)

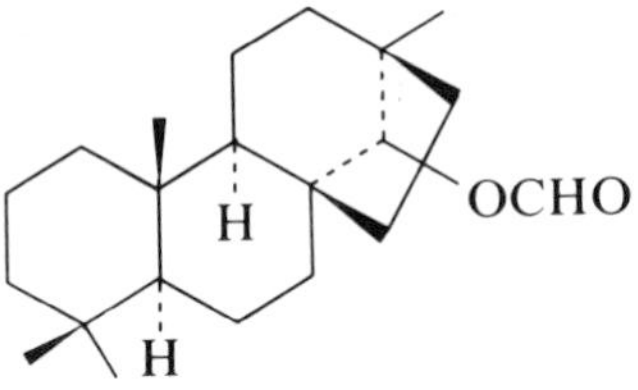

(LXXVII)

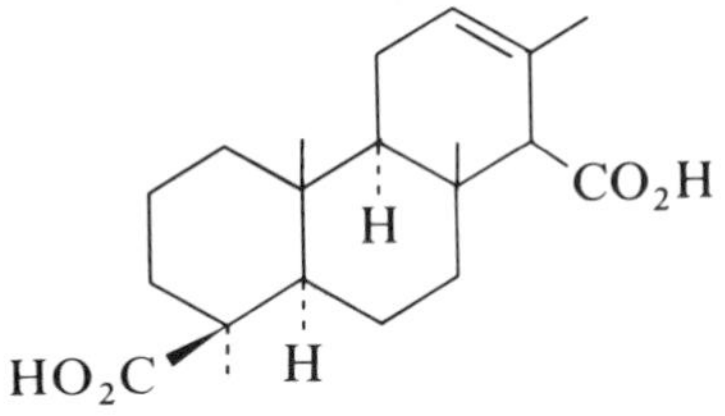

(LXXVIII)

FIG. 24. The acid-catalysed cyclization of the bicyclic diterpenes.

In contrast to the above cyclizations, treatment of agathic acid with formic acid yields isoagathic acid which was shown by a combination of spectroscopic and degradative techniques to possess structure (LXXVIII).

Another series of cyclizations is found in the ready ether formation reactions between C-8 and C-13. The naturally-occurring manoyl oxide, 13-epimanoyl oxide and keto-manoyl oxide are examples of this. The oxidation of manöol has been studied in some detail (Audier *et al.*, 1964; Mousseron-Canet *et al.*, 1963; Ohloff, 1958) and these reactions are also characterized by very ready cyclization between C-8 and C-13. The initial ketonic product (LXXIX) of potassium permanganate oxidation readily undergoes further hydroxylation with the formation of a stable ketal (LXXX). The C-13 epimer (LXXXII) of this ketal was produced by selective epoxydation of manöol at C-8, followed by cleavage of the vinyl group and internal opening of the epoxide (Mousseron-Canet *et al.*, 1963). Epoxydation is predominantly, but not entirely, from the α-face. Reduction of these epoxides affords sclareol and 8-episclareol. Sclareol also reveals several examples (Bory *et al.*, 1963; Scheidegger *et al.*, 1962) of this ready ether formation. For example, oxidation of the vinyl group gives a hydroxy-ketone which cyclizes extremely rapidly with the formation of an enol-ether.

The formation of the tetracyclic skeleton from the tricyclic skeleton has been less successful. Under base catalysed conditions the 7-keto-15,16-epoxide derived from isopimaric acid readily cyclizes to form the isohibaene carbon skeleton (Wenkert *et al.*, 1964).

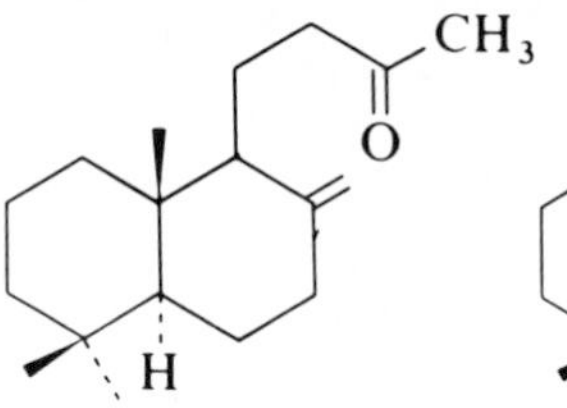

(LXXIX)

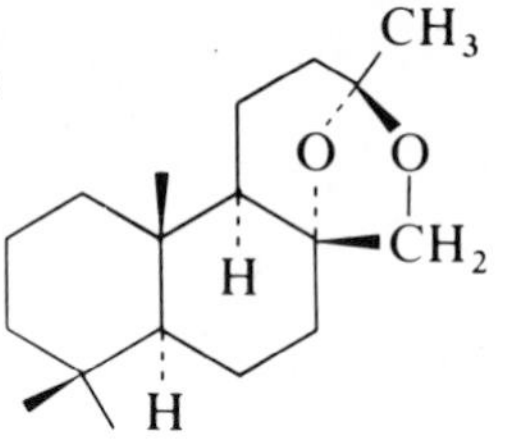

(LXXX)

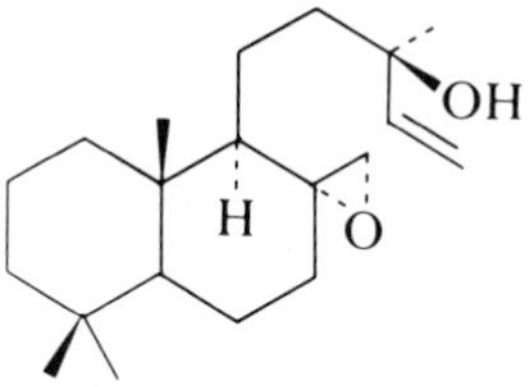

(LXXXI)

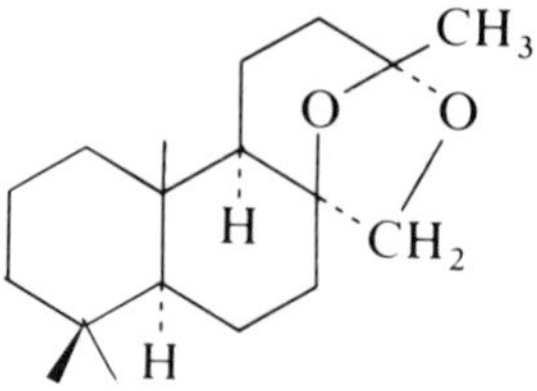

(LXXXII)

FIG. 25. The ketal formation from manöol.

The isomerization of the individual members of the tetracyclic series has been studied by various groups. Thus stachene is isomerized (Appleton *et al.*, 1966) by hydrogen chloride in chloroform to a mixture containing stachene, kaurene, isokaurene, isoatisirene and atisirene (cf. Fig. 5). Iodine catalysis can also bring about these rearrangements (Yoshikoshi *et al.*, 1967). It is also possible to initiate rearrangement using an epoxide as a leaving group.

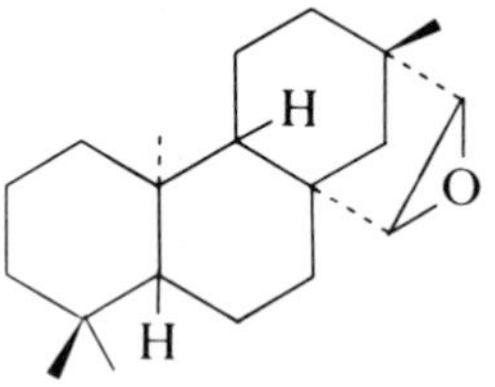

(LXXXIII)

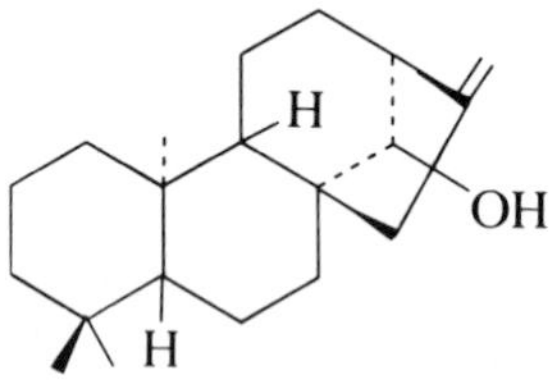

(LXXXIV)

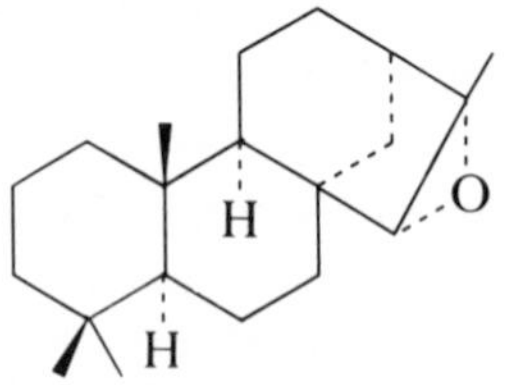

(LXXXV)

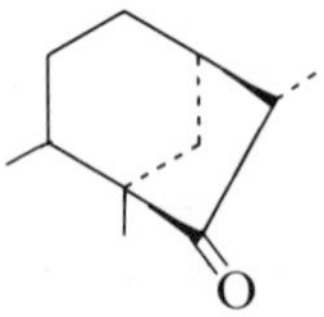

(LXXXVI)

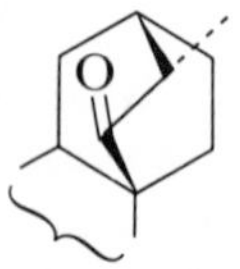

(LXXXVII)

FIG. 26. Some rearrangements of diterpene epoxides.

For example, treatment of stachene epoxide with boron trifluoride etherate affords the kaurene-14-ol (Hanson, 1967; Kapadi and Sukh Dev, 1965). Alternatively, treatment of 15α,16α-epoxyphyllocladane with boron trifluoride affords a mixture of phyllocladan-15-ketone and neoatiseran-15-one. The formation of the atiserane and neoatiserane skeleta requires a 1,3-hydride shift. Solvolysis (Appleton *et al.*, 1968) of the toluene-*p*-sulphonates of the epimeric 17-norkauran-16-ols generates acetates of 17-norkauran-12-ol, 17-noratisan-13-ol, 17-norphyllocladan-12-ol, 17-norphyllocladan-16-ol, 17-noratisan-16-ol and 17-norkauran-16-ol.

In view of the biological interest of the gibberellins there has been some work devoted to the contraction of ring B. Although the biosynthetic steps involve a 6-equatorial leaving group and the extrusion of C-7, this has not been successful *in vitro*.

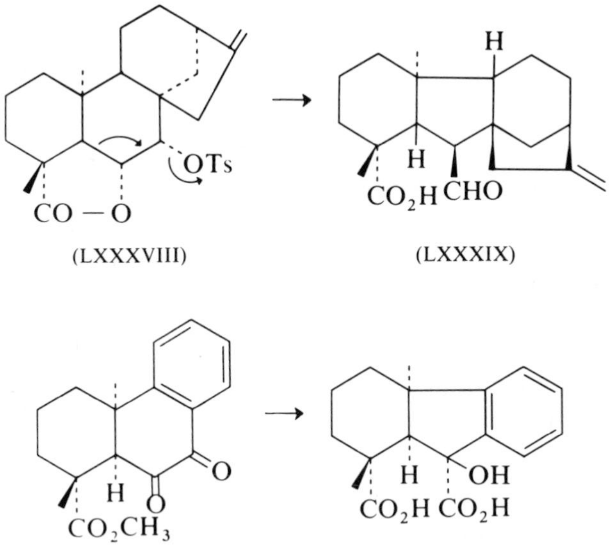

(LXXXVIII) (LXXXIX)

(XC) (XCI)

FIG. 27. The ring-contraction of some diterpenes.

However, treatment of the 7α-toluene-*p*-sulphonate (LXXXVIII) derived from 7-hydroxykaurenolide with strong base leads to the gibbane aldehyde (LXXXIX) (Galt and Hanson, 1965). Ring-contraction also occurs at a higher oxidation level. Thus, the α-diketone (XC) from desisopropyldehydroabietic acid afforded a gibbane hydroxy-acid (XCI) possessing the unnatural *cis* A/B fusion (Grove and Riley, 1961).

VI. The Total Synthesis of the Diterpenes

The synthesis of polycyclic substances has attracted considerable effort and ingenuity (Barltrop and Rogers, 1962). The diterpenoid substances provide a range of synthetic problems of graduated difficulty, ranging from the tricyclic aromatic compounds such as ferruginol to the complex polycyclic alkaloids and the labile gibberellins. Rather than attempt an exhaustive survey of total synthesis in this area, we shall take examples from several solutions to the various groups of synthetic problem posed by this series of compounds. For ease of stereochemical representation the diagrams drawn for this section are those of a single enantiomer. In most cases, however, the syntheses have involved racemates.

The stereochemical control of chemical reactions has had widespread ramifications in the synthesis of natural products. This is clearly illustrated in the synthesis of the diterpenes. Here the major problems have been concerned with the construction of ring junctions possessing a well-defined stereochemistry. The synthetic work has fallen into a number of stages culminating in the total synthesis of the tetracyclic diterpenes, the diterpene alkaloids and the gibberellins. The first stage involved the synthesis of compounds in which one ring, usually ring C, remained aromatic. These syntheses had as their goal compounds in the diterpenoid resin acid series. Experience in this area led to control over two major stereochemical features, the geminal substituents at C-4 and the A/B ring fusion. In the second stage ring C was reduced and the corresponding perhydrophenanthrenes alkylated at C-8. Here the angular methyl group at C-10 exerted control over the incoming group. Thus direct alkylation led to a new substituent at C-8 which was *trans* to the C-10 angular methyl group. Several of these perhydrophenanthrene derivatives have been cleaved to give total syntheses of the labdane diterpenes. The next stage involved the formation of ring D. These reactions involve cyclization between C-8 and C-13 which can, in certain circumstances, be a remarkably facile process. The final stage has involved the construction of special features such as the oxazolidine ring of the alkaloids.

Although there is a strong analogy with the biosynthesis of the diterpenes, the use of acid-catalysed multiple cyclizations of either completely aliphatic systems or ω-phenylalkenols has, until recently, given little control over the stereochemistry of the resultant ring junctions. However, recent studies have shown this to be a route to compounds of the podocarpane series.

The first stage involved methods which drew heavily on experience from the steroid field. The tricyclic system could be built up by constructing rings A and C and then forming ring B through a cyclization. Such cyclizations forming ring B are not always stereospecific. However, the presence of a methoxyl group, *para* to the newly formed bond, facilitates reversibility, and hence thermodynamic control, causing the *trans* isomer to predominate in the cyclization (Barltrop and Rogers, 1958). Alternatively, a ring extension reaction has been used to add ring A to a tetralone. Syntheses utilizing a tetralone afford greater control over the A/B fusion and the stereochemistry

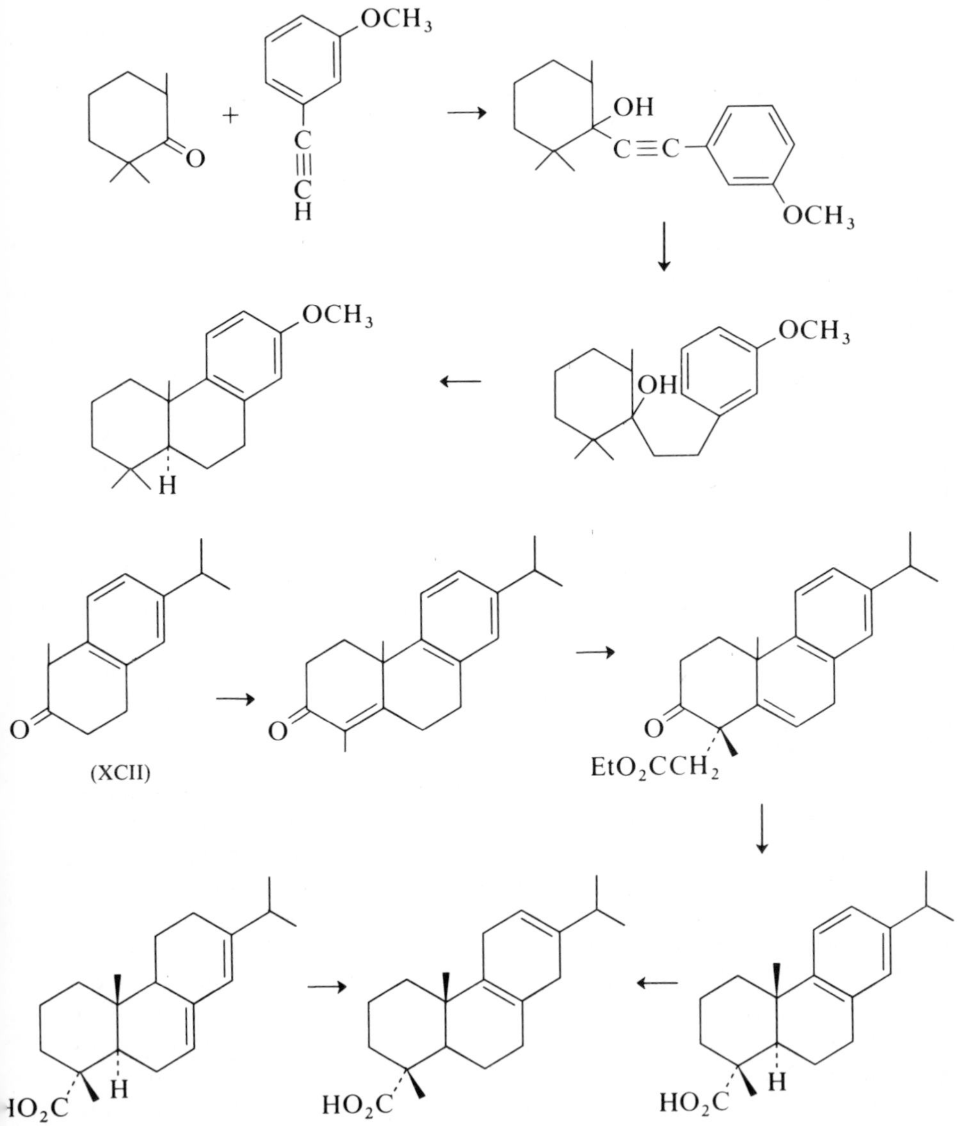

FIG. 28. Some syntheses of tricyclic skeleta.

at C-4. This is exemplified by a synthesis (Burgstahler and Worden, 1961; Stork and Schulenberg, 1956) of abietic acid (XCIII) from the tetralone (XCII).

In the second stage of the syntheses the aromatic ring C was reduced. Methods were developed to introduce substituents at C-8 oriented in such a manner as to form, on the one hand, ring D of the kaurene series in which the ring is on the opposite side of the molecule to the angular methyl group, and, on the other hand, to form ring D of the phyllocladene series. Two examples give an indication of the type of solution to this problem. Since

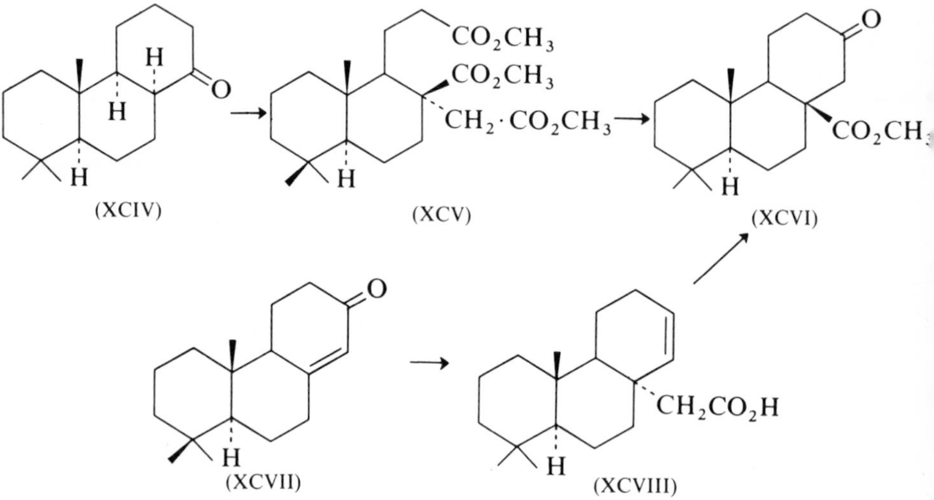

FIG. 29. Some syntheses of a phyllocladene degradation product.

the angular methyl group at C-10 directs *trans* alkylation at C-8, it was necessary in the synthesis (Turner and Shaw, 1960) of the degradation product (XCVI) of phyllocladene in which the carbomethoxyl group represented the residue of ring D, to cleave ring C after alkylation and to reclose the ring by a base-catalysed condensation to form the more stable *trans*-anti-*trans* backbone of phyllocladene. Thus, the α-methylene position of the ketone (XCIV) was blocked as its furfurylidene derivative and the ketone then alkylated with allyl bromide. Ozonolysis and esterification gave the tri-ester (XCV) which underwent Dieckmann cyclization to form the keto-ester (XCVI). In another approach (Church *et al.*, 1960), reduction of the unsaturated ketone (XCVII) gave a mixture of epimeric allylic alcohols. The predominant equatorial isomer gave a vinyl ether which on Claisen rearrangement and oxidation gave the acid (XCVIII). This acid, on ozonolysis and partial methylation, gave an anhydride ester which underwent a similar Dieckmann cyclization to form the keto-ester (XCVI). The type of method used in the construction of ring D is exemplified by the following synthesis of (±)-kaurene (Bell *et al.*, 1962). The aldehyde (XCIX) prepared by a Claisen

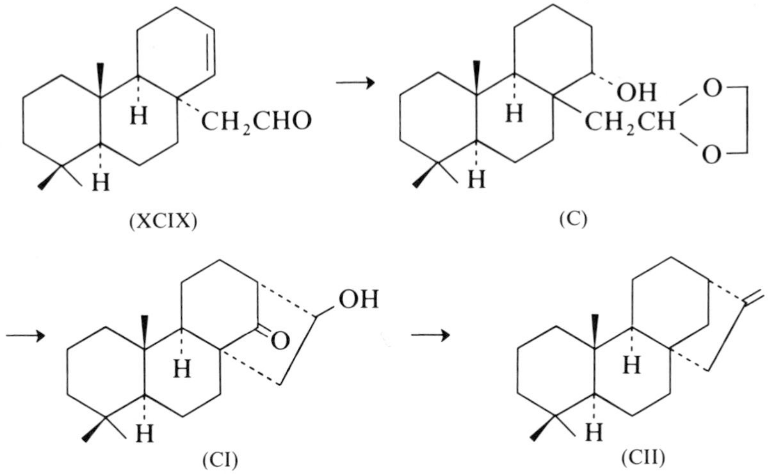

FIG. 30. The construction of ring D of (±)-kaurene.

rearrangement and oxidation procedure comparable with that described above, was protected as its ethylene acetal. Hydroboration gave the alcohol (C) which was oxidized to a ketone. Removal of the aldehyde protecting group permitted an internal aldol condensation leading to the tetracyclic hydroxy-ketone (CI). The 14-ketone was considerably more hindered than a ketone at C-16 and hence reacting the 14,16-diketone with excess methyl-enephosphorane gave (±)-14-ketokaurene which was subsequently reduced to (±)-kaurene (CII).

A method which has been used to construct the six-membered ring D of atisine used the following sequence. Aldol condensation of the diketone (CIII) gave a keto-alcohol which could be converted to the acetoxy-methane-sulphonate (CIV). This underwent a fragmentation reaction with the formation of the olefin (CV). Hydroboronation and formation of the terminal mesylate permitted alkylation at C-12 and formation of the atisine ring system (CVI) (Nagata *et al.*, 1963).

There are a number of alkaloid syntheses which have used photochemical methods to construct the heterocyclic ring. For example, photolysis (Edwards and ApSimon, 1962) of the acid–azide derived from podocarpic acid methyl ether gave a 25% yield of a δ-lactam in which the angular methyl group has been substituted. A similar reaction, using a C-20 acid azide, gave C-20 →

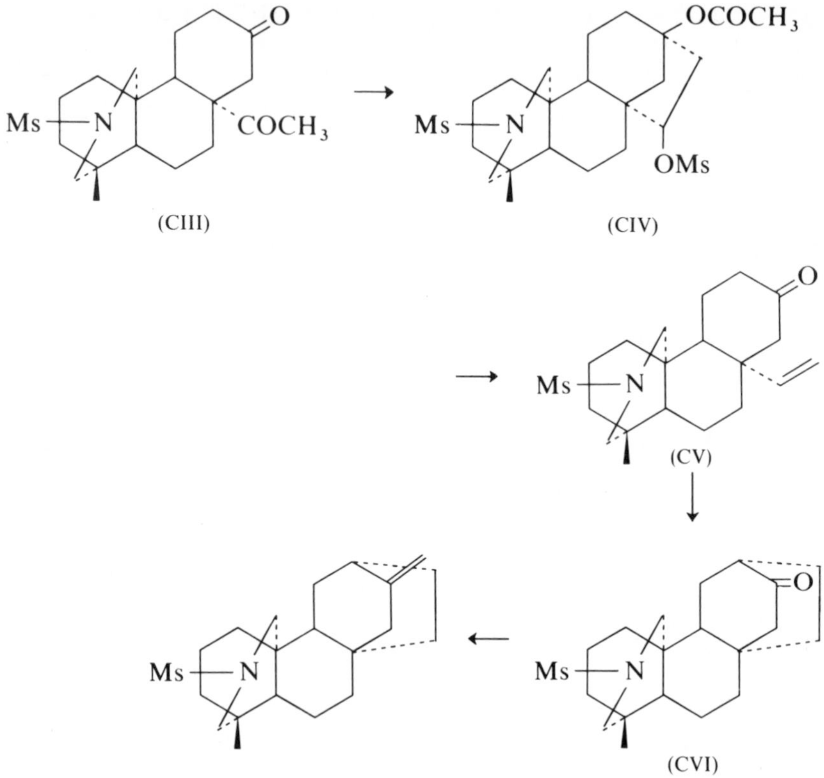

FIG. 31. The formation of the ring D of atisine.

C-19 lactam in a synthesis of the *Garrya* alkaloids. Photolysis of the *N*-iodo-amides derived from kauran-19-oic acids can also give 19 → 20-δ-lactones (Jefferies *et al.*, 1967).

A number of approaches have been developed to the synthesis of the gibberellins, and one has been carried through to completion (Mori *et al.*, 1969). The stages in these syntheses have involved, firstly, the synthesis of hexahydrofluorenones related to gibberone, gibberic acid and epigibberic acid. The next stage involved the substitution of ring A and the synthesis of a racemic diketo-ester (CX) from racemic epigibberic acid. This ester was available as a relay from gibberellic acid. Transformation of the optically active diketo-ester into a dienone utilized selective bromination stages. The lactone ring was reconstructed to give the isomerization product (CXII) of gibberellin A_1. The latter had previously been converted to gibberellin A_4.

These schemes can be exemplified by the synthesis of the gibberic acids. Annelation of the hydrindanone (CVII) with isopropenyl methyl ketone

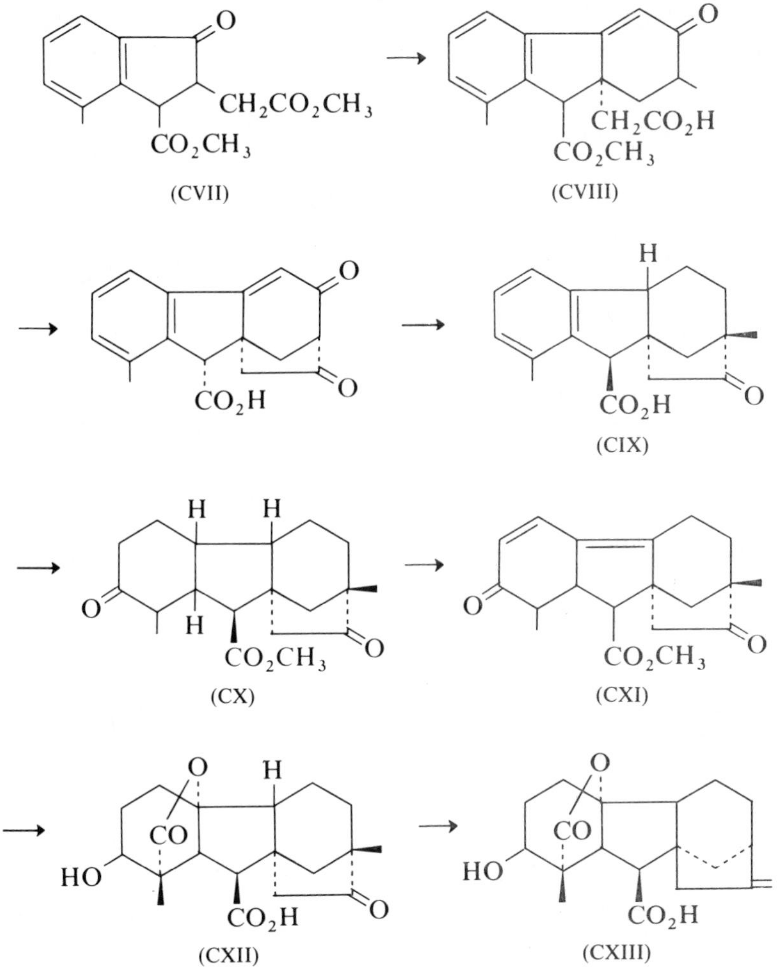

FIG. 32. Some stages in the synthesis of the gibberellins.

afforded the fluorenone (CVIII). Ring D was formed from this by acylation. Selective reduction then afforded (±)-gibberic and (±)-epigibberic acids (CIX). Ring A was activated by nitration and the nitro group converted through the amine to a phenol. Reduction then gave the racemic diketo-ester (CX). Successive bromination–dehydrobromination gave the dienone (CXI). Selective ketal formation blocked ring D, and ring A was then carboxymethylated, reduced, and the lactone ring formed. Epimerization of the ring A hydroxyl gave a degradation product of gibberellin A_1 (CXII) which was in turn converted to gibberellin A_4 (CXIII).

VII. Biological Properties of the Diterpenes

The diterpenes embrace a wide range of biological activity. Nevertheless, only a few compounds have had more than a passing interest. Amongst the best known substances are the gibberellin plant growth hormones (Brian *et al.*, 1960). The fungus *Gibberella fujikuroi* is responsible for a disease of rice seedlings known as bakanae disease. The active metabolites were extracted and eventually from one strain, gibberellic acid was isolated and purified. After considerable experimental effort its structure was shown to be (CXIV).

Investigation of its biological properties revealed effects that were typical of normal plant development and this stimulated the search for gibberellins in higher plants. At the same time, biosynthetic studies with the fungus revealed the existence of other gibberellins co-occurring with gibberellic acid. To date 34 gibberellins have been isolated. They possess the same

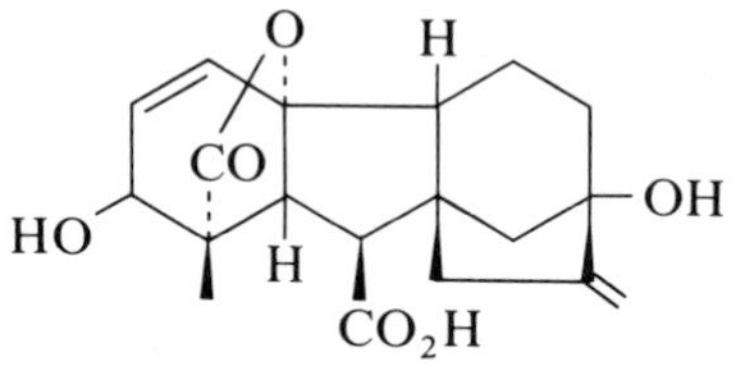

(CXIV)

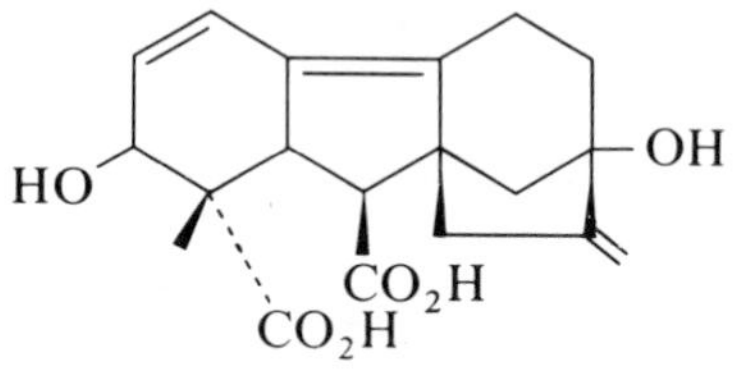

(CXV)

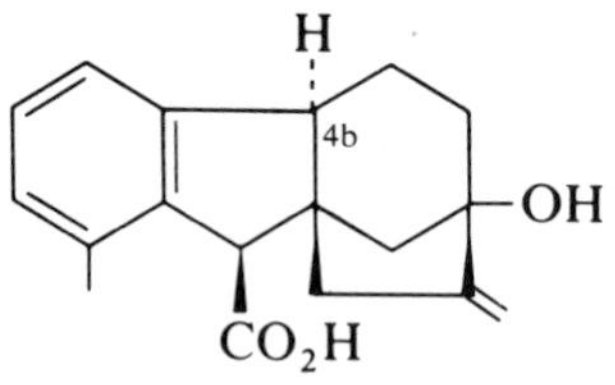

(CXVI)

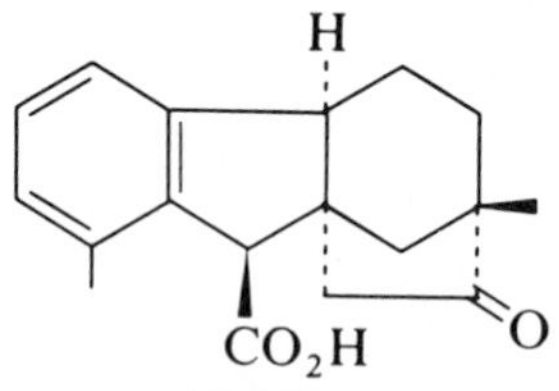

(CXVII)

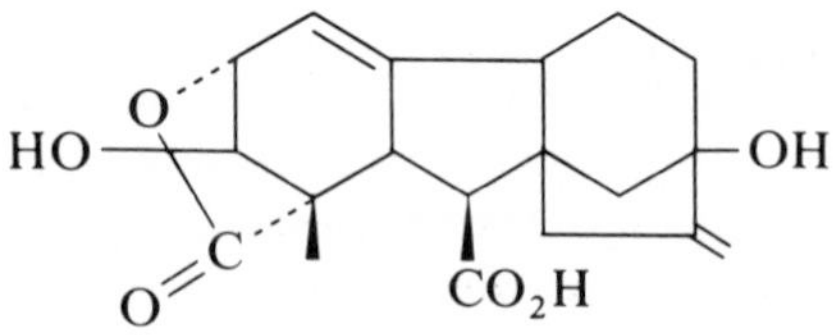

(CXVIII)

FIG. 33. Some gibberellin degradation products.

underlying perhydrofluorene skeleton and fall into two series: those C_{20} gibberellins in which the angular C-20 carbon atom remains at various oxidation levels, and those C_{19} gibberellins in which it is lost and a γ-lactone ring bridges ring A. The gibberellins show a range of activity in promoting cell extension, possibly in conjunction with auxins of the indolylacetic acid type. This level of activity differs with the plant species and individual gibberellin. Quite dramatic effects on the premature "bolting" of genetic dwarfs and on the vernalization of biennial plants have been observed. Gibberellic acid is produced commercially and is used in the malting process of beer and in horticulture.

Gibberellic acid (CXIV) is amongst the less stable gibberellins. In aqueous solutions at pH 6–8 it decomposes to gibberellenic acid (CXV), whilst in acid-solution at room temperature it gives allogibberic acid (Grove and Mulholland, 1960) (CXVI) (and its 4b epimer). At 100°C this aromatization reaction is accompanied by a Wagner–Meerwein rearrangement of ring D with the formation of gibberic acid (CXVII) and its 4b epimer (Cross, 1954; Grove *et al.*, 1960). In alkaline solution ring A undergoes a different rearrangement with the formation of a 1–3 γ-lactone (CXVIII). These rearrangement products lack the biological activity of gibberellic acid. Those gibberellins possessing a saturated ring A, do not undergo the aromatization reaction. However, on treatment (Cross *et al.*, 1961) with alkali, the axial hydroxyl group is epimerized to its equatorial isomer, possibly by a retro-aldol reaction. Another typical reaction of the gibberellins which lack the 7-hydroxyl group is hydration of the terminal methylene. Indeed, a number of naturally-occurring gibberellins such as gibberellin A_2 and A_{10} possess this feature.

In terms of quantity the resin acids are certainly amongst the natural products produced on a large scale. When the bark of a conifer is cut a semi-solid oleoresin is exuded. This is subjected to steam distillation to separate the monoterpenes of turpentine. A complex mixture of resin acids known as colophony or rosin remains behind, the exact composition of which depends upon the pine. It seems probable that the primary acidic constituents are pimaric acid, isopimaric acid, levopimaric acid, and neoabietic acid. On heating or treatment with acid, abietic acid is formed whilst on further heating this disproportionates to give dehydroabietic acid and di- and tetrahydroabietic acid. Apart from imparting a cohesive property to wood fibres, these, and particularly the phenolic resin acids, possess a mild antibiotic action and preserve the wood against microbial decay. Podocarpic acid and ferruginol have this action. Some tropical trees on decaying produce a fossil resin which is known as a copal. These copals, particularly from members of the Caesalpiniaceae, have provided a rich source of diterpenoid substances. It is possible that the resinous covering of plants in semi-arid regions which in a number of species contains diterpenoid substances also serves to protect the plant.

An antibacterial substance isolated from the *Basidiomycete*, *Pleurotus mutilus*, and known as pleuromutilin, has an interesting diterpenoid structure (Arigoni, 1968). It has not however attained any clinical significance.

Several diterpenoid substances have been shown to be tumour inhibitors. Included amongst these is the interesting quinone-methine (CXIX) isolated from *Taxodium distichum* (Kupchan *et al.*, 1968).

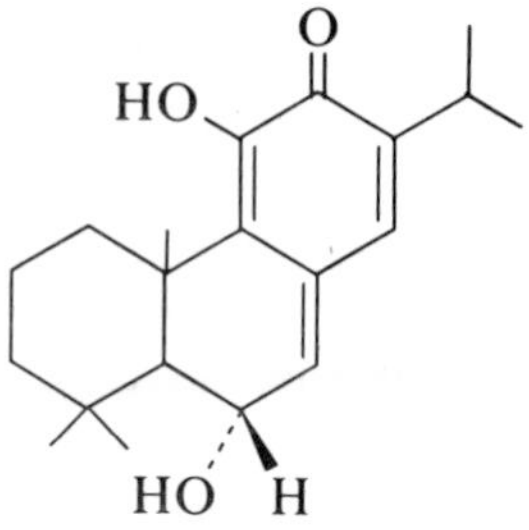

(CXIX)

A number of complex diterpenoid substances when esterified with nitrogenous residues are quite toxic. Thus ryanodine, which has some use as an insecticide, has a definite human toxicity, whilst the *Taxus* alkaloids, such as taxine-1, are responsible for the fatal properties associated with the yew tree (Della Casa de Marcano and Halsall, 1969). The aconite alkaloids isolated from various *Aconitum* species are highly toxic, causing fatal cardiac depression. The *Erythrophleum* alkaloids which are also toxic in small amounts have a digitalis-like action on the heart. Another group of diterpenoid poisons are the grayanotoxins isolated from various species of *Rhododendron*, *Kalmia*, and *Leucothoe*. These have been reported to possess a hypotensive action.

Amongst miscellaneous biological activity shown by the diterpenes are the bitter taste of many of the bicyclic compounds such as clerodin, andrographolide and columbin which form the bitter principles of various plant root extracts. In contrast the diterpenoid glycoside, stevioside, is amongst the sweetest known substances.

VIII. Typical Diterpenes

There are about five hundred naturally-occurring diterpenes, the majority of which are simple variations of oxidation pattern on a few basic skeleta. The following list of fifty compounds gives an indication of typical diterpenes and their physical constants.

Compound	Formula	M.P.	Rotation
1. Macrocyclic Diterpenes and their cyclization products			
Cembrene	$C_{20}H_{32}$	60°	+238°
Duvatrienediol-α-I	$C_{20}H_{34}O_2$	66°	+287°
Verticillol	$C_{20}H_{34}O$	105°	+168°
o-Cinnamoyl-taxicin-I	$C_{29}H_{36}O_7$	234°	+285°
2. Bicyclic Diterpenes, Manöol series			
Manöol	$C_{20}H_{34}O$	53°	+30°
13-epiManöol	$C_{20}H_{34}O$	138°	+51°
Manoyl oxide	$C_{20}H_{34}O$	29°	+20°
2-ketoManoyl oxide	$C_{20}H_{34}O_2$	77°	+40°
Sclareol	$C_{20}H_{36}O_2$	106°	−3°
3. Bicyclic Diterpenes, Labdanolic acid series			
Labdanolic acid	$C_{20}H_{36}O_3$	gum	−7°
Agathic acid	$C_{20}H_{30}O_4$	204°	+56°
Grindelic acid	$C_{20}H_{32}O_3$	101°	−102°
Marrubiin	$C_{20}H_{28}O_4$	160°	+36°
4. Bicyclic Diterpenes, rearranged Labdanes			
Kohlavenol	$C_{20}H_{34}O$	gum	—
Clerodin	$C_{20}H_{34}O_7$	171°	+20°
Columbin	$C_{20}H_{22}O_6$	180°	+45°
5. Tricyclic Diterpenes, Pimaranes			
Pimaradiene	$C_{20}H_{32}$	gum	
Sandaracopimaradiene	$C_{20}H_{32}O$	41°	−12°
Pimaric acid	$C_{20}H_{30}O_2$	212°	+60°
Isopimaric acid	$C_{20}H_{30}O_2$	164°	0°
Sandaracopimaric acid	$C_{20}H_{30}O_2$	168°	+20°
Darutigenol	$C_{20}H_{34}O_3$	160°	−11°
6. Tricyclic Diterpenes, rearranged Pimaranes			
Rimuene	$C_{20}H_{32}$	56°	+56°
Erythroxydiol Y	$C_{20}H_{34}O_2$	146°	+87°
Rosenonolactone	$C_{20}H_{28}O_3$	214°	−116°
Cassaic acid	$C_{20}H_{30}O_4$	203°	−126°
7. Tricyclic Diterpenes, Abietanes and rearranged Abietanes			
Podocarpic acid	$C_{17}H_{20}O_3$	194°	+144°
Ferruginol	$C_{20}H_{30}O$	175°	+41°
Totarol	$C_{20}H_{30}O$	132°	+41°
Abietic acid	$C_{20}H_{30}O_2$	175°	−116°

Compound	Formula	M.P.	Rotation
8. Tetracyclic Diterpenes, Kaurene-Phyllocladene series			
(−)-Kaurene	$C_{20}H_{32}$	50°	+80°
(−)-isoKaurene	$C_{20}H_{32}$	64°	−26°
(+)-Phyllocladene	$C_{20}H_{32}$	98°	+16°
(+)-isoPhyllocladene	$C_{20}H_{32}$	112°	+24°
Kaur-16-en-19-oic acid	$C_{20}H_{30}O_2$	181°	−112°
Steviol	$C_{20}H_{30}O_3$	215°	−95°
Kaur-16-en-3α,19-diol	$C_{20}H_{32}O_2$	185°	−66°
Abbeokutone	$C_{20}H_{32}O_3$	192°	−73°
7-Hydroxykaurenolide	$C_{20}H_{28}O_3$	188°	−25°
Cafestol	$C_{20}H_{28}O_3$	162°	−140°
Enmein	$C_{20}H_{26}O_6$	312°	−136°
Fujenal	$C_{20}H_{26}O_4$	170°	−74°
Garryfoline	$C_{22}H_{33}O_2N$	133°	−60°
9. Tetracyclic Diterpenes, Gibberellins			
Gibberellic acid	$C_{19}H_{22}O_6$	235°	+92°
Gibberellin A_9	$C_{19}H_{24}O_4$	211°	−22°
Gibberellin A_{13}	$C_{20}H_{26}O_7$	196°	−48°
10. Atiserene and related Alkaloids			
Atiserene	$C_{20}H_{32}$	58°	−40°
Atisine	$C_{22}H_{33}O_2N$	60°	+28°
Ajaconine	$C_{22}H_{33}O_3N$	172°	−119°
11. Trachylobane class			
Trachylobanic acid	$C_{20}H_{32}O_2$ (Me ester	112°	−41°)

REFERENCES

There are several thousand papers devoted to various aspects of diterpene chemistry. Space permits only a brief selection of these which has been made with a view to enabling the reader to obtain an entry to the various current developments in the field.

General Reviews

Simonsen, J. (1951). "The Terpenes", Vol. 3, Cambridge University Press, London.
Barton, D. H. R. (1949). *Quart. Rev.* **3**, 36.
Barltrop, J. A. and Rogers, N. A. J. (1961). *Prog. Org. Chem.* 96.
Fujita, E. (1964). *Bull. Inst. Chem. Res. (Kyoto)* **43**, 278; (1965) **44**, 239; (1966) **45**, 229.
Grigoreva, N. Y. and Kucherov, V. F. (1966). *Russ. Chem. Rev.* **35**, 850.
Grove, J. F. (1961). *Quart. Rev.* **15**, 56.
Hanson, J. R. (1968). "The Tetracyclic Diterpenes", Pergamon Press, Oxford.
Hanson, J. R. (1968). "*Advances in Phytochemistry*", **1**, 161.
Marion, L. (1963). "*Pure and Applied Chemistry*", **6**, 621.

de Mayo, P. (1959). "Chemistry of Natural Products", Vol. 3, Interscience Publishers, New York.

McCrindle, R. and Overton, K. H. (1969). *In* "Chemistry of Carbon Compounds" (E. H. Rodd, ed.), Vol. 2b, Elsevier, London.

McCrindle, R. and Overton, K. H. (1965). "Advances in Organic Chemistry. Methods and Results" (R. A. Raphael, ed.), Vol. 5, Interscience Publishers, New York.

Pelletier, S. W. (1964). *Experientia* **20**, 1.

Pelletier, S. W. (1967). *Quart. Rev.* **21**, 525.

Pinder, A. R. (1960). "Chemistry of Carbon Compounds" (E. H. Rodd, ed.), Vol. 4, Elsevier, London.

Stern, E. S. (1954). "The Alkaloids" (R. H. F. Manske, ed.), Vol. 4 and (1960) Vol. 7, Academic Press, New York and London.

Tsutsui, M. and Tsutsui, E. A. (1959). *Chem. Rev.* **59**, 1031.

Wiesner, K. and Valenta, Z. (1958). *Prog. Chem. Org. Nat. Prod.* 26.

In addition the reader will find useful summaries in the Annual and Specialist Reports of the Chemical Society.

Appleton, R. A., Fulke, J. W. B., Henderson, M. S. and McCrindle, R. (1967). *J. Chem. Soc.* 1943.

Appleton, R. A., Gunn, P. A. and McCrindle, R. (1968). *Chem. Comm.* 1131.

Appleton, R. A., McAlees, A. J., McCormick, A., McCrindle, R. and Murray, R. D. H. (1966). *J. Chem. Soc.* 2319.

Arigoni, D. (1968). *Pure Appl. Chem.* **17**, 331.

Audier, H., Bory, S. and Fetizon, M. (1964). *Bull. Soc. Chim. Fr.* 1351.

Baarschers, W. H., Horn, D. H. S. and Johnson, L. R. (1962). *J. Chem. Soc.* 4046.

Barltrop, J. A. and Rogers, N. A. J. (1958). *J. Chem. Soc.* 2566.

Barltrop, J. A., Bigley, D. B. and Rogers, N. A. J. (1960). *J. Chem. Soc.* 4613.

Barltrop, J. A. and Rogers, N. A. J. (1962). *Quart. Rev.* **16**, 117.

Barton, D. H. R. and Schmeidler, L. (1948). *J. Chem. Soc.* 1197.

Barton, D. H. R. and Elad, D. (1956). *J. Chem. Soc.* 2085.

Bell, R. A., Ireland, R. E. and Partyka, R. A. (1962). *J. Org. Chem.* **27**, 3741.

Birch, A. J., Rickards, R. W., Smith, H., Harris, A. and Whalley, W. B. (1959). *Tetrahedron* **7**, 241.

Bory, S., Fetizon, M. and Laszlo, P. (1963). *Bull. Soc. Chim. Fr.* 2310.

Bose, A. K. (1960). *Chem. Ind.* (*London*) 1104.

Brian, P. W., Grove, J. F. and MacMillan, J. (1960). *Prog. Chem. Org. Nat. Prod.* **18**, 350.

Briggs, L. H., Cain, B. F., Cambie, R. C. and Davis, B. R. (1962). *J. Chem. Soc.* 1840.

Briggs, L. H., Cain, B. F., Cambie, R. C., Davis, B. R., Rutledge, P. S. and Wilmshurst, J. K. (1963). *J. Chem. Soc.* 1345.

Burgstahler, A. W. and Worden, L. R. (1961). *J. Amer. Chem. Soc.* **83**, 2587.

Carmen, R. M. and Grant, P. K. (1961). *J. Chem. Soc.* 2187.

Church, R. F., Ireland, R. E. and Marshall, J. A. (1960). *Tetrahedron Lett.* **17**, 1.

Connolly, J. D., McCrindle, R., Murray, R. D. H. and Overton, K. H. (1966a). *J. Chem. Soc.* 273.

Connolly, J. D., McCrindle, R., Murray, R. D. H., Renfrew, A. J., Overton, K. H. and Melera, A. (1966b). *J. Chem. Soc.* 268.

Cross, B. E. (1954). *J. Chem. Soc.* 4670.

Cross, B. E., Grove, J. F. and Morisson, A. (1961). *J. Chem. Soc.* 2498.

Cross, B. E., Galt, R. H. B., Hanson, J. R. and Klyne, W. (1962). *Tetrahedron Lett.* 145.

Cross, B. E., Galt, R. H. B. and Hanson, J. R. (1963). *J. Chem. Soc.* 2944.
Della Casa de Marcano, D. P. and Halsall, T. G. (1969). *Chem. Comm.* 1282.
Djerassi, C.. Cais, M. and Mitscher, C. A. (1959). *J. Amer. Chem. Soc.* **81**, 2386.
Edwards, O. E. and Howe, R. (1959). *Can. J. Chem.* **37**, 760.
Edwards, O. E. and ApSimon, J. W. (1962). *Can. J. Chem.* **40**, 896.
Edwards, O. E. and Mootoo, B. S. (1969). *Can. J. Chem.* **47**, 1189.
Ellestad, G. A., Green, B., Harris, A., Whalley, W. B. and Smith, H. (1965). *J. Chem. Soc.* 7246.
Erdtman, H., Norin, T., Sumimoto, M. and Morrison, A. (1964). *Tetrahedron Lett.* 3879.
Galt, R. H. B. and Hanson, J. R. (1965). *J. Chem. Soc.* 1565.
Gaudemer, A., Polonsky, M. and Wenkert, E. (1964). *Bull. Soc. Chim. Fr.* 407.
Giles, J. A., Schumacher, J. N., Minns, S. S. and Bernasek, E. (1962). *Tetrahedron* **18**, 169.
Gopinath, K. W., Govindachari, T. R., Parthsarathy, P. C. and Viswanathan, N. (1961). *Helv. Chim. Acta.* **44**, 1040.
Grant, P. K. and Hodges, R. (1960a). *Tetrahedron* **8**, 261.
Grant, P. K. and Hodges, R. (1960b). *J. Chem. Soc.* 5274.
Grove, J. F., MacMillan, J., Mulholland, T. P. C. and Turner, W. B. (1960). *J. Chem. Soc.* 3049.
Grove, J. F. and Mulholland, T. P. C. (1960). *J. Chem. Soc.* 3007.
Grove, J. F. and Riley, B. J. (1961). *J. Chem. Soc.* 1105.
Halsall, T. G. and Cocker, J. D. (1956). *J. Chem. Soc.* 4262.
Halsall, T. G. and Cocker, J. D. (1957). *J. Chem. Soc.* 4401.
Hanson, J. R. (1963). *J. Chem. Soc.* 5061.
Hanson, J. R. (1967). *Tetrahedron* **23**, 793.
Harris, G. C. and Sanderson, T. F. (1948). *J. Amer. Chem. Soc.* **70**, 2079.
Hodges, R. and Reed, R. I. (1960). *Tetrahedron* **10**, 71.
Hosking, J. R. (1936). *Ber.* **69**, 760.
Hugel, G., Lods, L., Mellor, J. M., Theobald, D. W. and Ourisson, G. (1965). *Bull. Soc. Chim. Fr.* 2882.
Humber, L. G. and Taylor, W. T. (1955). *J. Chem. Soc.* 1044.
Ireland, R. E. and Schiess, P. W. (1963). *J. Org. Chem.* **28**, 6.
Jefferies, P. R., Ghisalberti, E. L. and Minchan, W. A. (1967). *Tetrahedron* **23**, 4463.
Jefferies, P. R., Rosich, R. S., White, D. E. and Woods, M. C. (1962). *Aust. J. Chem.* **15**, 521.
Jefferies, P. R., Rosich, R. S. and White, D. E. (1963). *Tetrahedron Lett.* 1793.
Kapadi, A. H. and Sukh Dev (1965). *Tetrahedron Lett.* 1255.
Klyne, W. (1953). *J. Chem. Soc.* 3072.
Kupchan, S. M., Karim, A. and Marcks, C. (1968). *J. Amer. Chem. Soc.* 90, 5923.
McCreadie, T. and Overton, K. H. (1968). *Chem. Comm.* 288.
Mori, K., Shiosaki, M., Itaya, N., Matsui, M. and Sumiki, Y. (1969). *Tetrahedron* **25**, 1293.
Mosettig, E., Quitt, P., Beglinger, U., Waters, J. A., Vorbrueggen, H. and Djerassi, C. (1961). *J. Amer. Chem. Soc.* **83**, 3163.
Mousseron-Canet, M., Millot, M. and Mani, M. (1963). *Isr. J. Chem.* **1**, 468.
Nagata, W., Sugasawa, T., Narusada, N., Wakabayashi, T. and Hayase, Y. (1963). *J. Amer. Chem. Soc.* **85**, 2342.
Ohloff, G. (1958). *Helv. Chim. Acta.* **41**, 845.
Overton, K. and Renfrew, A. J. (1967). *J. Chem. Soc.* 931.

Ruzicka, L. (1953). *Experientia* **9**, 357.
Ruzicka, L. and Bernold, E. (1941). *Helv. Chim. Acta.* **24**, 931.
Ruzicka, L., Seidel, C. F. and Engel, L. L. (1942). *Helv. Chim. Acta.* **25**, 621.
Ruzicka, L., Zwicky, C. and Jeger, O. (1948). *Helv. Chim. Acta.* **31**, 2143.
Schenk, H., Gutman, H., Jeger, O. and Ruzicka, L. (1952). *Helv. Chim. Acta.* **35**, 817.
Scheidegger, U., Schaffner, K. and Jeger, O. (1962). *Helv. Chim. Acta.* **45**, 400.
Scott, A. I., McCapra, F., Comer, F., Sutherland, S. A., Young, D. W., Sim, G. A. and Ferguson, G. (1964). *Tetrahedron* **20**, 1339.
Soucek, M. and Vlad, P. (1962). *Chem. and Ind.* 1946.
Stork, G. and Schulenberg, J. W. (1956). *J. Amer. Chem. Soc.* **78**, 250.
Turner, R. B. and Shaw, P. E. (1960). *Tetrahedron Lett.* **18**, 24.
Wenkert, E. (1955). *Chem. and Ind.* 282.
Wenkert, E., Peak, P. and Grant, P. K. (1961). *Chem. and Ind.* 1574.
Wenkert, E. and Kumasawa, Z. (1968). *Chem. Comm.* 140.
Wenkert, E., Jeffs, P. W. and Mahajan, J. R. (1964). *J. Amer. Chem. Soc.* **86**, 2218.
Wheeler, D. M. S., Fetizon, M. and Castine, W. H. (1967). *Tetrahedron* **23**, 3909.
Yoshikoshi, A., Kitadani, M. and Kitahara, Y. (1967). *Tetrahedron* **23**, 1175.

PART 2: THE SESTERTERPENES

I. Introduction

The sesterterpenes are a novel class of C_{25} terpene derived from geranylfarnesol pyrophosphate (I). The first representative of this class was described in 1965 (Argoni, 1965). Sesterterpenes have been isolated from insect protective waxes and from fungal sources. They form the phytotoxic principles of a number of *Helminthosporium* and *Cochliobolus* species. In the period that has elapsed since the description of the structure of gascardic acid (VII), derived from the insect *Gascardia madagascariensis*, about ten compounds of this type have been described. Both X-ray analysis and a wide combination of physical and chemical techniques have been used to determine the structures of these substances. Furthermore, a number of key stages in the biosynthesis of the sesterterpenes have been defined. Recently the acyclic C_{25} alcohols geranylfarnesol (Rios and Perez, 1969) and geranylnerolidol (Nozoe *et al.*, 1968) have been isolated from natural sources.

II. Biogenetic Relationship

Two related skeleta have been described. Biogenetically this relationship is shown in Fig. 1. Unlike the major diterpene pathway, the pyrophosphate group acts as the leaving group initiating cyclization to give an intermediate (II) containing an eleven-membered ring. A series of 1:2-shifts accompany a second cyclization leading to the carbon skeleton of gascardic acid. An alternative cyclization involving a 1:5-hydride shift leads to the unusual eight-membered ring system of the ophiobolins. This hydride shift has been confirmed by biosynthetic experiments (Canonica *et al.*, 1967, 1969; Nozoe *et al.*, 1967, 1960) using the appropriately labelled mevalonate.

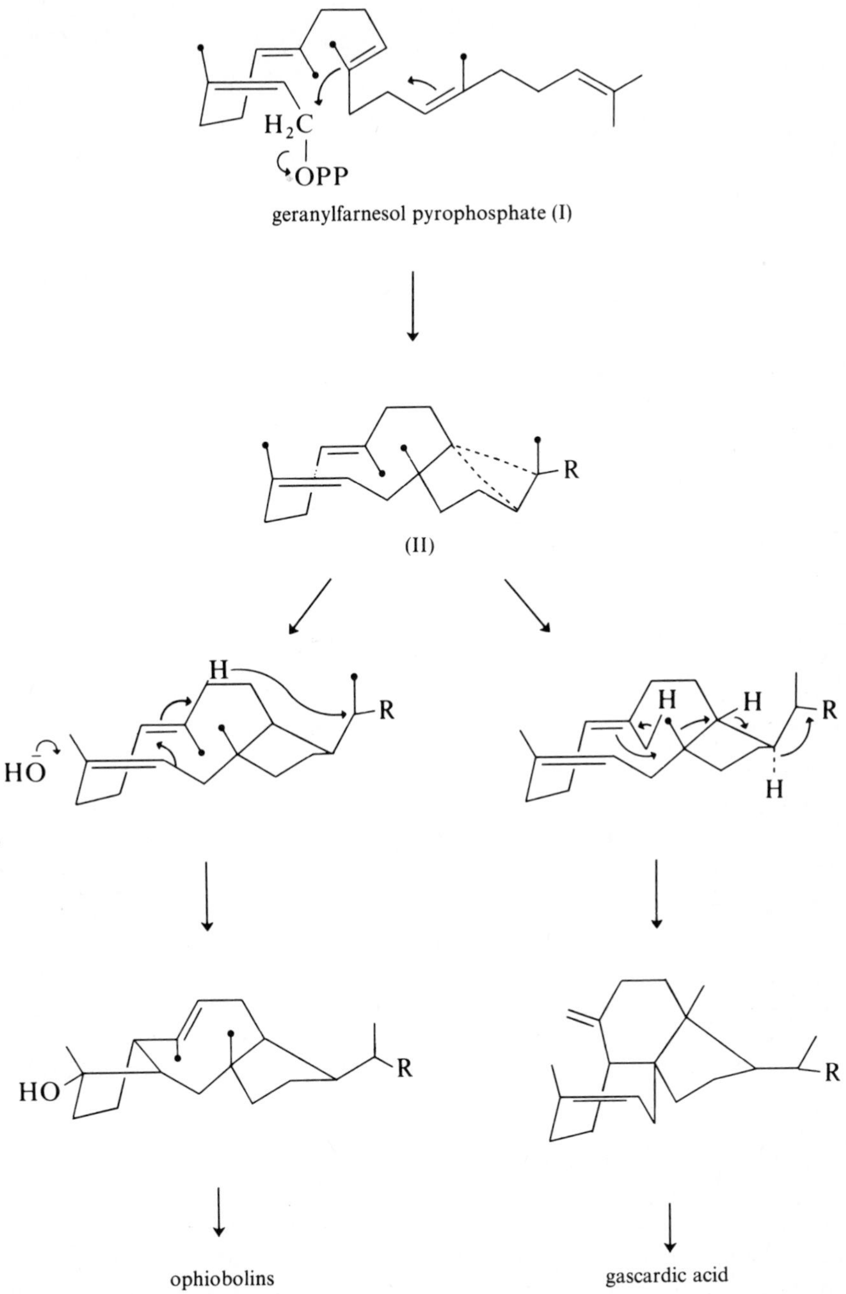

FIG. 1. Biogenetic relationship of the sesterterpenes.

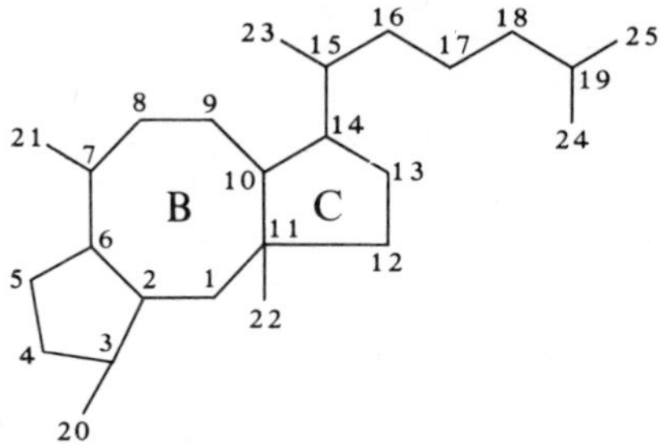

FIG. 2. Numbering of the sesterterpenes.

The numbering of the carbon skeleton is based on this mode of cyclization. A number of the fungal metabolites were studied independently by different workers who assigned them various trivial names. The identity of these compounds was subsequently established and consequently some of the trivial names have been abandoned. The accepted names are set out in Table 1 (Tsuda *et al.*, 1967).

TABLE 1

NOMENCLATURE OF THE OPHIOBOLINS

Literature names	Trivial names	Systematic names
Cochliobolin Ophiobolin Ophiobalin Cochliobolin A	Ophiobolin A	Ophiobola-7,18-dien-21-al-3α-ol-5-one 14α,17-oxide
Zizanin Ophiobolosin A Zizanin B Cochliobolin B	Ophiobolin B	Ophiobola-7,18-dien-21-al-3α,14α-diol-5-one
Zizanin A	Ophiobolin C	Ophiobola-7,18-dien-21-al-3α-ol-5-one
Cephalonic acid	Ophiobolin D	Ophiobola-3,6,18-trien-8β-ol-21-oic acid

III. Structural Determination

The study of these compounds has involved a combination of physical and chemical techniques. The n.m.r. spectra of the series have been particularly useful in revealing the location of the methyl groups relative to the centres of unsaturation and to the oxygen functions whilst the C_8 side chain is a fragment which is readily lost in the mass spectrum. X-ray analysis of a bromomethoxy derivative of ophiobolin A (Nozoe *et al.*, 1965), the bromoacetate of methyl cephalonate, and of the *p*-bromobenzoate of ceroplastol (Iitaka *et al.*, 1968) revealed the structure and absolute configuration of these substances. Nevertheless, fruitful chemical work has also been recorded for a number of the sesterterpenes. As an example of this work we shall describe some aspects of the chemistry of ophiobolin A (Canonica *et al.*, 1966; Nozoe *et al.*, 1966) and its relationship to other compounds.

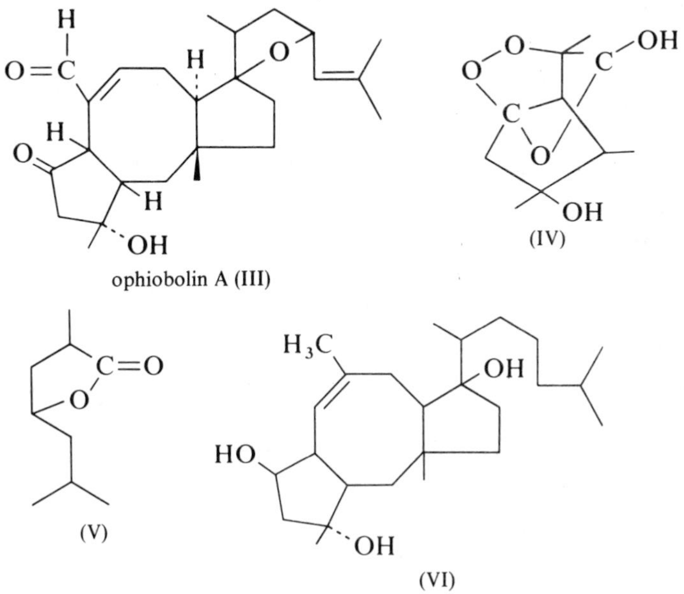

FIG. 30. Aspects of the degradation of ophiobolin A.

The n.m.r. spectrum of ophiobolin A (III) revealed two tertiary C-methyl groups, one secondary methyl and two attached to an olefin. It also showed two protons adjacent to a carbonyl group, an ether proton, two olefinic protons, and an aldehyde. The hydroxyl group of ophiobolin A was readily eliminated generating an $\alpha\beta$-unsaturated ketone. Furthermore the new double bond carried a methyl group and only one proton. This established the relationship of the hydroxyl group to the cyclopentanone. Lithium aluminium hydride reduction of ophiobolin A gave two stereoisomeric triols. One of these was re-oxidized to ophiobolin A. The other was oxidized by chromium trioxide to give a γ-lactone, thus establishing the relationship of the aldehydic and ketonic functions. Furthermore, the primary alcohol derived from the aldehyde could be re-oxidized with manganese dioxide to an $\alpha\beta$-unsaturated aldehyde. This reagent is specific for allylic alcohols and this therefore established the position of one of the double bonds. The relationship of the aldehyde and ketone functions was revealed by a number of other reactions. For example, hydrogenation of ophiobolin A gave a tetrahydro-derivative. The aldehyde group then existed predominantly in the enolic form, possibly stabilized by hydrogen bonding with the cyclopentanone. This reacted with oxygen to give a cyclic peroxide (IV) which was hydrolysed to form an enolic nor-β-diketone. With hydrazine a pyridazine is formed between these two centres.

The structure of the side chain was established by several degradations. The n.m.r. spectrum of ophiobolin A showed signals from only one ether-type proton. Hence, one terminus of the ether must be tertiary and the other secondary. Furthermore, decoupling experiments showed this to be allylic to an olefin. This relationship was established by the hydrogenolysis of the ether. The presence of a C_8 fragmentation in the mass spectra of a number of derivatives suggested the length of the side chain. This was confirmed by nitric acid oxidation of tetrahydro-ophiobolin A, when the heptanoic acid lactone (V) was isolated. Oxidation of the hydrogenolysed product gave 2,6-dimethylheptanoic acid.

Ophiobolin B lacks the side chain ether, possessing in its place a C-14 hydroxyl group. This in turn is absent from ophiobolin C. The relationship between ophiobolin A and B was established in one instance by the formation of a common hydrogenolysis product (VI), and in another by the partial synthesis of ophiobolin B. Thus ophiobolin A was reduced to the corresponding triols. Reduction with lithium in liquid ammonia led to hydrogenolysis of the allylic ether. Oxidation with chromium trioxide in pyridine regenerated the carbonyl functions of ophiobolin B.

The sesterterpenes isolated from insect wax that possess the ophiobolane skeleton, have been interrelated. The structure of ceroplastol 1 (VIII) was

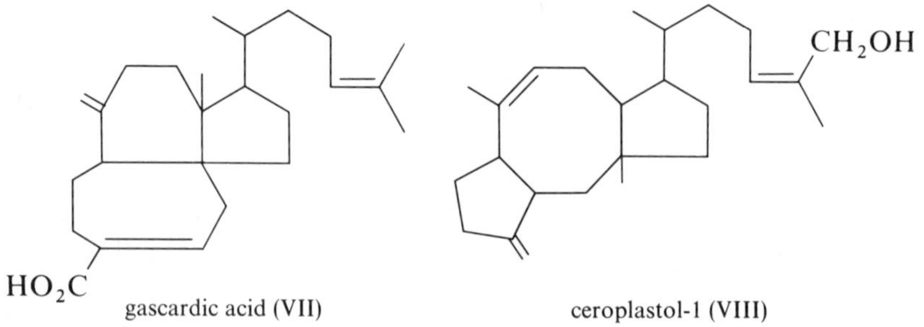

gascardic acid (VII) ceroplastol-1 (VIII)

FIG. 4. Some insect sesterterpenes.

determined (Iitaka *et al.*, 1968) by X-ray analysis of its *p*-bromobenzoate. Ceroplasteric acid is related to it. Ceroplastol 11 and albolic acid (Rios and Quijano, 1969) are double bond isomers of these two, possessing a tetrasubstituted double bond on ring A in place of the terminal methylene group. This relationship was established chemically.

The biosynthetic work on the sesterterpenes is reviewed elsewhere. It is characterized by the very successful use of multiply-labelled mevalonates. The carbon skeleton possesses a number of interesting synthetic problems and, no doubt, this will be the subject of some future investigations.

IV. Typical Sesterterpenes

Compound	Formula	M.P.	Rotation
Gascardic acid	$C_{25}H_{38}O_2$	124°	+145°
Ophiobolin A	$C_{25}H_{36}O_4$	182°	+301°
Ophiobolin B	$C_{25}H_{38}O_4$	175°	+300°
Ophiobolin C	$C_{25}H_{38}O_3$	121°	+363°
Ophiobolin D	$C_{25}H_{36}O_4$	140°	+76°
Ophiobolin F	$C_{25}H_{42}O$	81°	+23°

REFERENCES

OPHIOBOLINS

Tsuda, K., Nozoe, S., Morisaki, M., Hirai, K., Itai, A., Okuda, S., Canonica, L., Fiecchi, A., Galli-Kienle, M. and Scala, A. (1967). *Tetrahedron Lett.* 3369.

STRUCTURE DETERMINATION

Nozoe, S., Morisaki, M., Tsuda, K., Iitaka, Y., Takahashi, N., Tamura, S., Ishibashi, K. and Shirasaka, M. (1965). *J. Amer. Chem. Soc.* **87**, 4968.

Canonica, L., Fiecchi, A., Galli-Kienle, M. and Scala, A. (1966). *Tetrahedron Lett.* 1211; 1329.

Nozoe, S., Hirai, K. and Tsuda, K. (1966). *Tetrahedron Lett.* 2211.

Itai, A., Nozoe, S., Tsuda, K., Okuda, S., Iitaka, Y. and Nakayama, Y. (1967). *Tetrahedron Lett.* 4111.

Nozoe, S., Itai, A., Tsuda, K. and Okuda, S. (1967). *Tetrahedron Lett.* 4113.

Nozoe, S., Morisaki, M., Fukushima, K. and Okuda, S. (1968). *Tetrahedron Lett.* 4457.

BIOSYNTHESIS

Canonica, L., Fiecchi, A., Galli-Kienle, M., Ranzi, B. M., Scala, A., Salvatori, T. and Pella, E. (1967). *Tetrahedron Lett.* 3371.

Canonica, L., Fiecchi, A., Galli-Kienle, M., Ranzi, B. M. and Scala, A. (1967). *Tetrahedron Lett.* 4657; 1968, 275.

Nozoe, S., Morisaki, M., Tsuda, K. and Okuda, S. (1967). *Tetrahedron Lett.* 3365; 1968, 2347.

Insect Substances

Iitaka, Y., Watanabe, I., Harrison, I. T. and Harrison, S. (1968). *J. Amer. Chem. Soc.* **90**, 1092.
Rios, T. and Perez, S. (1969). *Chem. Comm.* 214.
Rios, T. and Quijano, L. (1969). *Tetrahedron Lett.* 1317; 2929.
Arigoni, D. (1965). Lecture, The Chemical Society, Nottingham.

5

THE TRITERPENOIDS

J. D. CONNOLLY and K. H. OVERTON

University of Glasgow

I. Introduction

The triterpenoids form a very large group of naturally occurring substances, widely distributed throughout the plant kingdom. There are probably upward of 500 natural triterpenoids of established structure. A small but important group, which includes lanosterol, is of animal origin. Although the isolation of many well-known triterpenoids dates back to the last century, the first correct structures were not assigned until the time of the Second World War. Thus the parent substances β-amyrin, α-amyrin, and lupeol were correctly formulated respectively in 1937,[157] 1949[216] and 1951.[7]

Triterpenoids together with steroids have played an important part in laying the foundations of the "New Organic Chemistry". In particular they provided an excellent experimental basis for the principles of Conformational

Analysis.[22,115] More recently lanosterol has become the focus of interest in studies directed towards an understanding of steroid biosynthesis.[93] Two major advances have resulted from this, each of outstanding importance, for terpenoid chemistry and biochemistry in general. The first is embodied in the Biogenetic Isoprene Rule,[250,251,263] associated with the Zürich School of Ruzicka, Arigoni and Eschenmoser. This postulates that each class of terpenoids is formed from an acyclic precursor which is cyclized and further elaborated according to a limited number of well-defined stereo-electronic principles. It is the triumph of this hypothesis that it successfully accommodates in constitutional and configurational detail the great diversity of terpenoid structures found in Nature. The conceptual edifice of the Biogenetic Isoprene Rule is matched and complemented by the biochemical investigations of Bloch, Lynen, Cornforth and Popjak[93,79,205] which have elucidated in extraordinary detail the precise mechanism whereby lanosterol is synthesised in living systems.

We have attempted in the following survey to indicate the structural relationships among triterpenoids in terms of the Biogenetic Isoprene Rule, essentially after the manner adopted by Ruzicka[250] in his Faraday Lecture of 1958. Since the basic chemistry of triterpenoids has been surveyed in a number of reference works and reviews,[261,213,236,278,106,175,176,149] we have used our limited space to emphasize what appear to us to be some of the more important advances of the last five years.

II. Squalene

The hexa-ene all-trans squalene (1) is the immediate biological precursor of all triterpenoids. Its stereochemistry was established by an X-ray investigation[228] of its urea adduct and it has been synthesised recently by a number of ingeneous methods.[49,52] Its biosynthesis has been established in elaborate detail;[93] only the final step, the coupling of two molecules of farnesyl pyrophosphate, awaits mechanistic clarification.[240,246] *

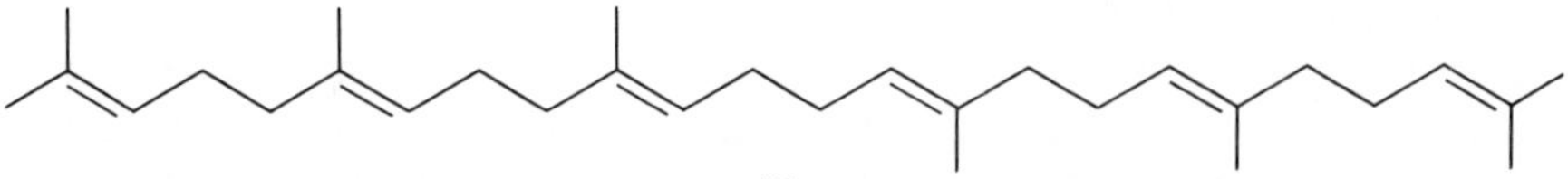

(1)

Squalene is transformed[91,275] into lanosterol (2) via cyclization of its terminal epoxide (4) (Fig. 1). Cycloartenol (3), the proximal cyclization product of squalene in plants, is similarly formed via squalene epoxide[244,245] and so are β-amyrin[90] and fusidic acid[141] (see below). It is assumed that other

* Several important recent publications[6a,65a,79a,114a,136a,246a,272a] advance this problem considerably.

3-hydroxy-triterpenoids arise analogously through squalene epoxide. The pentacyclic fern hydrocarbons (see section V) are the result of cyclization initiated by terminal protonation of squalene.

The Biogenetic Isoprene Rule postulates that the different types of tetra- and pentacyclic triterpenoids are formed according to the conformation that squalene (epoxide) adopts, presumably at an enzyme surface prior to cyclization. Each leads stereospecifically to a particular cyclization product according to the principles of the Biogenetic Isoprene Rule. We have, in the sequel classified triterpenoids according to this postulate.

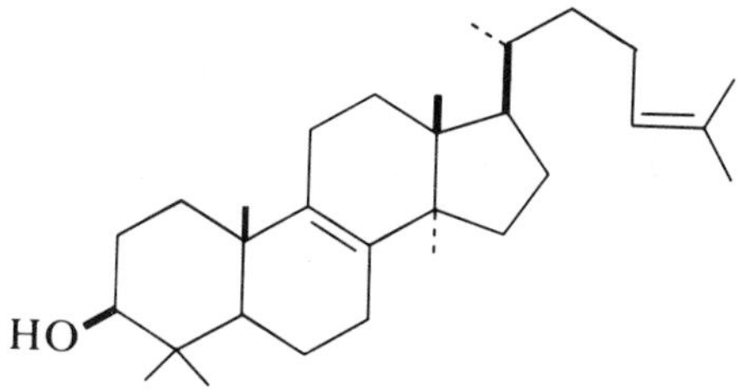

(2)

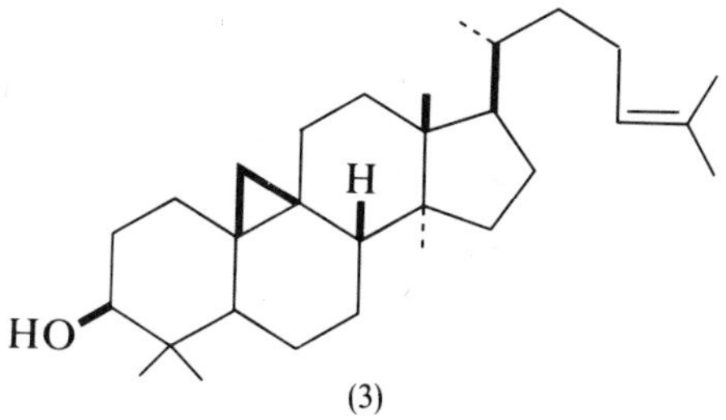

(3)

III. Cyclization of Squalene Epoxide in the Chair–Boat–Chair–Boat Sequence (Figure 1)

Cyclization of squalene epoxide (4) (Fig. 1) leads initially to the bridged cation (5) [an alternative pathway via a spiro-cation (12) has been proposed[274]] and then via (6) either directly to protosterol (7) or by a sequence of 1, 2-hydride and methyl shifts (6, arrows) to lanosterol (2), cycloartenol (3) and the cucurbitacins (11).

The intermediacy of the protosterol (7) in the biosynthesis of lanosterol was an essential postulate of the Biogenetic Isoprene Rule. It has recently been isolated[156] from *Cephalosporium caerulens* together with helvolic acid (9); moreover, it has been shown[179] that another micro-organism

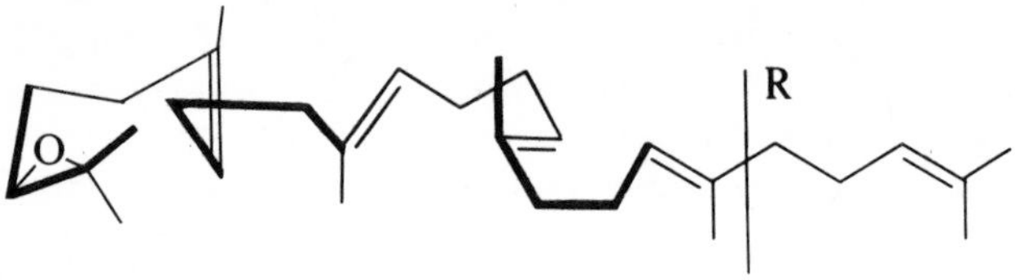

(4)

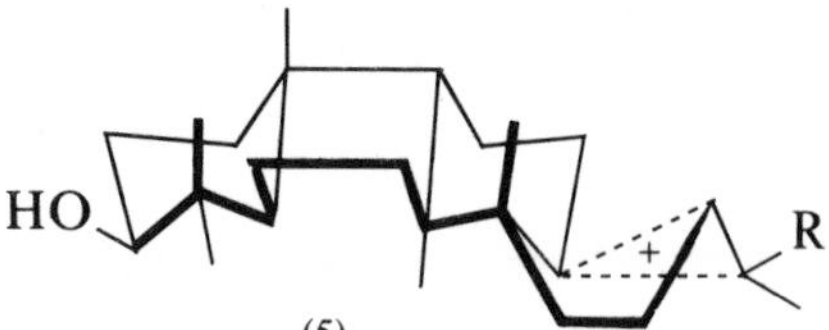

(5)

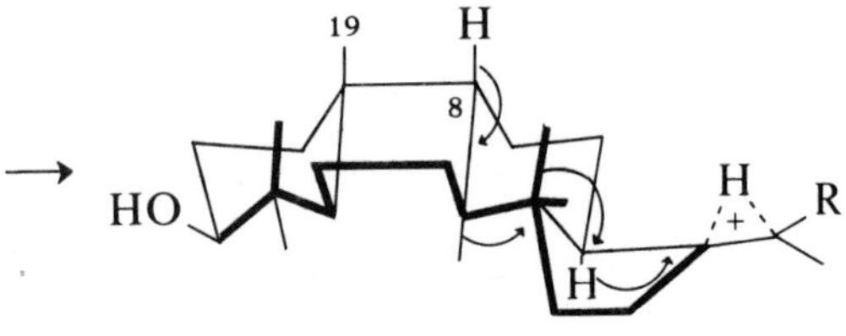

(6)

Lanosterol (2)
Cycloartenol (3)
Cucurbitacins (11)

Protosterol (7)
Fusidic Acid (8)
Helvolic Acid (9)
Cephalosporin P_1 (10)

FIG. 1.

incorporates radio-activity from MVA into (7). The $\Delta^{13,17}$-isomer has been converted by acid into dihydrolanosterol.[156] Alisol A (13)[223,177] is typical of a recently isolated group of biologically active protosterol derivatives.

Isolation of the antibiotics fusidic acid (8),[140] helvolic acid (9)[173,234] and cephalosporin Pl (10)[151] which lack one 4-methyl group of protosterol, preceded that of the simple protosterol (7) by a decade. It has been shown that fusidic acid, whose structure is supported by X-ray analysis,[88] is biosynthesised[141] from squalene 2, 3-oxide.

Cycloartenol (3), first formulated[39] in 1953 has recently assumed importance as a result of experiments[40,136,159,244,245] which demonstrate that in phytosterol biosynthesis it replaces lanosterol as the first-formed intermediate from squalene oxide. Cycloartenol can arise from the bridged cation (6) by 1,2-shift of the 8βH and proton loss from the C-19 methyl group, but since the migrating hydrogen and the cyclopropane ring formed are *syn* (both β), stabilization of the intermediate C-8 cation seems to be implied.

The cucurbitacins[197,291] have attracted considerable attention because of their varied oxygenation patterns and the initial difficulty encountered in establishing the rearranged skeleton common to them. The cucurbitane and lanostane series have recently been chemically inter-related[26,27] in an

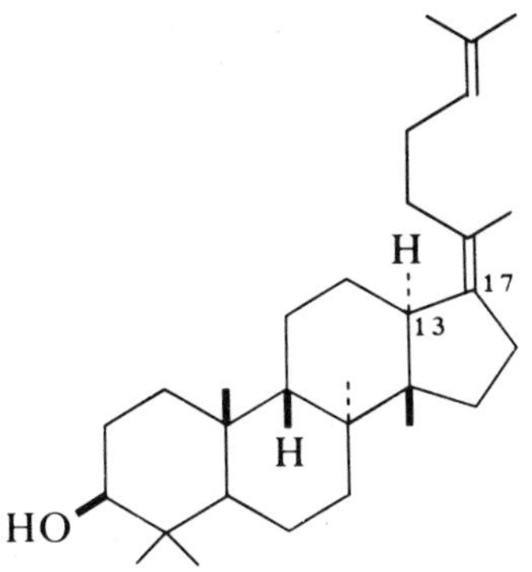

(7)

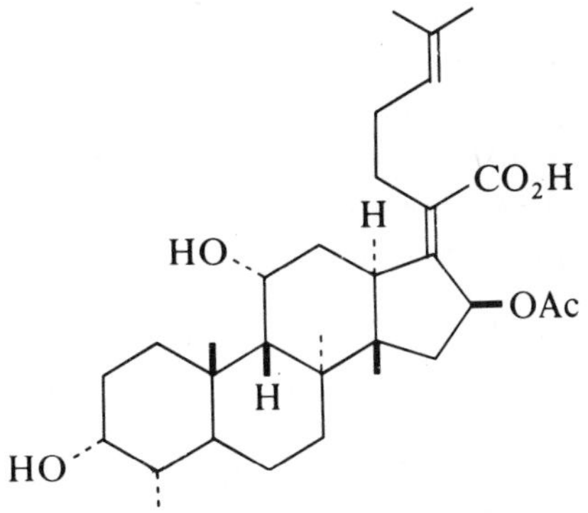

(8)

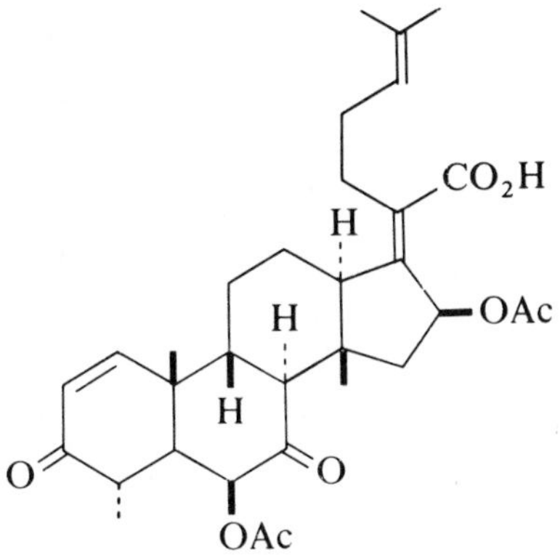

(9)

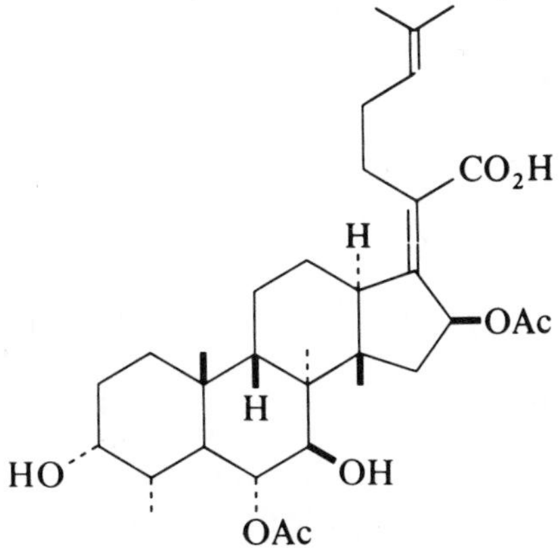

(10)

ingeneous way. Cucurbitacin (A) (11) was transformed into the cyclopropyl ketone (14) and this was opened with base to afford after oxidation the tetra-ketone (15), also obtained from eburicoic acid.

The Buxus alkaloids, typified by cyclobuxine (16),[59] are an interesting group of substances based on cycloartenol. The six terminal carbon atoms of the side chain have been removed, and nitrogen functions are attached at C-3 and C-20. Buxenine-G (17)[241] presumably results from cleavage of the cyclopropane to generate a seven-membered ring B.

The tricyclic C_{30} ketone malabaricol (18)[78,262] is of unusual interest since subsequent to elucidation of its structure it was discovered[276] that *non-enzymic* cyclization of squalene 2,3-oxide leads to the trienols (19) and (20). More recently (±)-malabaricanediol has been obtained[257] in 7% yield by picric acid-catalysed cyclization of the epoxydiol (21). Enzymes clearly play a major role in foiling this stereo-electronically preferred mode of cyclization during *in vivo* lanosterol synthesis. In the laboratory this has been effectively circumvented in the cyclization of (22) (see below). However, it is fascinating to add that the tricarbocyclic system of malabaricol has recently been obtained[273] *enzymically* when 2,3-oxido-squalene-lanosterol cyclase transformed 18,19-dihydrosqualene 2,3-oxide into (23).

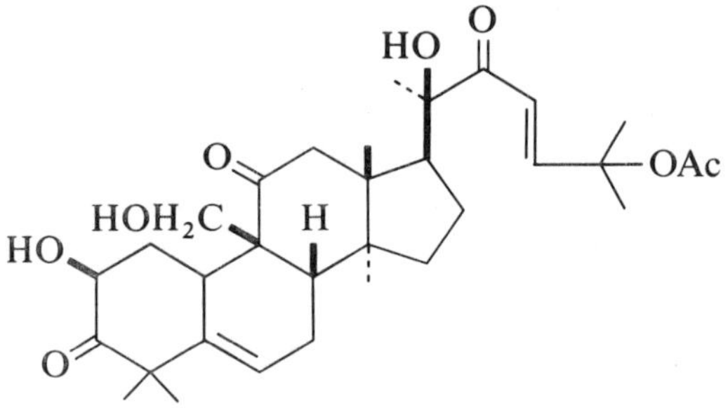

(11)

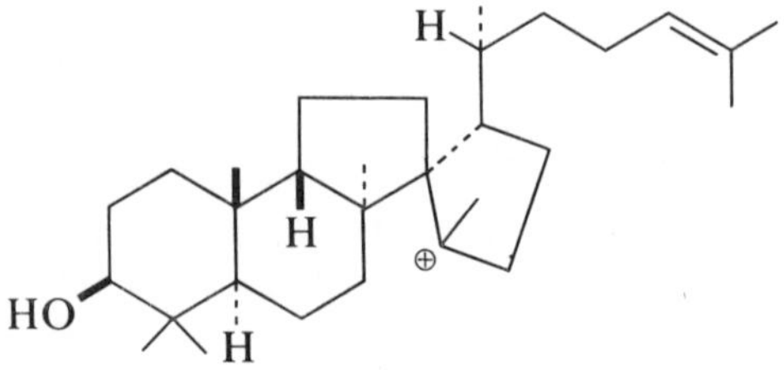

(12)

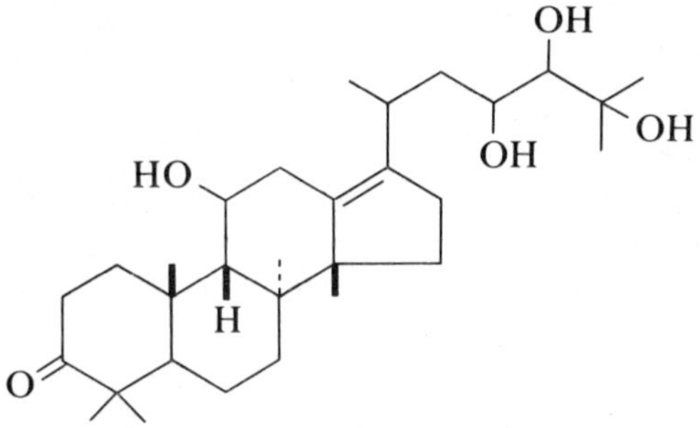

(13)

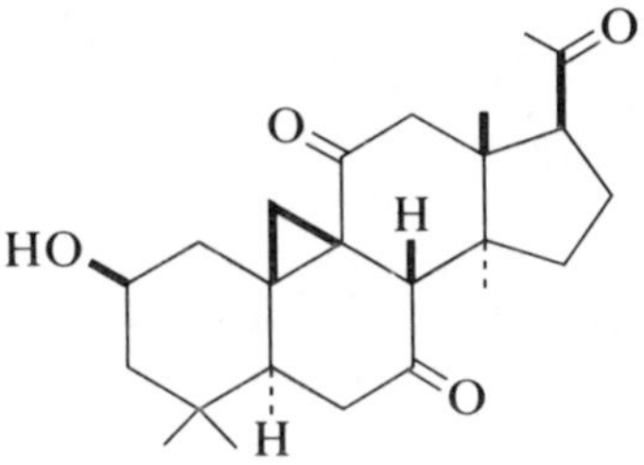

(14)

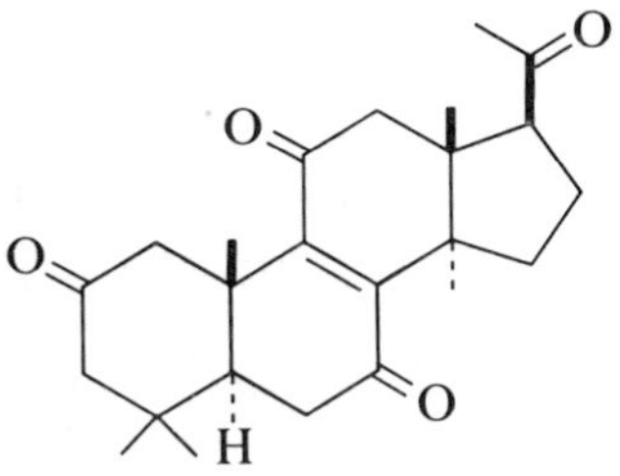

(15)

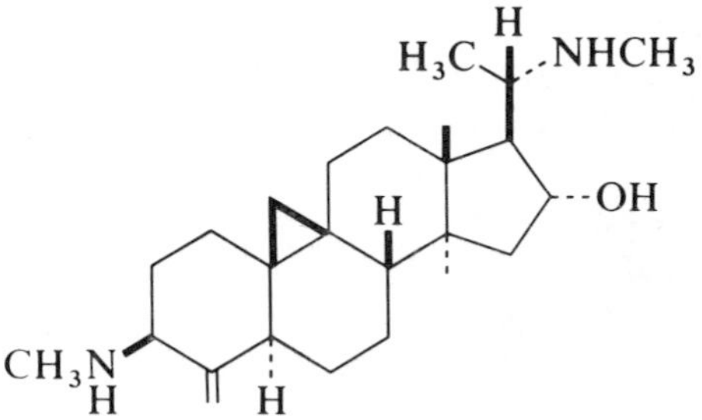

(16)

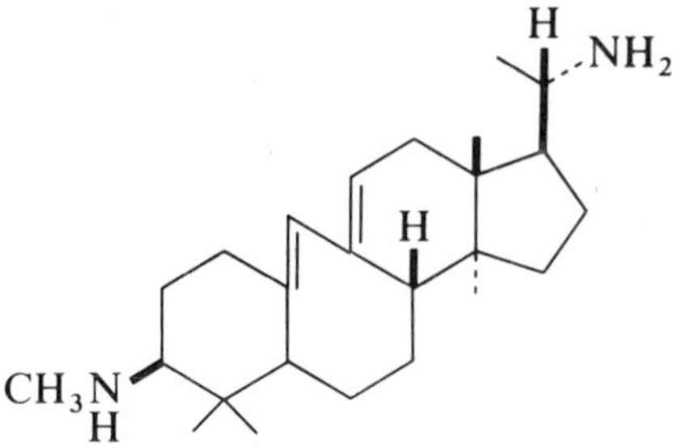

(17)

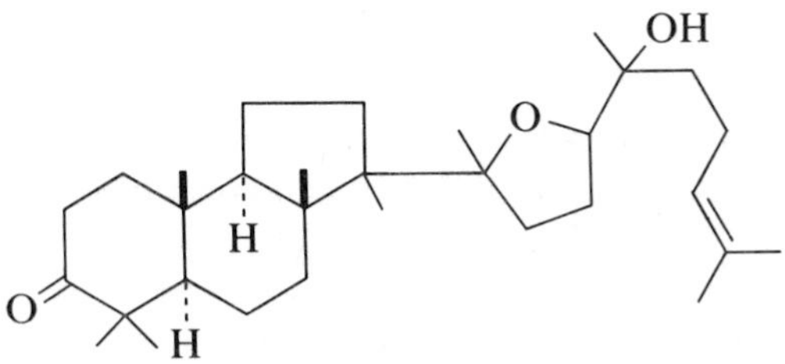

(18)

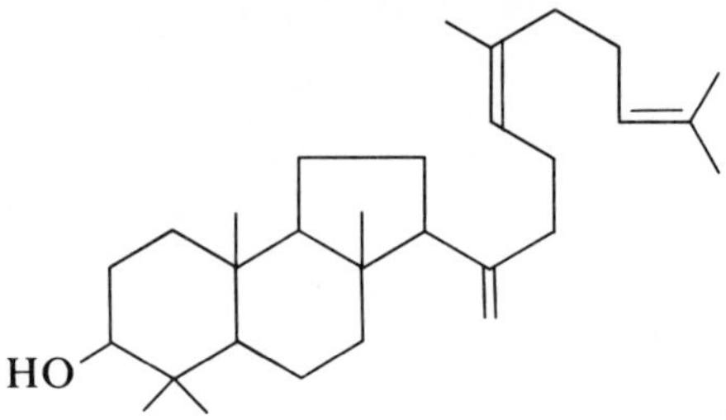

(19)

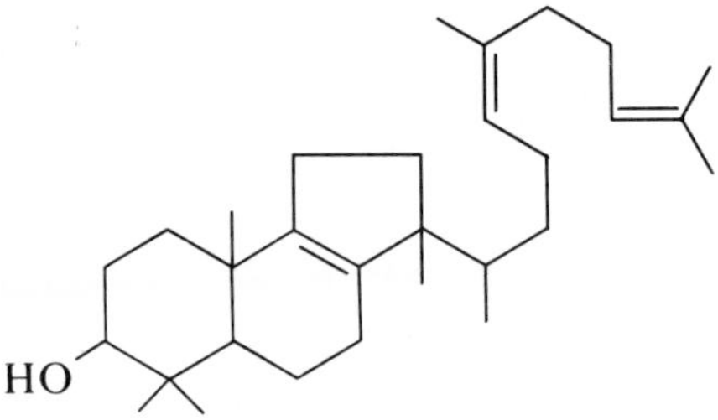

(20)

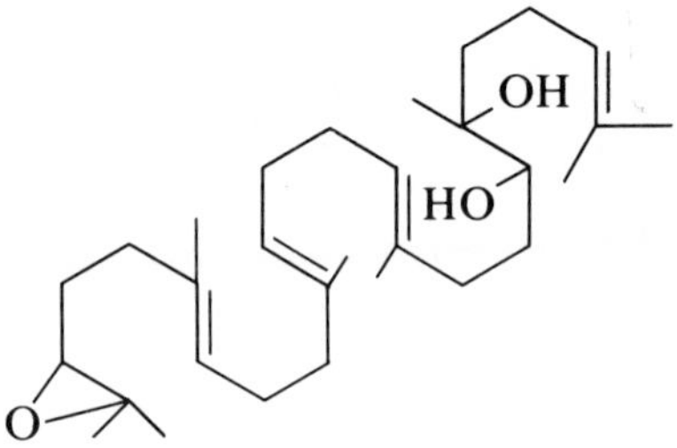

(21)

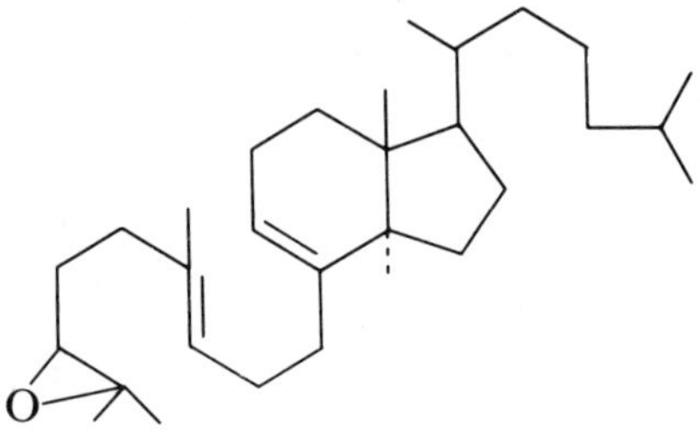

(22)

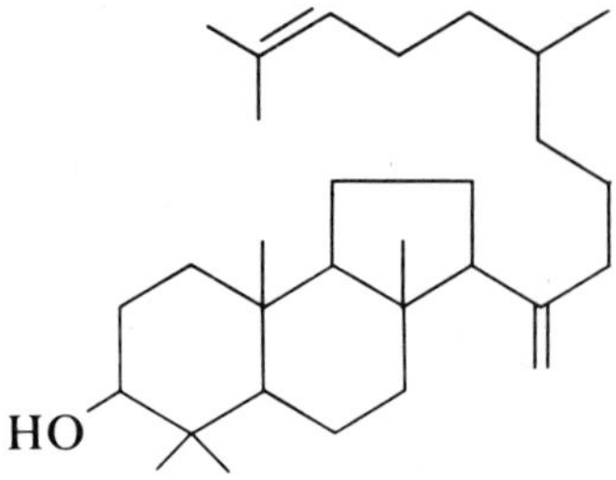

(23)

IV. Cyclization of Squalene Epoxide in the Chair–Chair–Chair–Boat Sequence

A. GROUP A (FIGURE 2)

Cyclization of squalene epoxide in the conformation (24) leads to cation (25) from which are derived the tetracyclic constituents of dammar resin, dammarene diols-I and -II (26), epimeric at C_{20}.[48,277] The ring-A cleaved dammarenolic acid (27)[11] was the first example of oxidative fission of C-3/C-4 in triterpenoids,[12,287] subsequently found in other groups (cf. nyctanthic,[12,287] roburic[209] and lansic acids[181]).

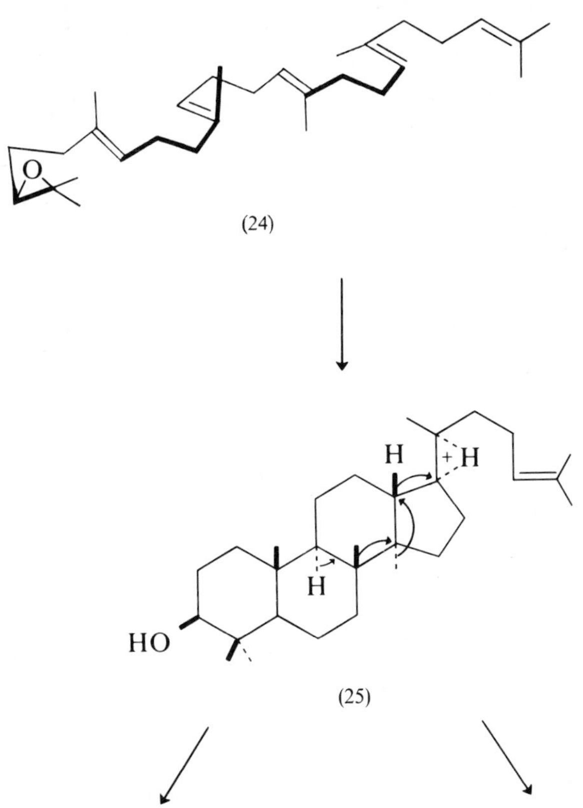

(24)

(25)

Euphol (28)
Tirucallol (29)
Cedrelone (33)
Simarolide (47)
Quassin (46)

Dammarenediols (26)

FIG. 2.

More important and numerous are the members of the euphane-tirucallane groups which are formed by rearrangement of the cation (25; arrows) by a process analogous to the rearrangement of cation (6) to lanosterol. Euphol (28) and tirucallol (29), differ in configuration at C-20, euphol having the same configuration as lanosterol.

A great deal of chemical interest has in recent years centred on the two groups known as limonoids and quassinoids which are notionally derivable from euphol or tirucallol. This area of terpenoid chemistry has been reviewed at length[110,85] and only a brief summary will here be given.

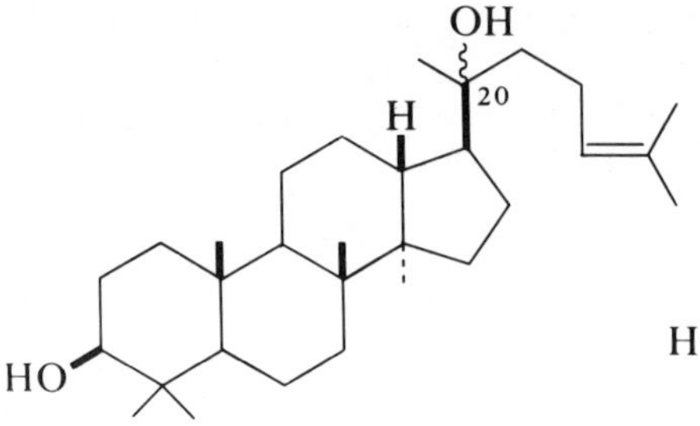

(26)

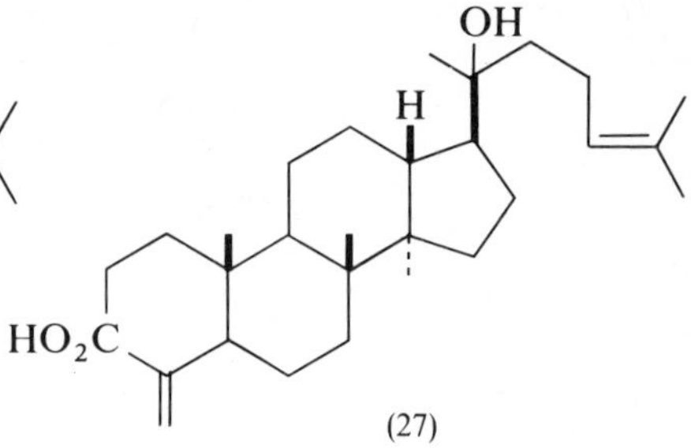

(27)

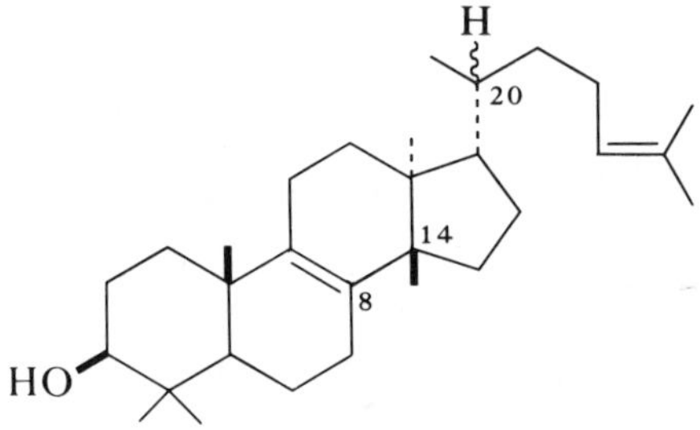

(28) 20βH
(29) 20αH

The limonoids[13,34] are in essence derived from the hypothetical intermediate apo-euphol (30), which arises from euphol (28) by skeletal rearrangement during which one methyl migrates from C-14 to C-8. This might be triggered by oxidative attack of the nuclear double bond of the Δ^7-isomer of euphol for instance via the epoxide. Laboratory analogies have been recently provided.[62,199] It is uncertain whether oxygenation of the side chain precedes or follows the apo-euphol rearrangement, since both melianone (31)[198] and grandifoliolenone (32)[82] have been isolated. However, oxidative modification eventually results in removal of the four terminal side chain carbon atoms and formation of a β-substituted furan ring.

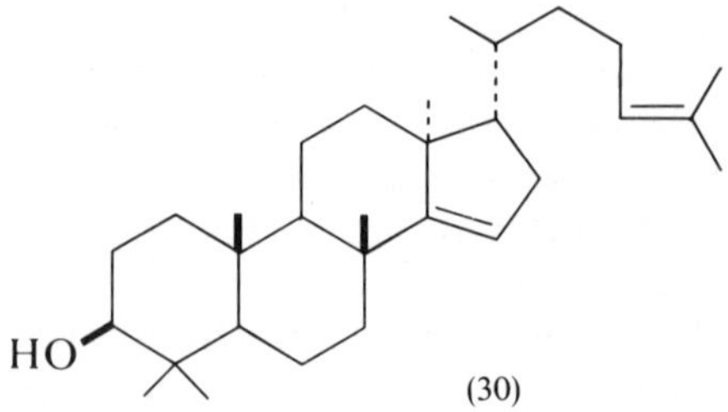

(30)

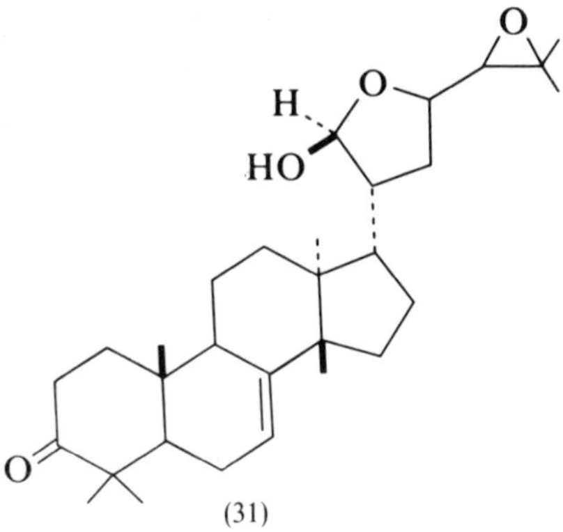

(31)

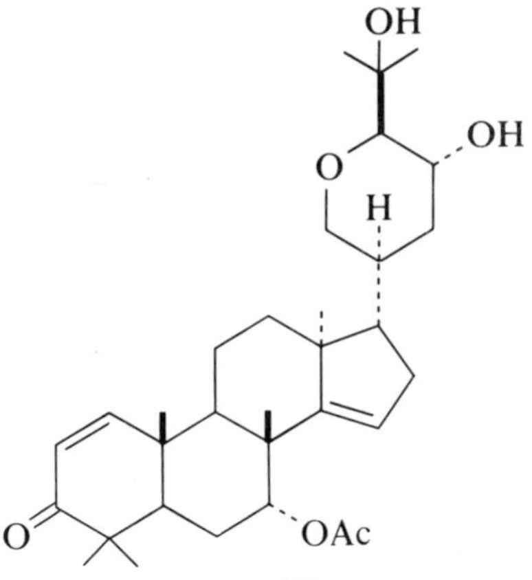

(32)

Cedrelone (33)[143,164] and gedunin (34)[6] are typical of the group which retains the apo-euphol skeleton intact. The ring C cyclopentene epoxide and $\alpha\beta$-oxido-δ-lactone represent oxidative modifications (cyclo-pentene $\rightarrow$ $\Delta^{\alpha\beta}$-cyclopentenone $\rightarrow$ $\Delta^{\alpha\beta}$-δ-lactone) typical of the limonoids. The ring-A enone and ring-B diosphenol systems, or close relatives, are also frequently found. The structures of cedrelone and gedunin are firmly based on X-ray analyses[146,266] of derivatives.

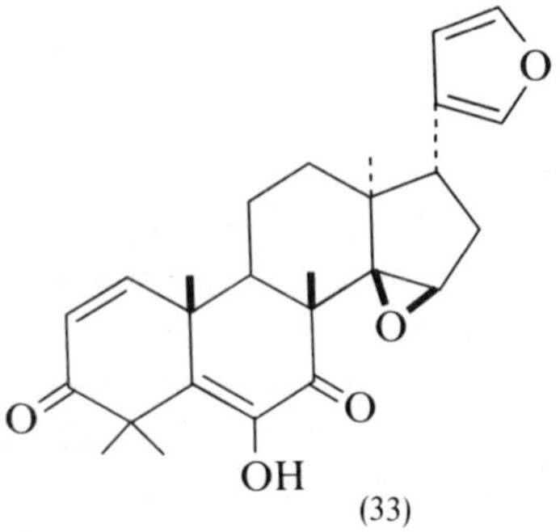

(33)

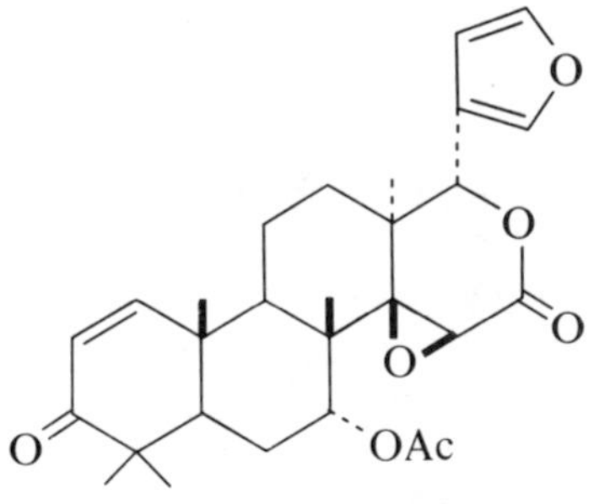

(34)

Gedunin, on treatment with alkali, suffers an interesting fragmentation[6] with formation of a δ-lactone (35) and furan-3-aldehyde. The reaction is typical of limonoids having the ring-D structure of gedunin and is termed the "merolimonol rearrangement"[13,34] from its original observation in the limonin series.

The other classes of limonoids each have one carbocyclic ring (A, B or C) cleaved. Limonin (36),[13,34] which gives its name to the whole class of tetranortriterpenoids has ring A cleaved between C-3 and C-4. An analogous cleavage occurs in dammarenolic acid (27).

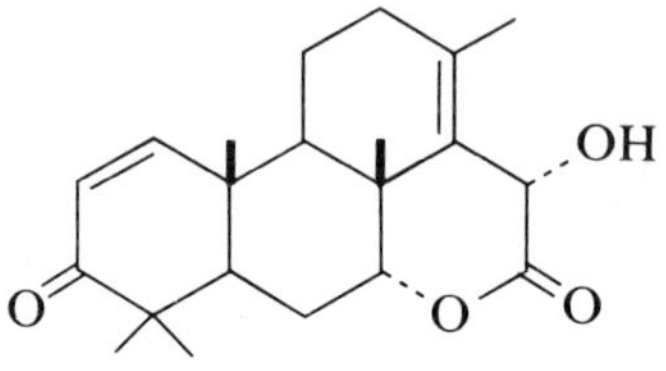

(35)

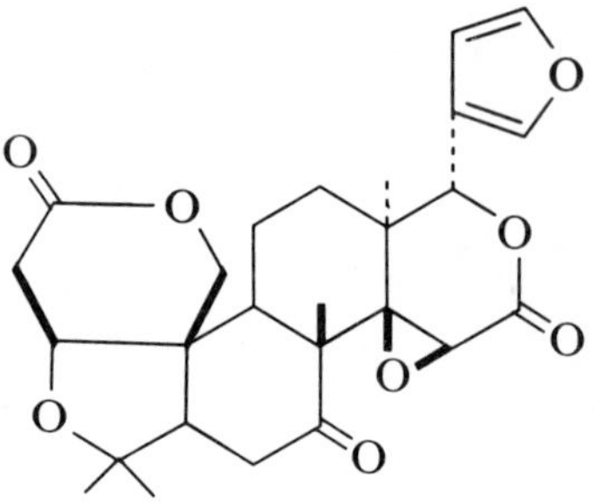

(36)

In the case of limonin the initially formed seco-hydroxy acid is however concealed by subsequent oxidation and cyclization [37 → 38 → 39 → 36].

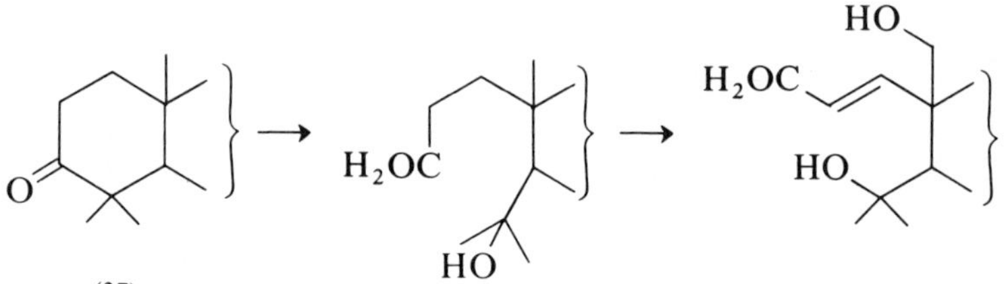

(37) (38) (39)

The chemical structure determination of limonin, which yielded a plethora of beautiful chemistry, is perhaps the last of the pre-X-ray era to deal with a substance of such complexity. A simultaneous independent X-ray analysis[15] supplied the configurational detail not revealed by the chemical work.

Cleavage of ring B, possibly by a similar process, leads to andirobin (40).[235] This cleavage also can be concealed in a subtle way, as is the case in the bicyclononanolides of which mexicanolide (41)[83] is typical. They probably arise in nature and have been made in the laboratory,[87] by Michael addition of C-2 to the conjugated diene-lactone system obtainable from an andirobin derivative [(42) → (41)]. The bicyclononandione system in mexicanolide cleaves[83] under mild basic conditions with extreme ease (as 41), arrows) to give (43).

Ring-C can be cleaved as in nimbin (45)[154,227] which very probably is formed as indicated in [(44) → (45)].

The C_{20} quassinoids, whose parent is quassin (46)[283] have a clear structural kinship with merogedunin (35). Incorporation experiments[220,221]

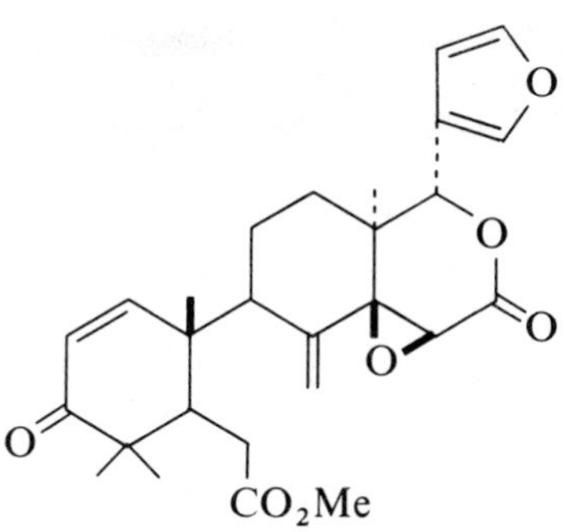

(40)

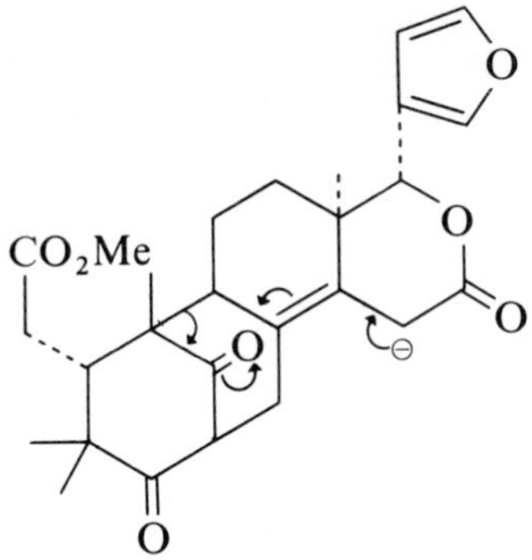

(41)

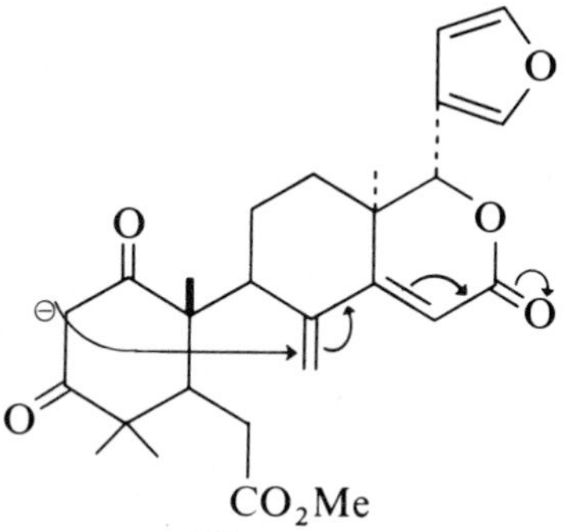

(42)

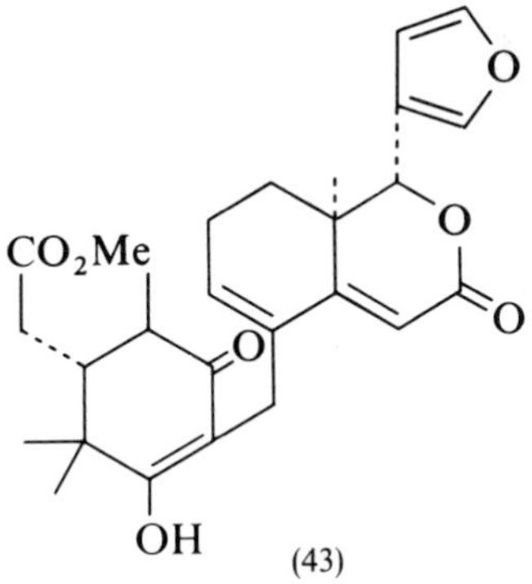

(43)

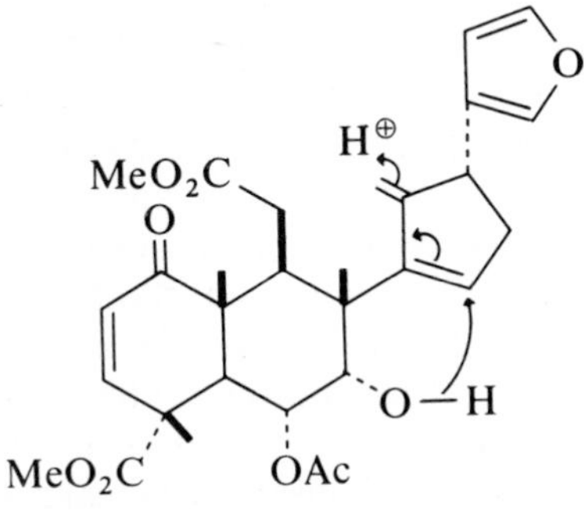

(44)

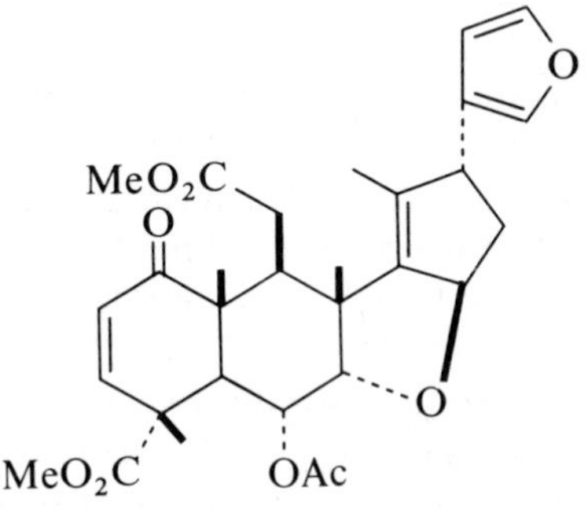

(45)

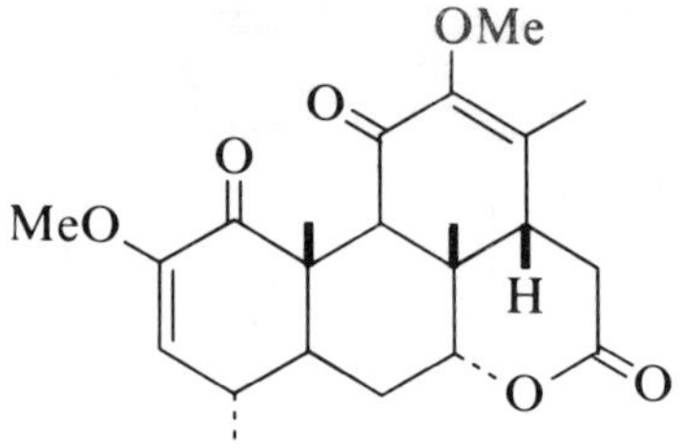

(46)

with differently labelled mevalonic acids are consonant with the suggestion that the quassinoids are derived *in vivo* from an *apo*-euphol precursor. While as a group they present fewer structural variations than the limonoids, the presence of numerous oxygen functions endows them with considerable chemical interest.

Simarolide (47)[60] and the corresponding diosphenol methyl ether[162] are related structurally both to limonoids and quassinoids.

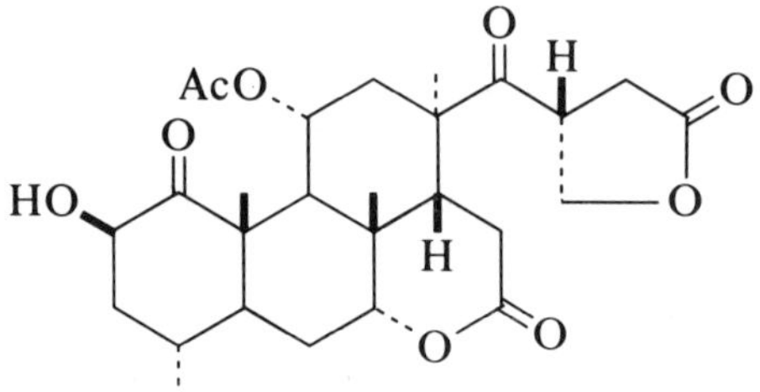

(47)

B. GROUP B (FIGURE 3)

The ion (48), derivable from (25) (Fig. 2) leads via (49) and (50) directly to allobetulin (51)[168] and germanicol (52)[36] and, by 1,2-hydride and methyl shifts to δ-amyrin (53)[224] and β-amyrin (54)[116,117] and to taraxerol (55),[38] multiflorenol (56),[255] glutinol (57)[137] and friedelin (58).[56] The rather unusual shionone (60),[267,268] ebelin lactone (61)[18] and baccharis oxide (64)[8] are derivable through cation (59).

Friedelin (58) occupies an important place in triterpene chemistry. Conversion[57,111] of the derived friedel-3-ene (62; arrows) with acid into olean-13(18)-ene (63) was the first and perhaps most spectacular of the "backbone rearrangements" which have come to be known as "friedo"-rearrangements. In this case seven stereospecific hydride or methyl migrations are involved, and the sequence of events is almost precisely the reverse of that postulated for the formation of friedelin from the cation (50).

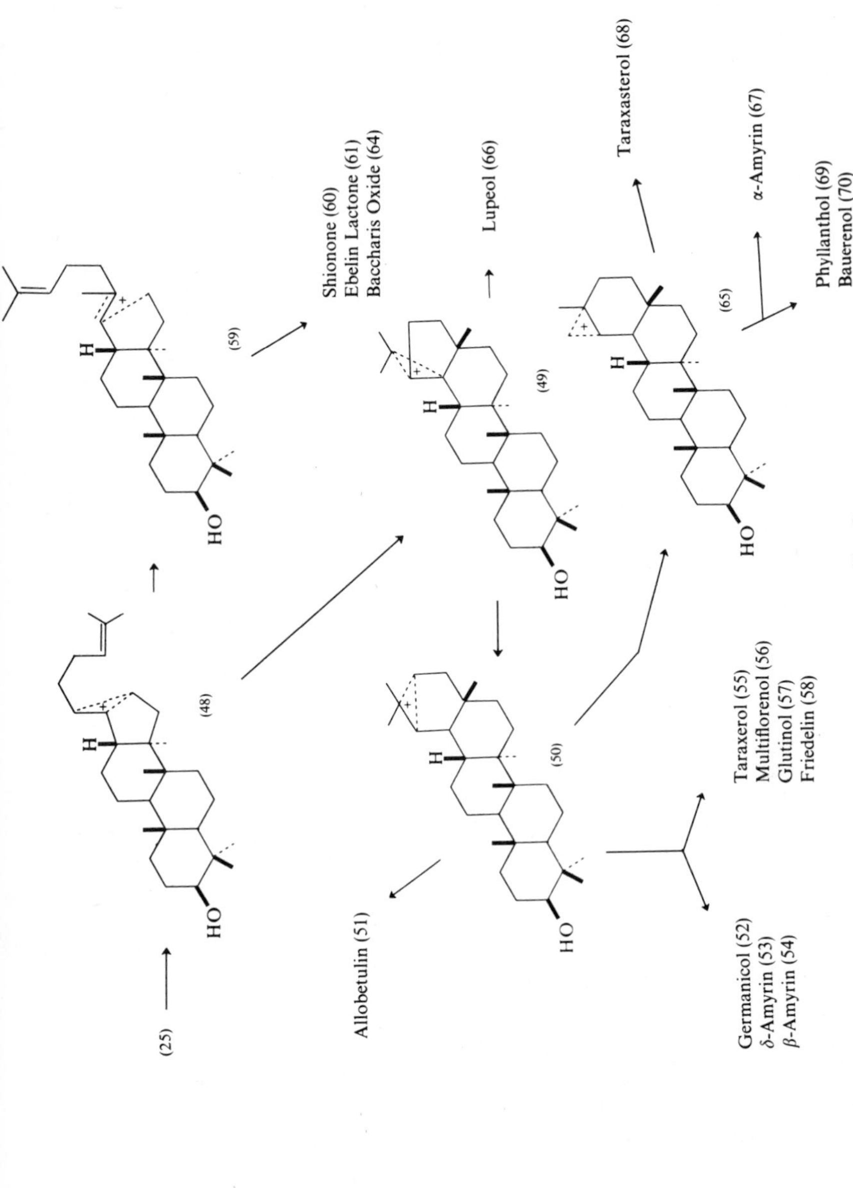
(25)
(48)
(59)
Shionone (60)
Ebelin Lactone (61)
Baccharis Oxide (64)
(49)
Lupeol (66)
(50)
Allobetulin (51)
Germanicol (52)
δ-Amyrin (53)
β-Amyrin (54)
Taraxerol (55)
Multiflorenol (56)
Glutinol (57)
Friedelin (58)
(65)
Taraxasterol (68)
α-Amyrin (67)
Phyllanthol (69)
Bauerenol (70)

FIG. 3.

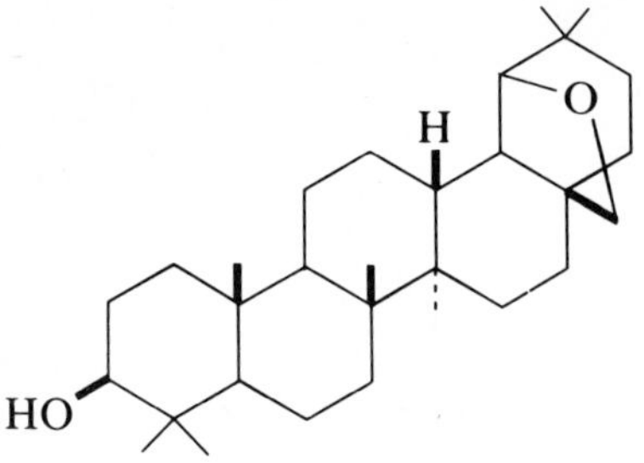
(51)

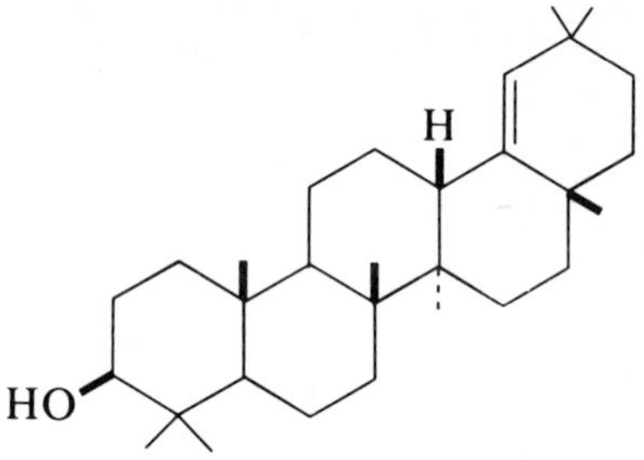
(52)

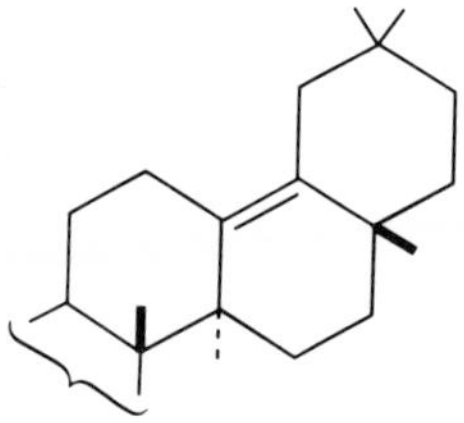
(53)

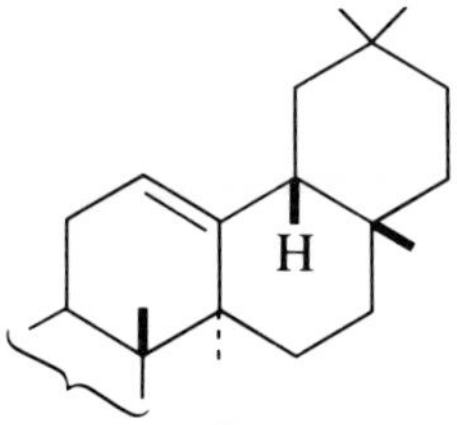
(54)

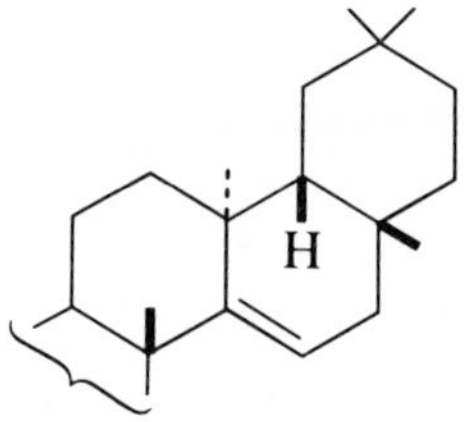
(55)

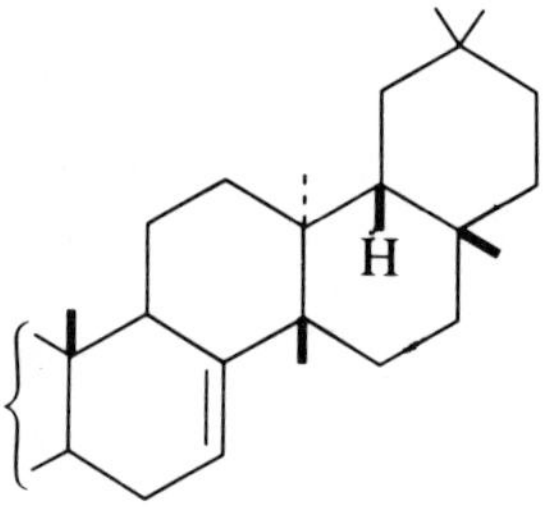
(56)

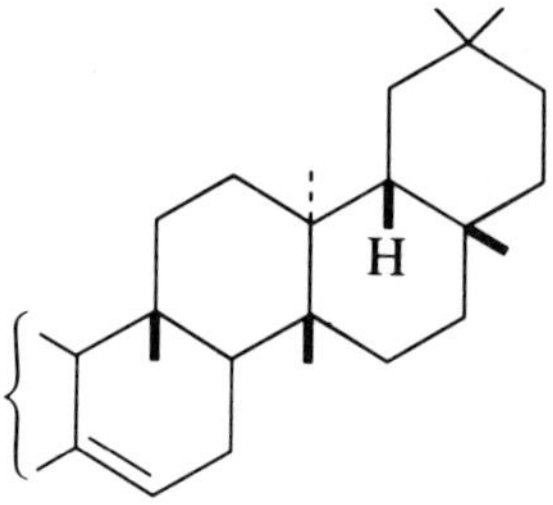
(57)

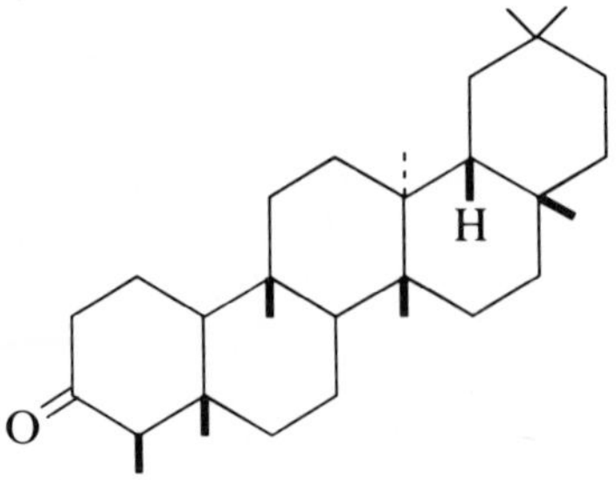
(58)

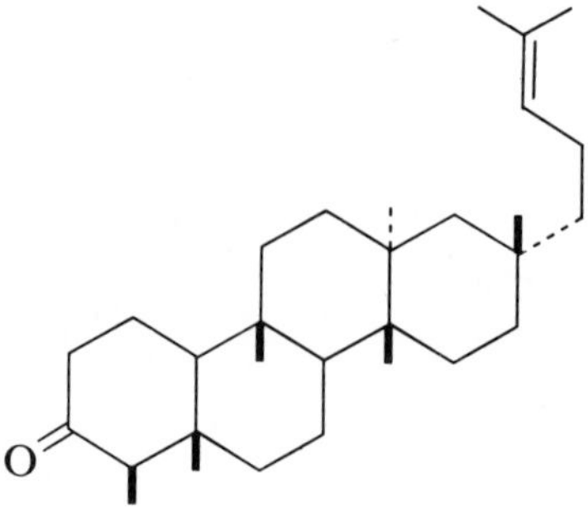
(60)

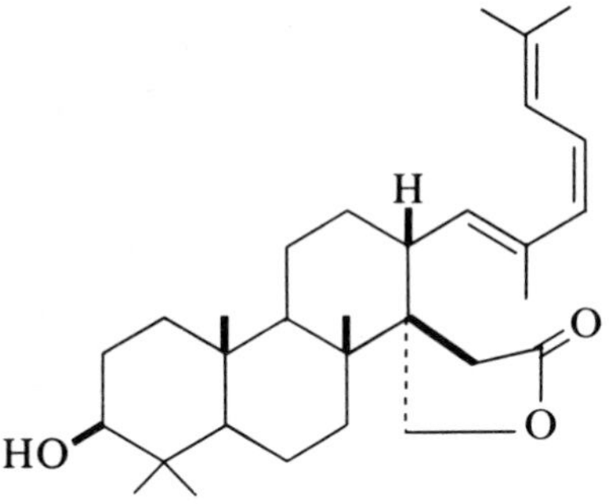

(61)

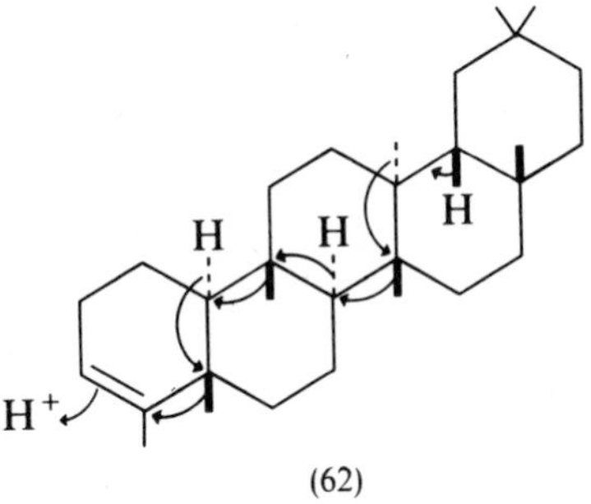

(62)

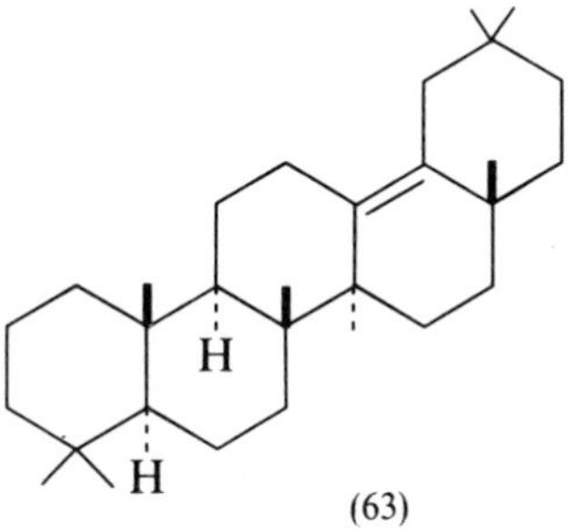

(63)

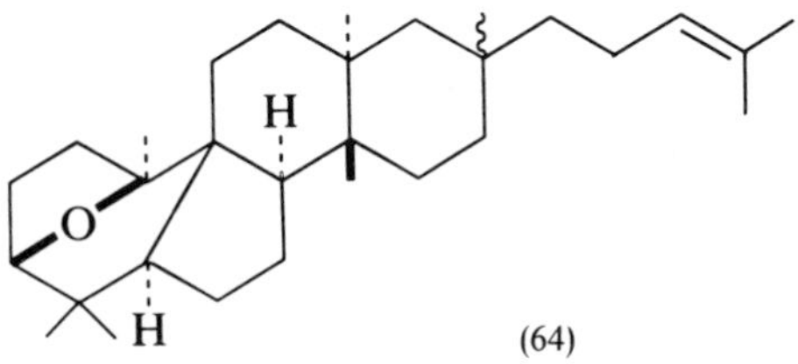

(64)

The large groups based on lupeol (66)[127] and α-amyrin (67)[89] are derivable, as shown, via the ions (49), (50) and (65). Phyllanthol (69)[33] and bauerenol (70)[195] represent an alternative fate for the cation (65), and are thus two members of the "friedo"-series related to α-amyrin. Taraxasterol (68)[130] is directly derivable from cation (65).

The β-amyrin (oleanane) group includes a very large number of naturally occurring members which cover an impressive variety of functional groups.

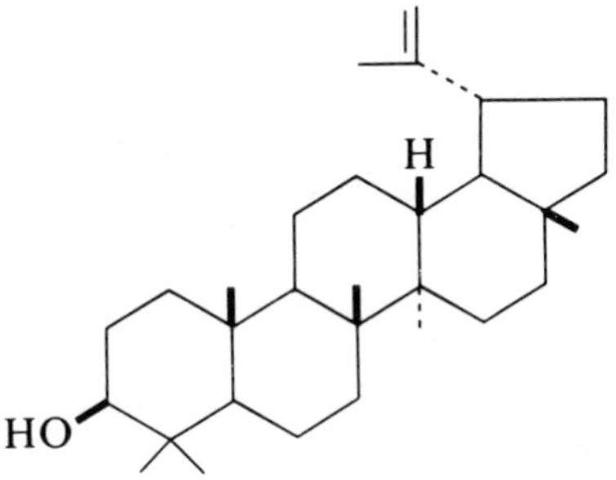

(66)

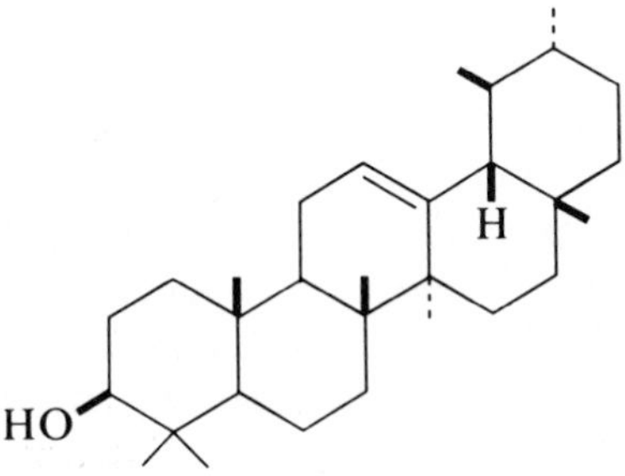

(67)

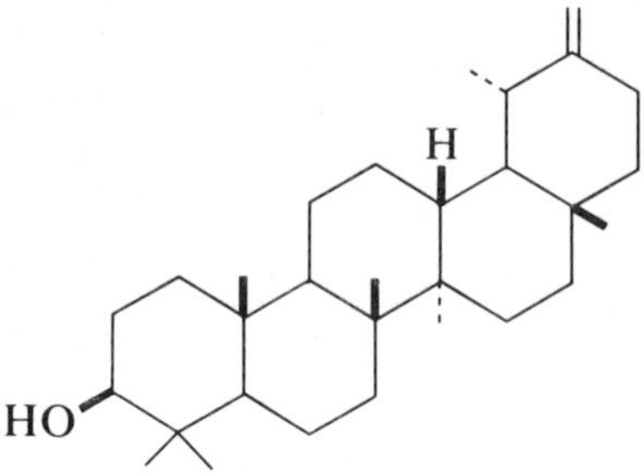

(68)

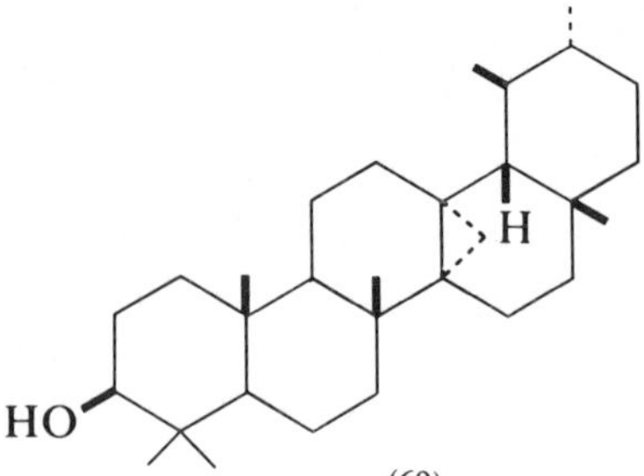

(69)

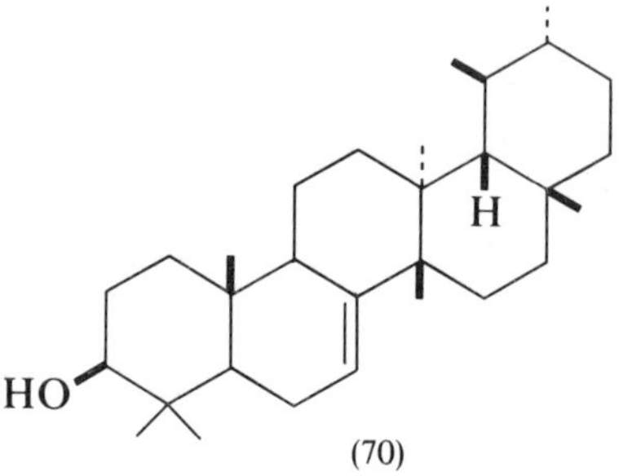

(70)

There are at present known examples of oxygen functions attached to all nuclear positions except, apparently, C-7 and C-11; of the angular methyl groups only that at C-8 has not been found oxidized.

V. Cyclization of Squalene in the Chair–Chair–Chair–Chair–Chair (Fig. 4), Chair–Chair–Chair–Chair–Boat (Fig. 5) and Chair–Boat–Chair–Chair–Boat (Fig. 6) Sequences

The triterpenoids considered up to this point are all oxygenated at C-3 and very probably arise, as does lanosterol, via squalene terminal epoxide. A small group of triterpene hydrocarbons have made their appearance in recent years which are formed by terminal protonation of squalene. It has recently been shown[28] that fern-9-ene (71) incorporates labelled squalene but does not incorporate labelled 2,3-oxidosqualene.

The relevant conformations of squalene are the C—C—C—C—C (72), C—C—C—C—B (73) and C—B—C—C—B (74) sequences (Figs. 4, 5, 6). Conformation (72) leads to cation (75) from which are derived diploptene (76)[5] and its congeners. The first pentacyclic triterpenoid from an animal source, the protozoan tetrahymanol (gammaceran-3β-ol) (78)[281] is accessible

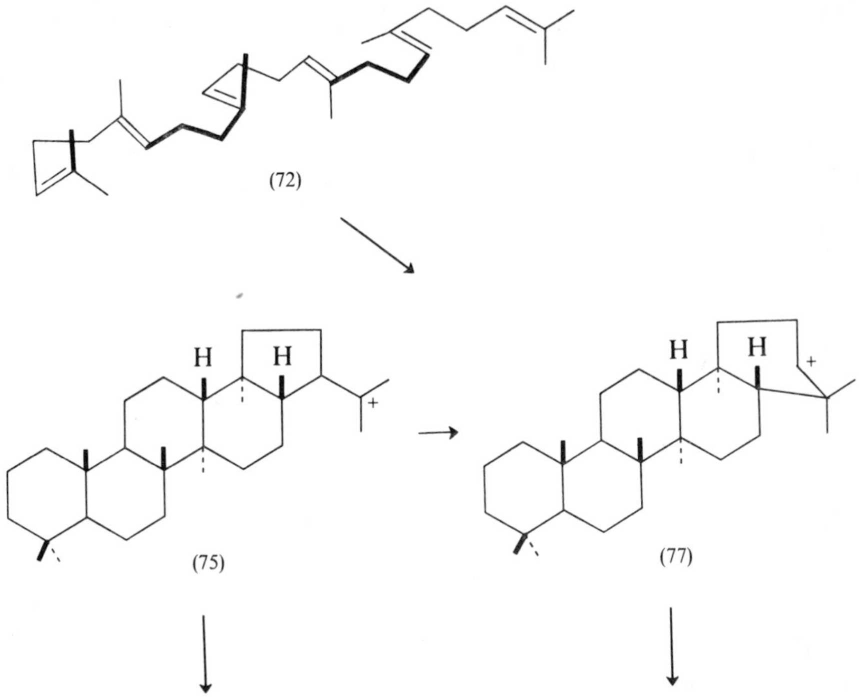

Diploptene (76)

Tetrahymanol (78)

FIG. 4.

through the cation (77). Experiments[69] with labelled squalene, D_2O and ^{18}O-enriched water demonstrate that tetrahymanol is formed from squalene by proton-induced cyclization and hydroxylation from the medium.

Conformation (73) of squalene leads via (79) directly to moretenol (80)[219] and, presumably by concerted rearrangement, to neomotiol (81), the fernenes (82),[226,2,3] adianene (83)[3] and filicene (84)[3] and their congeners.

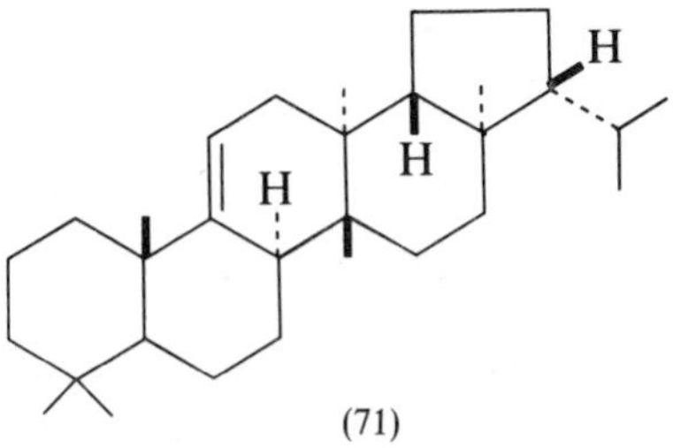

(71)

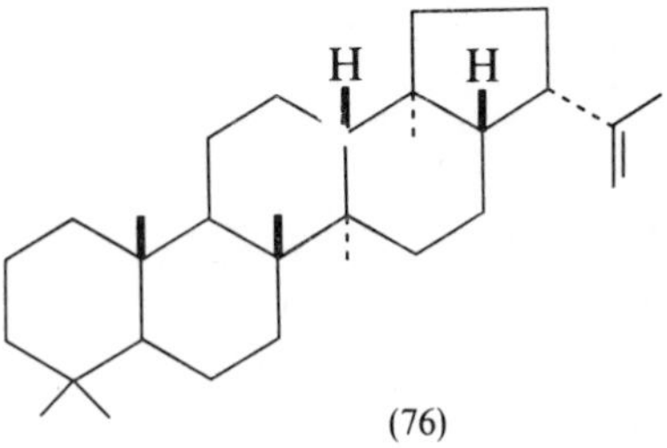

(76)

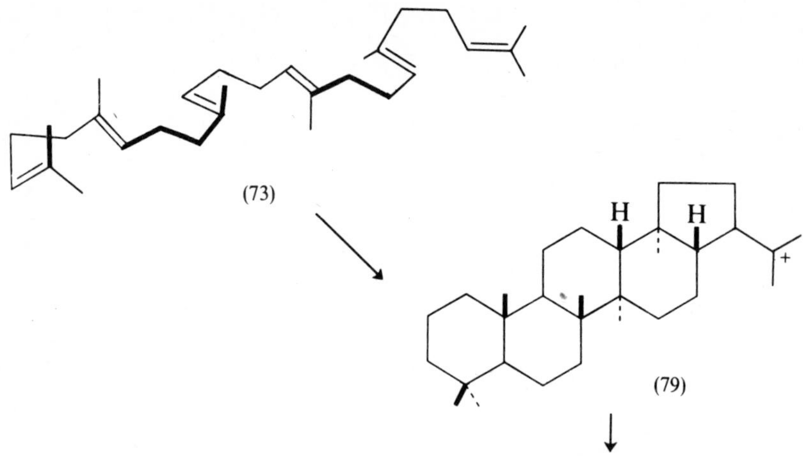

(73)

(79)

Moretenol (80)
Neomotiol (81)
Fernenes (82)
Adianene (83)
Filicene (84)

FIG. 5.

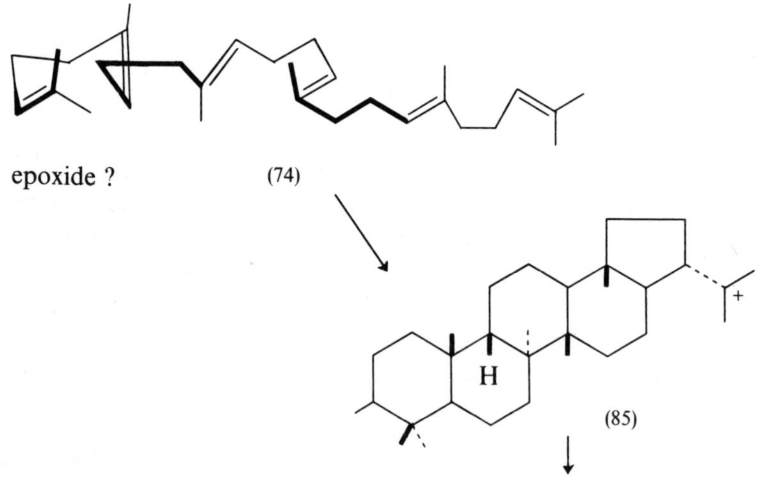

epoxide ?

(74)

(85)

Arborinol (86)

FIG. 6.

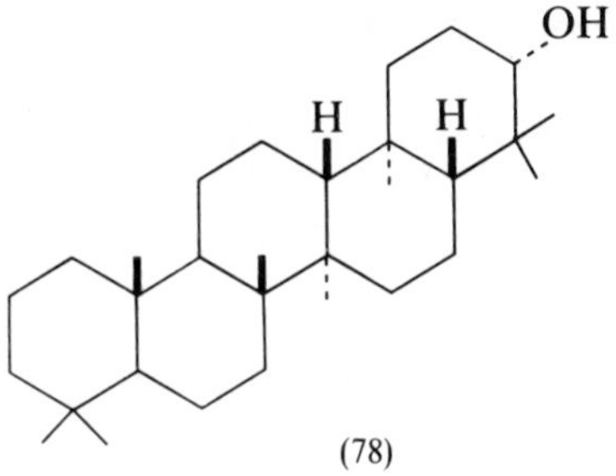

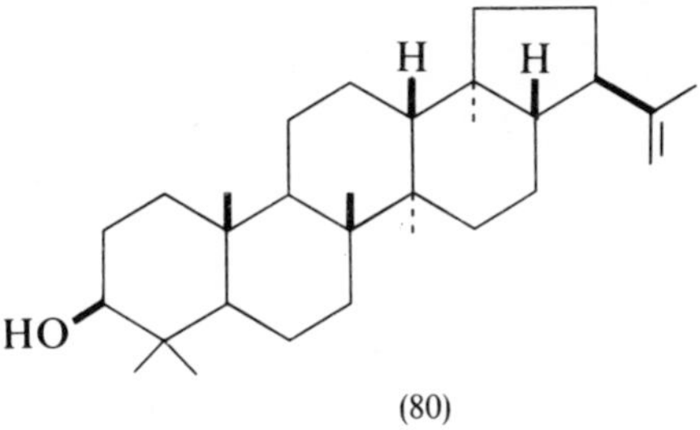

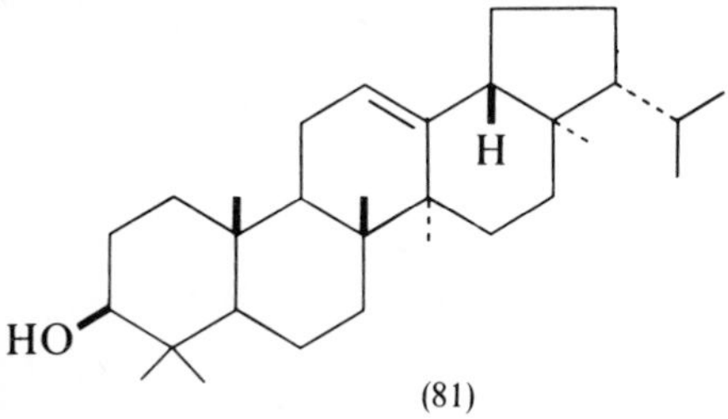

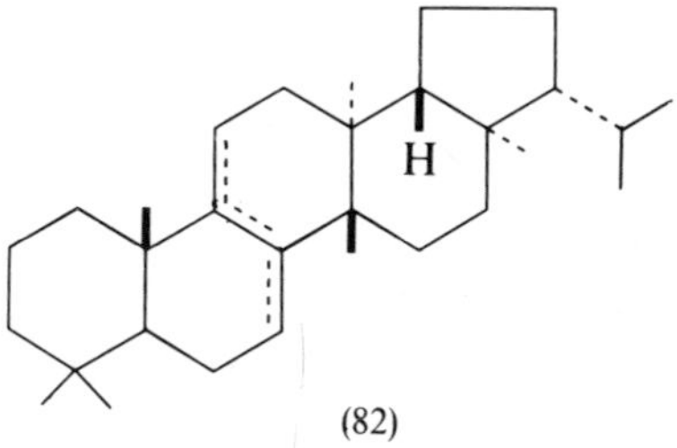

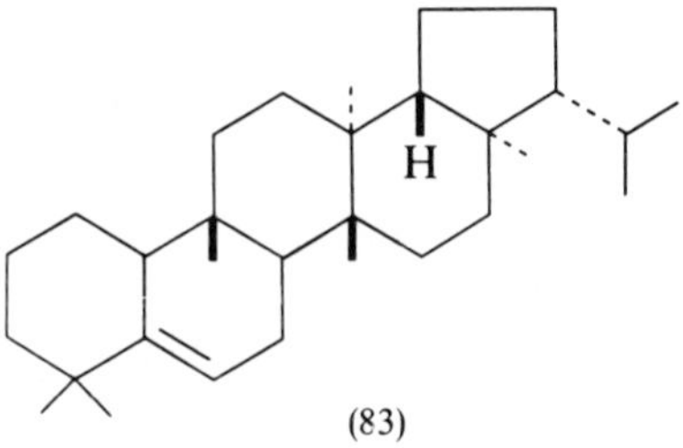

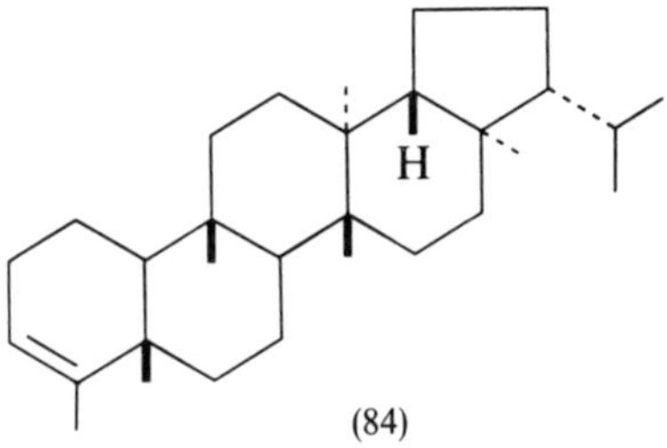

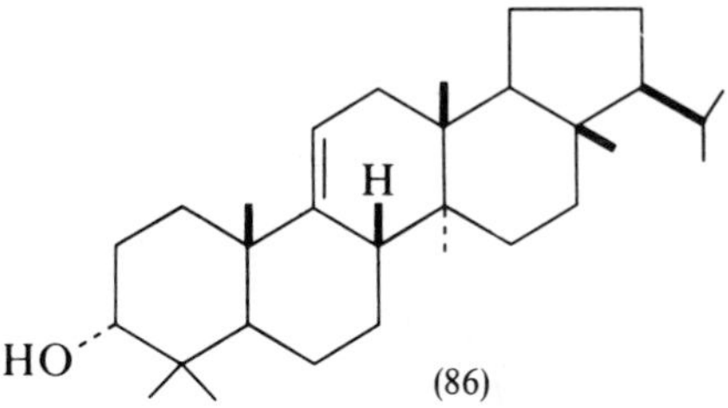

Arborinol (86)[180] is at present the only representative of the group derivable from conformation (74) via cation (85). However, it is likely that a series corresponding to the fernene-filicene group but based on the cation (85) will make its appearance in the future.

VI. Cyclization of Squalene Simultaneously from Both Ends

A small group of triterpenoids are the products of simultaneous oxidative attack at both ends of squalene [(72)⟶(87)] (Fig. 7). The first member of the 1,2-didecalyl ethane group was onocerin (88).[32] Its unique sym-

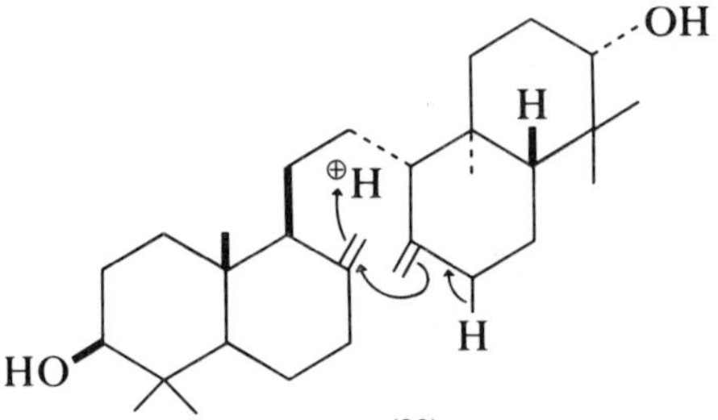

(88)

metry was demonstrated by showing that removal of either oxygen function (via Wolff–Kishner reduction of the acylated hydroxy-ketone)

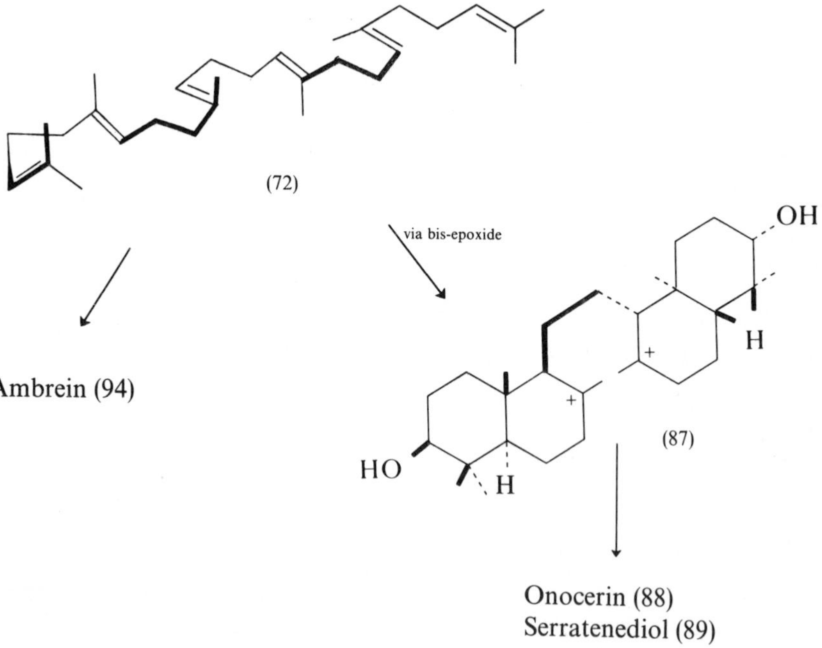

FIG. 7.

resulted in the *same* alcohol. The appealing and simplifying symmetry of onocerin led to several total syntheses (see p. 231), the earliest of natural polycyclic triterpenoids.

The pentacyclic serratanes, exemplified by serratenediol (89)[169] are a rapidly growing group. The carbon skeleton can in principle arise from an α-onocerin precursor. In fact cyclization[282] of α-onocerin diacetate with BF_3 in chloroform gives a mixture of γ-onocerin diacetate (90) and the isomeric serratenediol diacetates (91) and (92).

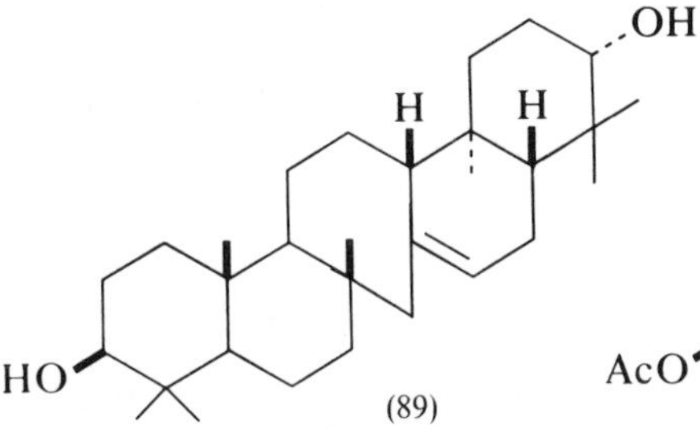

(89)

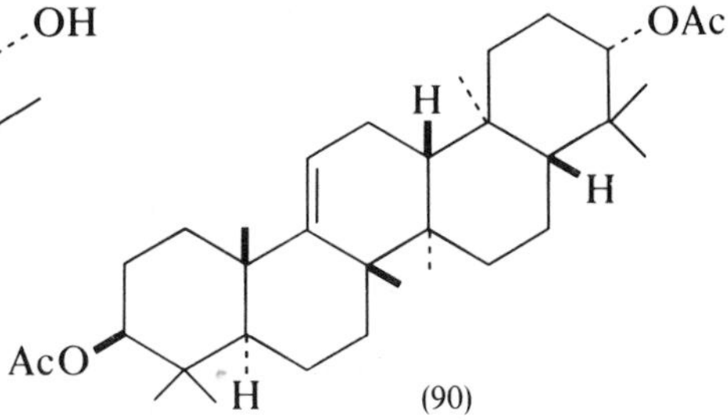

(90)

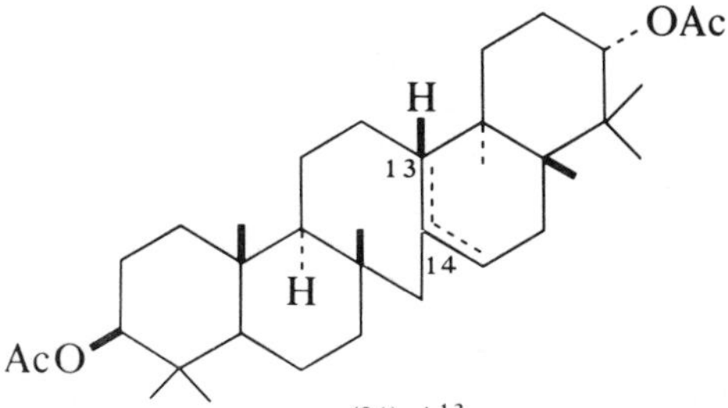

(91) Δ^{13}—
(92) Δ^{14}—

The bicyclic (!) triterpenoid lansic acid (93)[181] is an interesting onocerin derivative in which both terminal rings are cleaved after the manner of nyctanthic acid (see p. 214).

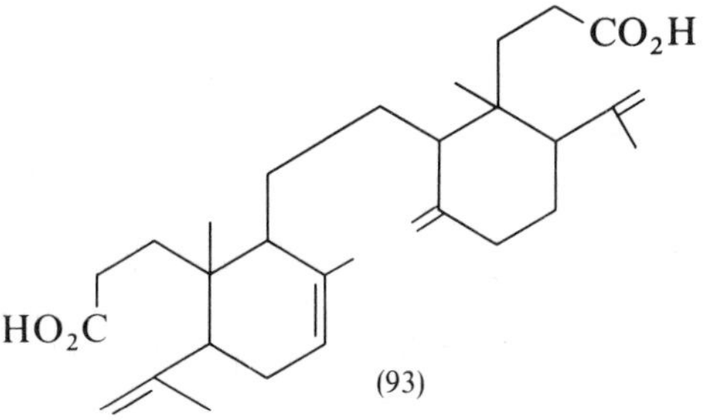

(93)

The long-known and extensively studied ambrein (94)[202,252] is clearly related in its genesis to onocerin.

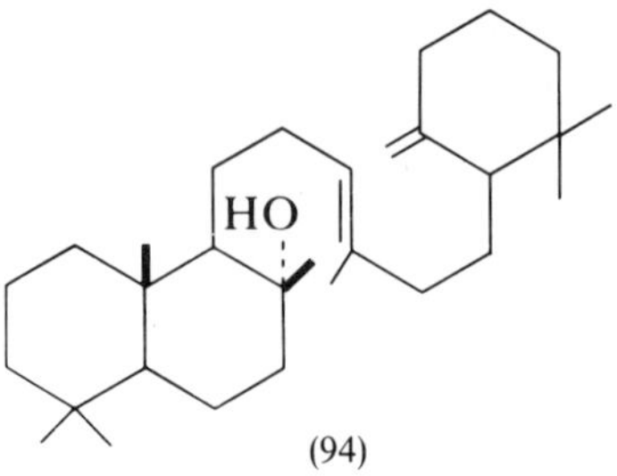

(94)

VII. Laboratory Synthesis of Triterpenoids

The synthetic challenge presented by triterpenoids has been met in two distinct ways. On the one hand methods for the assembly of polycarbocyclic systems, developed in the steroid field since the early 1950's have been extended to solve problems of increased complexity. The steric control and good yields achieved in complex reactions are impressive and well exemplified in the synthesis of germanicol.[171]

The entirely different approach of biogenetically patterned laboratory synthesis, notably developed by van Tamelen's school,[269] adds further incentive to that of chemical synthesis. This is to attempt to delineate the functions of the enzymes at work in the synthesis of terpenoids. Thus, using the substrates utilized by living systems, the aim is to see how far they can be transformed into natural metabolites without the agency of enzymes. This approach is closely complemented by studies which attempt to assess how far the enzymes responsible for the elaboration of the natural substrates will accept unnatural substrates.[270]

Even in terms of synthetic feasibility, biogenetically patterned syntheses may in the future compare favourably with conventional multi-stage laboratory synthesis.

A. Squalene

A stereoselective total synthesis[94] of squalene was achieved by Cornforth, Cornforth and Mathew as early as 1959. More recent syntheses of squalene have relied on the coupling of two C_{15} units in emulation of the biosynthetic pathway. Thus the difarnesyl sulphide (95) obtainable from farnesyl chloride was converted[52] by triphenylphosphine into farnesyl nerolidyl sulphide (96).

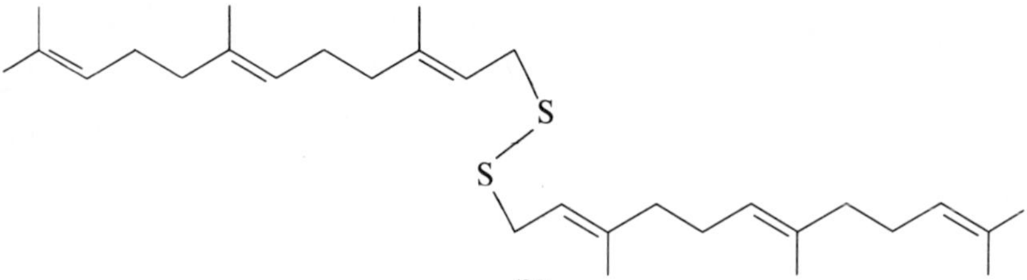

(95)

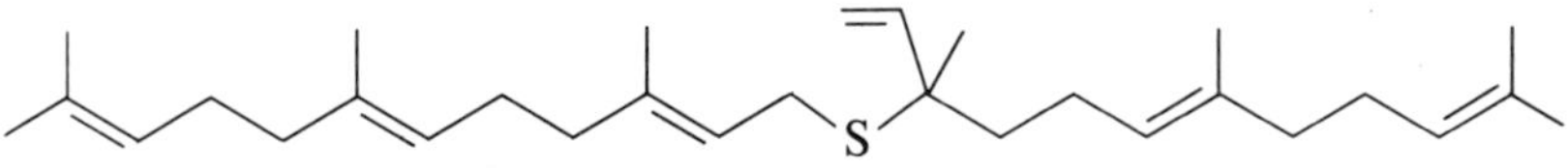

(96)

The sulphonium ylid obtained from (96) with benzyne underwent an intramolecular five-centre rearrangement to 12-phenylthiosqualene in high yield, and this with lithium in ammonia afforded squalene.

In an even simpler approach,[49] the anion of *tt*-farnesyl phenyl thioether was alkylated with *tt*-farnesyl bromide and the product (97) desulphurized with lithium in ethylamine.

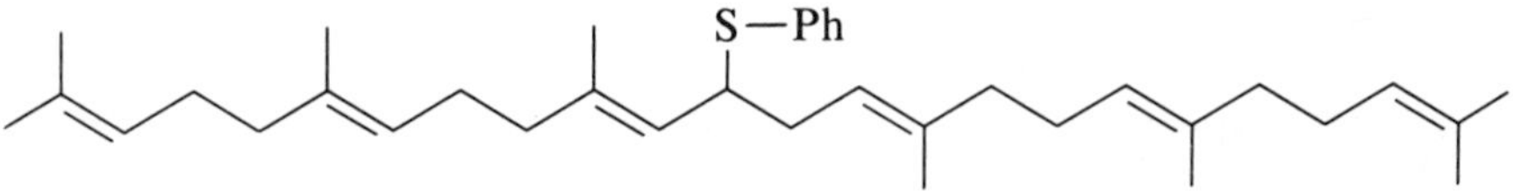

(97)

B. Onocerin

The first polycyclic triterpenoid to be synthesised was a α-onocerin (88). Its symmetry clearly marked it out as an attractive target for early total synthesis. The first recorded synthesis[264] used as its key step anodic coupling of the resolved acid (98) to the symmetrical dione (99) and this was converted into α-onocerin diacetate (100) by decarboxylation of the unsaturated diacid (101). In a later synthesis[271] the bicyclic fragment (102) to be coupled, was obtained by acid catalysed cyclization of the epoxide (103). This is in fact an early example of polyene cyclization promoted by terminal epoxide protonation, a process that plays a major role in biological terpene synthesis, and has been intensively explored in more recent times (see p. 233).

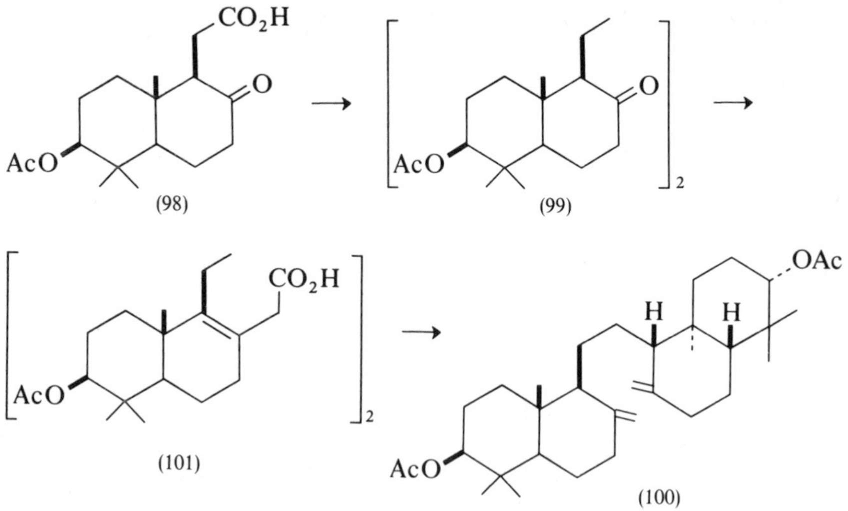

(98) (99) (101) (100)

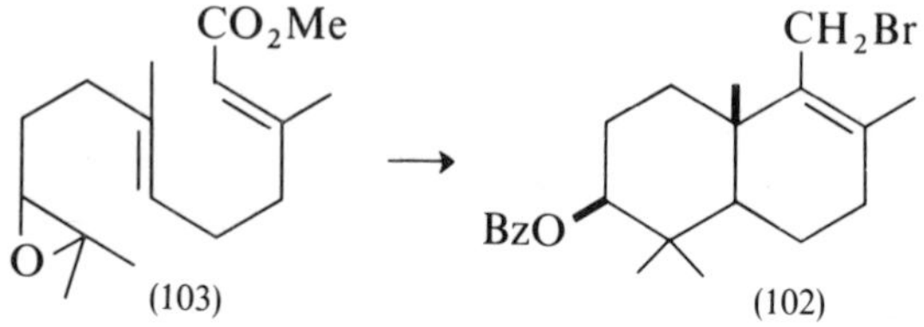

(103) (102)

C. Unsymmetrical Polycyclic Triterpenoids

The more formidable synthetic challenge presented by unsymmetrical triterpenoids has been successfully met only very recently. A total synthesis of (±)-germanicol (104) proceeded[171] through the tricyclic keto-acetate (105), conceived as a general intermediate which embodies the features of

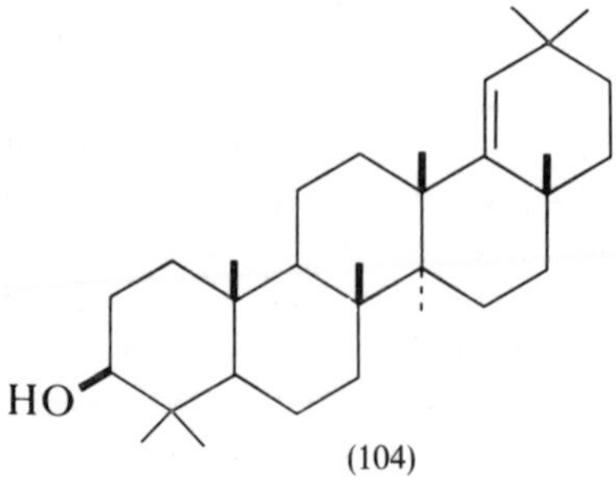

(104)

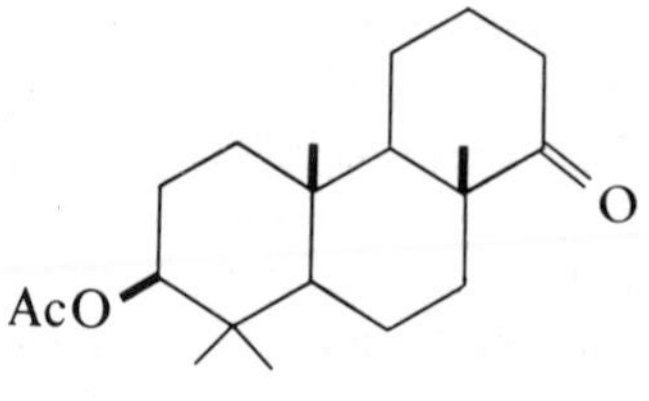

(105)

rings A, B and C of many triterpenoids. Rings D and E were introduced by conjugate addition of benzyl Grignard reagent to the methylene ketone (106), followed by PPA cyclization to (107).

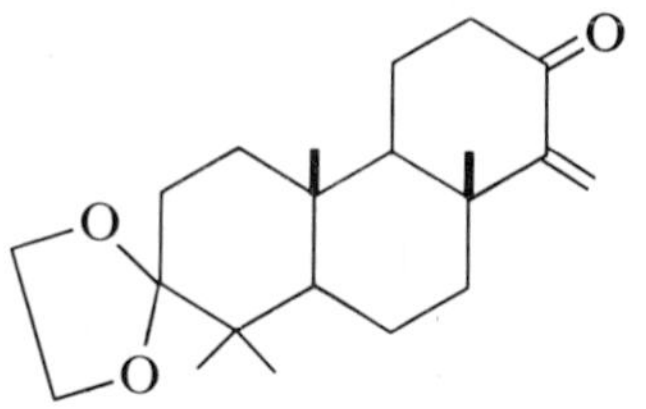

(106)

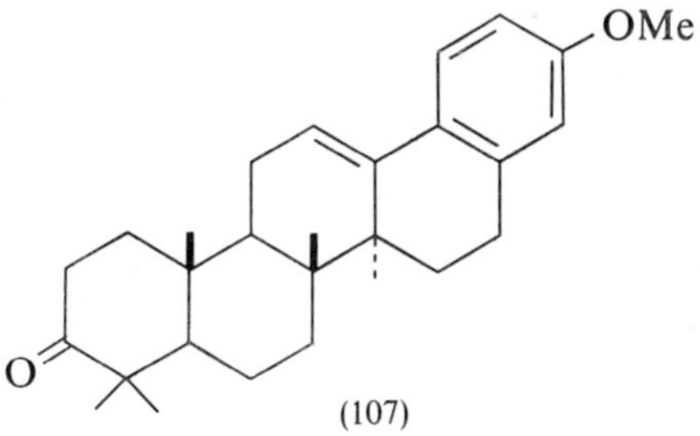

(107)

A different method of assembling the pentacarbocyclic system was carried to a successful conclusion in the synthesis of alnusenone (108).[172] The selected route started from condensation of (109) with (110) and proceeded through the diether (111).

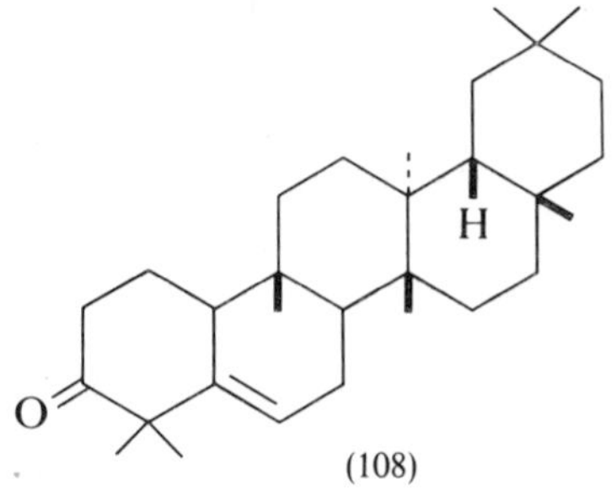

(108)

D. Formation of Tetracarbocyclic Triterpenoids through Biogenetic-type Cyclization

Van Tamelen and his colleagues have recently effected[272] cyclization of the triene epoxide (112) with $SnCl_4$ in CH_3NO_2 to the isoeuphenol system (113) in 35% yield. This most advanced example of *in vitro* synthesis complements the monumental studies in the same laboratory of enzyme-promoted polyene cyclizations. A pre-made ring D was incorporated into (112), to promote formation of a six- rather than a five-membered ring C (see p. 212) and also so that in future studies in the C_{30} series [$R{=}CH_2{-}CH{=}C(CH_3)_2$] the side chain double bond might be insulated from possible interaction with cationic sites generated in the nucleus during cyclization. The monocyclic epoxide (112) was synthesised by condensing (114) and (115).

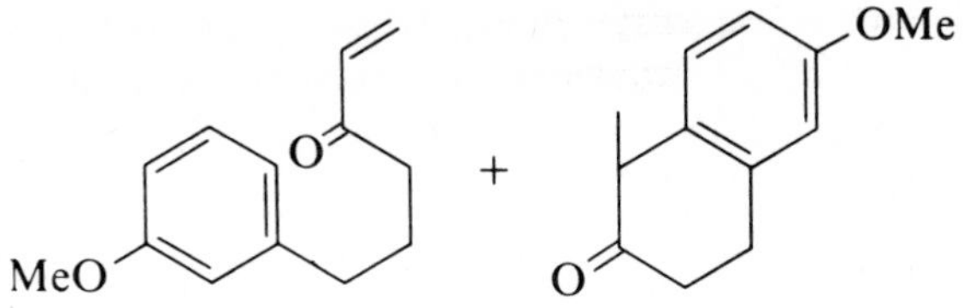

(109) (110)

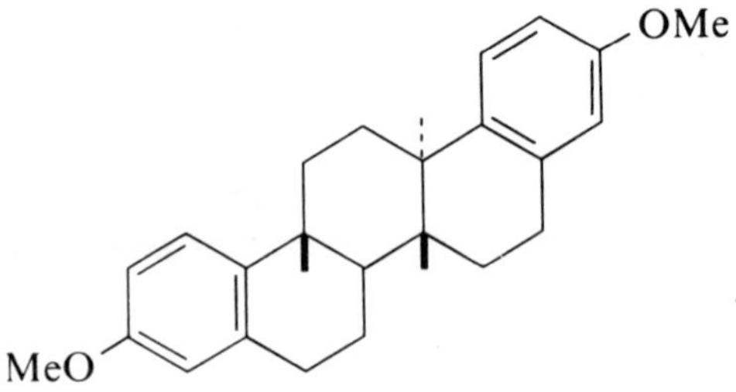

(111)

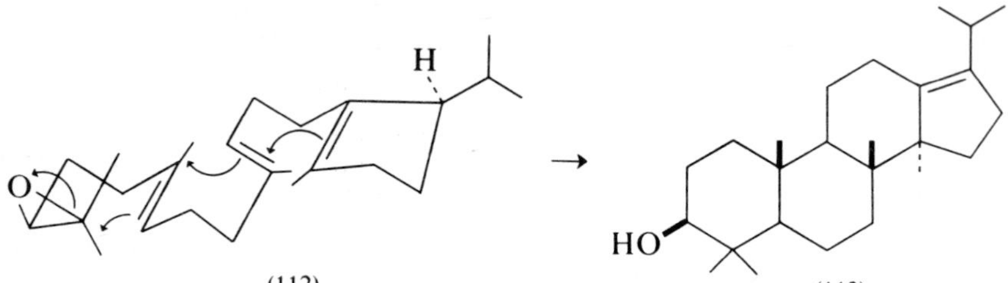

(112) (113)

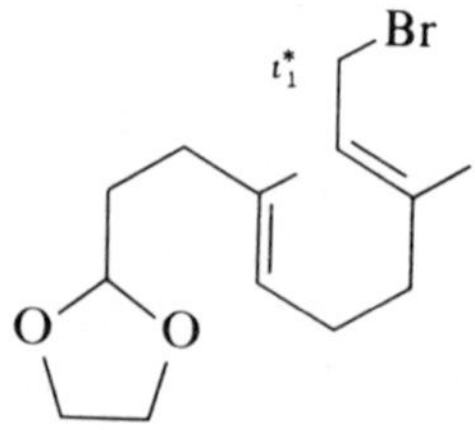

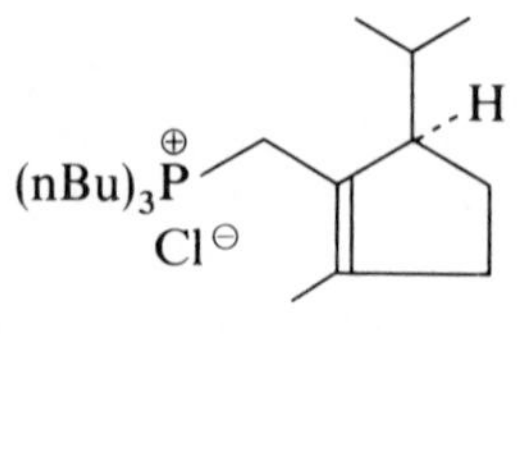

(114) (115)

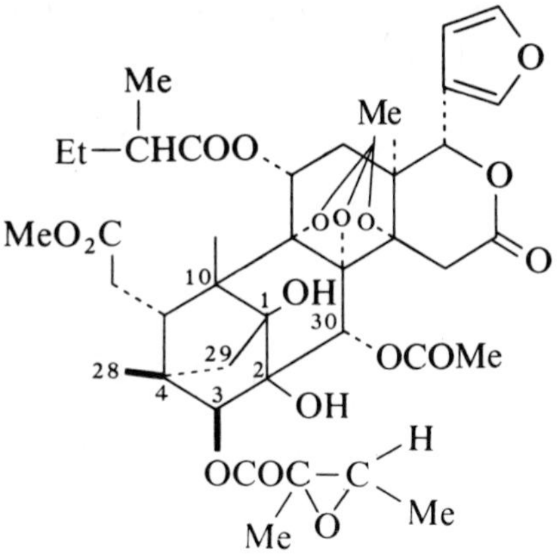

(116)

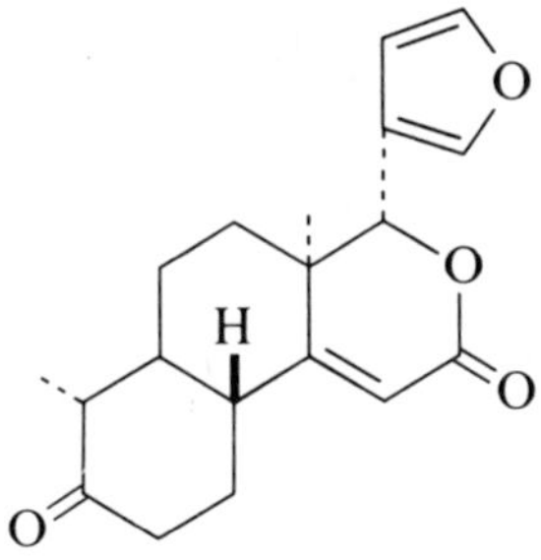

(117)

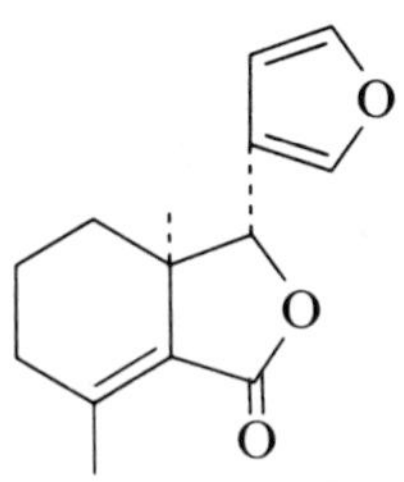

(118)

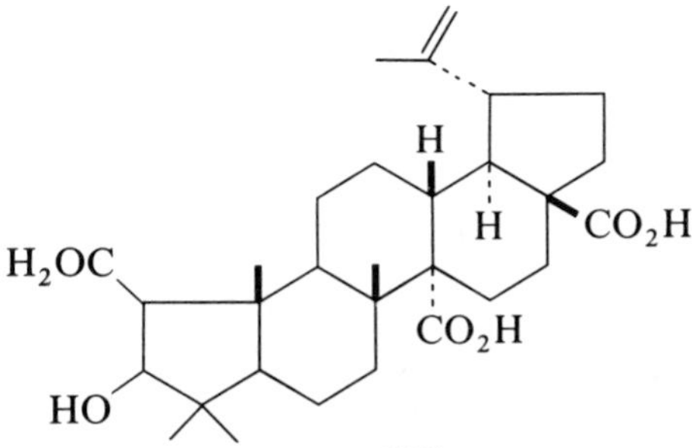

(119)

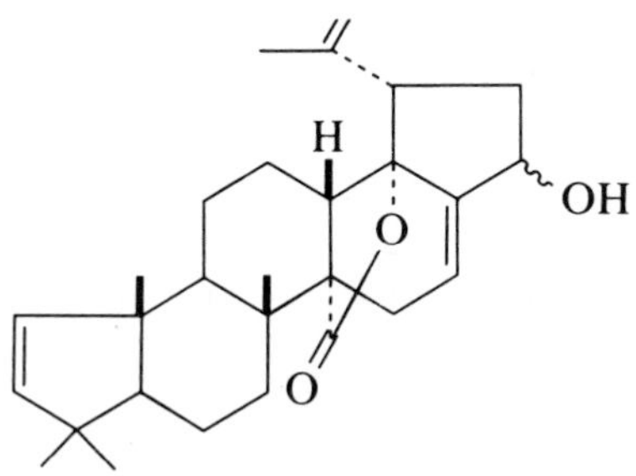

(120)

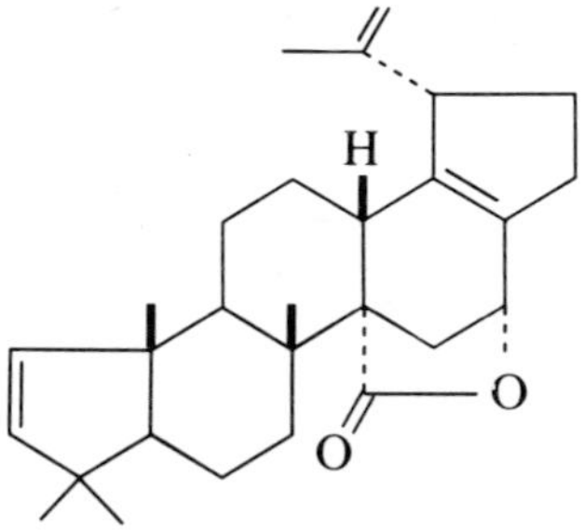

(121)

VIII. Tables of Representative Compounds

TABLE 1
PROTOSTANE GROUP

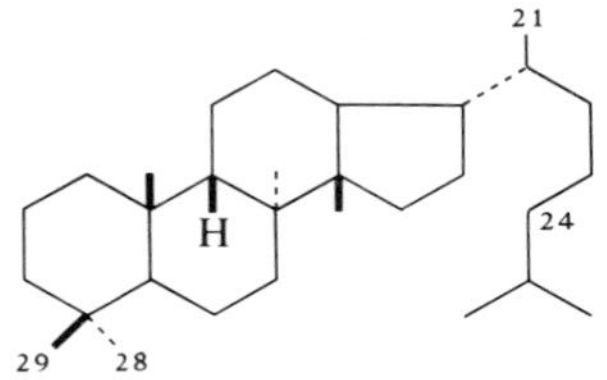

No.	Compound	Ref. No.	Molecular formula	C=C	—OH	CO	—CO_2H		M.P.	$[\alpha]_D^*$
1	Protosterol A	156	$C_{30}H_{50}O$	13(17), 24	3β				105	+10
2	Protosterol B	156	$C_{30}H_{50}O$	17(20), 24	3β				118	+16
3	Fusidic acid	140	$C_{31}H_{48}O_6$	17(20), 24	3α, 11α, 16βAc		21	29-nor	193	−9
4	Cephalosporin Pl	151	$C_{33}H_{50}O_8$	17(20), 24	3α, 6αAc, 7β, 16βAc		21	29-nor	147	+28
5	Helvolic acid	173, 234	$C_{33}H_{44}O_8$	1, 17(20), 24	6βAc, 16βAc	3, 7	21	29-nor	226	−132

* $[\alpha]_D$'s normally refer to solutions in $CHCl_3$, unless marked by an asterisk (*), when reference should be made to the original paper.

TABLE 2
MALABARICANE GROUP

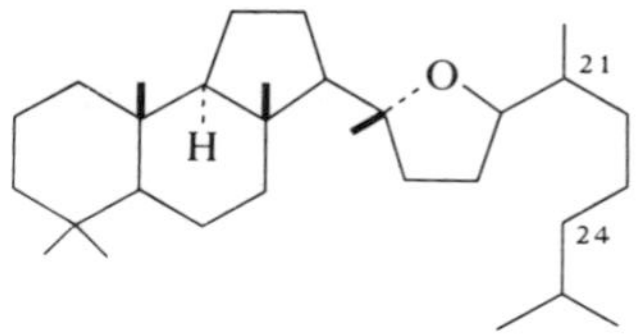

No.	Compound	Ref. No.	Molecular formula	C=C	—OH	CO	—CO_2H		M.P.	$[\alpha]_D$
6	Malabaricol	78, 262	$C_{30}H_{50}O_3$	24	21	3			70	+36
7	Malabaricanediol	78	$C_{30}H_{52}O_3$	24	3, 21				—	+23
8	Epoxymalabaricanediol	78	$C_{30}H_{50}O_4$	24	21	3		21, 24-oxide	144	+25

TABLE 3
LANOSTANE GROUP

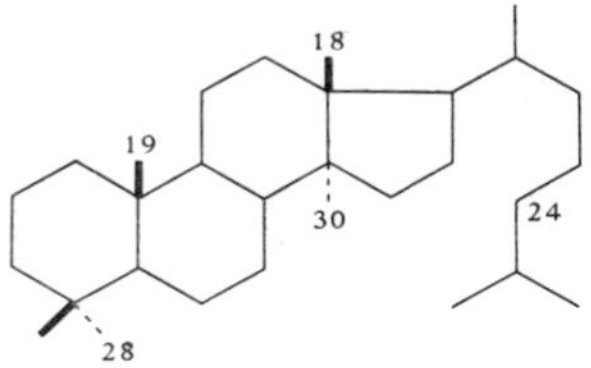

No.	Compound	Ref. No.	Molecular formula	C=C	—OH	CO	—CO_2H		M.P.	$[\alpha]_D$
9	Lanosterol	100, 138	$C_{30}H_{50}O$	8, 24	3β				140	+58
10	Parkeol	201	$C_{30}H_{50}O$	9(11), 24	3β			8β-H	160	+77
11	Agnosterol	203	$C_{30}H_{48}O$	7, 9(11), 24	3β				169	+69
12	Trametenolic acid B	150	$C_{30}H_{48}O_3$	8, 24	3β		21		258	+43*
13	Pinicolic acid A	66	$C_{30}H_{46}O_3$	8, 24		3	21		202	+68
14	Holothurinogenin I	73	$C_{30}H_{44}O_4$	7, 9(11)	3β			18, 20-lacetone, 22, 25-oxide	286	−9
15	Obtusifoldienol	142	$C_{31}H_{52}O$	8	3β			24-methylene	140	+72
16	Eburicoic acid	98	$C_{31}H_{50}O_3$	8	3β		21	24-methylene	292	+50
17	Polyporenic acid A	101	$C_{31}H_{50}O_4$	8	3α, 12α		26	24-methylene	200	+74
18	Abieslactone	211	$C_{31}H_{48}O_3$	9(11), 24	3αMe			26, 23-lactone	253	−187
19	Polyporenic acid C	53	$C_{31}H_{46}O_4$	7, 9(11)	16α	3	21	24-methylene	276	+6*
20	Lophenol	104	$C_{28}H_{48}O$	7	3β			29, 30-bisnor	151	+5
21	Macdougallin	187	$C_{28}H_{48}O_2$	8	3β, 6α			28, 29-bisnor	174	+72
22	Citrostadienol	215	$C_{30}H_{50}O$	7	3β			29, 30-bisnor; 24-ethylidene	164	+24

TABLE 4
CYCLOARTANE GROUP

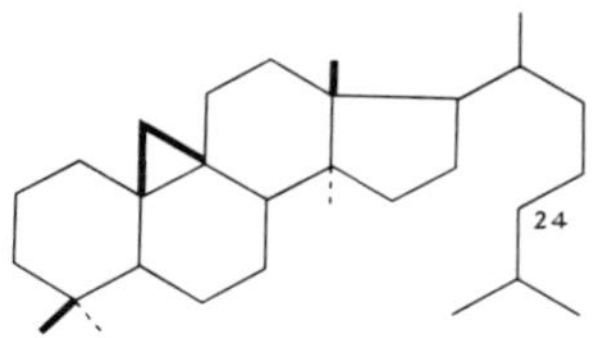

No.	Compound	Ref. No.	Molecular formula	C=C	—OH	CO	—CO_2H		M.P.	$[\alpha]_D$
23	Cycloartanol	260	$C_{30}H_{52}O$		3β				101	+45
24	Cycloartenol	39	$C_{30}H_{50}O$	24	3β				115	+54
25	Cycloartenone	21	$C_{30}H_{48}O$	24		3			109	+23
26	Mangiferolic acid	95	$C_{30}H_{48}O_3$	24	3β		26		183	+49
27	Cimigenol	97	$C_{30}H_{48}O_5$		3β, 15, 25			16β, 23-oxide; 16α, 24-oxide	—	—
28	Ambonic acid	96	$C_{31}H_{48}O_3$			3	26	24-methylene	150	+9
29	Cyclolaudenol	160	$C_{31}H_{52}O$	25	3β			24β-methyl	125	+46
30	Cycloeucalenol	99	$C_{30}H_{50}O$		3β			28-nor, 24-methylene	140	+45
31	Cycloneolitsin	247	$C_{33}H_{56}O$	25	3βMe			23, 23-dimethyl	174	+63

TABLE 5
CUCURBITANE GROUP

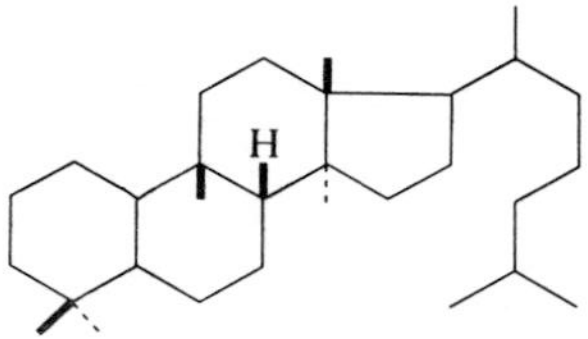

No.	Compound	Ref. No.	Molecular formula	C=C	—OH	CO	—CO_2H		M.P.	$[\alpha]_D$
32	Bryogenin	50a	$C_{30}H_{48}O_3$	5	3β	11, 24			157	+181
33	Gratiogenin	279a	$C_{30}H_{48}O_4$	5	3β, 25	11		20, 24-oxide	203	+175
34	Cucurbitacin F	218	$C_{30}H_{46}O_7$	5, 23	2, 3, 16, 20, 25	11, 22			245	+38*
35	Cucurbitacin D	189	$C_{30}H_{44}O_7$	5, 23	2, 16, 20, 25	3, 11, 22			152	+46
36	Cucubitacin I	189	$C_{30}H_{42}O_7$	1, 5, 23	2, 16, 20, 25	3, 11, 22			149	−52*
37	Cucurbitacin P	194	$C_{30}H_{48}O_7$	5	2, 3, 16, 20, 25	11, 22			159; 212	

TABLE 6
DAMMARANE GROUP

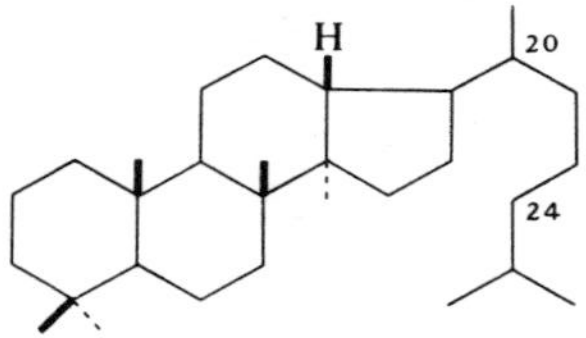

No.	Compound	Ref. No.	Molecular formula	C=C	—OH	CO	—CO_2H		M.P.	$[\alpha]_D$
38	Dammarenediol I	48, 277	$C_{30}H_{52}O_2$	24	3β, 20			C_{20} epimers	144	+27
39	Dammarenediol II	48, 277	$C_{30}H_{52}O_2$	24	3β, 20				133	+33
40	Octotillol	285	$C_{30}H_{52}O_3$		3β, 25			20, 24-oxide	200	+28
41	Protopanaxatriol	258	$C_{30}H_{52}O_4$	24 or 25	3β, 6α, 12β, 20				—	—
42	Panaxatriol	135	$C_{30}H_{52}O_4$		3β, 6α, 12β			20, 25-oxide	239	+14
43	Dammadienol	219a	$C_{30}H_{50}O$	20, 24	3β				138	+47
44	Agliaol	259	$C_{30}H_{50}O_2$	20	3β			24, 25-oxide	114	+53
45	Dipterocarpol	207	$C_{30}H_{50}O_2$	24	20	3			136	+66
46	Dammarenolic acid	11	$C_{30}H_{50}O_3$	4(29), 24	20		3	3, 4-seco	142	+43
47	Carnaubadiol	20	$C_{31}H_{54}O_2$	25	3β, 20			24Me	168	+34

TABLE 7
EUPHANE GROUP

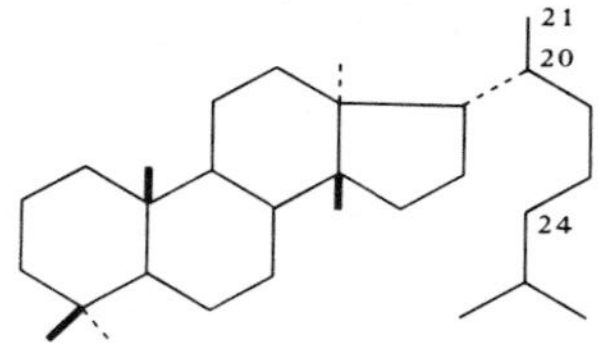

No.	Compound	Ref. No.	Molecular formula	C=C	—OH	CO	—CO_2H		M.P.	$[\alpha]_D$
48	Butyrospermol	102, 200	$C_{30}H_{50}O$	7, 24	3β			20βH	113	−12
49	Tirucallol	14, 17	$C_{30}H_{50}O$	8, 24	3β			20αH	135	+5
50	Euphol	29, 286	$C_{30}H_{50}O$	8, 24	3β			20βH	116	+32
51	Flindissol	51	$C_{30}H_{48}O_3$	7, 24	3α, 21			21, 23-oxide; 20αH	198	−46
52	Schinol	182	$C_{30}H_{48}O_3$	7, 24	3α		26	20αH	147	−40
53	α-Elemolic acid	51, 98a	$C_{30}H_{48}O_3$	8, 24	3α		21	20αH	222	−21
54	Bourjotinolone A	54	$C_{30}H_{48}O_4$	7	23. 25	3		21, 24-oxide; 20αH	176	−34
55	Turreanthin	44	$C_{32}H_{50}O_5$	7	21, 3βAc			21, 23-oxide; 24, 25-oxide	220	+3
56	Melianone	198	$C_{30}H_{46}O_4$	7	21	3		21, 23-oxide; 24, 25-oxide	233	−62
57	Masticadienoic acid	35	$C_{30}H_{46}O_3$	7, 24		3	26	20αH	178	−76
58	β-Elemonic acid	51	$C_{30}H_{46}O_3$	8, 24		3	21	20αH	217	+37
59	Euphorbol	217	$C_{31}H_{52}O$	8	3β			20αH; 24-methylene	127	±0

TABLE 8
CEDRELONE GROUP

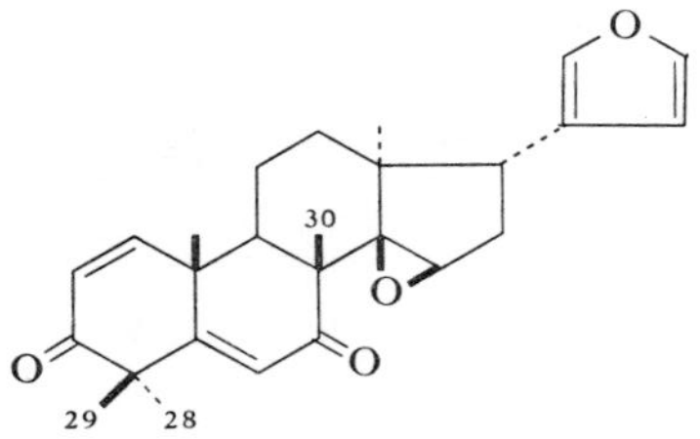

No.	Compound	Ref. No.	Molecular formula	C=C	—OH	CO	—CO_2H	M.P.	$[\alpha]_D$
60	Cedrelone	143, 164	$C_{26}H_{30}O_5$		6			204	−64
61	Anthothecol	47	$C_{28}H_{32}O_7$		6, 11αAc			225	−63

TABLE 9
GEDUNIN GROUP

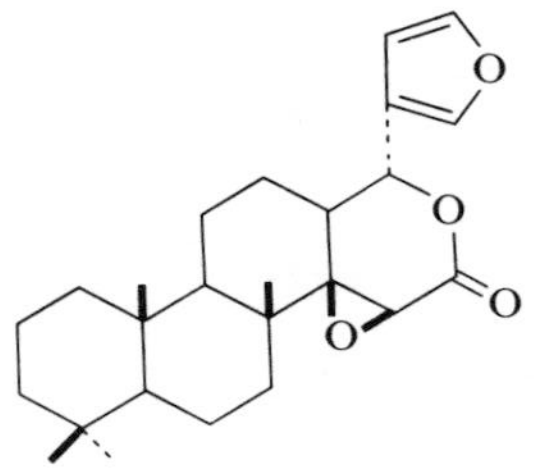

No.	Compound	Ref. No.	Molecular formula	C=C	—OH	CO	—CO_2H	M.P.	$[\alpha]_D$
62	Khivorin	45	$C_{32}H_{42}O_{10}$		1αAc, 3αAc, 7αAc			263	−42
63	Gedunin	6	$C_{28}H_{34}O_7$	1	7αAc	3		218	−44

TABLE 10
LIMONIN GROUP

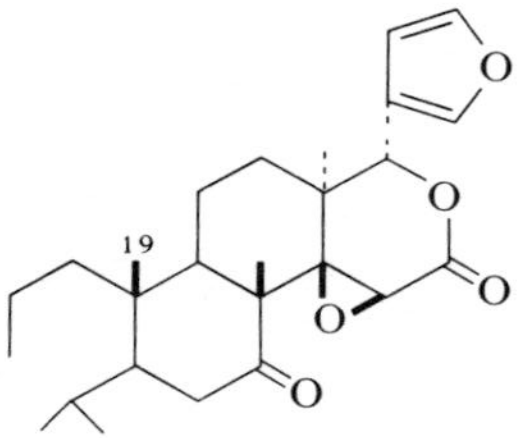

No.	Compound	Ref. No.	Molecular formula	C=C	—OH	CO	—CO_2H		M.P.	$[\alpha]_D$
64	Veprisone	144	$C_{27}H_{34}O_8$				$3CO_2Me$	1, 4-oxide	181	−18
65	Nomilin	34	$C_{28}H_{34}O_9$		1Ac			3, 4-lactone	279	−96*
66	Ichangin	109	$C_{26}H_{32}O_9$		1, 4			3, 19-lactone	212	—
67	Obacunone	34, 190	$C_{26}H_{30}O_7$	1				3, 4-lactone	231	−50
68	Limonin	13, 34	$C_{26}H_{30}O_8$					3, 19-lactone; 1, 4-oxide	305	−129*
69	Limonin diosphenol	163	$C_{26}H_{28}O_9$	5	6			3, 19-lactone; 1, 4-oxide	285	−199*

TABLE 11
ANDIROBIN GROUP

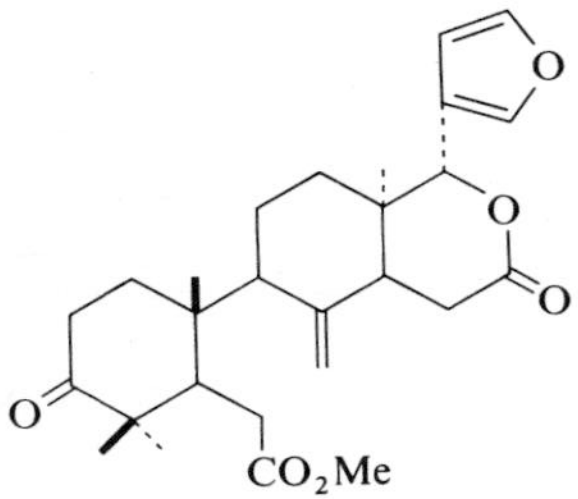

No.	Compound	Ref. No.	Molecular formula	C=C	—OH	CO	—CO_2H		M.P.	$[\alpha]_D$
70	Andirobin	235	$C_{27}H_{32}O_7$	1				14, 15β-oxide	197	—
71	Methyl angolensate	46, 71, 86	$C_{27}H_{34}O_7$					1α, 14β-oxide	201	−48

TABLE 12
SWIETENINE GROUP

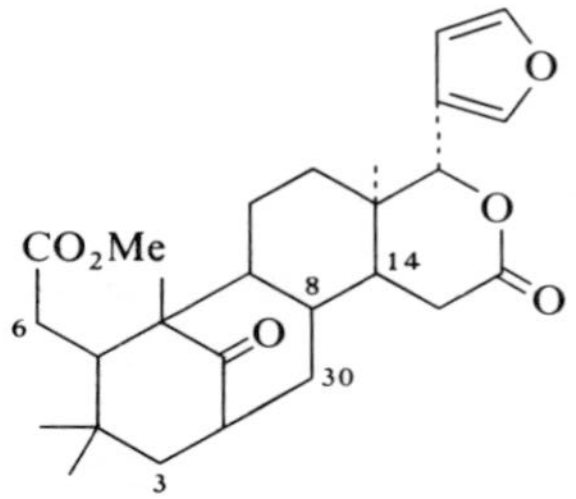

No.	Compound	Ref. No.	Molecular formula	C=C	—OH	CO	—CO_2H		M.P.	$[\alpha]_D$
72	Swietenolide	70, 84	$C_{27}H_{34}O_8$	8(14)	3β, 6				225	−136
73	Swietenine	81, 208	$C_{32}H_{40}O_9$	8(30)	6, 3βTg				276	−167
74	Mexicanolide	83	$C_{27}H_{32}O_7$	8(14)		3			227	−90
75	Carapin	10	$C_{27}H_{32}O_7$	14		3			178	+64
76	Xylocarpin	233	$C_{29}H_{36}O_9$		3βAc			8, 30-α-oxide	212	−88
77	Utilin (116)	155	$C_{41}H_{52}O_{17}$						278	−356

TABLE 13
NIMBIN GROUP

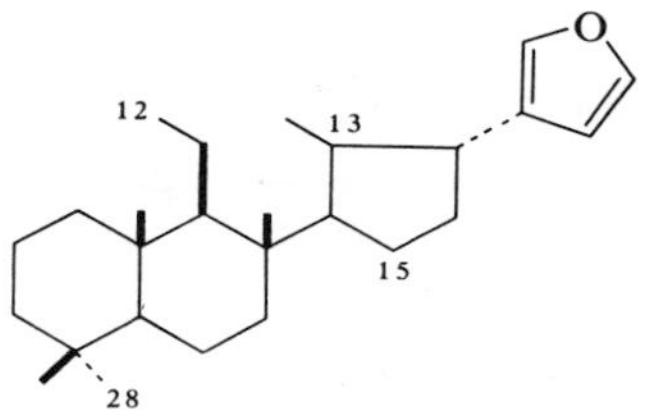

No.	Compound	Ref. No.	Molecular formula	C=C	—OH	CO	$—CO_2H$		M.P.	$[\alpha]_D$
78	Nimbin	154, 227	$C_{30}H_{36}O_9$	2, 13	6αAc	1	$12CO_2Me$ $28CO_2Me$	7α, 15α-oxide	205	+168
79	Nimbolin B	134	$C_{39}H_{46}O_{10}$	13	1αAc, 3αAc, 7αCin, 12			6α, 28-oxide; 12, 15-oxide	245	−93
80	Simarolide (47)	60	$C_{27}H_{36}O_9$						270	+74

TABLE 14

QUASSIN GROUP

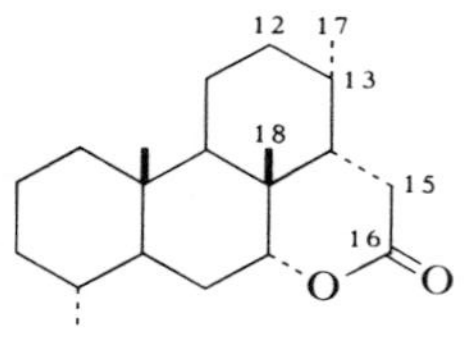

No.	Compound	Ref. No.	Molecular formula	C=C	—OH	CO	—CO_2H		M.P.	$[\alpha]_D$
81	Chaparrin	165	$C_{20}H_{28}O_7$	3	1β, 2α, 12α, 18	11			308	+14*
82	Glaucarubin	232, 238	$C_{25}H_{36}O_{10}$	3	1β, 2α, 11α, 12α, 15βR			R=$C_5H_9O_2$; 11, 18-oxide	255	+69*
83	Neoquassin	283	$C_{22}H_{30}O_6$	2, 12	2Me, 12Me	1, 11		16OH	231	+41
84	Chaparrinone	238a	$C_{20}H_{26}O_7$	3	1β, 11α, 12α	2		11, 18-oxide	242	−47*
85	Glaucarubolone	238	$C_{20}H_{26}O_8$	3	1β, 11α, 12β, 15β	2		11, 18-oxide	258	−34*
86	Quassin	283	$C_{22}H_{28}O_6$	2, 12	2Me, 12Me	1, 11			222	+35

TABLE 15
CEDRONINE GROUP

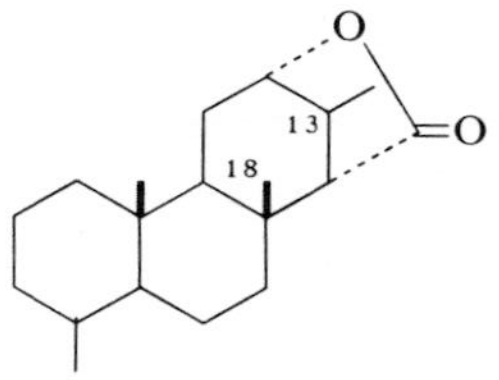

No.	Compound	Ref. No.	Molecular formula	C=C	—OH	CO	—CO₂H		M.P.	[α]D
87	Cedronyline	293	$C_{19}H_{26}O_7$	3	1, 2, 7, 11			13β, 18-oxide	267	+17*
88	Cedronine	293	$C_{19}H_{24}O_7$		1, 7, 11	2		13β, 18-oxide	280	−13*
89	Samaderine C	293	$C_{19}H_{24}O_7$	3	1, 2, 11	7		13β, 18-oxide	268	+59*
90	Odoratin (117)	72	$C_{20}H_{24}O_3$						223	+155
91	Fraxinellone (118)	80, 237	$C_{14}H_{18}O_3$						116	−44*

TABLE 16

No.	Compound	Ref. No.	Molecular formula	C=C	—OH	CO	—CO_2H	M.P.	$[\alpha]_D$
92	Shionone (60)	267, 268	$C_{30}H_{50}O$					162	−56
93	Ebelin lactone (61)	18	$C_{30}H_{46}O_3$					185	−14
94	Baccharis oxide (64)	8	$C_{30}H_{52}O$					149	+42

TABLE 17
Lupane Group

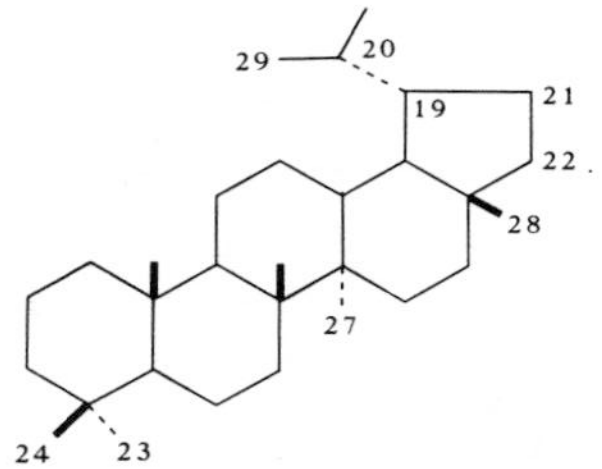

No.	Compound	Ref. No.	Molecular formula	C=C	—OH	CO	—CO_2H		M.P.	$[\alpha]_D$
95	Monogynol A	74	$C_{30}H_{52}O_2$		3, 20				234	+24
96	Lupeol	127	$C_{30}H_{50}O$	20(29)	3β				215	+33*
97	Betulin	124	$C_{30}H_{50}O_2$	20(29)	3β, 28				261	+20
98	Betulinal	125	$C_{30}H_{48}O_2$	20(29)	3β	28			193	+19
99	Canaric acid	68	$C_{30}H_{48}O_2$	4(23), 20(29)			3	3, 4-seco	216	+57
100	Betullinic acid	126	$C_{30}H_{48}O_3$	20(29)	3β		28		321	+12
101	Alphitolic acid	147	$C_{30}H_{48}O_4$	20(29)	2α, 3β		28		—	—
102	Glochidone	139	$C_{30}H_{46}O$	1, 20(29)		3			165	+73
103	Thurberogenin	210	$C_{30}H_{46}O_3$	20(29)	3β			28, 21-lactone	285	+11
104	Melaleucic acid	75, 148	$C_{30}H_{46}O_5$	20(29)	3β		27, 28		364	+19*
105	Ceanothic acid (119)	214	$C_{30}H_{46}O_5$						346	+38*
106	Emmolactone (120)	113	$C_{28}H_{38}O_3$						256	+123
107	Isodeoxyemmolactone (121)	190	$C_{28}H_{38}O_2$						190	−25

TABLE 18
ALLOBETULIN GROUP

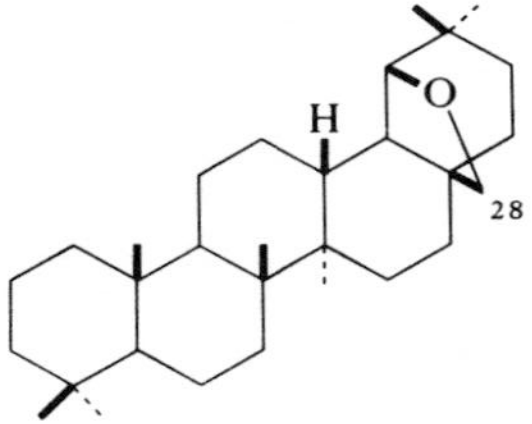

No.	Compound	Ref. No.	Molecular formula	C=C	—OH	CO	—CO_2H		M.P.	$[\alpha]_D$
108	Allobetulin	168	$C_{30}H_{50}O_2$		3				267	+51
109	Apoallobetulin	174	$C_{30}H_{48}O$	3(5)				5(4 → 3)abeo	207	+68
110	Oxyallobetulin	168	$C_{30}H_{48}O_3$		3	28			346	+47
111	Δ^2-Oxyallobetulene	24	$C_{30}H_{46}O_2$	2		28			363	+80

TABLE 19
OLEANANE GROUP

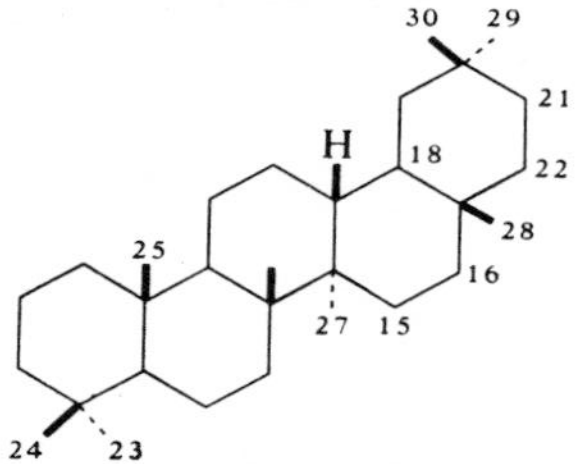

No.	Compound	Ref. No.	Molecular formula	C=C	—OH	CO	—CO_2H	M.P.	$[\alpha]_D$
112	β-Amyrin	116, 117	$C_{30}H_{50}O$	12	3β			200	+88
113	δ-Amyrin	224	$C_{30}H_{50}O$	13(18)	3β			214	−52
114	Germanicol	36	$C_{30}H_{50}O$	18	3β			181	+7
115	Maniladiol	151	$C_{30}H_{50}O_2$	12	3β, 16β			221	+68
116	Erythrodiol	119	$C_{30}H_{50}O_2$	12	3β, 28			242	+76
117	Primulagenin A	161	$C_{30}H_{50}O_3$	12	3β, 16α, 28			256	+52
118	Soyasapogenol-B	65	$C_{30}H_{50}O_3$	12	3β, 21α, 24			260	+83
119	Soyasapogenol-D	65	$C_{31}H_{52}O_3$	13(18)	3β, 21α, 24			299	−61
120	Barringtogenol	107	$C_{30}H_{50}O_4$	12	2α, 3β, 23 28			291	+18
121	Protoaescigenin	192	$C_{30}H_{50}O_6$	12	3β, 16α, 21α, 22β, 24, 28			310	+31*

TABLE 19—OLEANANE GROUP (*continued*)

No.	Compound	Ref. No.	Molecular formula	C=C	—OH	CO	—CO_2H		M.P.	$[\alpha]_D$
122	Soyasapogenol-C	65	$C_{30}H_{48}O_2$	12, 21	3β, 24				241	+63
123	Nyctanthic acid	12, 287	$C_{30}H_{48}O_2$	4(23), 12			3	3, 4-seco	235	+86
124	α-Boswellic acid	121	$C_{30}H_{48}O_3$	12	3α		24		289	+115
125	Oleanolic acid	166	$C_{30}H_{48}O_3$	12	3β		28		310	+80
126	Katonic acid	185	$C_{30}H_{48}O_3$	12	3α		29		287	+47
127	Morolic acid	23	$C_{30}H_{48}O_3$	18	3β		28		273	+33
128	Saikogenin-D	191	$C_{30}H_{48}O_4$	11, 13(18)	3β, 16α, 23, 28				266	−48*
129	Maslinic acid	64	$C_{30}H_{48}O_4$	12	2α, 3β		28		269	+43*
130	Bredemolic acid	280	$C_{30}H_{48}O_4$	12	2β, 3α		28		292	+101*
131	Sumaresinolic acid	108	$C_{30}H_{48}O_4$	12	3β, 6β		28		299	+54
132	Echinocystic acid	128	$C_{30}H_{48}O_4$	12	3β, 16α		28		310	+35
133	Siaresinolic acid	129	$C_{30}H_{48}O_4$	12	3β, 19α		28		280	+98*
134	Machaerinic acid	105	$C_{30}H_{48}O_4$	12	3β, 21β		28		—	—
135	Hederagenin	120	$C_{30}H_{48}O_4$	12	3β, 23		28		334	+82*
136	Queretaroic acid	103	$C_{30}H_{48}O_4$	12	3β, 30		28		323	—
137	Bayogenin	114	$C_{30}H_{48}O_5$	12	2β, 3β, 23		28		330	+98*
138	Entagenic acid	19, 35a	$C_{30}H_{48}O_5$	12	3, 15, 16		28		315	—
139	Acacic acid	284	$C_{30}H_{48}O_5$	12	3, 16, 21		28		276	—
140	Terminolic acid	184	$C_{30}H_{48}O_6$	12	2α, 3β, 6β, 23		28		347	+42*
141	Tomentosic acid	249	$C_{30}H_{48}O_6$	12	2α, 3β, 19β, 23		28		330	+64*

TABLE 19—OLEANANE GROUP (*continued*)

No.	Compound	Ref. No.	Molecular formula	C=C	—OH	CO	—CO_2H		M.P.	$[\alpha]_D$
142	Polygalacic acid	248	$C_{30}H_{48}O_6$	12	2β, 3β, 16α, 23		28		305	+60
143	Aegicerin	243	$C_{30}H_{46}O_3$	12	3β	16		15β, 28-oxide	256	−24
144	Cyclamigenin-B	112	$C_{30}H_{46}O_4$	12	3β	16, 30		13β, 28-oxide	—	—
145	Dumortierigenin	106	$C_{30}H_{46}O_4$	12	3β, 22α			28, 15β-lactone	295	−19
146	Glycyrrhetic acid	37	$C_{30}H_{46}O_4$	12	3β	11	30		300	+98
147	Gypsogenin	188	$C_{30}H_{46}O_4$	12	3β	23	28		274	+91
148	Rhemmanic acid	31	$C_{35}H_{52}O_5$	12	22βR	3	28	R = angeloyloxy	300	+84
149	Bassic acid	186	$C_{30}H_{46}O_5$	5, 12	2α, 3β, 23		28		319	+82
150	Quillaic acid	112	$C_{30}H_{46}O_5$	12	3β, 16α	23	28		293	+56*
151	Icterogenin	30	$C_{35}H_{52}O_6$	12	22βR, 24	3	28	R = angeloyloxy	241	+64*
152	Medicagenic acid	107	$C_{30}H_{46}O_6$	12	2β, 3β		23, 28		350	−13*
153	Phytolaccogenin	265	$C_{30}H_{46}O_7$	12	2β, 3β, 23		28, 30		318	—
154	Glabrolide	67	$C_{30}H_{44}O_4$	12	3β	11		30, 22β-lactone	365	+77
155	Albigenin	19a	$C_{29}H_{46}O_2$	13(18)	3β	16		28-nor	228	−114
156	Eupteleogenin	222, 229	$C_{29}H_{42}O_4$	20(30)	3β			29-nor; 11α, 12α-oxide; 28, 13β-lactone	272	+83
157	β-Peltoboykinolic acid	225	$C_{30}H_{48}O_3$	12	3β		27		222	+114

TABLE 20
TARAXERANE GROUP

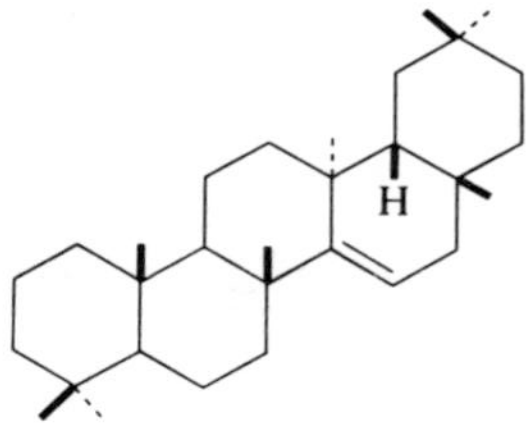

No.	Compound	Ref. No.	Molecular formula	C=C	—OH	CO	—CO_2H	M.P.	$[\alpha]_D$
158	Taraxerene	61	$C_{30}H_{50}$					239	+1
159	Taraxerol	38, 58	$C_{30}H_{50}O$		3β			285	+3
160	Sawamilletin	1	$C_{31}H_{52}O$		3βMe			278	+8
161	Taraxerone	58, 254	$C_{30}H_{48}O$			3		249	+12
162	Myricolal	212	$C_{30}H_{48}O_2$		3β	28		288	—

TABLE 21

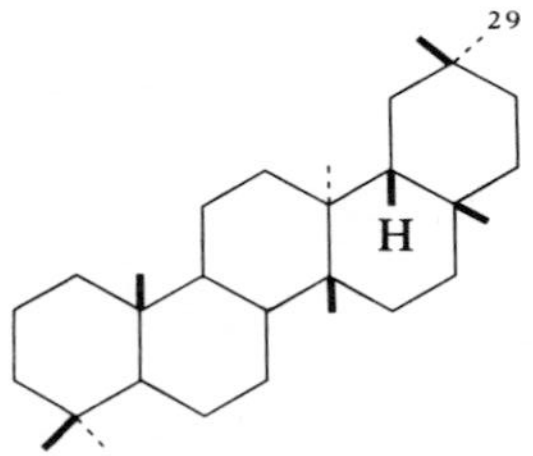

No.	Compound	Ref. No.	Molecular formula	C=C	—OH	CO	—CO_2H	M.P.	$[\alpha]_D$
163	Multiflorenol	255	$C_{30}H_{50}O$	7	3β			190	−28
164	Bryonolic acid	50	$C_{30}H_{46}O_3$	8	3β		29	305	+20

TABLE 22
GLUTINANE GROUP

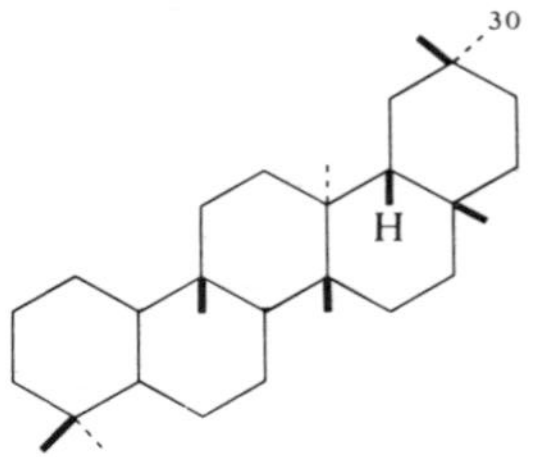

No.	Compound	Ref. No.	Molecular formula	C=C	—OH	CO	$-CO_2H$		M.P.	$[\alpha]_D$
165	Dendropanoxide (campanulin)	183	$C_{30}H_{50}O$					3β, 5β-oxide	207	+68
166	Glutinol	137	$C_{30}H_{50}O$	5	3β				211	+62
167	Celastrol	153	$C_{29}H_{38}O_4$	1(10), 3, 5, 7	3	2	30	23- or 24-nor (167, 168)	205	—
168	Pristimerin	152, 153	$C_{30}H_{40}O_4$	1(10), 3, 5, 7	3	2	$30CO_2Me$		220	—

TABLE 23
FRIEDELANE GROUP

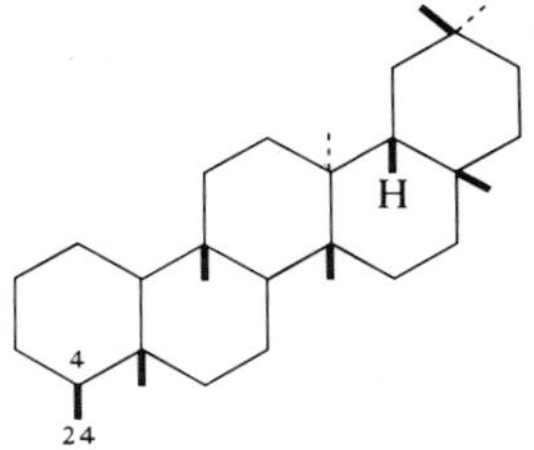

No.	Compound	Ref. No.	Molecular formula	C=C	—OH	CO	—CO_2H		M.P.	$[\alpha]_D$
169	Friedelan-3α-ol	288	$C_{30}H_{52}O$		3α				302	+18
170	Friedelin	56	$C_{30}H_{50}O$			3			265	−28
171	Cerin	92	$C_{30}H_{50}O_2$		2β	3			254	−41
172	Friedelan-2, 3-dione	178	$C_{30}H_{48}O_2$			2, 3			282	+25
173	Apetalactone	145	$C_{30}H_{50}O_3$		28			3, 4-seco; 3, 4-lactone	336	+38
174	Putranjivic acid	76	$C_{30}H_{50}O_2$	4(24)			3	3, 4-seco	179	−15

TABLE 24
TARAXASTANE GROUP

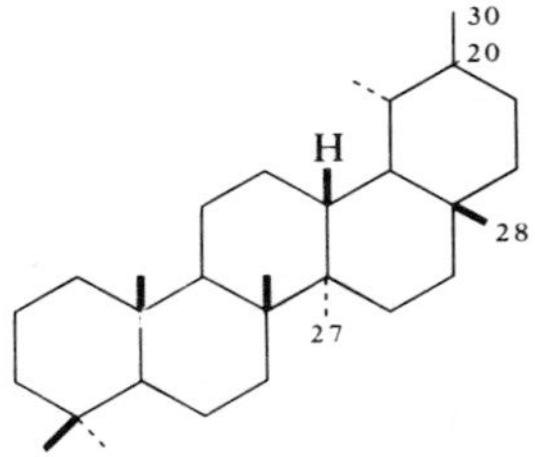

No.	Compound	Ref. No.	Molecular formula	C=C	—OH	CO	—CO_2H		M.P.	$[\alpha]_D$
175	ψ-Taraxastandiol	133	$C_{30}H_{52}O_2$		3β, 20				272	−11
176	Faradiol	253	$C_{30}H_{50}O_2$	20	3β, 12				237	+45
177	Arnidiol	253	$C_{30}H_{50}O_2$	20(30)	3β, 12				257	+83
178	Taraxasterol	130	$C_{30}H_{50}O$	20(30)	3β				227	+97
179	ψ-Taraxasterol	132	$C_{30}H_{50}O$	20	3β				221	+50
180	Phyllirigenin	77	$C_{30}H_{48}O_4$		3β, 27			28, 20-lactone	341	+23

TABLE 25
URSANE GROUP

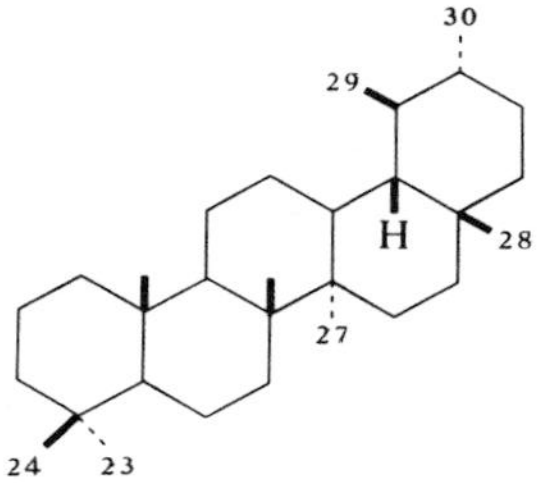

No.	Compound	Ref. No.	Molecular formula	C=C	—OH	CO	—CO_2H		M.P.	$[\alpha]_D$
181	α-Amyrin	89	$C_{30}H_{50}O$	12	3β				186	+83
182	Brein	196	$C_{30}H_{50}O_2$	12	3β, 16β				223	+66
183	Uvaol	292	$C_{30}H_{50}O_2$	12	3β, 28				223	+51
184	α-Amyrenone	118	$C_{30}H_{48}O$	12		3			126	+119
185	β-Boswellic acid	43	$C_{30}H_{48}O_3$	12	3α		24		232	+107
186	Ursolic acid	123	$C_{30}H_{48}O_3$	12	3β		28		292	+72*
187	Ifflaconic acid	256	$C_{30}H_{48}O_3$	12	3α		30		260	+88
188	Commic acid E	279	$C_{30}H_{48}O_5$	12	1β, 2β, 3β		23		333	+104*
189	Asiatic acid	239	$C_{30}H_{48}O_5$	12	2α, 3β, 23		28		305	+51*
190	Vanguerolic acid	25	$C_{30}H_{46}O_3$	12, 18	3β		28	30βMe	273	+306
191	Roburic acid	209	$C_{30}H_{48}O_2$	4(23), 12			3	3, 4-seco	182	+78
192	α-Peltoboykinolic acid	225	$C_{30}H_{48}O_3$	12	3β		27		236	+121
193	Dammar acid	55	$C_{32}H_{48}O_6$	12			23-CO_2Me 28-CO_2Me	2, 3-seco; 2, 3-lactone	236	+117

TABLE 26

No.	Compound	Ref. No.	Molecular formula	C=C	—OH	CO	$-CO_2H$		M.P.	$[\alpha]_D$
194	Phyllanthol	33	$C_{30}H_{50}O$		3β			13α, 27α-cyclo	234	+43

TABLE 27

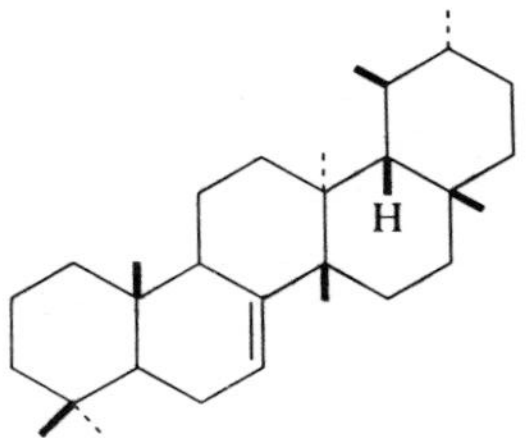

No.	Compound	Ref. No.	Molecular formula	C=C	—OH	CO	—CO_2H	M.P.	$[\alpha]_D$
195	Baurenol	195	$C_{30}H_{50}O$		3β			208	−30

TABLE 28

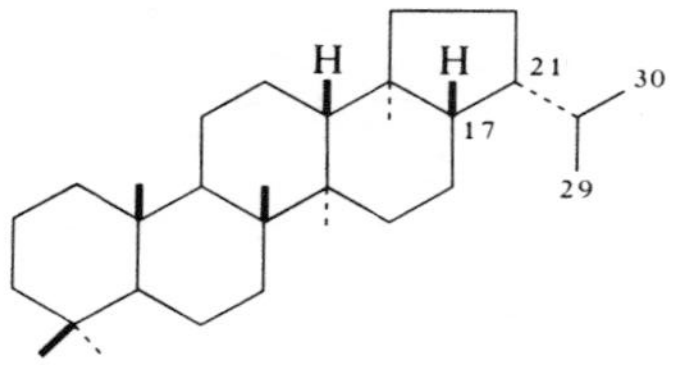

No.	Compound	Ref. No.	Molceular formula	C=C	—OH	CO	—CO_2H		M.P.	$[\alpha]_D$
196	Diplopterol (hydroxyhopane)	4	$C_{30}H_{52}O$		22				256	+44
197	Zeorin	167	$C_{30}H_{52}O_2$		6α, 22				253	+54
198	Leucotylin	289	$C_{30}H_{52}O_3$		6α, 16, 22				333	+49
199	Hydroxyhopanone	16	$C_{30}H_{50}O_2$		22	3			255	+64
200	Leucotylic acid	290	$C_{30}H_{50}O_4$		16β, 22		23		260	+330
201	Adiantone	42	$C_{29}H_{48}O$			22		29-nor	224	+81

TABLE 29

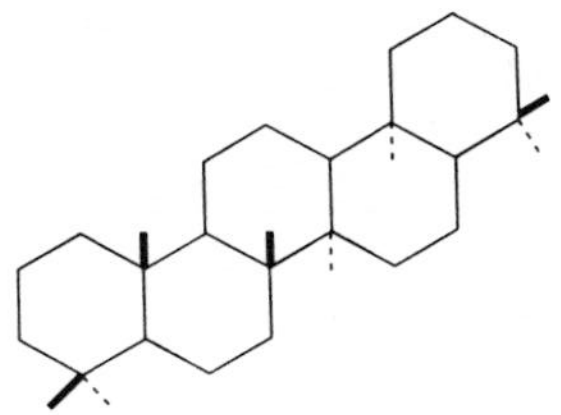

No.	Compound	Ref. No.	Molecular formula	C=C	—OH	CO	—CO_2H	M.P.	$[\alpha]_D$
202	Tetrahymanol	281	$C_{30}H_{52}O$		3β			315	—

TABLE 30

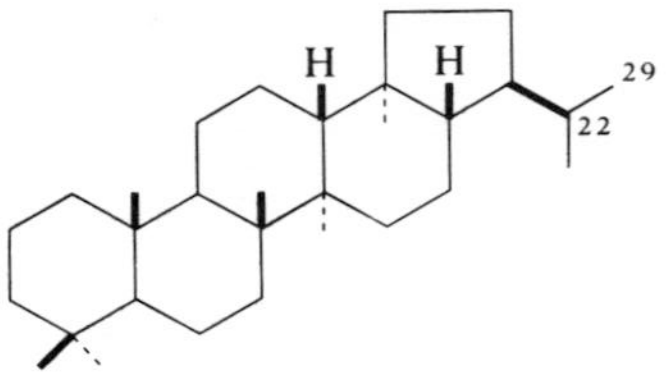

No.	Compound	Ref. No.	Molecular formula	C=C	—OH	CO	—CO_2H	M.P.	$[\alpha]_D$
203	Moretenol	219	$C_{30}H_{50}O$	22(29)	3β			237	+27

TABLE 31

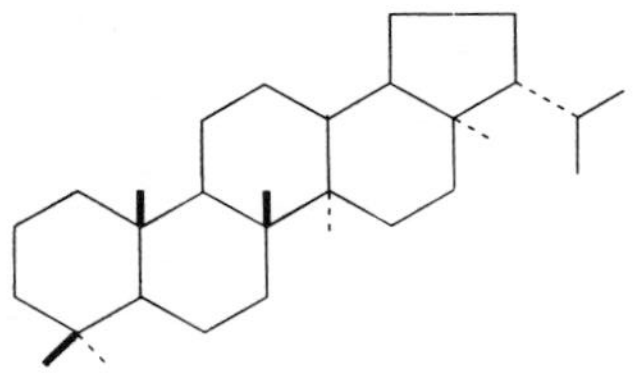

No.	Compound	Ref. No.	Molecular formula	C=C	—OH	CO	—CO_2H	M.P.	$[\alpha]_D$
204	Neomotiol	226	$C_{30}H_{50}O$	12	3β			228	−24

TABLE 32

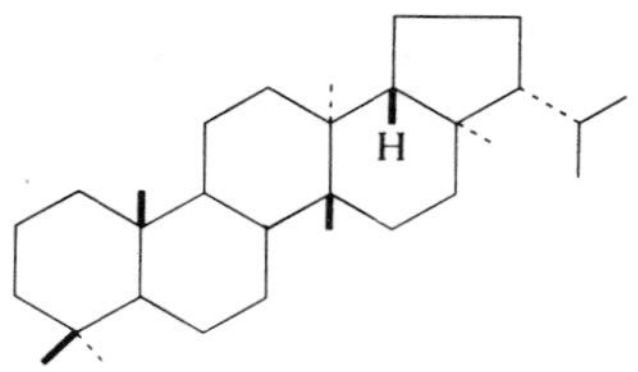

No.	Compound	Ref. No.	Molecular formula	C=C	—OH	CO	—CO_2H	M.P.	$[\alpha]_D$
205	Fernene	2, 3	$C_{30}H_{50}$	9(11)				171	−17
206	Motiol	226	$C_{30}H_{50}O$	7	3			218	−44
207	Fernenol	193	$C_{30}H_{50}O$	9(11)	3			194	−24
208	Arundoin	230	$C_{31}H_{52}O$	9(11)	3βMe			243	−5
209	Motidiol	226	$C_{30}H_{50}O_2$	7	2, 3			233	−22
210	Davallic acid	204	$C_{30}H_{48}O_2$	9(11)			24	283	+94

TABLE 33

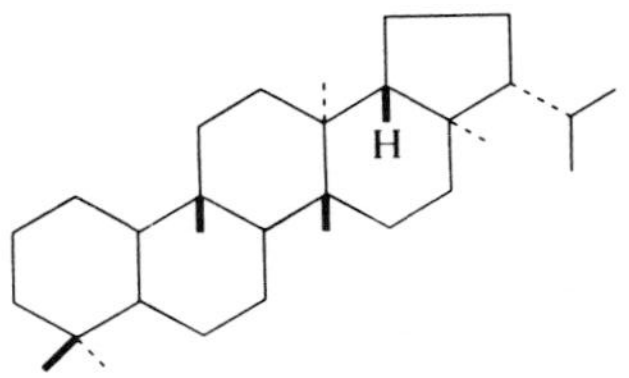

No.	Compound	Ref. No.	Molecular formula	C=C	—OH	CO	—CO_2H	M.P.	$[\alpha]_D$
211	Adianene	3	$C_{30}H_{50}$	5				192	+51
212	Simiarenol	9	$C_{30}H_{50}O$	5	3β			210	+51
213	Adianenediol	226	$C_{30}H_{50}O_2$	5	2, 3			231	+3

TABLE 34

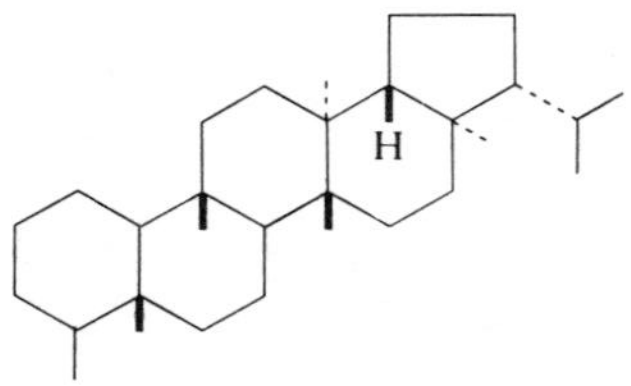

No.	Compound	Ref. No.	Molecular formula	C=C	—OH	CO	—CO_2H		M.P.	$[\alpha]_D$
214	Filicene	3	$C_{30}H_{50}$	3					230	+50
215	Filicene epoxide (adiantoxide)	41	$C_{30}H_{50}O$					3α, 4α-oxide	231	+47

TABLE 35

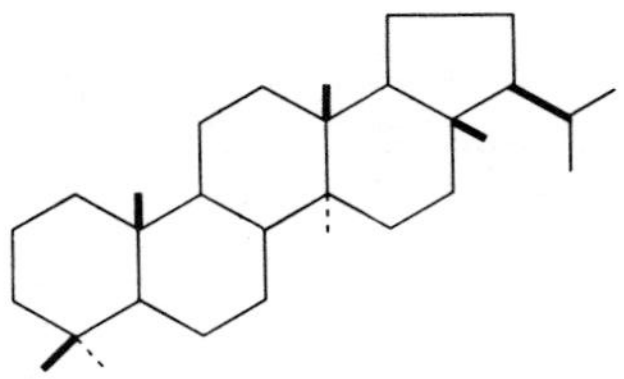

No.	Compound	Ref. No.	Molecular formula	C=C	—OH	CO	—CO_2H	M.P.	$[\alpha]_D$
216	Arborinol	180	$C_{30}H_{50}O$	9(11)	3α			275	+34
217	Isoarborinol	231	$C_{30}H_{50}O$	9(11)	3β			299	+47

TABLE 36

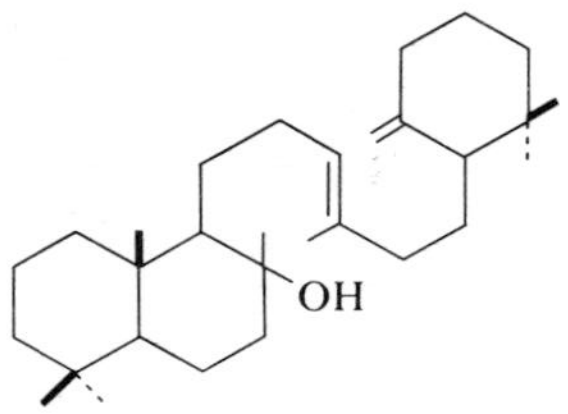

No.	Compound	Ref. No.	Molecular formula	C=C	—OH	CO	$-CO_2H$	M.P.	$[\alpha]_D$
218	Ambrein (94)	202, 252	$C_{30}H_{52}O$					84	+14*

TABLE 37

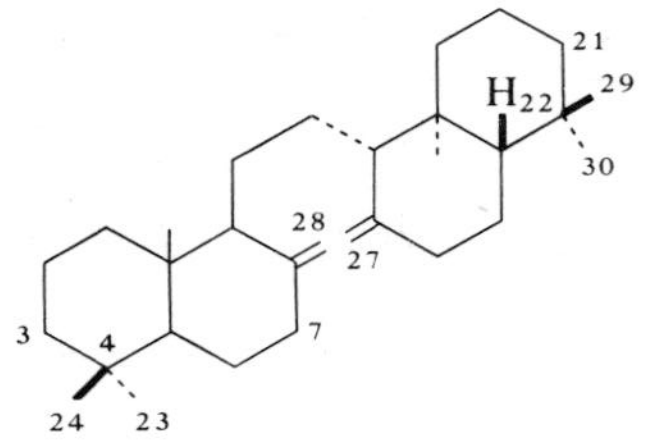

No.	Compound	Ref. No.	Molecular formula	C=C	—OH	CO	—CO_2H		M.P.	$[\alpha]_D$
219	α-Onocerin	32	$C_{30}H_{50}O_2$		3, 21				203	+18
220	Lansic acid	181	$C_{30}H_{46}O_4$	7, 14(27), 4(23), 22(29)			3, 21	3, 4-seco; 21, 22-seco	184	−7
221	Clavatol	252a	$C_{36}H_{58}O_8$		3βAc, 8βAc, 14αAc, 21βAc			27, 28-bisnor	279	—

TABLE 38

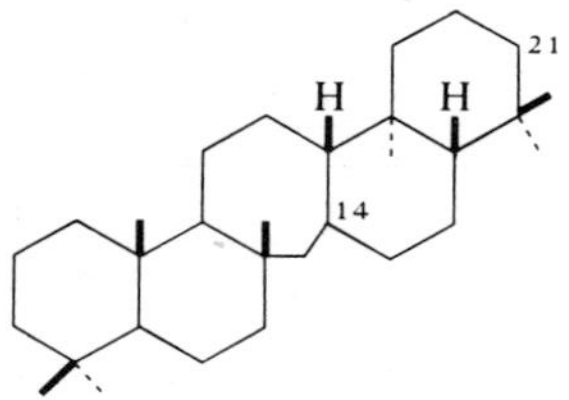

No.	Compound	Ref. No.	Molecular formula	C=C	—OH	CO	—CO_2H	M.P.	$[\alpha]_D$
222	Tohogenol	170	$C_{30}H_{52}O_3$		3β, 14β, 21α			244	—
223	Serratenediol	169	$C_{30}H_{50}O_2$	14	3β, 21α			305	−19

REFERENCES

1. Abe, S. and Obara, T. (1959). *J. Chem. Soc. Jap.* **80**, 1487.
2. Ageta, H., Iwata, K. and Natori, S. (1963). *Tetrahedron Lett.* 1447.
3. Ageta, H., Iwata, K. and Natori, S. (1964). *Tetrahedron Lett.* 3413.
4. Ageta, H., Iwata, K. and Otake, Y. (1963). *Chem. and Pharm. Bull.* (*Japan*) **11**, 407.
5. Ageta, H., Iwata, K. and Yonezawa, K. (1963). *Chem. and Pharm. Bull.* (*Japan*) **11**, 408.
6. Akisanya, A., Bevan, C. W. L., Halsall, T. G., Powell, J. W. and Taylor, D. A. H. (1961). *J. Chem. Soc.* 3705.

6a. Altman, L. J., Kowerski, R. C. and Rilling, H. C. (1971). *J. Amer. Chem. Soc.* **93**, 1782.

7. Ames, T. R., Halsall, T. G. and Jones, E. R. H. (1951). *J. Chem. Soc.* 450.
8. Anthonsen, J., Bruun, T., Hemmer, E., Holme, D., Lamvik, A., Sunde, E. and Sorensen, N. A. (1970). *Acta Chem. Scand.* **24**, 2479.
9. Aplin, R. T., Arthur, H. R. and Hui, W. H. (1966). *J. Chem. Soc.* (*C*) 1251.
10. Arene, E. O., Bevan, C. W. L., Powell, J. W. and Taylor, D. A. H. (1965). *Chem. Commun.* 302.
11. Arigoni, D., Barton, D. H. R., Bernasconi, R., Djerassi, C., Mills, J. S. and Wolff, R. (1959). *Proc. Chem. Soc.* 306.
12. Arigoni, D., Barton, D. H. R., Bernasconi, R., Djerassi, C., Mills, J. S. and Wolff, R. E. (1960). *J. Chem. Soc.* 1900.
13. Arigoni, D., Barton, D. H. R., Corey, E. J., Jeger, O., Caglioti, L., Sukh Dev, Ferrini, P. G., Glazier, E. R., Melera, A., Pradhan, S. K., Schaffner, K., Sternhell, S., Templeton, J. F. and Tobinga, S. (1960). *Experientia* **16**, 41.
14. Arigoni, D., Jeger, O. and Ruzicka, L. (1955). *Helv. Chim. Acta* **38**, 222.
15. Arnott, S., Davie, A. W., Robertson, J. M., Sim, G. A. and Watson, D. G. (1961). *J. Chem. Soc.* 4183.
16. Baddely, G. V., Halsall, T. G. and Jones, E. R. H. (1961). *J. Chem. Soc.* 3891.
17. Barbour, J. B., Warren, F. L. and Wood, D. A. (1951). *J. Chem. Soc.* 2537.
18. Barclay, G. A., Eade, R. A., Simes, H. V., Simes, J. J. H. and Taylor, J. C. (1963). *Chem. Ind.* (*London*) 1206.
19. Barna, A. K. (1956). *Naturwiss* **43**, 250.

19a. Barna, A. K. and Raman, S. P. (1962). *Tetrahedron* **18**, 155.

20. Barnes, C. S., Galbraith, M. N., Ritchie, E. and Taylor, W. C. (1965). *Aust. J. Chem.* **18**, 1411.
21. Barton, D. H. R. (1951). *J. Chem. Soc.* 1444.
22. Barton, D. H. R. (1953). *J. Chem. Soc.* 1027.
23. Barton, D. H. R. and Brooks, C. J. W. (1951). *J. Chem. Soc.* 257.
24. Barton, D. H. R., Carruthers, W. and Overton, K. H. (1956). *J. Chem. Soc.* 788.
25. Barton, D. H. R., Cheung, H. T., Daniels, P. J. L., Lewis, K. G. and McGhie, J. F. (1962). *J. Chem. Soc.* 5163.
26. Barton, D. H. R., Giacopello, D., Manitto, P. and Struble, D. L. (1969). *J. Chem. Soc.* (*C*) 1047.
27. Barton, D. H. R., Garbers, C. F., Giacopello, D., Harvey, R. G., Lessard, J. and Taylor, D. R. (1969). *J. Chem. Soc.* (*C*) 1050.
28. Barton, D. H. R., Gosden, A. F., Mellows, G. and Widdowson, D. A. (1969). *Chem. Commun.* 184.
29. Barton, D. H. R., McGhie, J. F., Pradhan, M. K. and Knight, S. A. (1955). *J. Chem. Soc.* 876.

30. Barton, D. H. R. and de Mayo, P. (1954). *J. Chem. Soc.* 887.
31. Barton, D. H. R., de Mayo, P. and Orr, J. C. (1956). *J. Chem. Soc.* 4160.
32. Barton, D. H. R. and Overton, K. H. (1955). *J. Chem. Soc.* 2639.
33. Barton, D. H. R., Page, J. E. and Warnhoff, E. W. (1954). *J. Chem. Soc.* 2715.
34. Barton, D. H. R., Pradhan, S. K., Sternhell, S. and Templeton, J. F. (1961). *J. Chem. Soc.* 255.
35. Barton, D. H. R. and Seoane, E. (1956). *J. Chem. Soc.* 4150.
35a. Barua, A. K., Chakrabarti, P., Pal, S. K. and Das, B. C. (1970). *J. Indian Chem. Soc.* **47**, 195.
36. Beaton, J. M., Johnston, J. D., McKean, L. C. and Spring, F. S. (1953). *J. Chem. Soc.* 3660.
37. Beaton, J. M. and Spring, F. S. (1955). *J. Chem. Soc.* 3126.
38. Beaton, J. M., Spring, F. S., Stevenson, R. and Stewart, J. L. (1955). *J. Chem. Soc.* 2123.
39. Bentley, H. R., Henry, J. A., Irvine, D. S. and Spring, F. S. (1953). *J. Chem. Soc.* 3673.
40. Benveniste, P., Hewlins, M. J. E. and Fritig, B. (1969). *Eur. J. Biochem.* **9**, 526.
41. Berti, G., Bottari, F. and Marsili, A. (1969). *Tetrahedron* **25**, 2939.
42. Berti, G., Bottari, F., Marsili, A., Lehn, J. M., Witz, P. and Ourisson, G. (1963). *Tetrahedron Lett.* 1283.
43. Beton, J. L., Halsall, T. G. and Jones, E. R. H. (1956). *J. Chem. Soc.* 2904.
44. Bevan, C. W. L., Ekong, D. E. U., Halsall, T. G. and Toft, P. (1967). *J. Chem. Soc.* (*C*) 820.
45. Bevan, C. W. L., Powell, J. W. and Taylor, D. A. H. (1963). *J. Chem. Soc.* 980.
46. Bevan, C. W. L., Powell, J. W., Taylor, D. A. H., Toft, P. and Welford, M. (1967). *J. Chem. Soc.* (*C*) 163.
47. Bevan, C. W. L., Rees, A. H. and Taylor, D. A. H. (1963). *J. Chem. Soc.* 983.
48. Biellman, J. F. (1966). *Tetrahedron Lett.* 4803.
49. Biellman, J. F. and Ducep, J. B. (1969). *Tetrahedron Lett.* 3707.
50. Biglino, G., Cattel, L., Caputo, O. and Nobili, G. (1969). *Gazz. Chim. Ital.* **99**, 830.
50a. Biglino, G., Lehn, J.-M. and Ourisson, G. (1963). *Tetrahedron Lett.* 1651.
51. Birch, A. J., Collins, D. J., Muhammad, S. and Turnbull, J. P. (1963). *J. Chem. Soc.* 2762.
52. Blackburn, G. M., Ollis, W. D., Smith, C. and Sutherland, I. O. (1969). *Chem. Commun.* 99.
53. Bowers, A., Halsall, T. G., Jones, E. R. H. and Lemin, A. J. (1953). *J. Chem. Soc.* 2548.
54. Breen, G. J. W., Ritchie, E., Sidwell, W. T. L. and Taylor, W. C. (1966). *Aust. J. Chem.* **19**, 455.
55. Brewis, S., Halsall, T. G., Harrison, H. R. and Hodder, O. J. R. (1970). *Chem. Commun.* 891.
56. Brownlie, G., Spring, F. S., Stevenson, R. and Strachan, W. S. (1955). *Chem. Ind.* (*London*) 686, 1156.
57. Brownlie, G., Spring, F. S., Stevenson, R. and Strachan, W. S. (1956). *J. Chem. Soc.* 2419.
58. Brooks, C. J. W. (1955). *J. Chem. Soc.* 1675.
59. Brown, K. S. and Kupchan, S. M. (1962). *J. Amer. Chem. Soc.* **84**, 4592.
60. Brown, W. A. C. and Sim, G. A. (1962). *Proc. Chem. Soc.* 293.
61. Bruum, T. (1954). *Acta Chem. Scand.* **8**, 1291.
62. Buchanan, J. G. St. C. and Halsall, T. G. (1969). *Chem. Commun.* **48**, 242.

63. Burbage, M. B., Jewers, K., Eade, R. A., Harper, P. and Simes, J. J. H. (1971). *Chem. Commun.* 195.
64. Caglioti, L., Cainelli, G. and Minutilli, F. (1961). *Gazz. Chim. Ital.* 1387.
65. Cainelli, G., Britt, J. J., Arigoni, D. and Jeger, O. (1958). *Helv. Chim. Acta.* **41**, 2053.
65a. Campbell, R. V. M., Crombie, L. and Pattenden, G. (1971), *Chem. Commun.* 218.
66. Canonica, L., Fedeli, E. and Fiecchi, A. (1959). *Gazz. Chim. Ital.* **89**, 818.
67. Canonica, L., Russo, G. and Bonati, A. (1966). *Gazz. Chim. Ital.* **96**, 772.
68. Carman, R. M. and Cowley, D. (1965). *Aust. J. Chem.* **18**, 213, 1493.
69. Caspi, E. J., Grieg, J. B., Zander, J. M. and Mandelbaum, A. (1969). *Chem. Commun.* **28**, 210.
70. Chakrabarthy, T., Connolly, J. D., McCrindle, R., Overton, K. H. and Schwarz, J. C. P. (1968). *Tetrahedron* **24**, 1503.
71. Chan, W. R., Magnus, K. E. and Mootoo, B. S. (1967). *J. Chem. Soc.* (*C*) 171.
72. Chan, W. R., Taylor, D. R. and Aplin, R. T. (1966). *Chem. Commun.* 576.
73. Chanley, J. D., Mezzetti, T. and Sobotka, H. (1966). *Tetrahedron* **22**, 1857.
74. Chatterjee, S. K., Anand, N. and Dhan, M. L. (1959). *J. Sci. Ind. Res.* (*India*) **18B**, 262.
75. Chopra, C. S., Cole, A. R. H., Theiberg, K. J. L., White, D. E. and Arthur, H. R. (1965). *Tetrahedron* **21**, 1529.
76. Chopra, G. R., Jain, A. C. and Seshadri, T. R. (1969). *Indian J. Chem.* **7**, 1179.
77. Chopra, C. S., White, D. E. and Melrose, G. J. H. (1965). *Tetrahedron* **21**, 2585.
78. Chowla, A. and Sukh Dev. (1967). *Tetrahedron Lett.* 4837.
79. Clayton, R. B. (1965). *Quart. Rev. Chem. Soc.* **19**, 168.
79a. Coates, R. M. and Robinson, W. H. (1971). *J. Amer. Chem. Soc.* **93**, 1785.
80. Coggon, P., McPhail, A. T., Storer, R. and Young, D. W. (1969). *Chem. Commun.* 828.
81. Connolly, J. D., Henderson, R., McCrindle, R., Overton, K. H. and Bhacca, N. S. (1965). *J. Chem. Soc.* 6935.
82. Connolly, J. D. and McCrindle, R. (1967). *Chem. Commun.* 1193; (1971) *J. Chem. Soc.* (*C*) 1715.
83. Connolly, J. D., McCrindle, R. and Overton, K. H. (1968). *Tetrahedron* **24**, 1489, 1497.
84. Connolly, J. D., McCrindle, R., Overton, K. H. and Warnock, W. D. C. (1968). *Tetrahedron* **24**, 1507.
85. Connolly, J. D., Polonsky, J. and Overton, K. H. (1970). *Advan. Phytochem.* **2**, 385.
86. Connolly, J. D., Thornton, I. M. S. and Taylor, D. A. H. (1970). *Chem. Commun.* 1205.
87. Connolly, J. D., Thornton, I. M. S. and Taylor, D. A. H. (1971). *Chem. Commun.* 17.
88. Cooper, A. and Hodgkin, D. C. (1968). *Tetrahedron* **24**, 909.
89. Corey, E. J. and Cantrall, E. W. (1959). *J. Amer. Chem. Soc.* **81**, 1745.
90. Corey, E. J. and de Montellano, P. R. O. (1967). *J. Amer. Chem. Soc.* **89**, 3362.
91. Corey, E. J., Russey, W. E. and de Montellano, P. R. O. (1966). *J. Amer. Chem. Soc.* **88**, 4750.
92. Corey, E. J. and Ursprung, J. J. (1955). *J. Amer. Chem. Soc.* **77**, 3667, 3668.
93. Cornforth, J. D. (1969). *Quart. Rev. Chem. Soc.* **23**, 125.
94. Cornforth, J. W., Cornforth, R. H. and Mathew, K. K. (1959). *J. Chem. Soc.* 2539.

95. Corsano, S. and Mincione, E. (1965). *Tetrahedron Lett.* 2377.
96. Corsano, S. and Minicione, E. (1968). *Chem. Commun.* 739.
97. Corsano, S. and Piancatelli, G. (1969). *Gazz. Chim. Ital.* **99**, 1140.
98. Cort, L. A., Gascoigne, R. M., Holker, J. S. E., Ralph, B. J., Robertson, A. and Simes, J. J. H. (1954). *J. Chem. Soc.* 3713.
98a. Cotterell, G. P., Halsall, T. G. and Wriglesworth, M. J. (1970). *J. Chem. Soc.* (*C*) 739.
99. Cox, J. S. G., King, F. E. and King, T. J. (1959). *J. Chem. Soc.* 514.
100. Curtis, R. G., Fridrichsons, J. and Mathieson, A. McL. (1952). *Nature, Lond.* **170**, 321.
101. Curtis, R. G., Heilbron, I., Jones, E. R. H. and Woods, G. F. (1953). *J. Chem. Soc.* 457.
102. Dawson, M. C., Halsall, T. G., Jones, E. R. H., Meakins, G. D. and Phillips, P. C. (1956). *J. Chem. Soc.* 3172.
103. Djerassi, C., Henry, J. A., Lemin, A. J., Rios, T. and Thomas, G. H. (1956). *J. Amer. Chem. Soc.* **78**, 3783.
104. Djerassi, C., Krakower, G. W., Lemin, A. J., Liu, L. H., Mills, J. S. and Villotti, R. (1958). *J. Amer. Chem. Soc.* **80**, 6284.
105. Djerassi, C. and Lippman, A. E. (1955). *J. Amer. Chem. Soc.* **77**, 1825.
106. Djerassi, C., Robinson, C. H. and Thomas, D. B. (1956). *J. Amer. Chem. Soc.* **78**, 5685.
107. Djerassi, C., Thomas, D. B., Livingston, A. L. and Thompson, C. R. (1957). *J. Amer. Chem. Soc.* **79**, 5292.
108. Djerassi, C., Thomas, G. H. and Jeger, O. (1955). *Helv. Chim. Acta* **38**, 1304.
109. Dreyer, D. L. (1966). *J. Org. Chem.* **31**, 2279.
110. Dreyer, D. L. (1968). *Prog. Chem. Org. Nat. Prod.* 190.
111. Dutler, H., Jeger, O. and Ruzicka, L. (1955). *Helv. Chim. Acta* **38**, 1268.
112. Dorchai, R. O. and Thomson, J. B. (1965). *Tetrahedron Lett.* 2223.
113. Eade, R. A., Ellis, J. and Simes, J. J. H. (1966). *Chem. Commun.* 246.
114. Eade, R. A., Simes, J. J. H. and Stevenson, B. (1963). *Aust. J. Chem.* **16**, 900.
114a. Edmond, J., Popjak, G., Wong, S. M. and Williams, V. (1971). *Fed. Proc.* **30**, 1106.
115. Eliel, E. L., Allinger, N. A., Angyal, S. J. and Morrison, G. A. (1965). "Conformational Analysis", p. 256, Interscience Publishers, New York.
116. Elsevier's Encyclopaedia of Organic Chemistry, (1940). Vol. 14, pp. 528, 532. (Amsterdam.)
117. Elsevier Supplement. (1952). p. 946.
118. Ibid. 532; 1073S.
119. Ibid. 539; 974S.
120. Ibid. 548; 1010S.
121. Ibid. 559; 981S.
122. Ibid. 564; 1052S.
123. Ibid. 565; 1092S.
124. Ibid. 568; 1133S.
125. Ibid. 569; 1137S.
126. Ibid. 571; 1142S.
127. Ibid. 575; 1115S.
128. Ibid. 587; 1029S.
129. Ibid. 591; 1043S.
130. Ibid. 606; 1159S.

131. Ibid. 974S.
132. Ibid. 1161S.
133. Ibid. 1163S.
134. Ekong, D. E. U., Fakunle, C. O., Fasina, A. K. and Okogun, J. I. (1969). *Chem. Commun.* 1166.
135. Elyakov, G. B., Dzizenko, A. K. and Elkin, Y. N. (1966). *Tetrahedron Lett.* 141.
136. Eppenberger, U., Hirth, L. and Ourisson, G. (1969). *Euro J. Biochem.* **8**, 180.
136a. Epstein, W. W. and Rilling, H. C. (1970). *J. Biol. Chem.* **245**, 4597.
137. Fischer, F. G. and Seiler, N. (1961). *Ann.* **644**, 162.
138. Fridrichsons, J. and Mathieson, A. McL. (1953). *J. Chem. Soc.* 2159.
139. Ganguly, A. K., Govindachari, T. R., Mohamed, P. A., Rahimtulla, A. D. and Viswanathan, N. (1966). *Tetrahedron* **22**, 1513.
140. Godtfredsen, W. O., von Daehne, W., Vangedal, S., Marquet, A., Arigoni, D. and Melera, A. (1965). *Tetrahedron* **21**, 3505.
141. Godtfredsen, W. O., Lorck, H., van Tamelen, E. E., Willett, J. D. and Clayton, R. B. (1968). *J. Amer. Chem. Soc.* **90**, 208.
142. Gonzalez, A. G. and Breton, J. L. (1959). *An. Real Soc. Espan. Fis. Quim.* **55B**, 93.
143. Gopinath, K. W., Govindachari, T. R., Parthasarathy, P. C., Viswanathan, N., Arigoni, D. and Wildman, W. C. (1961). *Proc. Chem. Soc.* 446.
144. Govindachari, T. R., Joshi, B. S. and Sundararajan, V. N. (1964). *Tetrahedron* 2985.
145. Govindachari, T. R., Prokash, D. and Viswanathan, N. (1968). *J. Chem. Soc.* (*C*) 1323.
146. Grant, I. J., Hamilton, J. A., Hamor, T. G., Robertson, J. M. and Sim, G. A. (1963). *J. Chem. Soc.* 2506.
147. Guise, G. B., Ritchie, E. and Taylor, W. C. (1962). *Aust. J. Chem.* **15**, 314.
148. Hall, S. R. and Maslen, E. N. (1965). *Acta Cryst.* **18**, 265.
149. Halsall, T. G. and Aplin, R. T. (1964). *Progr. Chem. Org. Natur. Prod.* **22**, 153.
150. Halsall, T. G., Hodges, R. and Sayers, G. C. (1959). *J. Chem. Soc.* 2036.
151. Halsall, T. G., Jones, E. R. H., Lowe, G. and Newall, C. E. (1966). *Chem. Commun.* 685.
152. Ham, P. J. and Whiting, D. A. (1970). *Chem. Ind.* (*London*) 1379.
153. Harada, R., Kakisawa, H., Kobayashi, S., Musya, M., Nakanishi, K. and Takahashi, Y. (1962). *Tetrahedron Lett.* 603.
154. Harris, M., Henderson, R., McCrindle, R., Overton, K. H. and Turner, D. W. (1968). *Tetrahedron* **24**, 1517.
155. Harrison, H. R., Hodder, O. J. R., Bevan, C. W. L., Taylor, D. A. H. and Halsall, T. G. (1970). *Chem. Commun.* 1388.
156. Hattori, J., Igarashi, H., Iwasaki, S. and Okuda, S. (1969). *Tetrahedron Lett.* 1023.
157. Haworth, R. D. (1937). *Ann. Rep.* **34**, 327.
158. Heintz, R. and Benveniste, P. (1970). *Phytochemistry* **9**, 1499.
159. Heintz, R., Schaefer, P. C. and Benveniste, P. (1970). *Chem. Commun.* 946.
160. Henry, J. A., Irvine, D. S. and Spring, F. S. (1955). *J. Chem. Soc.* 1607.
161. Hensens, O. D. and Lewis, K. G. (1965). *Tetrahedron Lett.* **51**, 4639.
162. Hikino, H., Ohta, T. and Takemoto, T. (1970). *Chem. Pharm. Bull.* (*Japan*) **18**, 1082.
163. Hirose, Y. (1963). *Chem. Pharm. Bull.* (*Japan*) **11**, 535.
164. Hodges, R., McGeachin, S. G. and Raphael, R. A. (1963). *J. Chem. Soc.* 2515.
165. Hollands, T. R., de Mayo, P., Nisbet, M. and Crabbé, P. (1965). *Can. J. Chem.* **43,** 2996; 3008.

166. Huneck, S. (1963). *Tetrahedron* **19**, 479.
167. Huneck, S. and Lehn, J.-M. (1963). *Bull. Soc. Chim. Fr.* 1702.
168. Ikan, R. and McLean, J. (1960). *J. Chem. Soc.* 893.
169. Inubushi, Y., Sano, T. and Tsuda, Y. (1964). *Tetrahedron Lett.* 1303.
170. Inubushi, Y., Tsuda, Y. and Sano, T. (1965). *Chem. Pharm. Bull.* (*Japan*) **13**, 750.
171. Ireland, R. E., Baldwin, S. W., Dawson, D. J., Dawson, M. I., Dolfini, J. E., Newbould, J., Johnson, W. S., Brown, M., Crawford, R. J., Hudrlik, P. F., Rasmussen, G. H. and Schmiegel, K. K. (1970). *J. Amer. Chem. Soc.* **92**, 5743.
172. Ireland, R. E. and Welch, S. C. (1970). *J. Amer. Chem. Soc.* **92**, 7232.
173. Iwasaki, S., Sair, M. I., Igarashi, H. and Okuda, S. (1970). *Chem. Commun.* 1119.
174. Jarolim, V., Hejno, K. and Sorm, F. (1963). *Collect. Czech. Chem. Commun.* **28**, 2443.
175. Jeger, O. (1950). *Progr. Chem. Org. Natur. Prod.* **7**, 1.
176. Jones, E. R. H. and Halsall, T. G. (1955). *Progr. Chem. Org. Natur. Prod.* **12**, 44.
177. Kamiya, K., Murata, T. and Nishikawa, M. (1970). *Chem. Pharm. Bull.* (*Japan*) **18**, 1362.
178. Kane, V. V. and Stevenson, R. (1960). *J. Org. Chem.* **25**, 1394.
179. Kawaguchi, A. and Okuda, S. (1970). *Chem. Commun.* 1012.
180. Kennard, O., di Sanseverino, L. R., Vorbruggen, H. and Djerassi, C. (1965). *Tetrahedron Lett.* 3433.
181. Kiang, A. K., Tan, E. L., Lim, F. Y., Habaguchi, K., Nakanishi, K., Fachan, L. and Ourisson, G. (1967). *Tetrahedron Lett.* 3571.
182. Kier, L. B., Lehn, J. M. and Ourisson, G. (1963). *Bull. Soc. Chim. Fr.* 911.
183. Kimura, K., Hashimoto, Y. and Agata, I. (1960). *Chem. Pharm. Bull.* (*Japan*) **8**, 1145.
184. King, F. E. and King, T. J. (1956). *J. Chem. Soc.* 4469.
185. King, F. E. and Morgan, J. W. W. (1960). *J. Chem. Soc.* 4738.
186. King, T. J. and Yardley, J. P. (1961). *J. Chem. Soc.* 4308.
187. Knight, J. C., Wilkinson, D. I. and Djerassi, C. (1966). *J. Amer. Chem. Soc.* **88**, 790.
188. Kochetkov, N. K., Khorlin, A. J. and Ovodov, Ju. S. (1963). *Tetrahedron Lett.* 477.
189. de Kock, W. T., Enslin, P. R., Norton, K. B., Barton, D. H. R., Sklarz, B. and Bothner-By, A. A. (1963). *J. Chem. Soc.* 3828.
190. Kubota, T., Matsuura, T., Tokoroyama, T., Kamikawa, T. and Matsumoto, T. (1961). *Tetrahedron Lett.* 325.
191. Kubota, T., Tonami, F. and Hinoh, H. (1966). *Tetrahedron Lett.* 701.
192. Kuhn, R. and Low, I. (1963). *Ann.* **669**, 183.
193. Kundu, S. K., Chatterjee, A. and Rao, A. S. (1966). *Tetrahedron Lett.* 1043.
194. Kupchan, S. M., Smith, R. M., Aynehchi, Y. and Maruyama, M. (1970). *J. Org. Chem.* **35**, 2891.
195. Lahey, F. N. and Leeding, M. V. (1958). *Proc. Chem. Soc.* 342.
196. Laird, W., Spring, F. S. and Stevenson, R. (1960). *J. Amer. Chem. Soc.* **82**, 4108.
197. Lavie, D. and Glotter, E. (1972). *Progr. Chem. Org. Natur. Prod.* 29, *in press.*
198. Lavie, D., Jain, M. K. and Kirson, J. (1967). *J. Chem. Soc.* (*C*) 1347.
199. Lavie, D. and Levy, E. C. (1968). *Tetrahedron Lett.* 2097.
200. Lawrie, W., Hamilton, W., Spring, F. S. and Watson, H. S. (1956). *J. Chem. Soc.* 3272.
201. Lawrie, W., Spring, F. S. and Watson, H. S. (1956). *Chem. Ind.* (*London*) 1458.
202. Lederer, E., Marx, F., Mercier, D. and Perot, G. (1946). *Helv. Chim. Acta* **29**, 1354.
203. Lewis, D. A. and McGhie, J. F. (1956). *Chem. and Ind.* (*London*) 550.

204. Lin, Y-Y., Kakisawa, H., Shiobara, Y. and Nakanishi, K. (1965). *Chem. Pharm. Bull. (Japan)* **13**, 986.
205. Lynen, F. (1965). *Umschau* **65**, 321.
206. McCrindle, R. and Overton, K. H. (1969). *In* "Rodd's Chemistry of Carbon Compounds" (S. Coffey, ed.), 2nd Ed., Vol. II, Part C, p. 406, Elsevier, Amsterdam.
207. McLean, J. and Watts, W. E. (1960). *J. Org. Chem.* **25**, 1263.
208. McPhail, A. T. and Sim, G. A. (1966). *J. Chem. Soc.* (*B*) 318.
209. Mangoni, L. and Belardini, M. (1963). *Tetrahedron Lett.* 921.
210. Marx, M., Leclercq, J., Tursch, B. and Djerassi, C. (1967). *J. Org. Chem.* **32**, 3150.
211. Matsunaga, S., Okada, J. and Uyeo, S. (1965). *Chem. Commun.* 525.
212. Matyukhina, L. G. and Ryabinin, A. A. (1960). *Dokl. Acad. Sci. USSR, Earth Sci. Sect.* **131**, 316.
213. de Mayo, P. (1959). "The Higher Terpenoids", p. 64. Interscience Publishers, New York.
214. de Mayo, P. and Starratt, A. N. (1962). *Can. J. Chem.* **40**, 788.
215. Mazur, Y., Weizmann, A. and Sondheimer, F. (1958). *J. Amer. Chem. Soc.* **80**, 6293.
216. Meisels, A., Jeger, O. and Ruzicka, L. (1949). *Helv. Chim. Acta* **32**, 1075.
217. Menard, E., Wyler, H., Hiestand, A., Arigoni, D., Jeger, O. and Ruzicka, L. (1955). *Helv. Chim. Acta* **38**, 1517.
218. van der Merwe, K. J., Enslin, P. R. and Pachler, K. (1963). *J. Chem. Soc.* 4275.
219. Miller, C. H., Rawson, J. W. L., Ritchie, E., Shannon, J. S. and Taylor, W. C. (1965). *Aust. J. Chem.* **18**, 226.
220. Moron, J. and Polonsky, J. (1968). *Tetrahedron Lett.* 385.
221. Moron, J., Rondest, J. and Polonsky, J. (1966). *Experientia* **22**, 511.
222. Murata, T., Imai, S., Imanishi, M., Goto, M. and Morita, K. (1965). *Tetrahedron Lett.* 3215.
223. Murata, T. and Miyamoto, M. (1970). *Chem. Pharm. Bull (Japan)* **18**, 1354.
224. Musgrave, O. C., Stark, J. and Spring, F. S. (1952). *J. Chem. Soc.* 4393.
225. Nagai, M., Izawa, K. and Inoue, T. (1969). *Chem. Pharm. Bull. (Japan)* **17**, 1438.
226. Nakamura, S., Yamada, T., Wada, H., Inoue, Y., Goto, T. and Hirata, Y. (1965). *Tetrahedron Lett.* 2017.
227. Narayanan, C. R., Pachapurkar, R. V., Pradhan, S. K. and Shah, V. R. (1964). *Indian J. Chem.* **2**, 108.
228. Nicolaides, N. and Laves, F. (1954). *J. Amer. Chem. Soc.* **76**, 2596.
229. Nishikawa, N., Kamiya, N., Murata, T., Toinira, Y. and Nitta, I. (1965). *Tetrahedron Lett.* 3223.
230. Nishimoto, K., Ito, M., Natori, S. and Ohmoto, T. (1965). *Tetrahedron Lett.* 2245.
231. Nishimoto, K., Ito, M., Natori, S. and Ohmoto, T. (1966). *Chem. Pharm. Bull. (Japan)* **14**, 97.
232. Nyburg, S. C., Walford, G. L. and Yates, P. (1965). *Chem. Commun.* 203.
233. Okorie, D. A. and Taylor, D. A. H. (1970). *J. Chem. Soc.* (*C*) 211.
234. Okuda, S., Nakayama, Y. and Tsuda, K. (1966). *Chem. Pharm. Bull. (Japan)* **14**, 436.
235. Ollis, W. D., Ward, A. D., de Oliviera, H. M. and Zelnik, R. (1970). *Tetrahedron* **26**, 1637.
236. Ourisson, G., Crabbé, P. and Rodig, O. R. (1964). "Tetracyclic Triterpenes", Hermann, Paris.
237. Pailer, M., Schader, G., Spiteller, G. and Fenzl, W. (1965). *Monatsh Chem.* **96**, 1324.

238. Polonsky, J., Fouquey, C. and Gaudemer, A. (1964). *Bull. Soc. Chim. Fr.* 1827.
238a. Polonsky, J. and Zylber, N. B. (1965). *Bull. Soc. Chim. Fr.* 2793.
239. Polonsky, J. and Zylber, J. (1961). *Bull. Soc. Chim. Fr.* 1586.
240. Popjak, G., Edmond, J., Clifford, K. and Williams, V. (1969). *J. Biol Chem.* **244**, 1897.
241. Puckett, R. T., Sim, G. A., Abushanab, E. and Kupchan, S. M. (1966). *Tetrahedron Lett.* 3815.
242. Rangaswami, S. and Samba-murthi, K. (1961). *Proc. Ind. Acad. Sci.* **54**, 132.
243. Rao, K. V. (1963). *Chem. Ind.* (*London*) 1523.
244. Rees, H. H., Goad, L. J. and Goodwin, T. W. (1968). *Tetrahedron Lett.* 723.
245. Rees, H. H., Goad, L. J. and Goodwin, T. W. (1968). *Biochem. J.* **104**, 417.
246. Rilling, H. C. and Epstein, W. W. (1970). *J. Biol. Chem.* **245**, 4597.
246a. Rilling, H. C., Poulter, C. D., Epstein, W. W. and Larsen, B. (1971). *J. Amer. Chem. Soc.* **93**, 1783.
247. Ritchie, E., Senior, R. G. and Taylor, W. C. (1969). *Aust. J. Chem.* **22**, 2371.
248. Rondest, J. and Polonsky, J. (1963). *Bull. Soc. Chim. Fr.* 1253.
249. Row, L. R. and Subba Rao, G. S. R. (1962). *Tetrahedron* **18**, 827.
250. Ruzicka, L. (1959). *Proc. Chem. Soc.* 341.
251. Ruzicka, L., Eschenmoser, A., Arigoni, D. and Jeger, O. (1955). *Helv. Chim. Acta* **38**, 1890.
252. Ruzicka, L. and Lardon, F. (1946). *Helv. Chim. Acta* **29**, 912.
252a. Sano, T., Fujimoto, T. and Tsuda, Y. (1970). *Chem. Commun.* 1274.
253. Santer, J. O. and Stevenson, R. (1962). *J. Org. Chem.* **72**, 3204.
254. Sasaki, S., Aoyagi, S. and Hsu, H. Y. (1965). *Chem. Pharm. Bull.* (*Japan*) **13**, 87.
255. Sengupta, P. and Khastgir, H. N. (1963). *Tetrahedron* **19**, 123.
256. Shannon, J. S. (1963). *Aust. J. Chem.* **16**, 683.
257. Sharpless, K. B. (1970). *J. Amer. Chem. Soc.* **92**, 6999.
258. Shibata, S., Tanaka, O., Soma, K., Tida, Y., Ando, T. and Nakamura, H. (1965). *Tetrahedron Lett.* 207.
259. Shiengthong, D., Verasarn, A., NaNonggai-Suwanrath, P. and Warnhoff, E. W. (1965). *Tetrahedron* **21**, 917.
260. Shimizu, M. and Ohta, G. (1960). *Chem. Pharm. Bull.* (*Japan*) **8**, 108.
261. Simonsen, J. and Ross, W. C. J. (1957). "The Terpenes", Vol. V, Cambridge University Press, London.
262. Sobti, R. R. and Sukh Dev, (1968). *Tetrahedron Lett.* 2215.
263. Stork, G. and Burgstahler, A. W. (1955). *J. Amer. Chem. Soc.* **77**, 5068.
264. Stork, G., Davies, J. E. and Meisels, A. (1963). *J. Amer. Chem. Soc.* **85**, 3419.
265. Stout, G. H., Malofsky, B. M. and Stout, V. F. (1964). *J. Amer. Chem. Soc.* **86**, 957.
266. Sutherland, S. A., Sim, G. A. and Robertson, J. M. (1962). *Proc. Chem. Soc.* 222.
267. Takahashi, T., Moriyama, Y., Tanahashi, T. and Ourisson, G. (1967). *Tetrahedron Lett.* 2991.
268. Takahashi, T., Tsuyuki, T., Hoshino, T. and Ito, M. (1967) *Tetrahedron Lett.* 2997.
269. van Tamelen, E. E. (1968). *Acc. Chem. Res.* **1**, 111.
270. van Tamelen, E. E. and Fried, J. H. (1970). *J. Amer. Chem. Soc.* **92**, 7206.
271. van Tamelen, E. E. and Hessler, E. J. (1966). *Chem. Commun.* 411.
271a. van Tamelen, E. E. and Schwartz, M. A. (1971). *J. Amer. Chem. Soc.* **93**, 1780.
272. van Tamelen, E. E., Milne, G. M., Suffness, M. I., Rudler Chauvin, M. C., Anderson, R. J. and Achini, R. S. (1970) *J. Amer. Chem. Soc.* **92**, 7202.

273. van Tamelen, E. E., Sharpless, K. B., Hanzlik, R., Clayton, R. B., Burlinghame, A. L. and Wszolek, P. C. (1967). *J. Amer. Chem. Soc.* **89**, 7150.
274. van Tamelen, E. E., Willett, J. D. and Clayton, R. B. (1967). *J. Amer. Chem. Soc.* **89**, 3371.
275. van Tamelen, E. E., Willett, J. D., Clayton, R. B. and Lord, K. E. (1966). *J. Amer. Chem. Soc.* **88**, 4752.
276. van Tamelen, E. E., Willett, J., Schwartz, M. and Nadean, R. (1966). *J. Amer. Chem. Soc.* **88**, 5937.
277. Tanaka, O., Tanaka, N., Ohsawa, T., Iitaka, Y. and Shibata, S. (1968). *Tetrahedron Lett.* 4235.
278. Templeton, W. (1969). "An Introduction to the Chemistry of the Terpenoids and Steroids," Butterworths, London.
279. Thomas, A. F. Heusler, K. and Muller, J. M. (1961). *Tetrahedron* **16**, 264.
279a. Tschesche, R., Biernoth, G. and Snatzke, G. (1964). *Ann.* **674**, 196.
280. Tschesche, R., Henckel, E. and Snatzke, G. (1963). *Tetrahedron Lett.* 613.
281. Tsuda, Y., Morimoto, A., Sano, T. and Inubushi, Y. (1965). *Tetrahedron Lett.* 1427.
282. Tsuda, Y., Sano, T., Kawaguchi, K. and Inubushi, Y. (1964). *Tetrahedron Lett.* 1279.
283. Valenta, Z., Gray, A. H., Orr, D. E., Papadopoulos, S. and Podesva, C. (1962). *Tetrahedron* **18**, 1433.
284. Varshney, I. P., Shamsuddin, K. M. and Beyler, R. E. (1965). *Tetrahedron Lett.* 1187.
285. Warnhoff, E. W. and Halls, C. M. M. (1965). *Can. J. Chem.* **43**, 3311.
286. Warren, F. L. and Watling, K. H. (1958). *J. Chem. Soc.* 179.
287. Whitham, G. H. (1960). *J. Chem. Soc.* 2016.
288. Wrzeciono, U. *Rocz. Chem.* **37**, 1457.
289. Yosioka, I. and Nakanishi, T. (1963). *Chem. Pharm. Bull* (*Japan*) **11**, 1468.
290. Yosioka, I., Nakanishi, T. and Tsuda, E. (1966). *Tetrahedron Lett.* 607.
291. Zander, J. M. and Wigfield, D. C. (1970). *Chem. Commun.* 1599.
292. Zurcher, A., Jeger, O. and Ruzicka, L. (1954). *Helv. Chim. Acta* **37**, 2145.
293. Zylber, J. and Polonsky, J. (1964). *Bull. Soc. Chim. Fr.* 2016.

6

CAROTENOID CHEMISTRY

R. RAMAGE

Senior Lecturer, University of Liverpool

I. Introduction

There are dramatic differences between the chemistry of the tetraterpenes and that of the lower terpenoids described earlier. The carotenoids, by far the predominant class of tetraterpenes, do not have the fascinating three-dimensional structures of triterpenes and steroids, with subtle differences in reactivity being explicable in conformational terms. This is a result of the most important hybridization in the carotenoid skeleton being sp^2, in the form of a polyene system. Although this simplifies one aspect of the stereochemistry of these natural products, it introduces a major complication in terms of the vast numbers of possible geometric isomers. In fact the double bonds exist mainly in the more stable *trans*-form (see Section V B).

The high degree of unsaturation in carotenoids has rendered them heat and light sensitive making carotenoids the most experimentally demanding group of terpenes. To compensate for this, the polyene chain is responsible for the intense colour of the carotenoids, which are pigments that occur in many plant and animal sources.

The hydrocarbon carotene was the first member of the class to be isolated, by Wackenroder in 1831 from carrot, as ruby red crystals, m.p. 168°C. It was thought to be a single compound until a century later when Kuhn and Lederer (1931a) separated it into three isomers, α-carotene (1), m.p. 188°C, β-carotene (2), m.p. 184°C, and γ-carotene (3), m.p. 178°C using the technique of chromatography, which had been discovered by Tswett (1903). The re-introduction of this forgotten separatory method has had important consequences in the development of organic chemistry. Among the other techniques used during the structural phase of carotenoid chemistry ca. 1930, and which found widespread use in other fields, were microhydrogenation (Kuhn and Müller, 1934) and the Kuhn-Roth C-methyl determination (Kuhn and Roth, 1933a). Also, the early work of Kuhn on model polyene systems led to a greater understanding of the relationship of structure with light absorption and led quite naturally to the more recent work of Bohlmann (1953) and Jones (1960) on poly-yne systems. The early structural phase of carotenoid chemistry was dominated by the work of Karrer, Kuhn and Zechmeister. A summary of the literature belonging to this period is given in the monograph by Karrer and Jucker (1948).

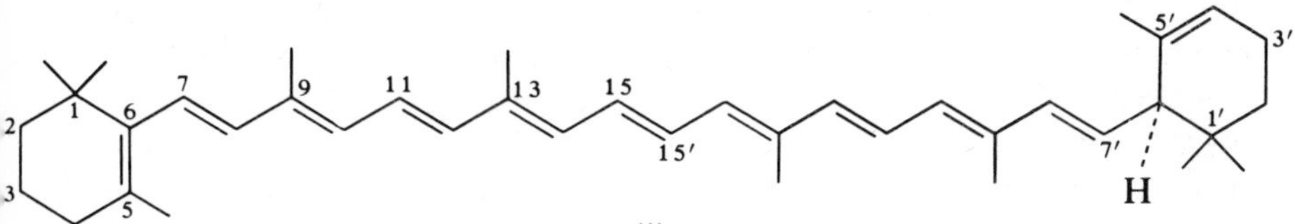

(1)

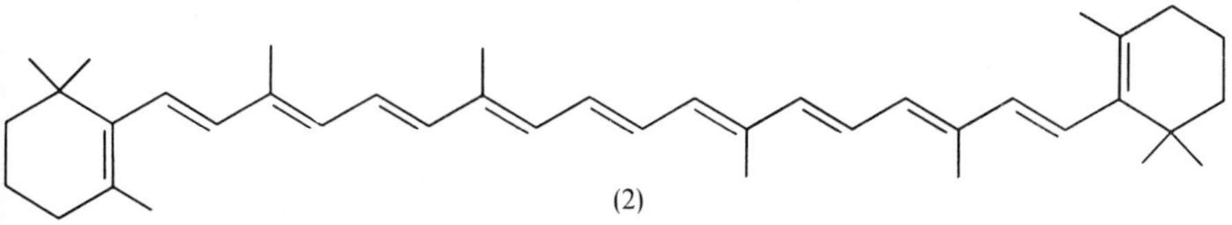

(2)

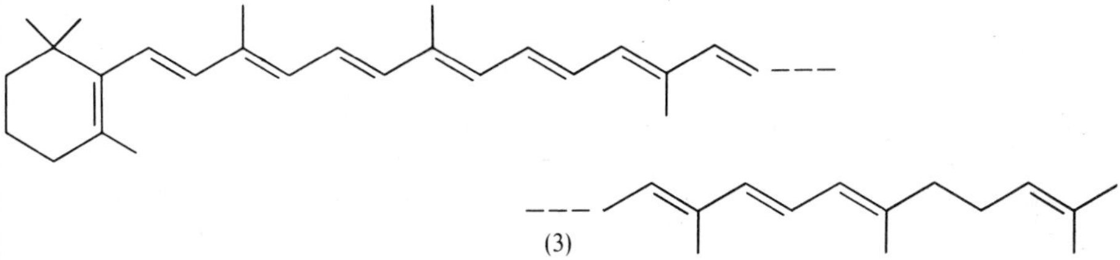

(3)

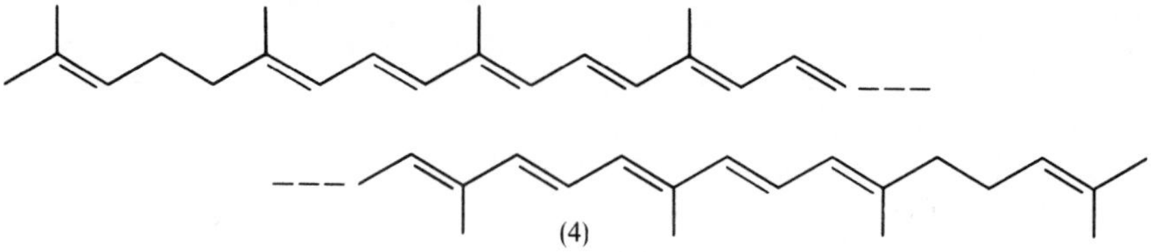

(4)

Following the classical structural elucidation studies, the synthesis of β-carotene (2) and lycopene (4) in 1950 (Karrer and Eugster, 1950a; Inhoffen, 1950a,b,; Milas *et al.*, 1950) gave new synthetic methods to the general organic chemist and set new standards in synthesis. An excellent review of this field has been given by Isler and Schudel (1963).

Recent work has been directed to biosynthetic studies (Goodwin, 1965) and also, using modern physical methods, to extending structural studies to rare carotenoids, principally by Weedon and Liaaen-Jensen.

II. Carotenoid Nomenclature

Those carotenoids which may be considered as basic structures, from which most others are derived, are α-carotene (1), β-carotene (2), γ-carotene (3), lycopene (4), δ-carotene (5) and ε-carotene (6). The numbering system is as shown in α-carotene (1) where the C-atoms of the main chain are numbered 1–15, beginning at the carbons carrying the *gem.* methyl groups. When two different ring systems are present the β-ionone ring carries the unprimed numbers, e.g. α-carotene (1). In the case of fully unsaturated aliphatic carotenoids, 1 and 1^1 are assigned to the same carbon atoms as in the corresponding cyclic structures. The term *retro*-carotenoid is used to depict the situation where both rings are joined to the polyene chain via exocyclic double bonds, e.g. rhodoxanthin (7).

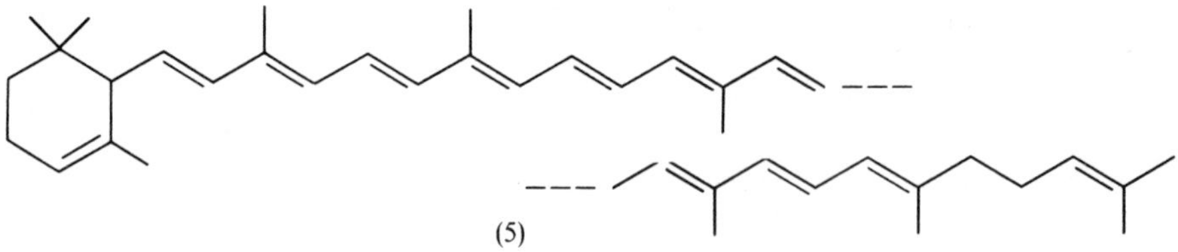

(5)

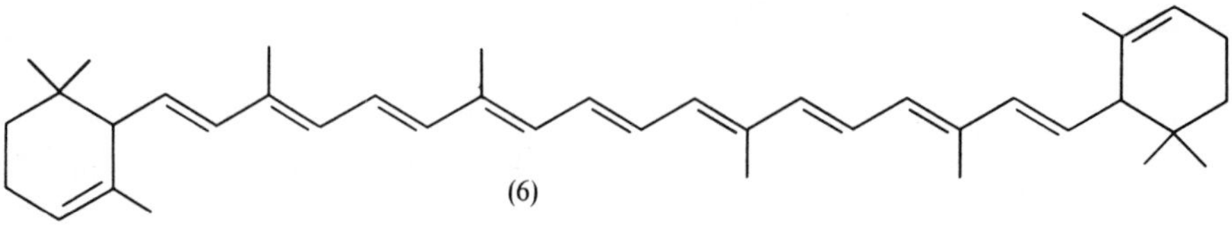

(6)

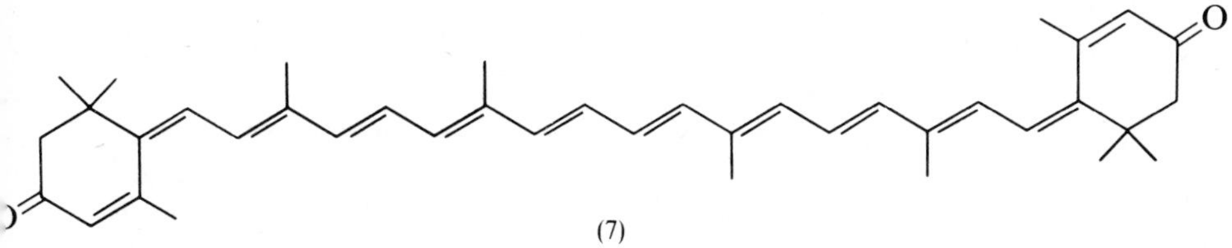

(7)

Carotenoids have long been sub-divided into two main groups: hydrocarbons (carotenes) and oxygen-containing derivatives (xanthophylls). The latter group are, however, *in vivo* transformation products of the parent carotenes and lend themselves to classification as such.

Although most of the double bonds in carotenoids have the *trans*-configuration, it is possible to set up an equilibrium between the *cis* and *trans* forms. The term "stereoisomeric set" includes all the possible *cis-trans* forms of a given carotenoid (Zechmeister, 1960).

III. Distribution and Isolation

A. Occurrence

1. *Higher plants*

The leaves of all green plants contain the same major carotenoids β-carotene (2), lutein (3,3′-dihydroxy-α-carotene) (8), violaxanthin (5,6,5′,6′-diepoxy-3,3′-dihydroxy-β-carotene) (9), and neoxanthin (10). No acyclic carotenoids are present in the photosynthetic tissues and the carotenoids are located in the grana of the chloroplast in the form of chromoproteins. Carotenoids are thought to occur as carotenoid-protein complexes in the living plant, since they are readily extracted after treatment with a polar solvent, which would denature the protein moiety. Evidence for such complexes is given by the difference in absorption spectra between the bound carotenoid and the isolated specimen.

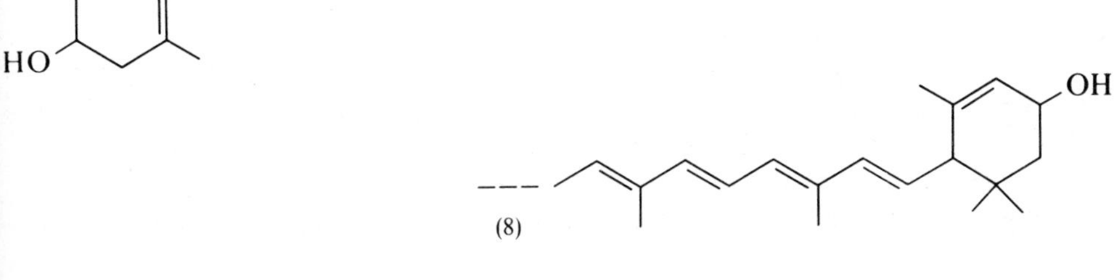

(8)

Carotenoids are also found in non-photosynthetic tissues of the fruit of some higher plants, e.g. lycopene (4) from tomato and capsanthin (11) present in red peppers. Highly oxidized xanthophylls are characteristic of carotenoids isolated from yellow flowers, whilst deep orange flowers have large amounts of β-carotene (2) or lycopene (4). Although carotene was first isolated from carrot, few roots contain significant amounts of carotenoids.

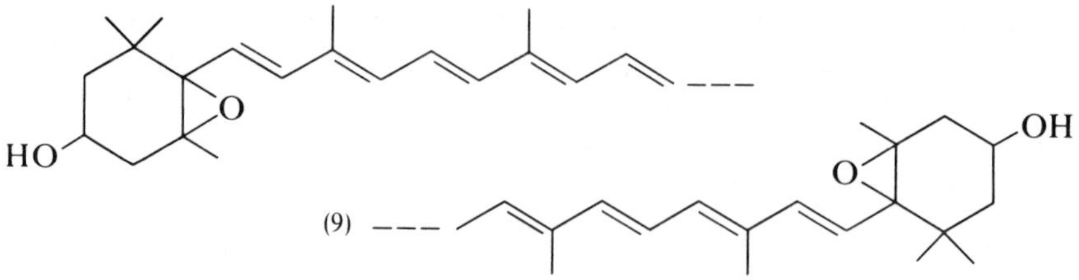

(9)

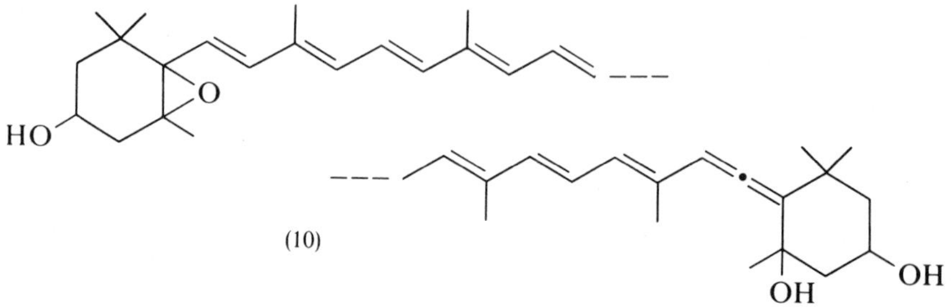

(10)

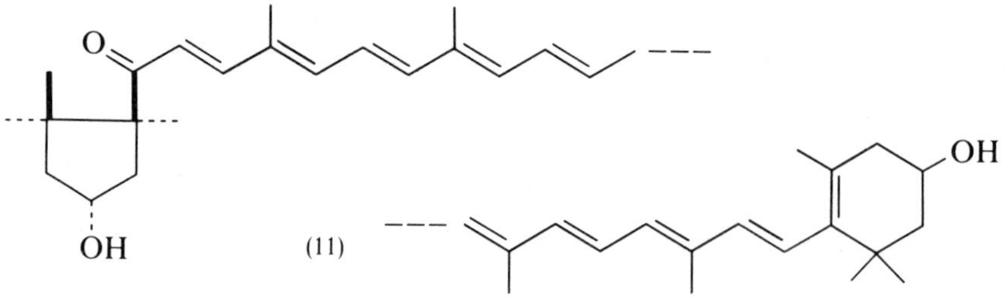

(11)

2. *Protista*

Since algae are photosynthetic they contain carotenoids in their chloroplasts. The distribution amongst the various classes of algae has been summarised by Goodwin (1965).

Photosynthetic bacteria synthesize carotenoids mainly of the acyclic type, whilst non-photosynthetic bacteria have been shown to produce novel C-50 carotenoids (Liaaen-Jensen, 1967).

Characteristic fungal carotenoids are frequently acidic, e.g. torularhodin (12). The distribution of pigments can vary with the age of culture e.g.

Rhizophlyctis rosea (Davies, 1961a,b). Differential distribution can occur in the two sexual forms of certain Phycomycetes. The asexual and female plants of the aquatic fungus *Allomyces* produce no carotenoids, whereas the male form accumulates γ-carotene (3) (Emerson and Fox, 1940). The + and − forms of *Blakeslea trispora* are approximately twenty times more efficient at producing β-carotene (2) when grown in a mixed culture compared with separate cultures. It has been found that trisporic-C acid (13) stimulates the production of β-carotene (2) in cultures of single strains of *Blakeslea trispora* (Caglioti *et al.*, 1966).

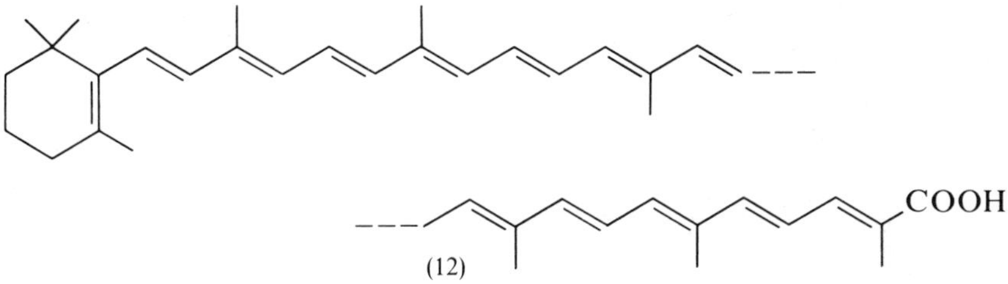

(12)

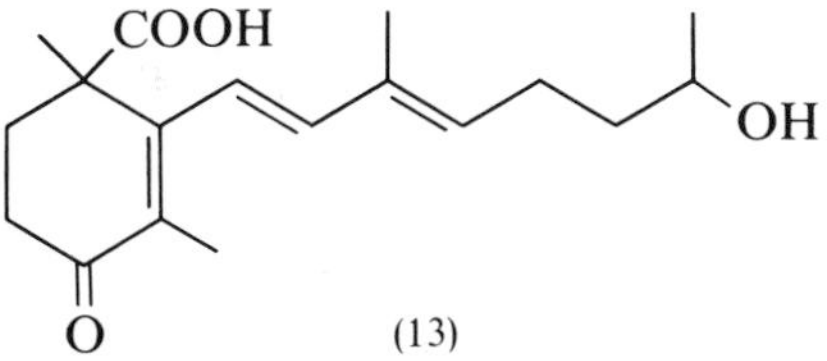

(13)

B. Isolation

The carotenoids are isolated, by solvent extraction from natural sources as a complex mixture and preliminary separation may be effected by partition between two immiscible solvents such as petroleum ether and 90% methanol. The carotenes and xanthophyll esters are soluble in the upper phase (epiphase) whereas the xanthophylls and carotenoid acids are soluble in the lower layer (hypophase). This procedure can be repeated on the epiphase pigments after saponification, or in some cases lithium aluminium hydride treatment, of the xanthophyll esters. Although this crude separation is efficient with carotenoids having two or more hydroxyl groups, the monohydroxy carotenoids tend to be distributed between the two phases. This procedure can be greatly improved using counter-current distribution with a Craig machine.

By far the most important technique for separating, purifying and isolating carotenoids is that of column chromatography, mentioned earlier in connection with Kuhn's separation of crude carotene into its constituent isomers.

The various types of chromatograms used in this field are summarized by Davies (1965). Relatively recent chromatographic techniques such as paper chromatography and thin-layer chromatography provide an efficient means of analytical separation. A full spectroscopic analysis, i.e. ultra-violet/visible and infra-red absorption spectra coupled with n.m.r. spectra provide the best means of characterisation of carotenoids. Proof of identity is usually effected by mixed chromatogram with an authentic sample using thin-layer or paper chromatography. This method is more reliable than the classical mixed melting point with the thermally labile carotenoids.

IV. Biological Function

Carotenoids are thought to be accessory pigments in photosynthesis, which is basically the photoreduction of carbon dioxide to an organic form such as carbohydrate. In green plants there is a concomitant liberation of oxygen from water.

$$CO_2 + H_2O \rightarrow (CHOH) + O_2$$

However, in photosynthetic bacteria water is replaced by hydrogen, hydrogen sulphide or H_2R where R is an organic residue. The pigments which sensitize these reactions are chlorophyll and bacteriochlorophyll or chloroviridin, respectively. Excitation of the blue or red absorption band of chlorophyll produces the same fluorescent state from which energy, liberated on returning to the ground state, may be used in chemical reaction.

In vitro, carotenoids do not fluoresce but *in vivo* they are able to sensitize the fluorescence of chlorophyll in green plants. The accessory pigments in blue-green and red algae are the phycobilins. In the diatom Nitzchia, light which would be predominantly absorbed by fucoxanthin (14) produced fluorescence of chlorophyll *a* with almost the same yield as light absorbed by chlorophyll itself (Dutton *et al.*, 1943).

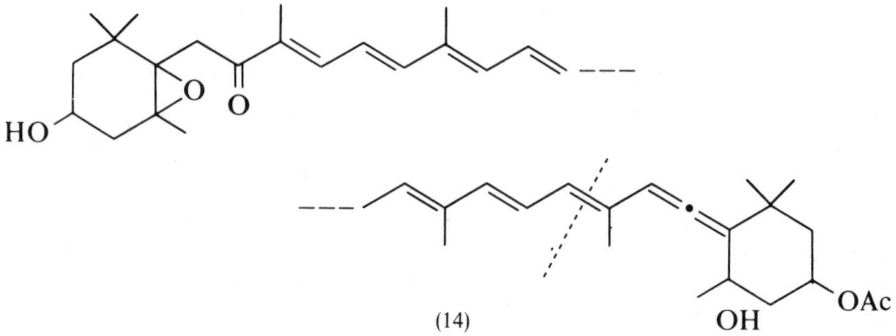

(14)

Carotenoids may also be concerned with other functions, e.g. photo-responses such as phototropism in higher plants or phototaxis in algae. It is thought that carotenoids may be involved in reproduction of Zygomycetes but, as in the case of photoresponses, the evidence is inconclusive (Burnett, 1965). What seems to be better established is that carotenoids can protect the cell from damage due to photo-oxidation catalysed by other light absorbing pigments such as chlorophylls. This was tested by comparing normal carotenoid-containing bacteria with mutants in which the carotenes are replaced by the more saturated colourless, phytoene (15). The protective action of carotenoids has also been indicated by inhibiting the normal carotenoid synthesis on addition of diphenylamine. This led to destruction of the bacteria on exposure to light and oxygen.

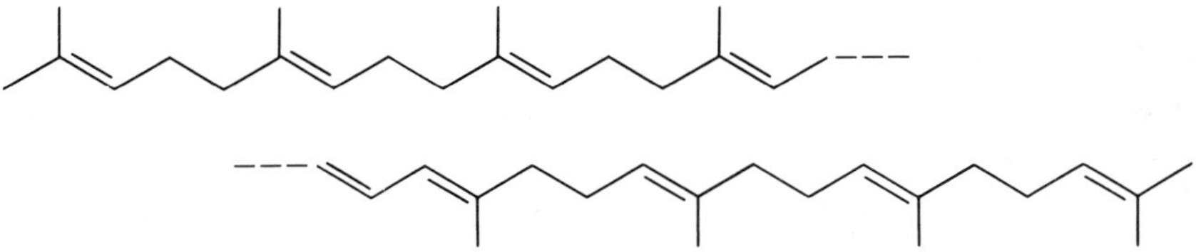

(15)

V. Structural Elucidation

A. Classical Chemical Methods

In a highly unsaturated molecule such as β-carotene (2) it was obviously crucial to determine the numbers of double bonds present. This was accomplished by Zechmeister (1928), who found that carotene $C_{40}H_{56}$ required eleven moles of hydrogen for complete saturation and thus showed carotene to be bicyclic. From this came the microhydrogenation technique of Kuhn and Möller (1934) which is now a standard analytical method. Although normally reliable, care must be exercised in interpreting results from carotenoids containing aromatic rings, or highly activated oxygen functions susceptible to hydrogenolysis.

Using oxidative methods the groups of Karrer and Kuhn gained insight into the structure of the carotenoid skeleton. Permanganate oxidation of β-carotene (2) afforded a mixture of 2:2-dimethyl-glutaric acid, 2:2-dimethylsuccinic acid, dimethylmalonic acid, acetic acid and a key degradation product, geronic acid which was shown to have the structure (16). This work by Karrer and his co-workers (1929a, 1930, 1931a) had two important consequences. Quantitative estimation of the geronic acid (16) produced on ozonolysis of β-carotene (2), and consideration of its structure, showed

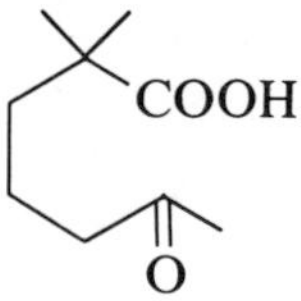

(16)

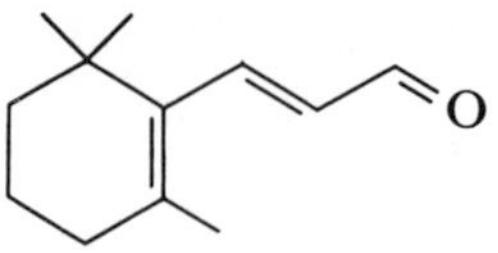

β-ionone

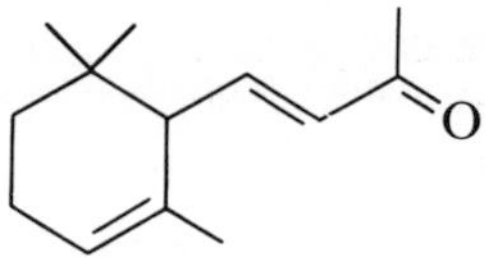

α-ionone

β-carotene (2) to have two *β*-ionone rings. Estimation of the acetic acid produced indicated the presence of four groupings of the type

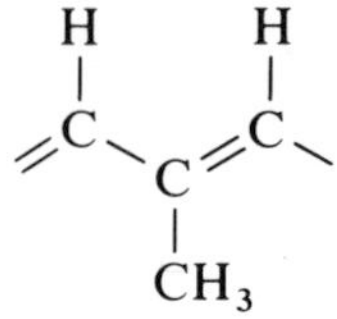

Kuhn and his co-workers (1929, 1931b, 1934) also used this oxidative method of C-methyl determination and evolved the present analytical procedure, which is used in so many modern biosynthetic studies. Karrer then proposed the correct structure, with the more stable *trans* configuration of the double bonds based on the tail to tail union of two C-20 units, whose structures were based on the isoprene rule.

Partial permanganate oxidation of *β*-carotene (2) (Karrer and Solmssen, 1937) yielded the *apo*carotenals (17) and (18), whereas chromic acid led to ring cleavage to give semi *β*-carotenone (19) and *β*-carotenone (20) (Kuhn and Brockmann, 1935). These latter degradation products have recently been isolated (Yokoyama and White 1968) from citrus plants.

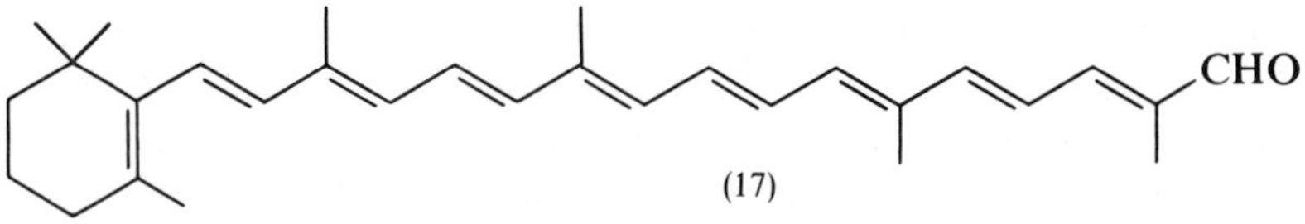

(17)

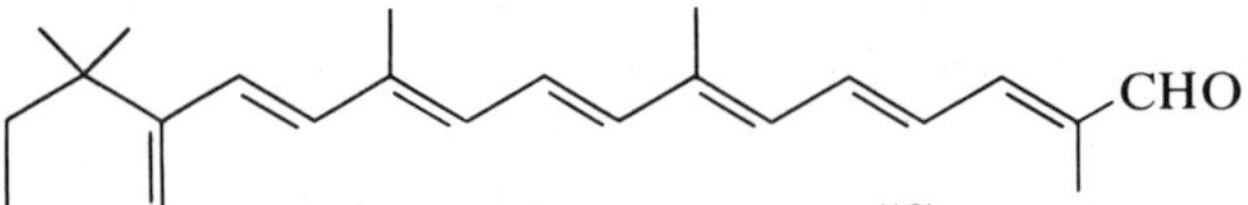

(18)

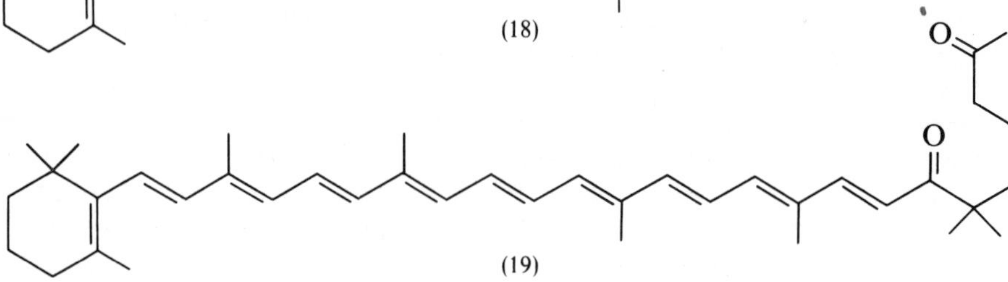

(19)

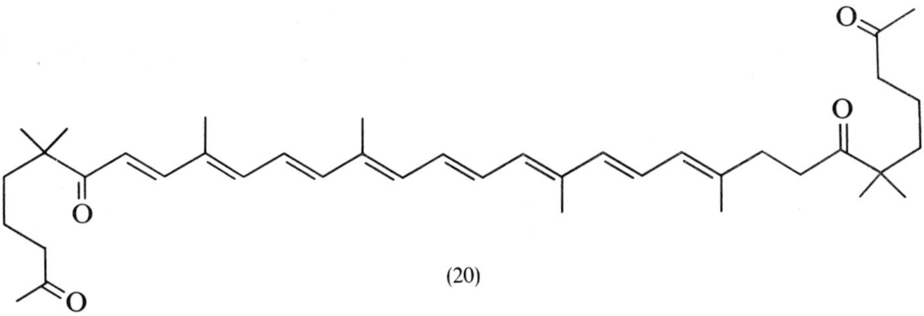

(20)

Oxidation of α-carotene (1) gave isogeronic acid (21) in addition geronic acid (16) and the other oxidation products of β-carotene (2). ε-Carotene (6) has both rings in the α-ionone form. The absolute configuration at C-6 in

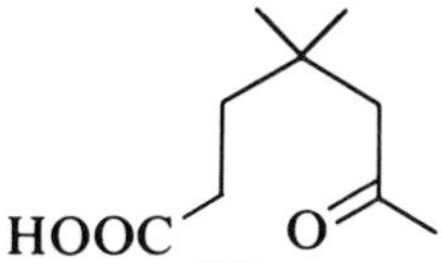

(21)

α-carotene (1) has been established (Eugster, 1969) by chemical correlation of the enantiomers of α-cyclogeranic acid with α-ionone and hence α-carotene (1) by synthesis. In both permanganate and chromic acid oxidation, α-carotene (1) is attacked preferentially at the electrophilic tetra-substituted double bond of the β-ionone residue. This leads to products isomeric with those obtained from β-carotene (2).

Ozonolysis of lycopene (4) and estimation of the acetone produced indicated two isopropylidene residues (Karrer, Bachmann, 1929b; Karrer *et al.*, 1931b). Succinic acid was also found to be produced by ozonolysis. Hydrogenation and C-methyl determination together with this oxidative evidence led to the acyclic structure (4) being proposed (Karrer, 1931). Evidence for the central polyene moiety of lycopene (4) came from partial oxidation with chromic acid which afforded 6-methylhept-5-en-2-one and lycopenal (22). Further oxidation of (22) gave bixindialdehyde (23) which was converted, via the dioxime and dinitrile, to norbixin (24) (Kuhn and Grundmann, 1932a; cf. Karrer and Jucker, 1948).

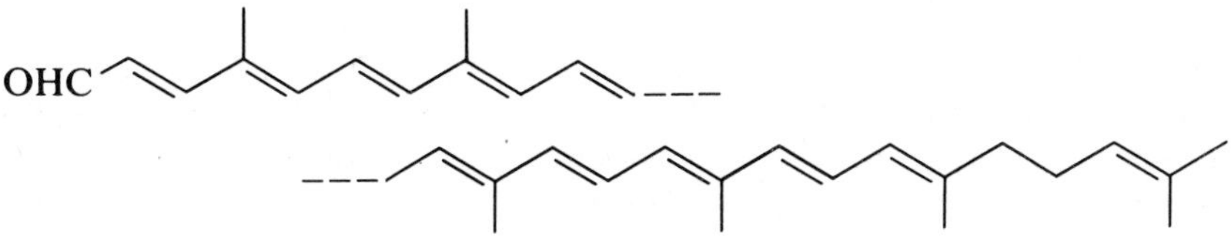

(22)

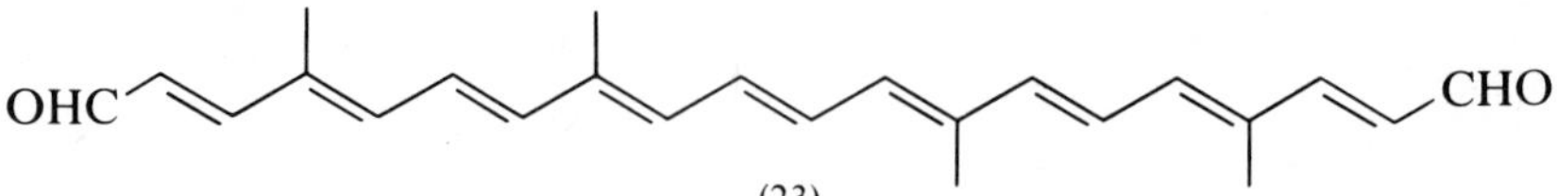

(23)

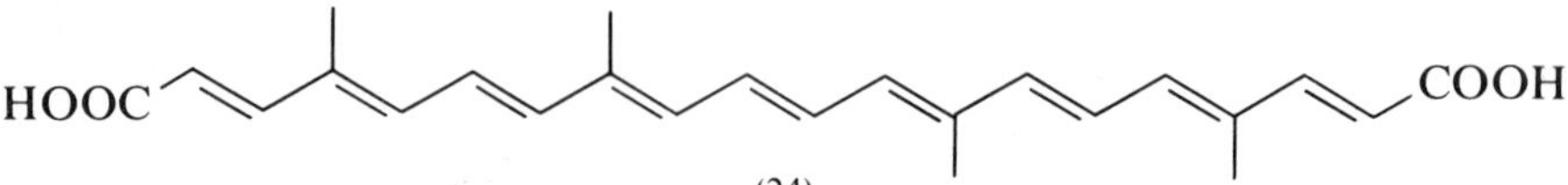

(24)

Oxidation of the xanthophyll, zeaxanthin (25), with ozone and permanganate produced 2:2-dimethyl succinic acid; obviously oxidation of the hydroxyl group substituted in the β-ionone ring caused extensive degradadation so that large fragments could not be isolated as in the case of β-carotene (2) (Karrer *et al.*, 1930). Partial oxidation of lutein (8) and zeaxanthin (25) afforded mainly α-citraurin (26) and β-citraurin (27) respectively. (Karrer *et al.*, 1938.)

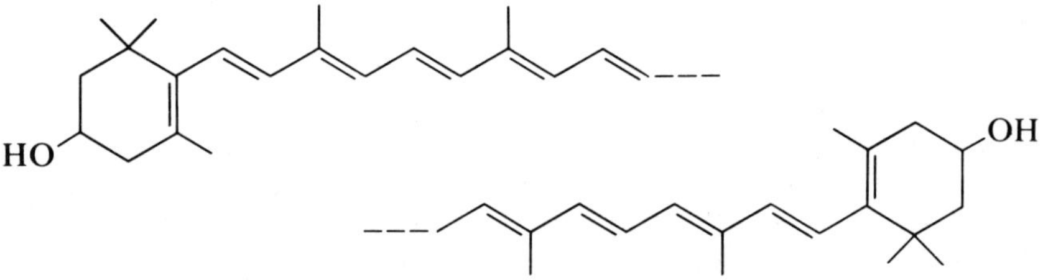

(25)

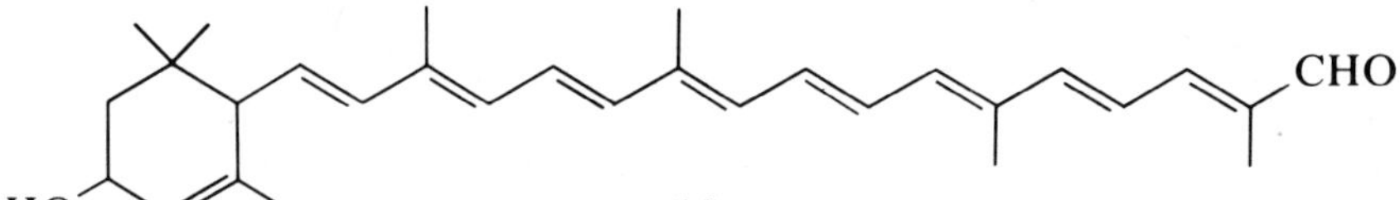

(26)

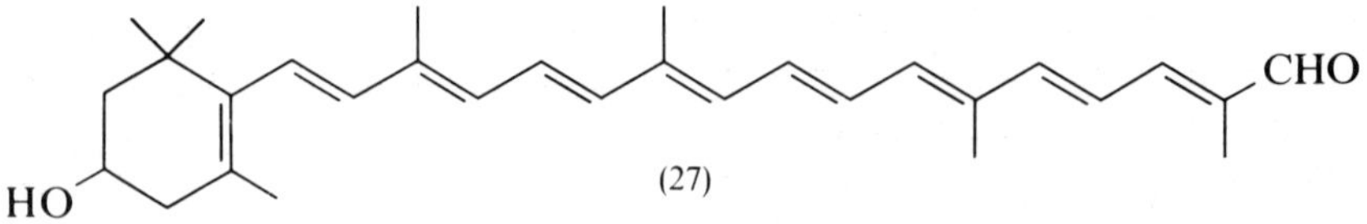

(27)

Further evidence, of a more tenuous kind, for the carotenoid skeleton came from thermolysis studies by Kuhn and Winterstein (1933b). It was found that thermal decomposition of β-carotene (2) led to the formation of 2,6-dimethylnaphthalene, a process which is thought to involve cyclization of the central polyene fragment followed by cleavage of the terminal residues as shown:

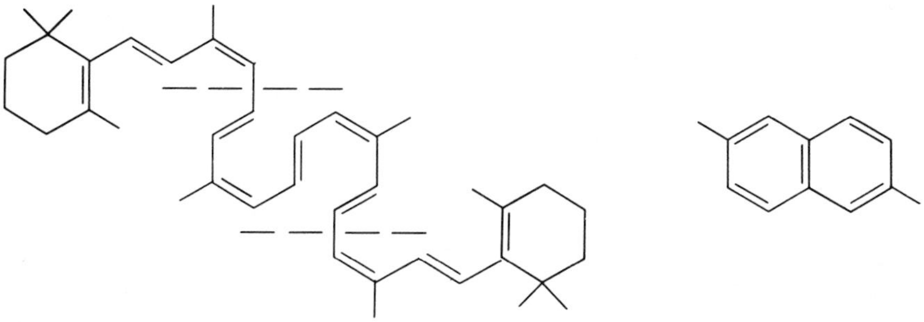

The identification of the functional groups in naturally occurring xanthophylls follows the usual procedures for aldehydes, ketones, alcohols, epoxides etc. and will be discussed in Section VI dealing with the various classes of carotenoids.

B. Ultra-violet/Visible Light Absorption

1. *All-trans conjugated systems*

As would be expected of molecules containing conjugated double bonds, the ultra-violet or visible spectra give the best indication of the nature of the polyene system, and is of a paramount importance in the analysis of the carotenoids (Davies, 1965). There are three regions in carotenoid absorption spectra. These are designated the λ_1 band in the visible region, which has intense absorption and gives rise to the characteristic colour of this class of natural products, the λ_2 band in the near ultra-violet (see Section B2), and the λ_3 band in the ultra-violet region. The λ_1 band usually consists of three peaks in the visible spectrum. The wavelengths of the three absorption bands in the visible spectrum increase with each extension to the conjugated system. In carotenoids containing two β-ionone residues, the band of shortest wavelength is reduced to an inflection. More drastic modification to the typical three-band spectrum occurs in the oxo-carotenoids when a carbonyl group is conjugated to the polyene system. In this case the spectrum is replaced by an almost symmetrical single band or a single band flanked by weak inflections.

When a double bond is moved out of conjugation with the polyene system, as in the conversion of a β-ionone system to an α-ionone system, there is the expected hypsochromic shift in the visible spectrum. This displacement to shorter wavelengths is also observed on cyclization at either end of the carotenoid skeleton, e.g. lycopene (4), γ-carotene (3), and β-carotene (2), indicating a reduced participation by the terminal cyclic double bond in the conjugated polyene system. Another important hypsochromic shift occurs on rearrangement of a 5,6 epoxide to the isomeric furanoid oxides. A simple example of this is the acid-catalysed transformation of α-carotene-5,6-epoxide (28) to flavochrome (29).

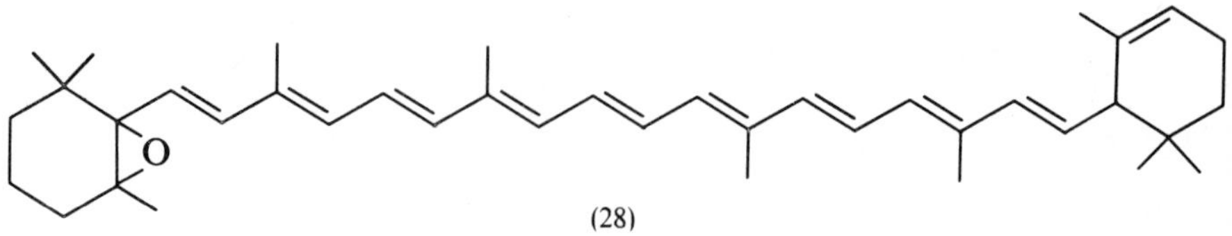

(28)

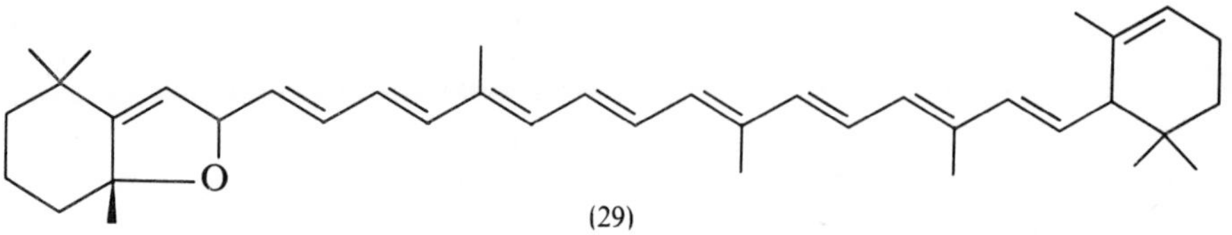

(29)

A few selected examples of the variation of visible absorption spectra with carotenoid structure are given in Table 1. All spectra given are from light petroleum or hexane solution.

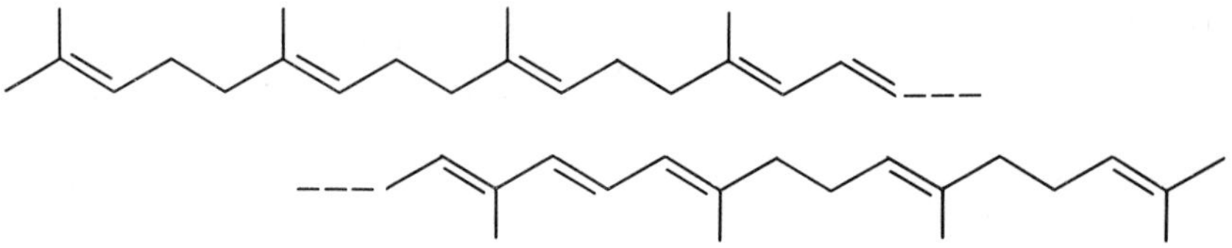

(30)

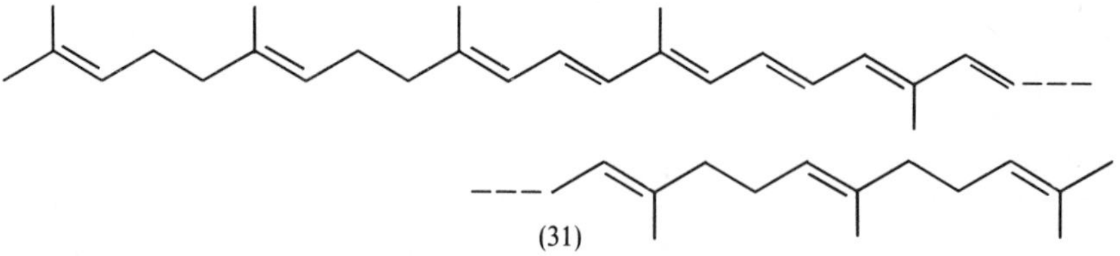

(31)

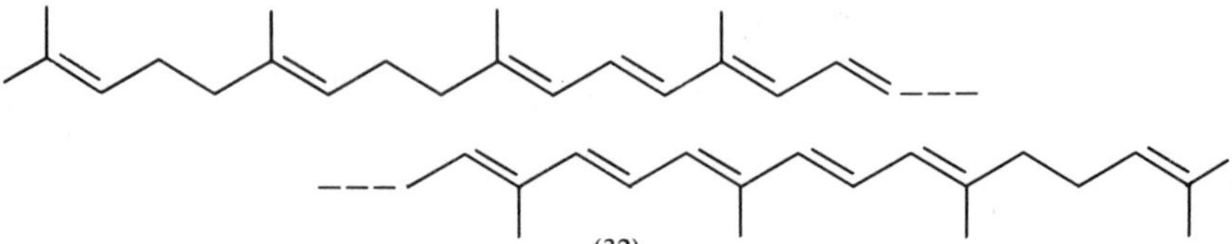

(32)

TABLE 1

VISIBLE SPECTRA OF SELECTED CAROTENOIDS

Polyene (number of conjugated double bonds)	Visible absorption maxima (mμ)			Reference
Phytoene (15) 3 conjugated double bonds	275	285	296	Davies, 1961a
Phytofluene (30) 5 conjugated double bonds	331	348	367	Davies, 1961a
ζ-Carotene (31) 7 conjugated double bonds	378	400	425	Davies, 1965
Neurosporene (32) 9 conjugated double bonds	416	440	470	Isler and Schudel, 1963b
Lycopene (4) 11 conjugated double bonds	446	472	505	Isler and Schudel, 1963b
Spirilloxanthin (33) 13 conjugated double bonds	468	499	534	Liaaen-Jensen, 1962a
γ-Carotene (3)	437	462	494	Isler and Schudel, 1963b
β-Carotene	425	451	482	Goodwin, 1955
Echinenone (34) (4-oxo-β-carotene)	—	458	—	Davies, 1965
Canthaxanthin (35) (4,4-dioxo-β-carotene)	—	466	—	Isler and Schudel, 1963b
α-Carotene-5,6-epoxide (28)	—	442	471	Goodwin, 1955
Flavochrome (29)	—	422	450	Karrer and Jucker, 1948

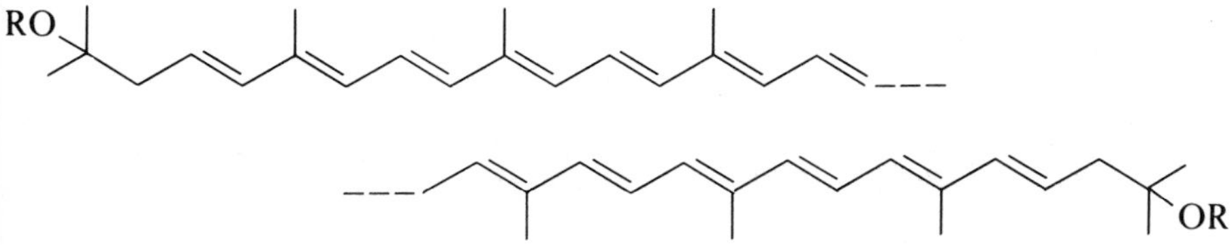

R = Me (33) R = H (37)

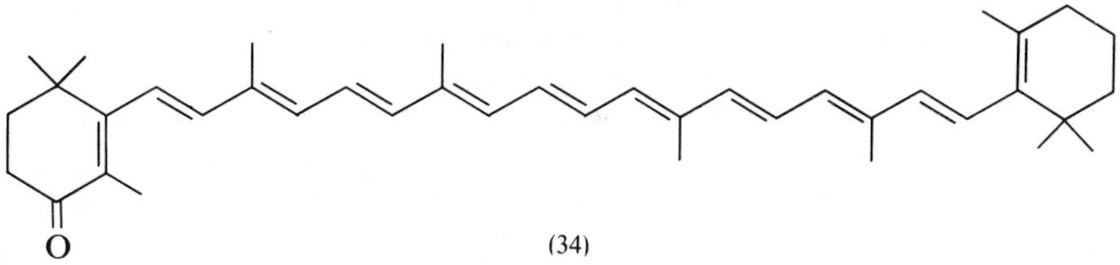

(34)

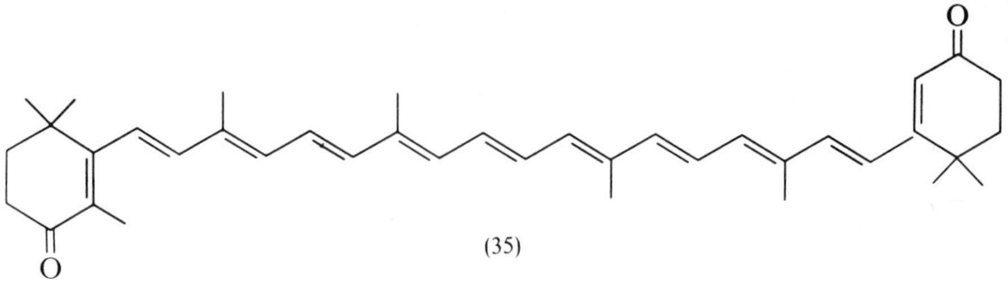

(35)

2. *Stereomutation of the carotenoid skeleton*

Gillam and El Ridi (1935, 1936) found that β-carotene (2), after repeated absorption on alumina, was transformed into an isomeric mixture. It was subsequently shown, however, that the transformation was independent of absorption processes (Zechmeister and Tuzson, 1938) and was spontaneous in solution. It was early realized that this was a cis $\rightleftharpoons$ trans equilibration of a double bond.

Rearrangement of all-trans carotenoids does not equilibrate each double bond to give poly-*cis* forms, but yields mainly mono-*cis* and di-*cis* forms. Not all the double bonds of β-carotene (2), for example, are equally sensitive to isomerization. Pauling (1939) proposed that the double bonds in carotenoids are divided into two types (cf. Fig. 1). A double bond is said to be

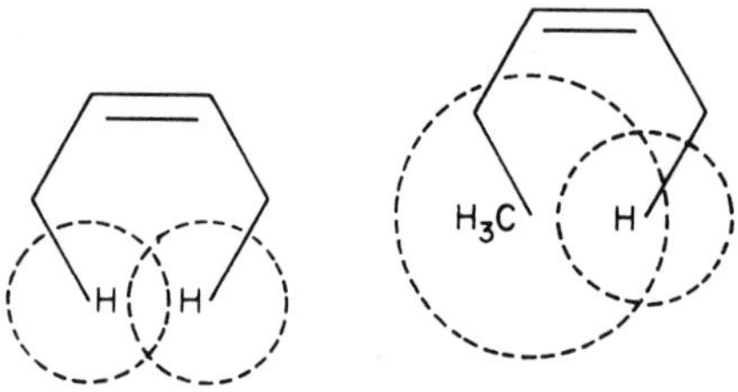

FIG. 1. Overlapping of hydrogen atoms in —CH—CH=CH—CH—, and of hydrogen and methyl in —CH—CH=CH—C(CH_3)— with *cis* configuration; from Pauling (1939).

"unhindered" when there is little steric hindrance in the *cis*-configuration, i.e. between two hydrogen atoms, and "hindered" when there is severe steric repulsions in the *cis*-form, for instance between a methyl group and a hydrogen (cf. Fig. 1). Into this second category fall all the trisubstituted double bonds of the carotenoid skeleton as well as the 15,15′-double bond isomer (36) which was the penultimate product in the synthesis of β-carotene (Inhoffen *et al.*, 1950c). On subjecting an all-*trans* carotenoid to stereoisomerization, there is competitive formation of both unhindered and hindered *cis* isomers.

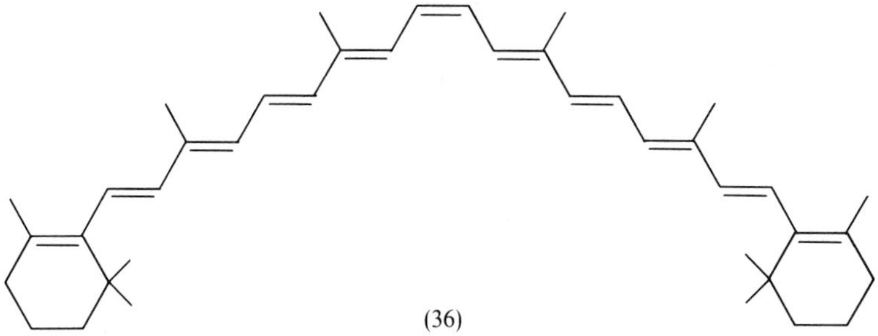

(36)

From steric considerations the former type predominate. Conversely, in stereomutation of synthetic *cis* isomers the "hindered" double bonds are most labile.

Stereomutation of a carotenoid takes place on solution (Zechmeister and Tuzson, 1938) but the process is usually slow at room temperature. The rate depends, markedly, on the structure of the carotenoid, for instance only 1–2 % of α-, β-, and γ-carotene undergo stereomutation in 24 h at room temperature, whereas, in the case of the acyclic carotenoids, the proportion is much greater, e.g. all-*trans* lycopene (4) (10 %), spirilloxanthin (33) (23 %) and α-bacterioruberin (37) (42 %) (Zechmeister, 1962; Liaaen-Jensen, 1962b). An equilibrium mixture of geometrical isomers from all-*trans* carotenoids is usually produced within 1 h in refluxing benzene or hexane.

Although cis-trans isomerization can be achieved by irradiation with light of wavelength close to the absorption band of the carotenoid, the most useful and rapid method of stereomutation involves exposure to light of a carotenoid solution in benzene, containing catalytic amounts of iodine.

The geometrical configuration of a carotenoid has a marked effect on the chromatographic behaviour and stereomutation can be used to characterize carotenoids, since each gives rise to a characteristic mixture of isomers which can be compared by chromatographic techniques.

With the formation of *cis* isomers, the colour intensity decreases and a characteristic new absorption in the ultra-violet region appears. The "cis-peak" (λ_2) occurs 142 ± 2 mμ below the long wavelength band in the visible spectrum, and is found between 320 and 380 mμ for normal, fully unsaturated carotenoids. Dale (1954) has analysed the empirical relationship of the minor bands, including the *cis*-peak, in polyene spectra and found that for an aliphatic polyene containing *n* conjugated double bonds, the minor band (λ_s) will lie close to the main absorption band of a corresponding polyene with n/s conjugated double bonds. The configuration of normal carotenoids may be determined by the change in ultra-violet/visible absorption spectra before, and after, iodine-catalysed stereomutation (cf. Table 2).

TABLE 2

CONFIGURATIONAL TYPES OF *NORMAL* CAROTENOIDS AS INDICATED BY SPECTROSCOPIC CHANGES UPON TREATMENT WITH IODINE; FROM ZECHMEISTER (1960)

Configuration	Shape of the curve in the *cis*-peak region	Change of the extinction upon iodine catalysis	
		in the *cis*-peak region	in the visible region
All-*trans*	Flat	Increase	Decrease
Central-mono *cis*	High peak	Decrease	Increase
Peripheral-mono *cis*	Moderate peak	Slight increase	Increase
Poly-*cis*	Flat	Increase	Strong increase

C. INFRA-RED SPECTROSCOPY

Infra-red spectroscopy is extremely important in determining the nature of the ancillary functional groups present in xanthophylls, e.g. hydroxyl, carbonyl, allene or acetylene. This will be exemplified in Section VI, where appropriate. In the application of infra-red to configurational studies of carotenoids, the important regions are ~7.25, ~10.0–10.6, and ~13 μ (Zechmeister, 1960). The peak at 7.25 μ has been assigned to a methyl-substituted *cis*-double bond of the polyene chain, due to C—H in-plane vibration. Out of plane vibrations of the C—H on vicinal carbons of a *trans*-disubstituted double bond cause the band at 10.0–10.6 μ, but this can be split in the spectra of some *cis*-isomers (Lunde and Zechmeister, 1955). The corresponding C—H out of plane vibrations of the central *cis*-double bond of (36) gives rise to a strong absorption at 12.84 μ.

D. THE N.M.R. OF CAROTENOIDS

Since the definitive work of Weedon, Jackson and their collaborators (Barber *et al.*, 1960), n.m.r. has proved to be invaluable for structural and synthetic work in carotenoid chemistry. The protons of normal carotenoids are of three types: methyl, methylene, and olefinic. The latter are found at relatively low field (2.0–4.5τ) and are of limited use in structural assignment due to complex spin-spin coupling and overlap of the signals. Although the methyl and methylene resonances occur in the same region of the spectrum, the methyl groups are all on fully substituted carbon atoms or on sp^2 hybridized carbons of the polyene chain. Thus there are no large splittings of the methyl peaks due to spin-spin coupling with a $C_{(\alpha)}$-proton and the methyl resonances appear as single peaks, readily distinguishable from the broad methylene resonances. It is, therefore, the methyl region ~8–9τ, which gives most information about the basic carotenoid skeleton.

Methyl groups attached to non-terminal double bonds of the polyene chain (38) are the most strongly de-shielded of the C-methyl groups and have peaks in the region 7.95–8.15τ. Structural modification can lead to a difference in shielding of the various "in chain" methyl groups. A methyl group at a *β*position with respect to a triple bond in a conjugated system (39) gives rise to a peak at slightly lower field (7.87–7.95τ) than those of the other "in chain" methyl groups.

It is found that methyl groups on terminal positions of polyene chains (40) are less deshielded than "in chain" groups and resonate ~8.20τ. "End of chain" methyl groups on a carbon *α*to a carbonyl group (41), are de-shielded to 7.98–8.13τ.

Lycopene (4) has peaks at 8.38 and 8.31τ attributed to the isopropylidene grouping (42), showing the lower deshielding effect of an isolated double bond. In spirilloxanthin (33), however, the terminal C-methyl groups are attached to saturated carbon and are found at 8.83τ; a position consistent with a paramagnetic shift due to the oxygen substituent on the methyl-bearing carbon.

The *β*end groups (43) associated with *β*-carotene (2) and many other carotenoids gives rise to two peaks at 8.30 and 8.97τ in the ratio (1:2). From relative position and intensity, the 8.30τ resonance is due to the olefinic 5-methyl group. This is relatively high field for an "end of chain" type (40) and close to the position for methyl groups on isolated double bonds, suggesting that the conjugation between the cyclic double bond and the polyene chain is partial. There is evidence in the visible absorption spectra (Section V B) which indicates that, in carotenoids having the *β*end group (43), steric hindrance between the *gem* dimethyl grouping and the polyene chain forces the cyclic double bond out of the plane of the polyene chain. The peak at 8.97τ is due to two magnetically equivalent C-methyl groups.

In the *α*end group (44) of *α*-carotene (1), however, the methyl groups at C-1 are no longer equivalent and exhibit two peaks at 9.08 and 9.17τ. The band at 8.30τ is obviously the olefinic methyl group.

These findings are summarized as follows:

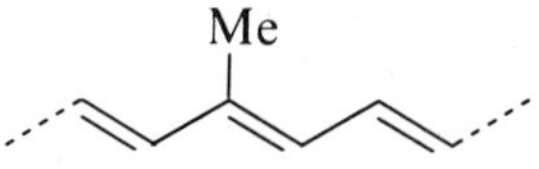

(7.95–8.15) (38)

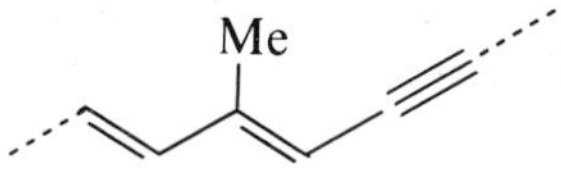

(7.87–7.95) (39)

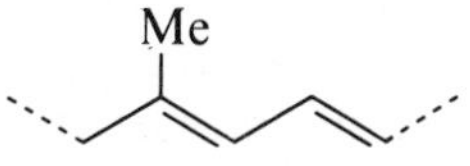

(8.20) (40)

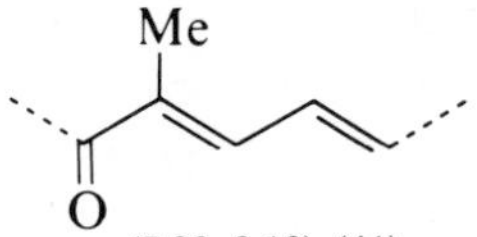

(7.98–8.13) (41)

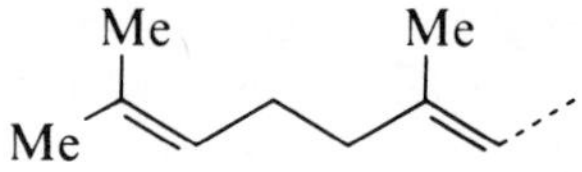

(8.38, 8.31) (42)

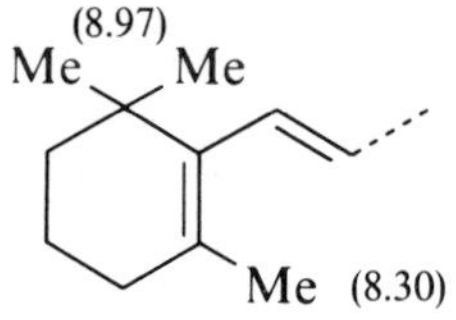

(43)

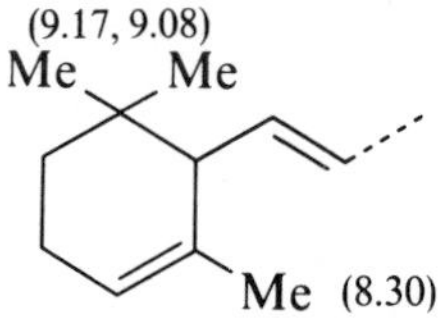

(44)

An instructive example of the application of n.m.r. to carotenoid structures is given in Fig. 2. A change of end-group dramatically alters the spectrum. Further examples of the use of n.m.r. spectroscopy will be given in Section VI.

E. Mass Spectroscopy

The instability, solvation and basic limitations of conventional micro-analysis, together with the very small quantities of material available for experimentation, made it difficult to deduce the correct molecular formula for an unknown carotenoid. This has largely been circumvented by the application of mass spectrometry to the structural elucidation of carotenoids. Molecular composition is given by accurate mass determination. Further evidence about structure may be obtained by consideration of the fragmentation processes, which occur on electron impact.

A study by Schwieter *et al.* (1965) showed that the isomeric carotenes and lycopene gave rise to M-92 and M-106 ions, which were thought to be produced by elimination of part of the central polyene chain of the carotenoid skeleton. The mechanism shown was invoked to account for the loss of masses 92 and 106, corresponding to toluene and m-xylene respectively.

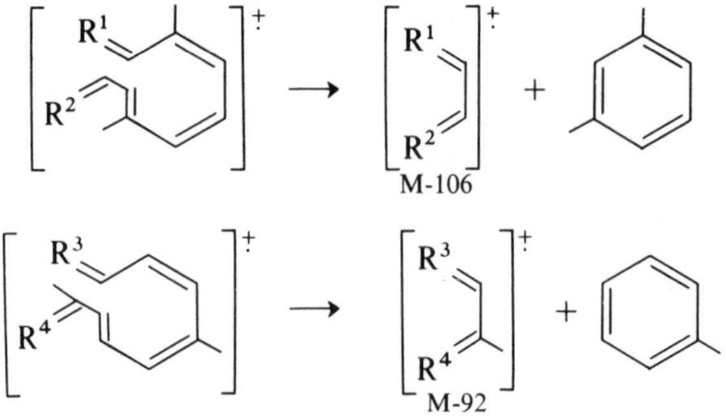

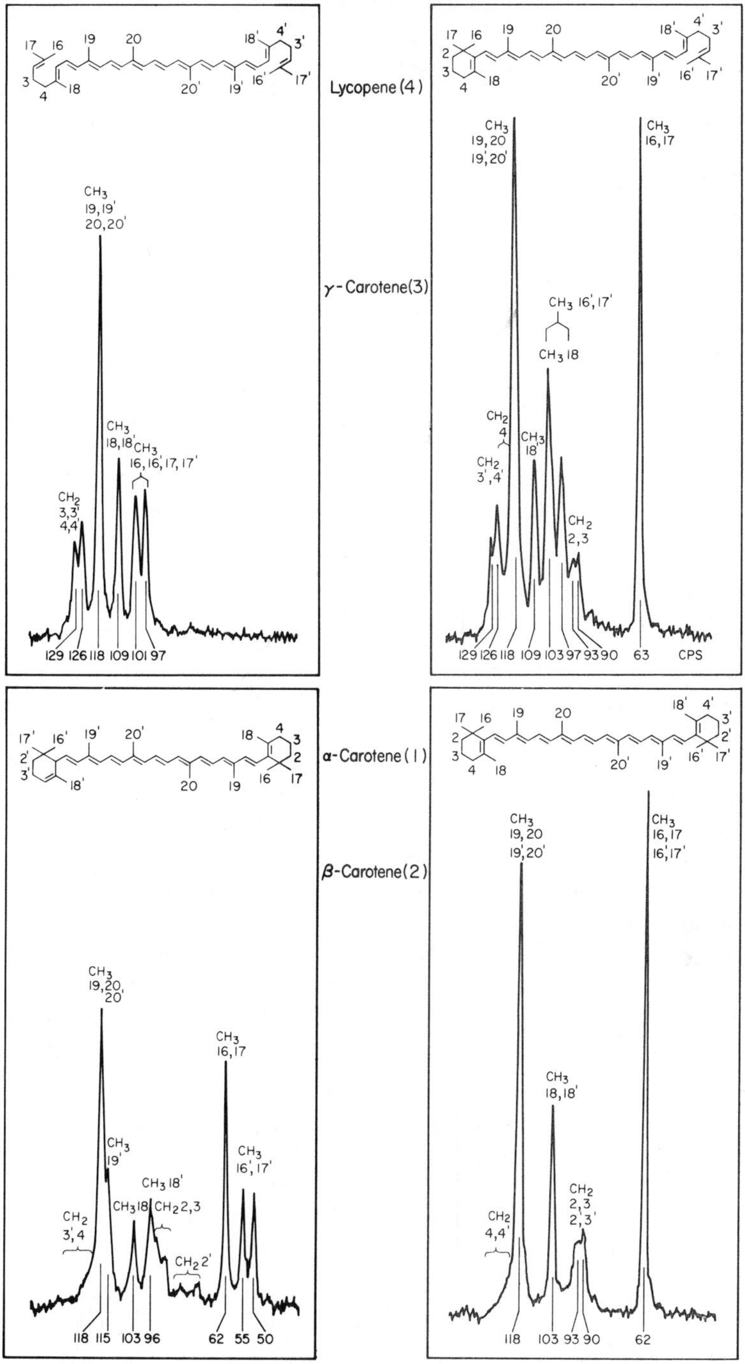

FIG. 2. Methyl and methylene region of the n.m.r. of lycopene, α, β, and γ-carotene at 60 Hz; Adapted from Schwieter *et al.* (1965). Solvent $CDCl_3$. Chemical shifts in Hz.

An analogy for this is the isolation of toluene and xylene from the thermolysis of lycopene (Kuhn and Winterstein, 1932). Later work has shown that the ratio of M-92/M-106 peaks varies with the number of double bonds in the polyene chain and, in fact, decreases as the number of double bonds increases from 9 to 13 (Enzell *et al.*, 1968). In this context it is interesting to compare the ratios for β-carotene and lycopene (Schwieter *et al.*, 1965) which are 9 and 0.3 respectively, again indicating diminished conjugation between the polyene chain and the terminal cyclic double bonds in β-carotene.

Perhaps the most useful fragmentation processes involve the end groups. For example, the α- end group (44) found in α, δ and ε-carotene show a peak at m/e 480 corresponding to M-56. A retro-Diels–Alder process was postulated (Schwieter *et al.*, 1965). By contrast, the acyclic end group (42) undergoes

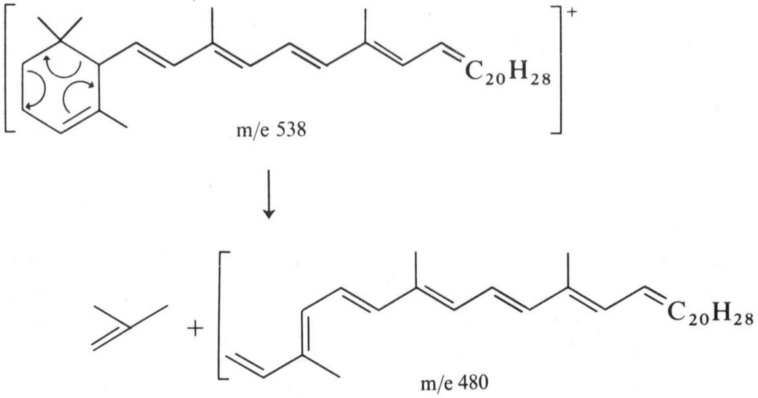

allylic fission to produce a M-69 peak at m/e 467 observed in δ- and γ-carotene and, as expected, lycopene.

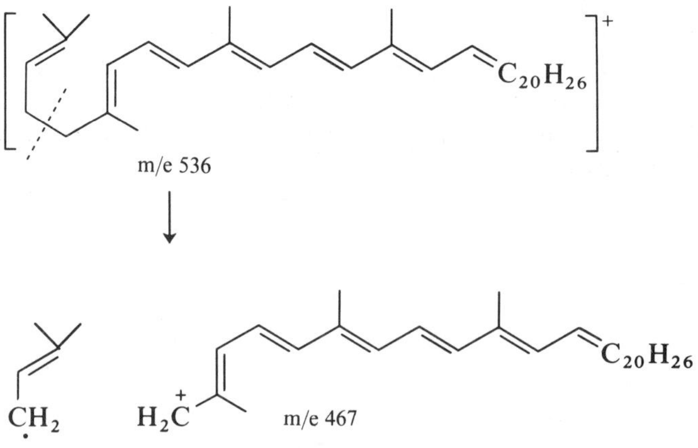

Good examples of how such a fragmentation process can indicate the positions of substituents in a carotenoid skeleton are the mass spectra of lycoxanthin (45) and lycophyll (46) (Cholnoky *et al.*, 1968). In addition to fragments corresponding to the loss of one and two H_2O in lycoxanthin (45) and lycophyll (46) respectively, important peaks were found at 483 (M-69) and 467 (M-85) in the former but only at 483 (M-85) in the latter. The (M-69) fragment indicates a lycopene end group (42) and the (M85) fragment indicates a hydroxylated acyclic end group (42).

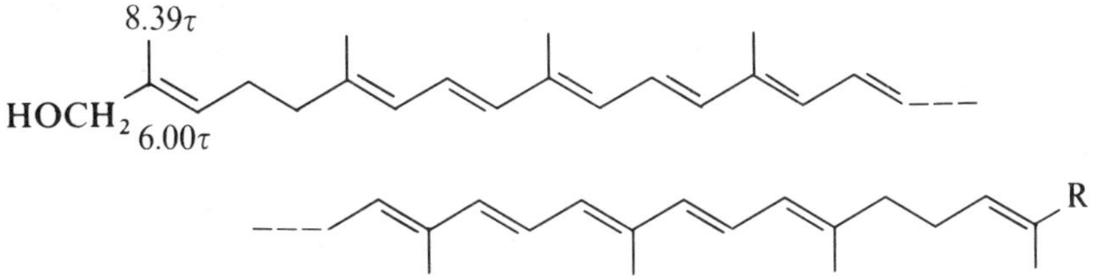

R=Me (45) R=CH_2OH (46)

The identification of tertiary hydroxyl functions in carotenoid structures is difficult due to the sluggishness with which they undergo esterification. An excellent technique has been devised by McCormick and Liaaen-Jensen (1966) which involves trimethylsilylation of the tertiary hydroxyl group followed by mass spectral measurement of the parent ion. Although other hydroxyl groups, e.g. secondary also react to give trimethylsilyl ethers, these can be cleaved selectively due to the slow hydrolysis of tertiary trimethylsilyl ethers. Thus any change in the parent ion, after hydrolysis, must be due to reaction at the tertiary hydroxyl groupings present in the unknown carotenoid.

Mass spectrometry offers a method of identifying the 3-hydroxy furanoid oxide (48) end group which is formed by acid catalysed rearrangement of the 3-hydroxy-5,6-epoxy end group of neoxanthin (10) and other natural carotenoids. The end group (48) gives rise to ions of m/e 181 and 221 (Baldas *et al.*, 1966; Aitzetmuller *et al.*, 1968) which have been attributed to structures (49) and (50) respectively.

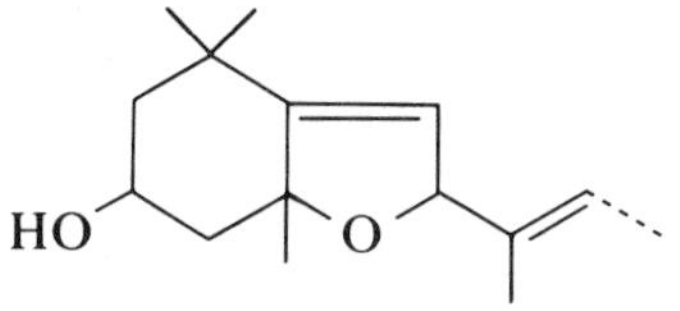

(48)

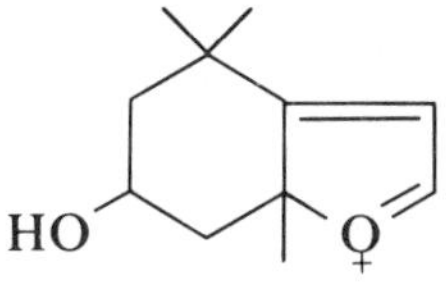

(49)

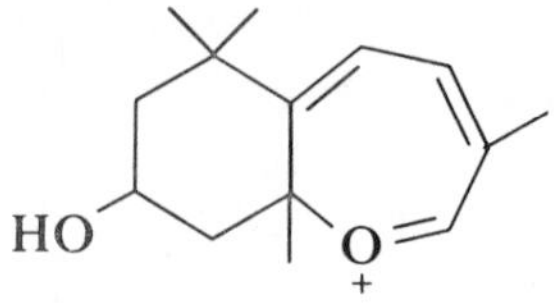

(50)

VI. Structural Relationships

A. Acyclic Carotenoids

By analogy with squalene (51), it was thought that lycopersene (52) would be the initial product of tail to tail union of two C-20 units. Karrer and Kramer (1944) synthesized lycopersene (52) from geranyl-geranyl bromide and sodium, but, so far, it has not been isolated from natural sources. In the *in vivo* cyclization of squalene 2:3-epoxide (47) leading to tetra- and penta-cyclic structures, the central C—C bond must be saturated, thus if the union of the two C-20 units is *oxidative* two observations can be explained. Firstly, the failure to isolate lycopersene (52) from natural sources, since phytoene (15) would be the initially formed carotenoid; and, secondly, the 15,15′ double bond would preclude full participation of all the π-bonds in a cyclization process leading to polycyclic structures with a C-40 skeleton. Phytoene (15) has been assigned a 15,15′-*cis* structure on the basis of infra red (12.84 μ) and its inability to complex with thiourea (Rabourn *et al.*, 1954, 1956). Stereomutation of natural phytoene gave a product identical to that synthesized by Weedon and his collaborators (Davis *et al.*, 1966).

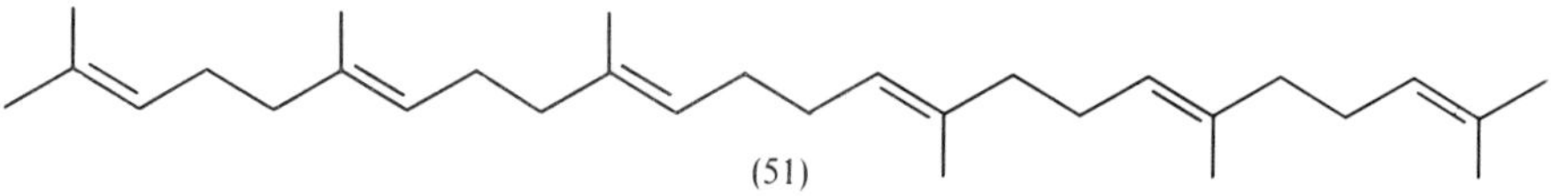

(51)

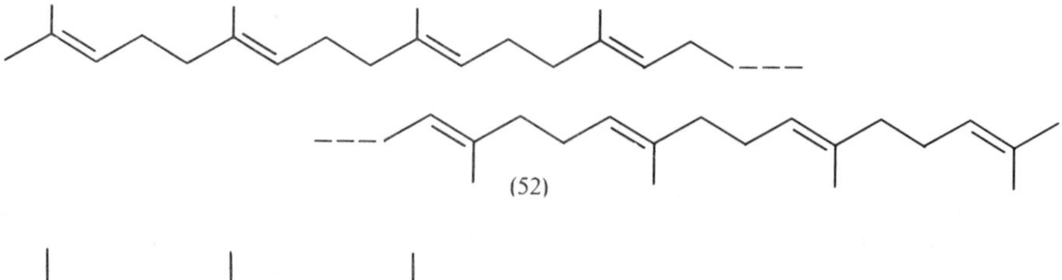

(52)

(47)

It now seems clear that phytoene (15) is oxidized, *in vivo*, to phytofluene (30), ζ-carotene (31), neurosporene (32) and lycopene (4); the driving force being extension of conjugation in each case. All-*trans* phytofluene (30) has been synthesized by a Wittig reaction between (53) and the aldehyde (54) (Davis *et al.*, 1966). The 15,15′*cis* isomer of ζ-carotene (31) was synthesised

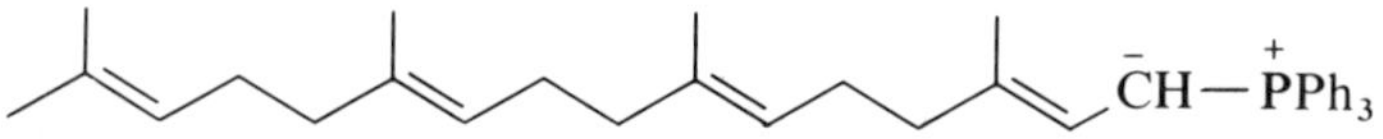

(53)

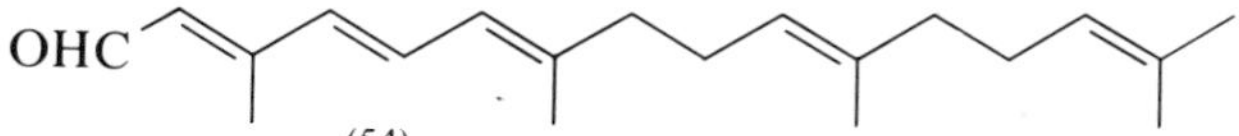

(54)

(Davis *et al.*, 1966) from (55), derived from farnesyl bromide, and (56) followed by *cis* reduction bond using Lindlar's catalyst. The ultra-violet spectrum showed a hypsochromic shift of 3 mμ, compared with the all-*trans* isomer, and had the expected infra-red peak for a 15,15′-*cis* double bond. Moreover, the synthetic product was identical with ζ-carotene (31) isolated from *Chlorella* mutants.

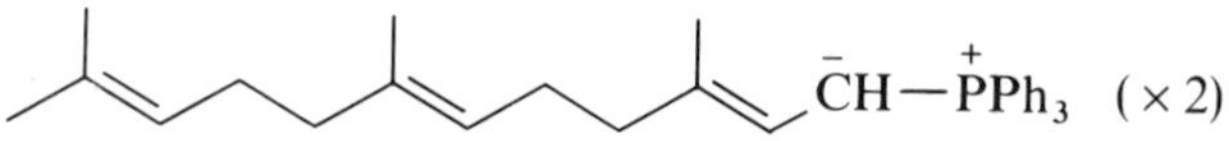

(55)

(56)

Chloroxanthin (57), isolated from a green mutant of the photosynthetic bacteria. *Rhodopseudomonas spheroides*, and rhodopin (58), ex protolithotropic bacterium (Liaaen-Jensen, 1965; Aasen and Jensen, 1967a) represent hydration products at a terminal double bond of neurosporene (32) and lycopene (4) respectively. It must be the terminal double bond which is hydrated, since the absorption spectrum of rhodopin (58) is the same as lycopene (4). *Rhodopseudomonas spheroides* is yellow-brown when grown anaerobically but rapidly becomes red on exposure to oxygen due to the oxidation of spheroidene (59) to spheroidenone (60) (Barber *et al.*, 1966).

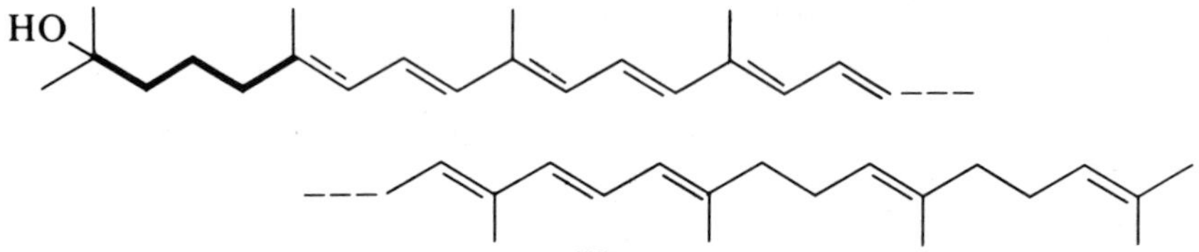

(57)

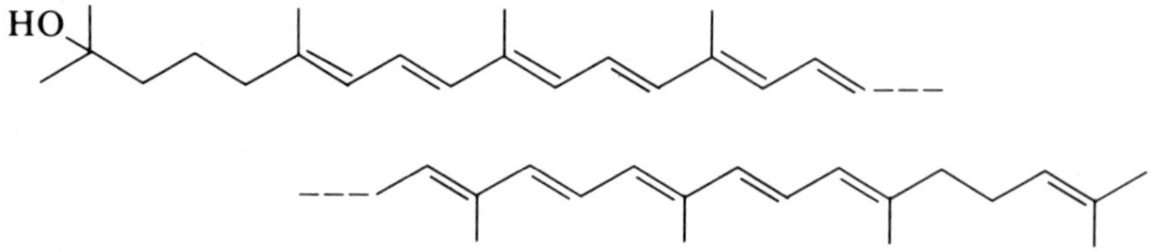

(58)

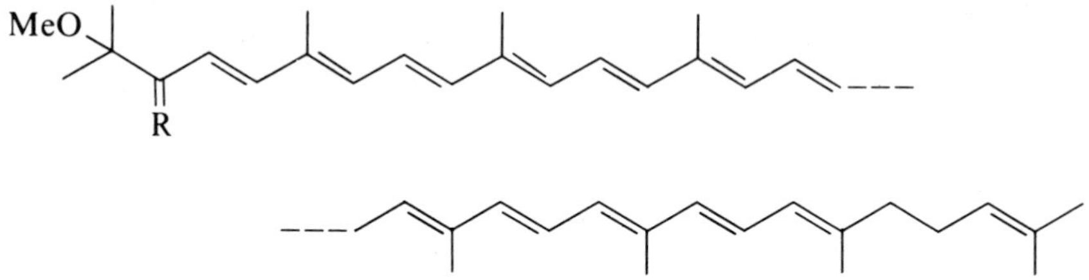

$R = H_2$ (59)
R = O (60)

Addition of the elements of water or methanol to both terminal bonds of lycopene (4), followed by the introduction of two further double bonds affords α-bacterioruberin (37) (Liaaen-Jensen, 1960) and spirilloxanthin (33) respectively. Spirilloxanthin (33) is the characteristic pigment of many purple photosynthetic bacteria and has been synthesized (Schneider and Weedon, 1967) from reaction of the phosphorane (61) with the dialdehyde (62).

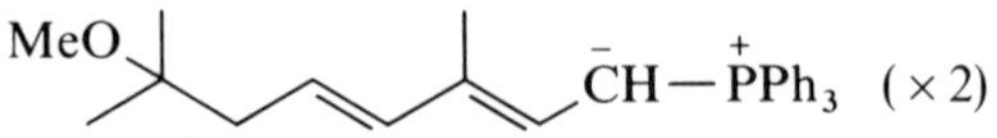

(61)

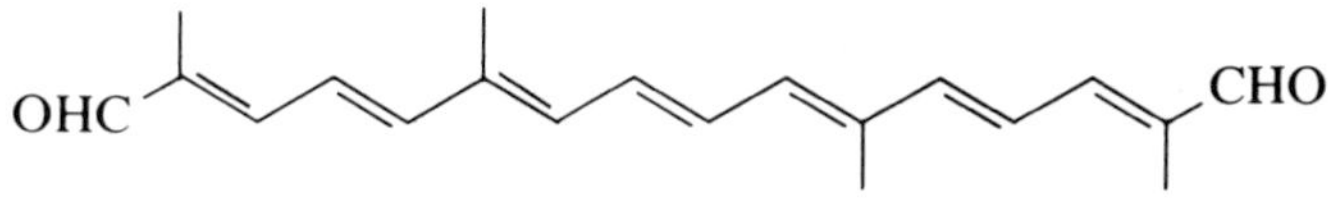

(62)

Functionalization of the acyclic carotenoid skeleton may also be achieved *in vivo* by oxidation at the allylic methyl groups. Lycoxanthin (45) and lycophyll (46) are oxidation products of lycopene (4). The absorption spectrum (in benzene) showed maxima at 521, 487, 458 indicating an undecaene chromophore. Mass spectra of the compounds indicated the likely positions of hydroxylation (Cholnoky *et al.*, 1968; Markham and Liaaen-Jensen, 1968) (see Section V E). Both lycoxanthin (45) and lycophyll (46), on oxidation, showed no change in the visible spectrum. This indicated that the hydroxyl functions were not allylic to the polyene chromophore, which was substantiated by the finding that neither (45) nor (46) were dehydrated by 0.01% HCl in chloroform; a reagent specific for the dehydration of allylic alcohols in the carotenoid field. The n.m.r. spectrum was unambiguous in placing the hydroxyl functions at the terminal positions. Oxidation at a non-terminal methyl group of rhodopin (58) and isomerism of the double bond adjacent to the carbonyl, occurs in photosynthetic purple sulphur bacteria (Aasen, and Liaaen-Jensen, 1967b) producing warmingone, for which structure (63) is one possibility given; the position of oxidation being uncertain.

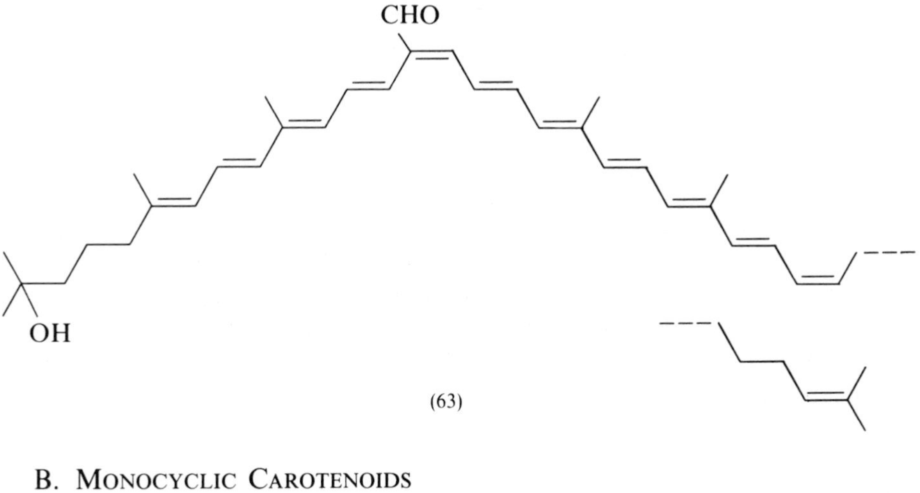

(63)

B. Monocyclic Carotenoids

It is thought that neurosporene (32), not lycopene (4), is the precursor of the cyclic carotenoids (Goodwin, 1963), but the evidence is not conclusive.

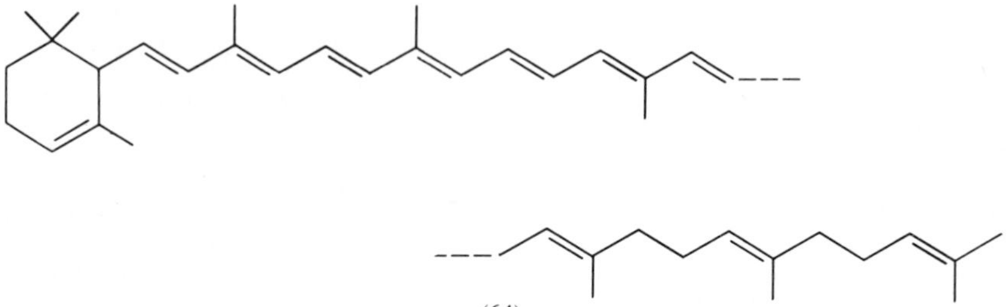

(64)

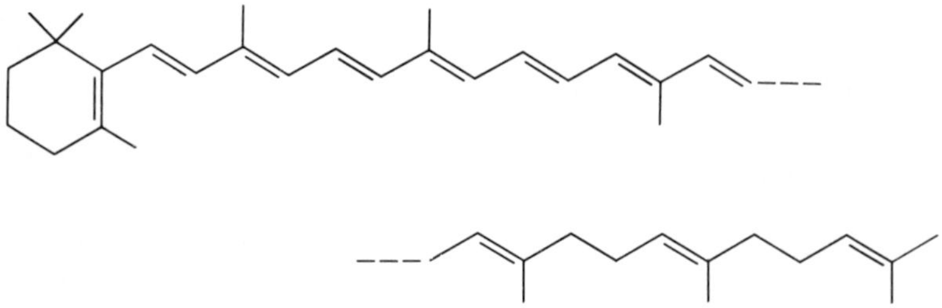

(65)

Cyclization would yield α-zeacarotene (64) and β-zeacarotene (65). There is strong biosynthetic evidence that the α- and β- end groups are not in equilibrium (Williams *et al.*, 1967). Oxidation of α- and β-zeacarotene would afford δ-carotene (5) and γ-carotene (3), respectively.

Most monocyclic carotenoids are oxidation products of γ-carotene (3). Again functionalization of γ-carotene follows the pattern of the acyclic carotenoids with addition to the terminal double bond, introduction of a further double bond to extend conjugation, and oxidation at allylic positions. Torulene (66) and torularhodin (12) represent examples of the latter two processes *in vivo*. Hertzberg and Liaaen-Jensen (1967) isolated γ-carotene (3) and 4-keto-γ-carotene (67) from *Mycobacterium phei* together with the

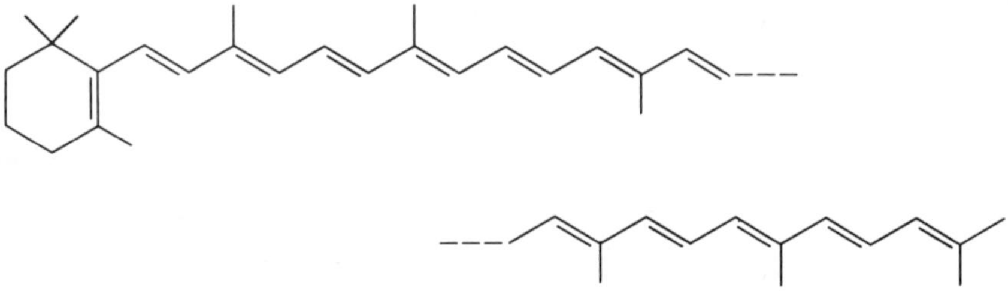

(66)

products of hydration at the terminal non-conjugated double bond (68) and deoxyflexixanthin (69) respectively. The α-ketol, flexixanthin (70) (Aasen and Liaaen-Jensen 1966a) isolated from flexibacteria is probably a product of further oxidation of (69), or saproxanthin (71) (Aasen and Liaaen-Jensen, 1966b). Pleixanthophyll (72) and 4-ketopleixanthophyll (73) are other γ-carotene derivatives having formal addition at the terminal double bond. The carbonyl grouping of (67) and (73) can be placed at the 4-position from infra-red carbonyl frequency at 6.03 μ and also the hypsochromic shift of 4 mμ in the visible spectrum after lithium aluminium hydride reduction of the carbonyl group (Leftwick and Weedon, 1966).

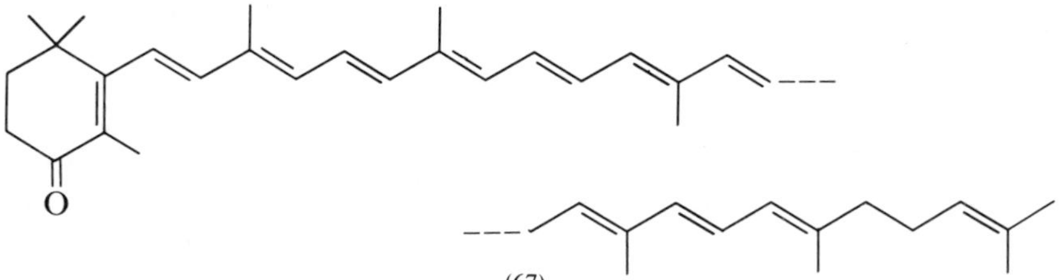

(67)

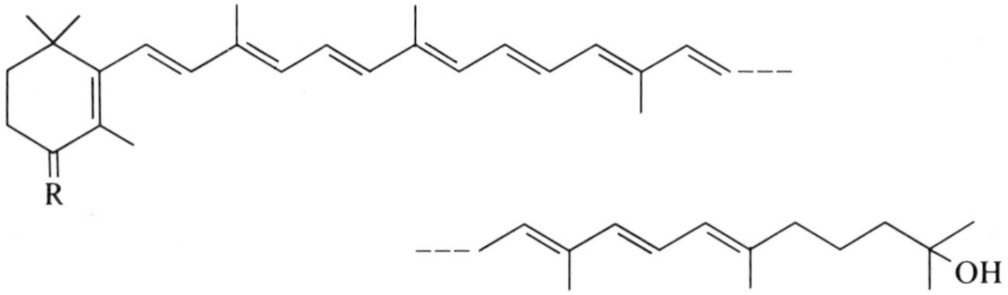

R = H_2 (68)
R = O (69)

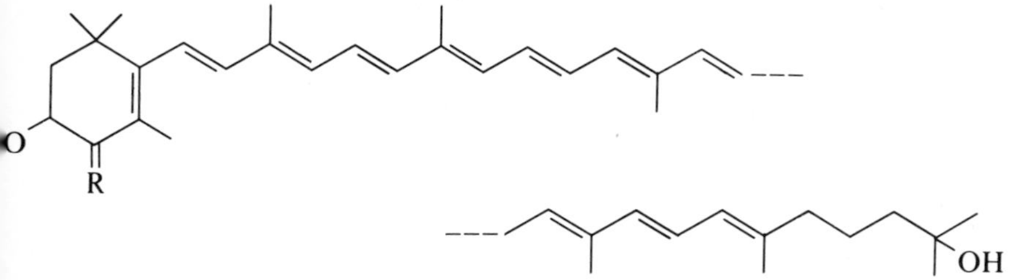

R = O (70)
R = H_2 (71)

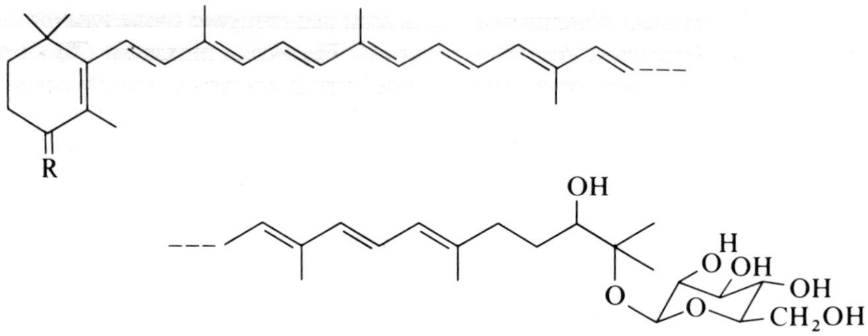

R = H_2 (72)
R = O (73)

Rubixanthin (74) and gazaniaxanthin (75) have been postulated to be isomeric about the 5′,6′ double bond (Brown and Weedon, 1968). Both give similar mixtures on iodine-catalysed stereomutation and the absence of a "*cis*-peak" in the visible spectrum of gazaniaxanthin (75) was attributed to slight stereochemical difference between the two double bond isomers. Perhaps the most salient feature to note at this stage, is the presence of a hydroxyl function at C-3 in saproxanthin (71) and rubixanthin (74), which, although not an allylic position, is a very common site of oxidation in the bicyclic carotenoids. This end group gives rise to a strong M-18 peak in the mass spectrum.

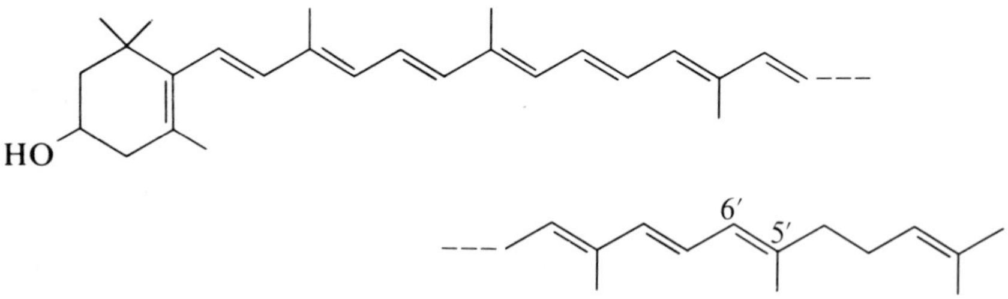

(*trans* 5′, 6′) (74)
(*cis* 5′, 6′) (75)

B. Bicyclic Carotenoids

1. *Carotenoids derived from α-carotene* (1).

Hydroxylation of α-carotene (1) in the 3,3′ positions gives lutein (8), a major carotenoid of all green plants. Treatment of (8) with 0.01 N methanolic hydrochloric acid at 40°C selectively afforded the allylic methyl ether

(Liaaen-Jensen, and Hertzberg, 1966). Carotenoids having a hydroxyl function allylic to the polyene chain are readily methylated by this reagent (Petracek and Zechmeister, 1956). Products from oxidation at the allylic methyl groups are also found in nature. Pyrenoxanthin (76) has been isolated from *Chlorella pyrenoidosa* (Yamamoto *et al.*, 1969) and siphonaxanthin (77) shown to be present in some siphonous green algae (Kleinig *et al.*, 1969).

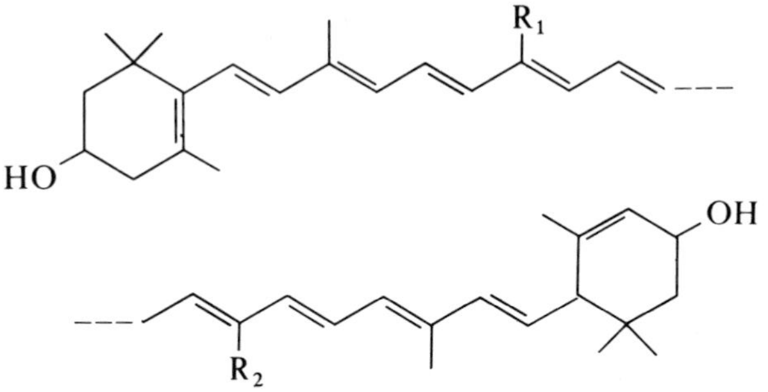

R_1 = Me, R_2 = CH_2OH } or reverse

(76)

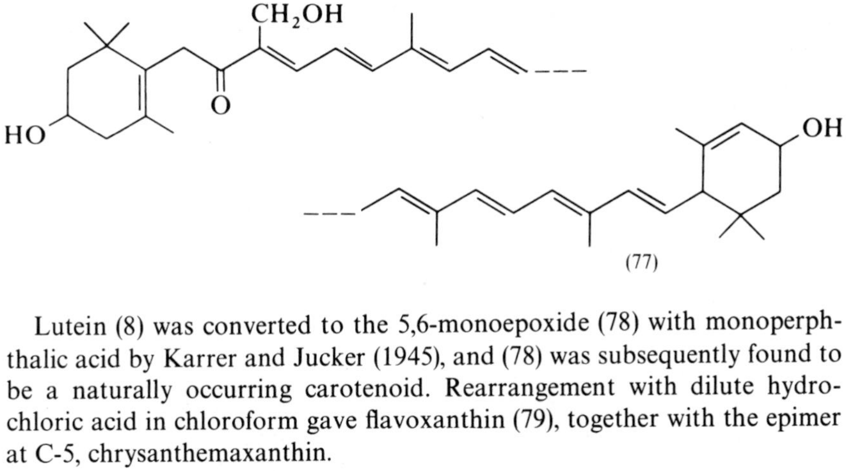

(77)

Lutein (8) was converted to the 5,6-monoepoxide (78) with monoperphthalic acid by Karrer and Jucker (1945), and (78) was subsequently found to be a naturally occurring carotenoid. Rearrangement with dilute hydrochloric acid in chloroform gave flavoxanthin (79), together with the epimer at C-5, chrysanthemaxanthin.

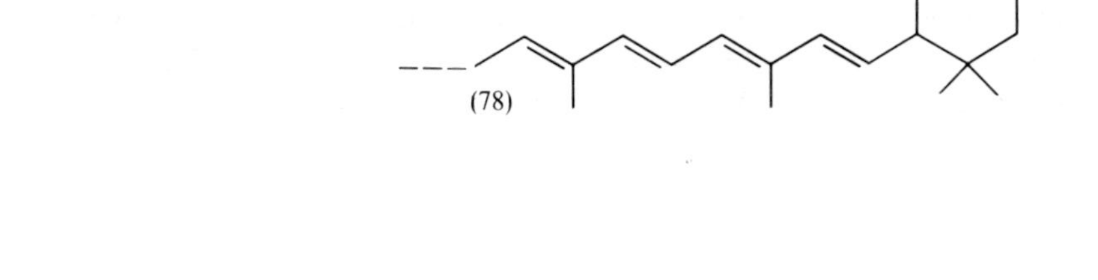

(78)

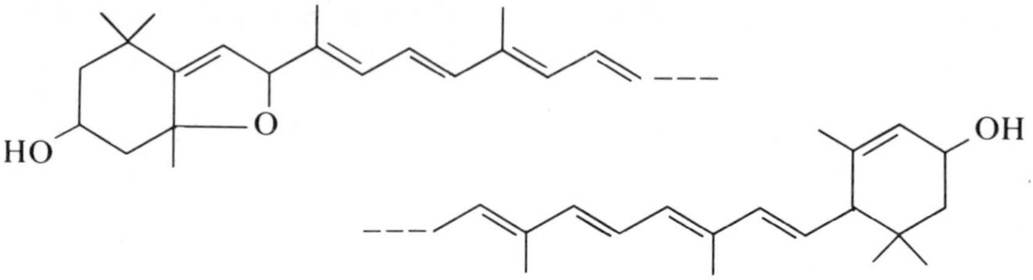

(79)

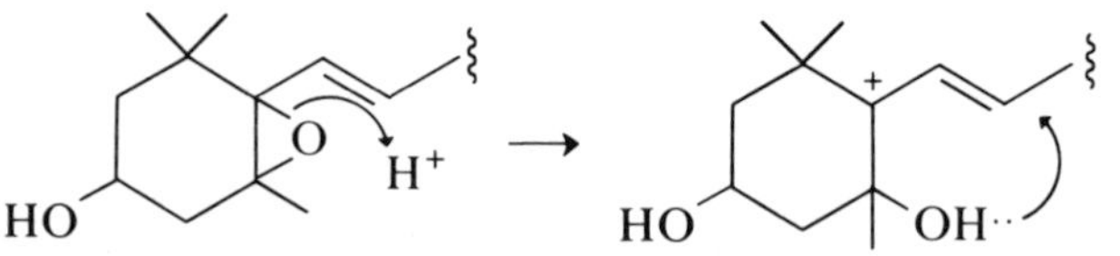

The effect of this transformation on the mass spectrum and visible absorption spectrum has been discussed in Section V. An alternative mode of neutralization of the intermediate carbonium ion would lead to the acetylenic carotenoid, monoadoxanthin (80), but this will be dealt with later in detail.

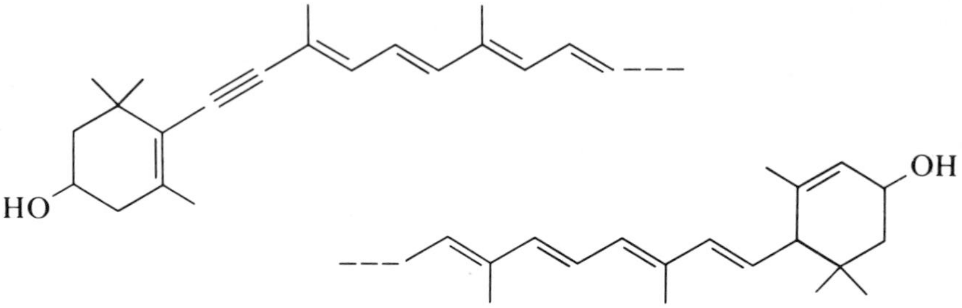

(80)

2. *Carotenoids derived from β-carotene* (2)

In this series also the favoured positions of hydroxylation are 3,3′ and 4,4′—e.g. isozeaxanthin (81) and zeaxanthin (25). The 4,4′-hydroxyl groups are differentiated by their greater reactivity towards hydrochloric acid in methanol (methyl ether formation) and hydrochloric acid in chloroform (dehydration). Oxidation of carotenoids having a hydroxyl group at the 4 or 4′ positions produces a change in the visible absorption spectrum (Table 1, Section V B) since both carbonyl groups formed are conjugated with the polyene system. For this reason, the infra-red carbonyl frequency is at 6.03 μ. In carotenoids having both rings hydroxylated, the symmetrical arrangement is most common.

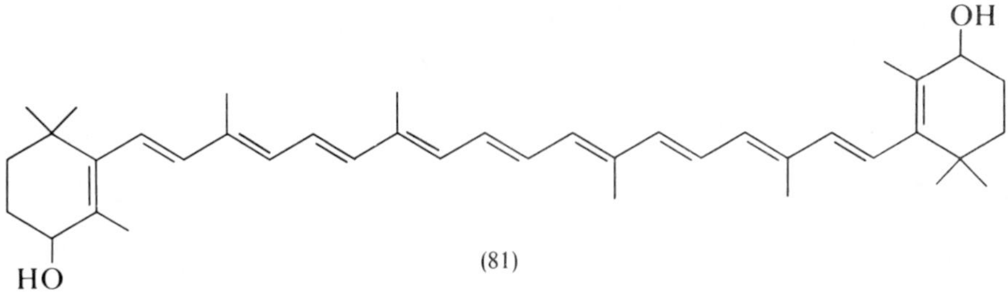

(81)

Zeaxanthin (25) gives rhodoxanthin (7) on vigorous oxidation with manganese dioxide (Entschel and Karrer, 1959) via the dihydrorhodoxanthin (82). This is an analogous case to the oxidation of 1,4-diketones to enedione

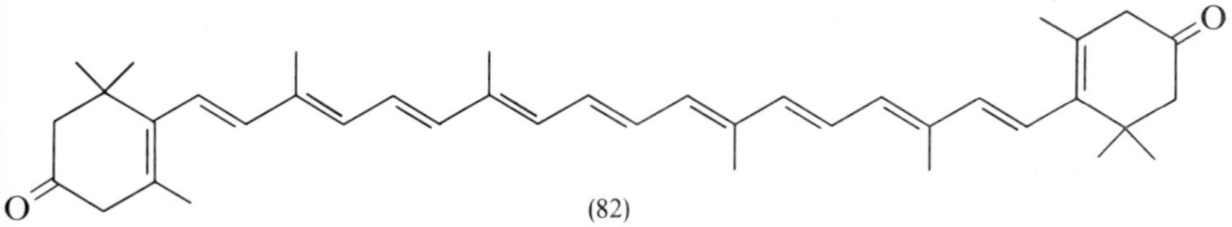

(82)

systems. Conversely, reduction of rhodoxanthin (7) with zinc and acetic acid in pyridine affords (82) which on Ponndorf reduction gives (±) zeaxanthin (25). Rhodoxanthin (7) has been synthesized (Mayer *et al.*, 1967) from the phosphorane (83) and the dialdehyde (84).

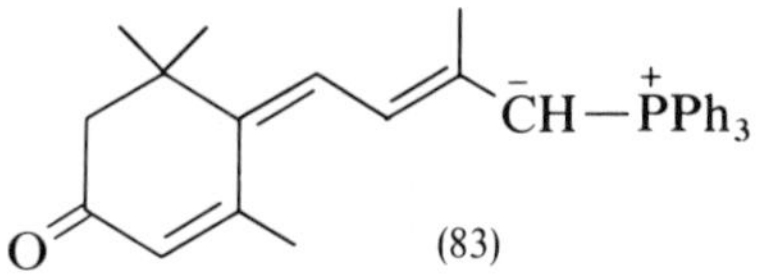

(83)

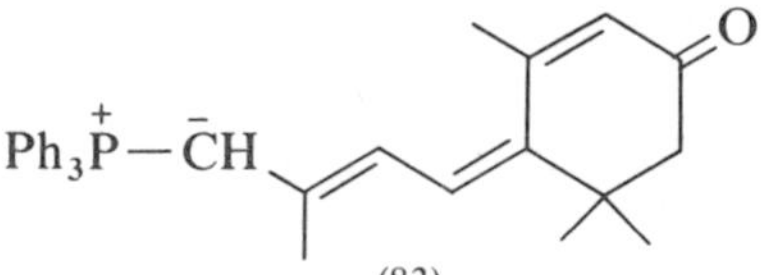

(83)

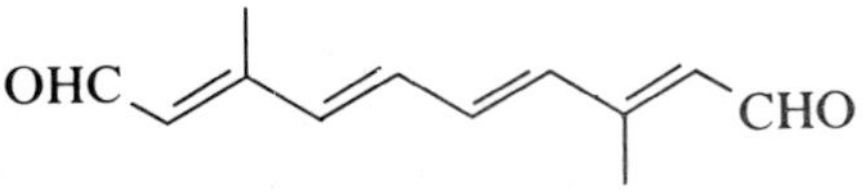

(84)

The α-ketol end group of flexixanthin (70) is also found in the bicyclic carotenoids, hydroxy-echinenone (85) and astaxanthin (86). Both may be considered to be oxidation products of cryptoxanthin and zeaxanthin (25) respectively or alternatively formed by hydroxylation of echinenone (34) and canthaxanthin (35) respectively. 3-Hydroxyechinenone (85) has been isolated from *Adonis annua* (Egger, 1965). Astaxanthin (86) is found in nature

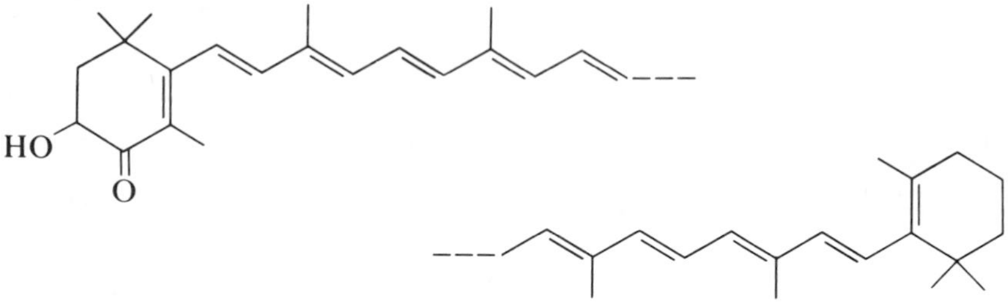

(85)

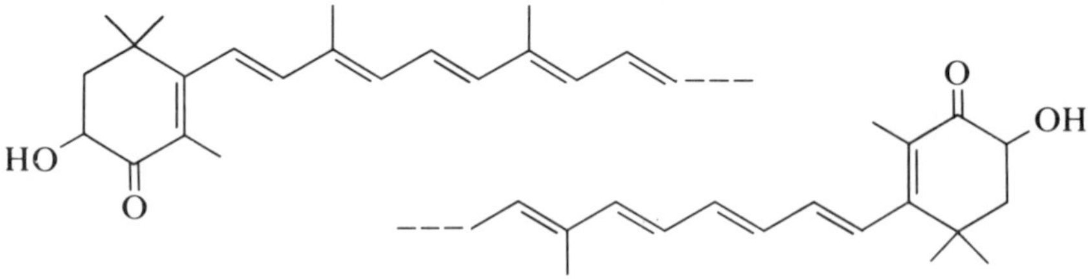

(86)

in the free form, as an ester, or part of a protein complex (Karrer and Jucker, 1948). Combination with the appropriate apoprotein from the lobster carapace, gives the characteristic blue colour (λ 630 mμ) of α-crustacyanin (Cheeseman *et al.*, 1966). Attempts to isolate astaxanthin (86) often led to the

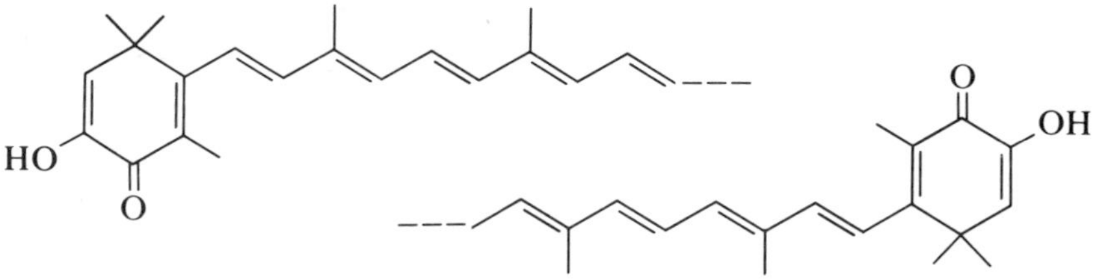

(87)

oxidation to astacene (87). Astacene (87) can be obtained by autoxidation of canthaxanthin (35), and has been transformed (Leftwick and Weedon, 1967) into astaxanthin (86) on reduction with potassium borohydride followed by selective allylic oxidation of the intermediate 3,4-diol system using manganese dioxide. 3-Hydroxy-echinenone (85) has been synthesized by a similar route from echinenone (34).

Astaxanthin (86) has been postulated to be an intermediate in the biosynthesis of the animal carotenoids actinioerythrin (88) and violerythrin (89). Actinioerythrin (88) is a red pigment, which on treatment with alkali

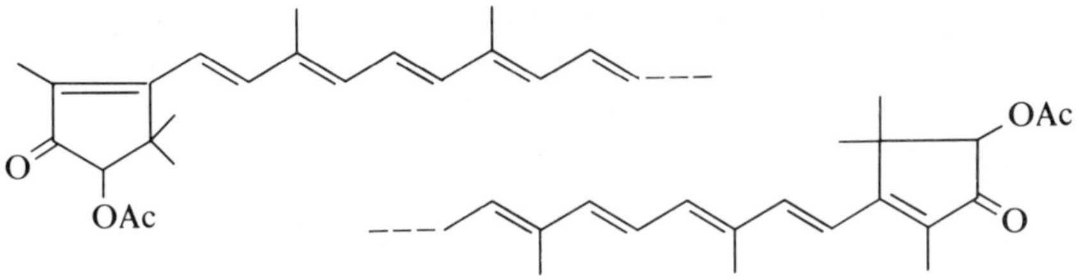

(88)

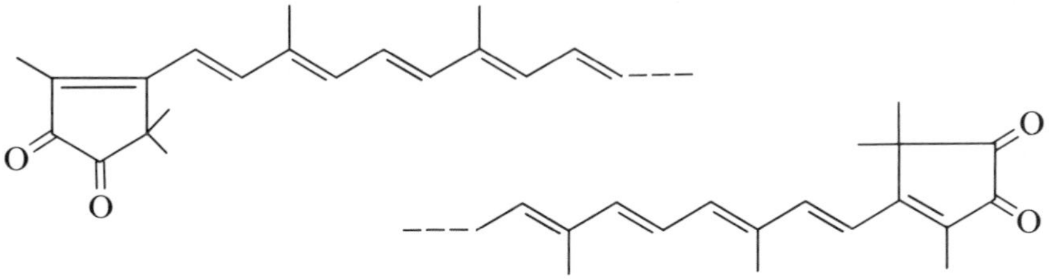

(89)

and aerial oxidation can be transformed into the blue pigment, violerythrin (89). The structure proof depended heavily on physical methods (Hertzberg and Liaaen-Jensen, 1968) and has been confirmed by synthesis (Holzel *et al.*, 1969). Astacene is thought to be further oxidized to the 1,2,3-triketo system which then suffers a benzylic acid rearrangement followed by oxidation of the α-hydroxy acid intermediate.

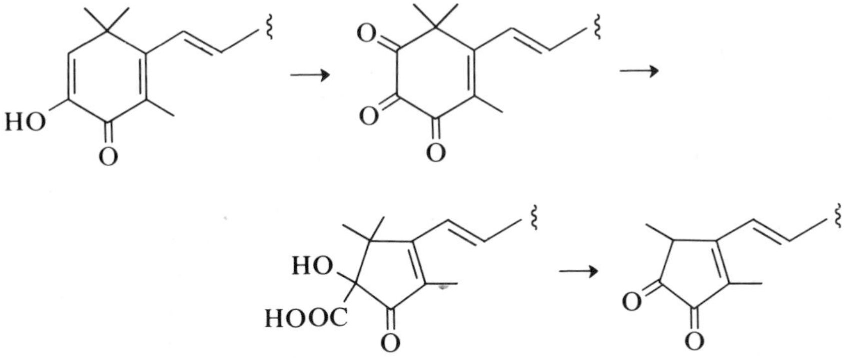

Capsanthin (11) and capsorubin (90) have been isolated from red peppers, *Capsicum annuum*, and structure elucidation (Barber *et al.*, 1961; Entschel and Karrer, 1960) showed them to have the novel ring-contracted structures.

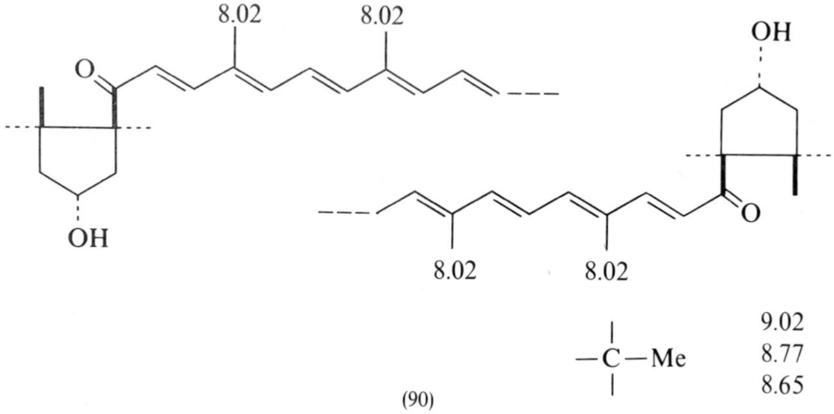

(90)

The visible spectra of capsanthin (474, 505 mμ) and capsorubin (444, 474, 506 mμ) indicated a decaenone and nonaenedione chromophore respectively. Oxidation of capsanthin (11) by the Oppenauer method gave a strong band at 5.7 μ in the infra-red suggesting the presence of a cyclopentanone moiety in the oxidation product. This gave the first clue to the existence of the cyclopentane ring in these carotenoids. The n.m.r. spectra showed the peaks expected for "in-chain" methyl groups at 8.02τ and in (11) the methyl resonances for the β- end group of zeaxanthin (25) at 8.95τ (6H) and 8.30τ (3H). In addition, the spectra of (11) and (90) also showed methyl singlets at 9.02, 8.77 and 8.65τ which were assigned methyl groups on the cyclopentane ring on the basis of chemical shift. The absolute configuration of capsanthin (11) and capsorubin (90) was determined by Faigle and Karrer (1961) by degrading these carotenoids to (−)-camphoronic acid. It was then proved by Cooper *et al.*, (1962) that the hydroxyl groups in the end groups are *trans* to the polyene chain. Thus the absolute stereochemistry is as shown for capsanthin (11) and capsorubin (90). The presence of antheraxanthin (91) and violaxanthin (9) in *Capsicum annuum* led to the postulate that rearrangement of these 5,6-epoxides was responsible for ring contraction to capsanthin (11) and capsorubin as shown.

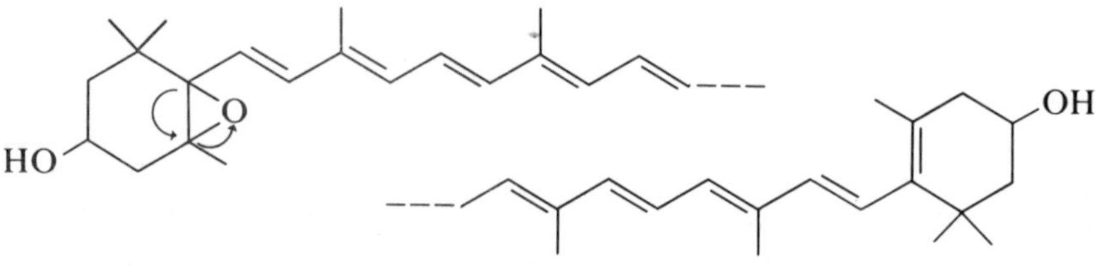

(91)

Thus we have seen that the 5,6-epoxide group can lead to furanoid oxides, e.g. (78) → (79) or, *in vivo*, to rearrangement with ring contraction, e.g. (91) → (11). There is yet another important series of carotenoids which must surely require 5,6-epoxides intermediates for biosynthesis. These are the naturally occurring allenic or acetylenic carotenoids, e.g. (10) or (80). The relationship between the 5,6-epoxides, allenes, and acetylenic compounds may be depicted thus.

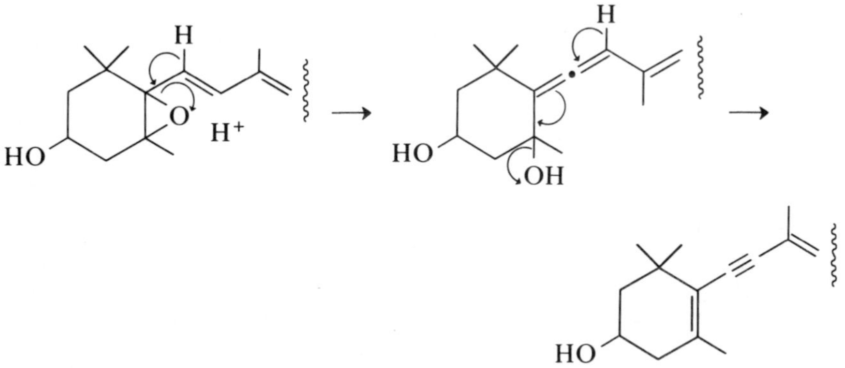

The chief carotenoid of the flowers of *Mimulus guttatus*. (Nietsche *et al.*, 1969), deepoxyneoxanthin (92) would therefore be produced from antheraxanthin (91), the hypothetical precursor of capsanthin (11). Acid-catalysed rearrangement of deepoxyneoxanthin (92) afforded diaxanthin (93); oxidation of which *in vivo* would yield pectenolone (94) having the familiar α-ketol system in one of the end groups (Cambell *et al.*, 1967). Similar

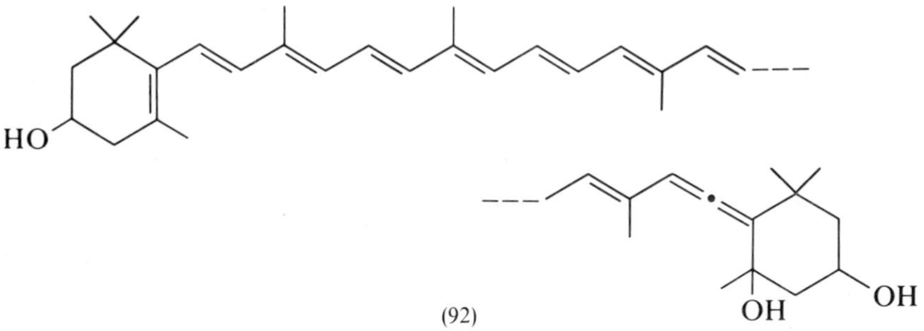

(92)

rearrangements would be expected starting from the bis-epoxide violaxanthin (9). The initial product would be neoxanthin (10), a major xanthophyll of photosynthetic tissues (Donohue *et al.*, 1966; Mallams *et al.*, 1967). Acid-catalysed rearrangement of neoxanthin (10) (Egger *et al.*, 1969) afforded

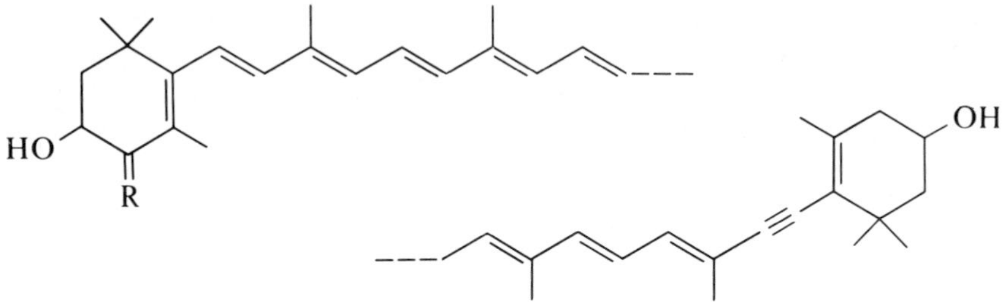

$(R{=}H_2)$ (93) $(R{=}O)$ (94)

diadinochrome (95), due to proton attack at both the 5,6-epoxy grouping and the allenic alcohol function. Diadinoxanthin (96), isolated from *Euglena* (Aitzetmuller *et al.*, 1968) must be produced *in vivo* from reaction specifically at the allenic function. A hypsochromic shift of 20 mμ was observed in the acid-catalysed rearrangement of diadinoxanthin to the furanoid oxide (95) indicating the isomerization of one 5,6-epoxy grouping in the carotenoid.

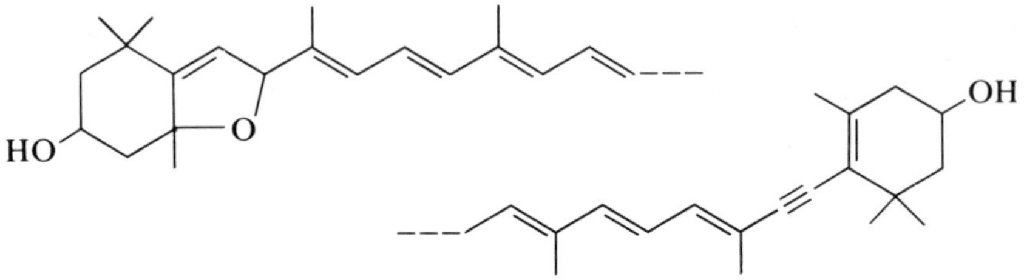

(95)

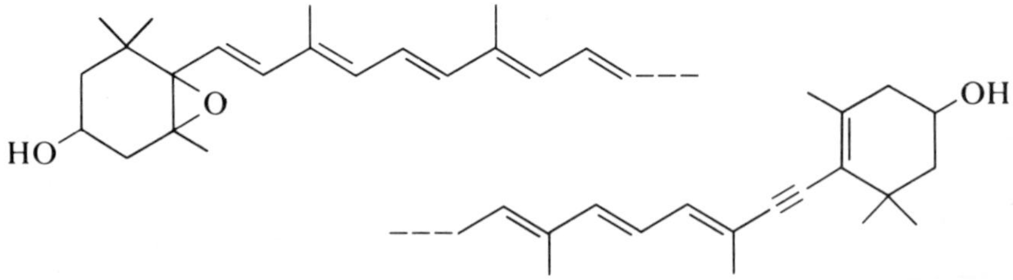

(96)

Further rearrangement of the 5,6-epoxy end group would lead to the di-acetylenic structure (97) assigned to alloxan, isolated from flagellates of the algal class *Cryptophyceae* (Mallams *et al.*, 1967b). The monoacetylene, monadoxanthin (80) was also obtained from this source. The triple bond in alloxan (97) gave a characteristic infra-red absorption at 4.63 μ. Cambell *et al.* (1967) isolated pectenoxanthin from the giant scallop *Pecten maximus* and assigned the same structure as for alloxan (97).

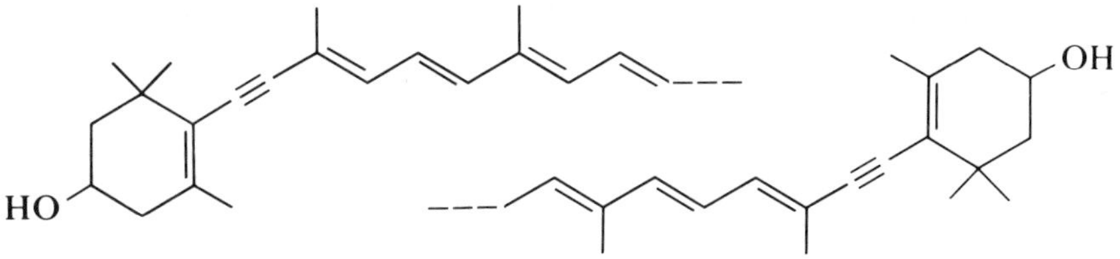

(97)

Fucoxanthin occurs in brown algae (*Phaeophyceae*) and is one of the most abundant carotenoids in nature. The structure of fucoxanthin (14) (Bonnett *et al.*, 1969) showed it to be an oxidation product of neoxanthin (10). Infra-red absorption at 5.22 μ was attributed to the allene system. The mass spectral fragmentation (M, M-18, M-18-18, M-18-60, M-18-18-60) indicated the composition $C_{42}H_{58}O_6$ and was consistent with two hydroxyl groups as well as an acetoxyl function being present in fucoxanthin (14). Zinc permanganate partial oxidation gave fragments formed by the cleavage indicated in structure (14). Oxidation of fucoxanthin by the Oppenauer method gave the C_{31} carotenoid, paracentrone (98) which occurs in trace amounts in the sea urchin, *Paracentrolus lividus* (Hora *et al.*, 1970). The mechanism involves the prior oxidation of the secondary hydroxyl grouping followed by fragmentation

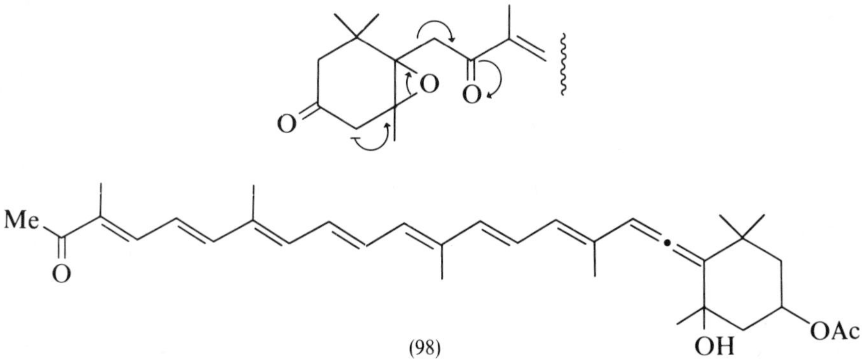

(98)

as shown. The use of seaweed meal as a feed ingredient for poultry initiated a study of the fate of fucoxanthin (14) (Jensen, 1966) and led to the isolation of iso-fucoxanthin (99).

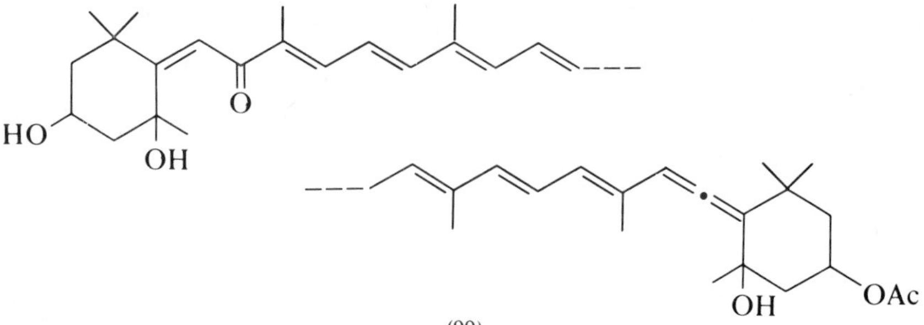

(99)

D. Carotenoids with Aromatic End Groups

Aryl carotenoids have been isolated from the sea sponge, *Reniera japonica* (Yamaguchi, 1960), green, brown and purple photosynthetic sulphur bacteria (Liaaen-Jensen *et al.*, 1964), (Liaaen-Jensen, 1965), (Aasen and Liaaen-Jensen, 1967c). These carotenoids must be derived by rearrangement of the carotenoid skeleton since they have two unique end groups (100) and

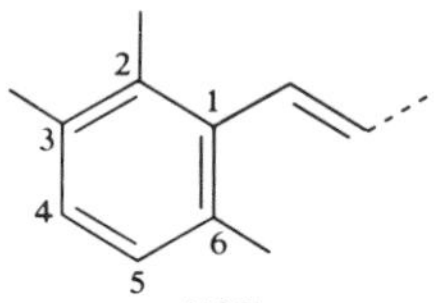

(100)

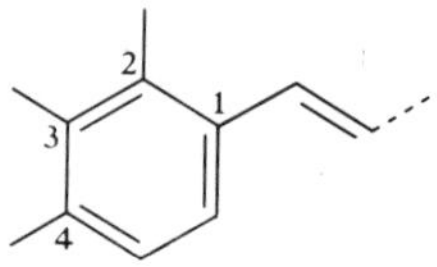

(101)

(101). Because of the methyl groups at C-2 and C-6, the aryl group (100) cannot be coplanar with the polyene system. This is reminiscent of the situation with the β end group (43), therefore (100) and (43) make the same contribution to the visible spectra of carotenoids. The end group (101) does not have the same steric inhibition to resonance and thus makes a similar contribution to the visible spectrum as the acyclic end group (42). Chlorobactene (102) is the major carotenoid of photosynthetic bacteria (Liaaen-Jensen *et al.*, 1964) and was found to have the same visible absorption characteristics as γ-carotene (3). To complicate matters further, no structural aid could be obtained from the closely similar melting points and infra-red spectra.

The n.m.r. proved to be the means of differentiating γ-carotene (3) and chlorobactene (102). The positions of the methyl peaks are shown, but particularly important are the positions and number of the aryl-methyl groups at 7.75 and 7.77τ.

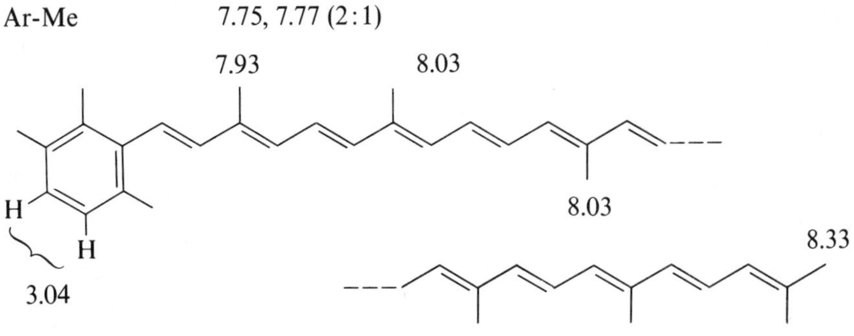

(102)

Photosynthetic brown bacteria contain isorenieratene (103) and β-isorenieratene (104) with only traces of β-carotene (2) and chlorobactene (102), (Liaaen-Jensen, 1965). The synthesis of these carotenoids, as well as

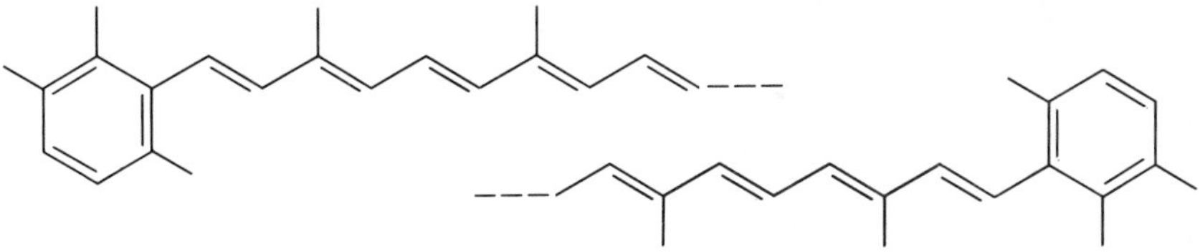
(103)

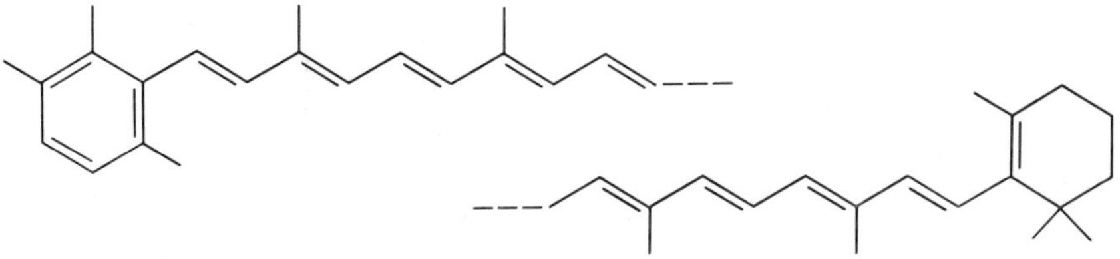

Ar-Me 7.73, 7.78 (2:1)

(104)

others in the series, involved reaction of the appropriately substituted benzyl phosphorane (105) with crocetindial (106), (Cooper *et al.*, 1963).

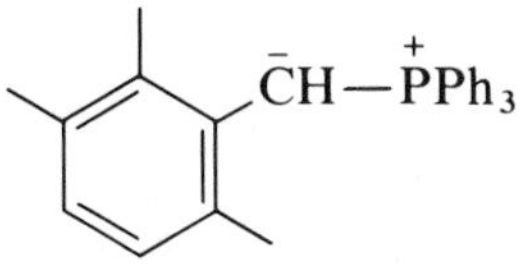

(×2)

(105)

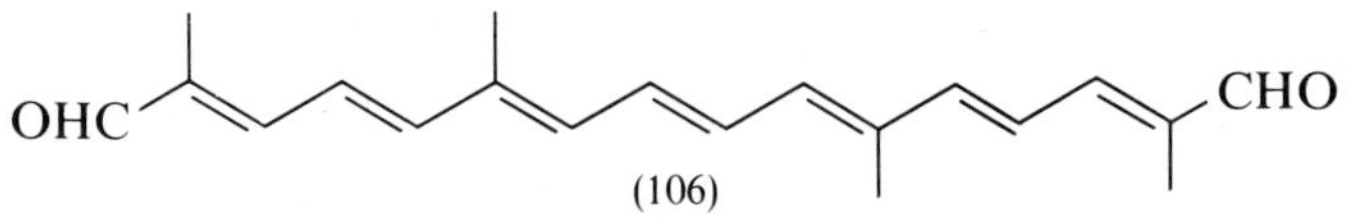

(106)

Carotenoids having the end group (101) are found in the purple photosynthetic sulphur bacteria. Okenone (107) had a carbonyl frequency at 6.10 μ and the visible spectrum confirmed extended conjugation to the carbonyl group (375, (460) 484, 516 mμ). The n.m.r. data, shown, indicated a spirilloxanthin end group and a tetrasubstituted aromatic ring, (Liaaen-Jensen, 1967a). Final proof of structure came from synthesis (Aasen and Liaaen-Jensen, 1967c). Renierapurpurin (108) and renieratene (109) also

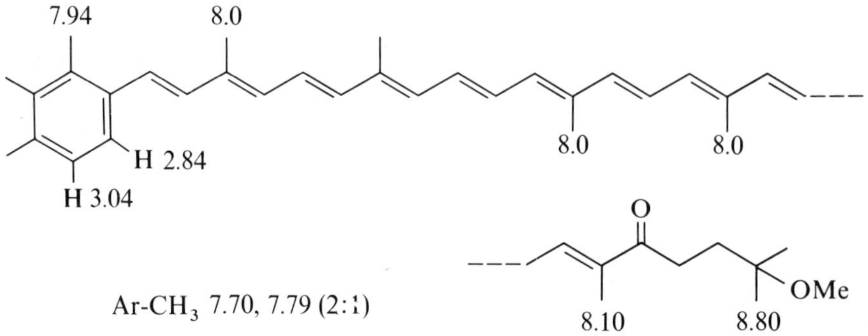

(107)

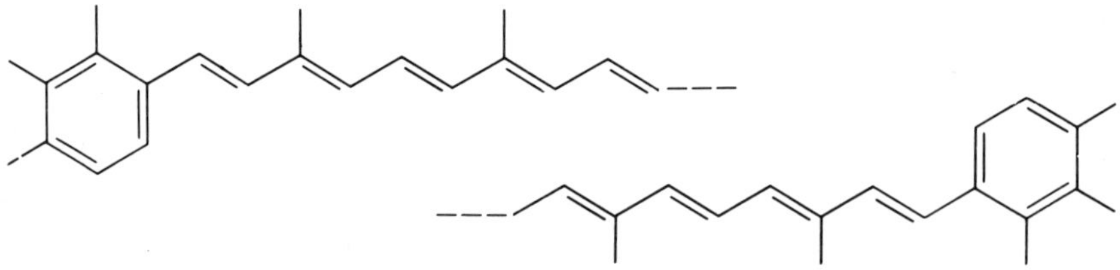

(108)

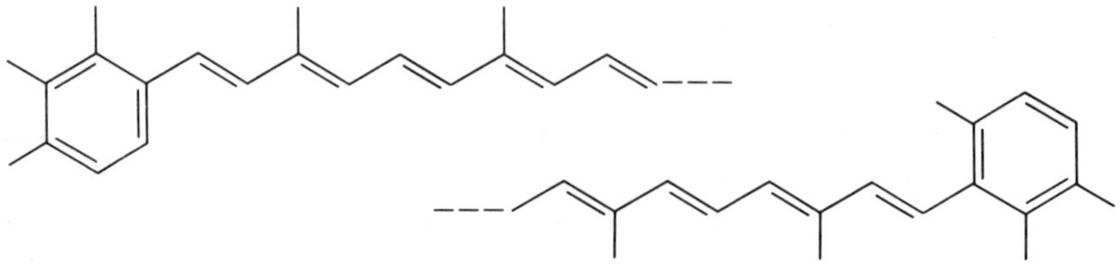

(109)

possess the end group (101) (Yamaguchi, 1960), although renieratene is not symmetrical having examples of both known aryl end groups.

The derivation of end groups (100) and (101) *in vivo* is a matter for conjecture, but it seems likely that 1,2-methyl migration occurs in a β- end group (43). If this occurs with the formation of the cyclohexadiene (110) then oxidation would lead to end group (100). Since the intermediate (110)

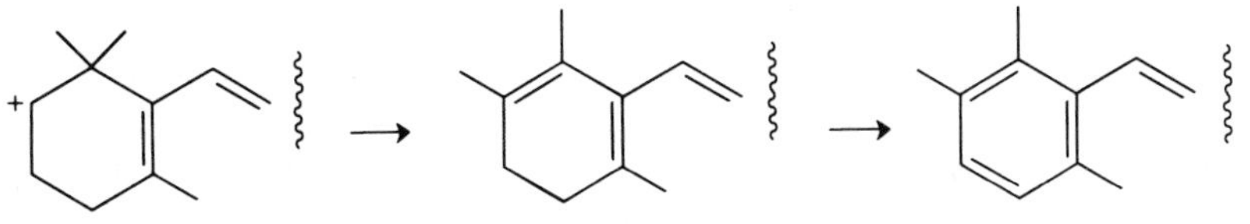

(43) (110) (100)

is a cross-conjugated system, double bond migration leading to (111) should be favoured from both steric and electronic standpoints. Ring opening of the cyclohexadiene (111) would give the hexatriene (112), which could undergo isomerization to (113)—possibly via a [1,7]-sigmatropic shift. Cyclization of (113) followed by oxidation would then produce the end group (101).

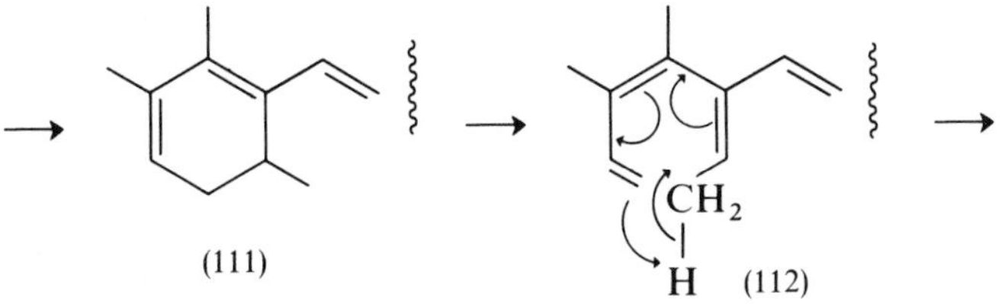

(111) (112)

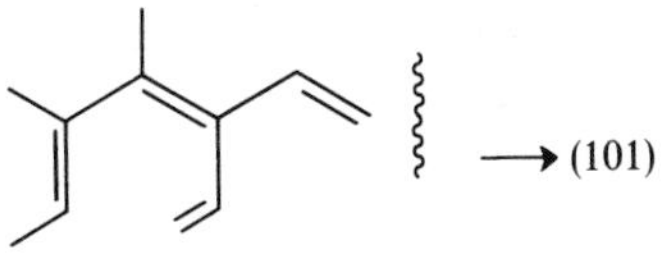

(113)

E. C-45 and C-50 Carotenoids

At the present time, few carotenoids having >C-40 skeleton are known. The C-45 carotenoid isolated by Norgard and Liaaen-Jensen (1969) from *Corynebacterium poinsettiae* was assigned the structure, 2-isopentenyl-3,4-dehydrorhodopin (114) largely on the basis of mass spectra of the carotenoid and its trimethylsilyl ether. The important fragmentations (M-69, M-55) are shown as are the methyl resonances from n.m.r. studies.

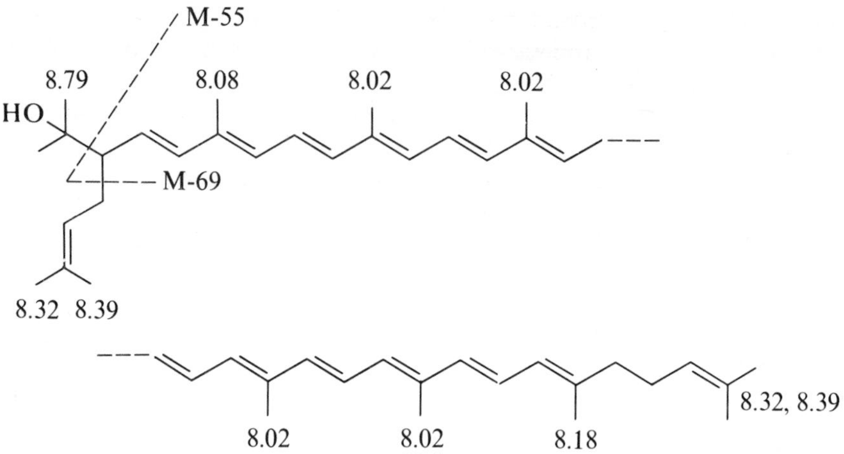

(114)

A bacterial carotenoid having a C-50 skeleton was isolated by Kelly and Liaaen-Jensen (1967) and one of the possible structures proposed was (115) in which isoprenoid moieties have added to each end of the normal acyclic

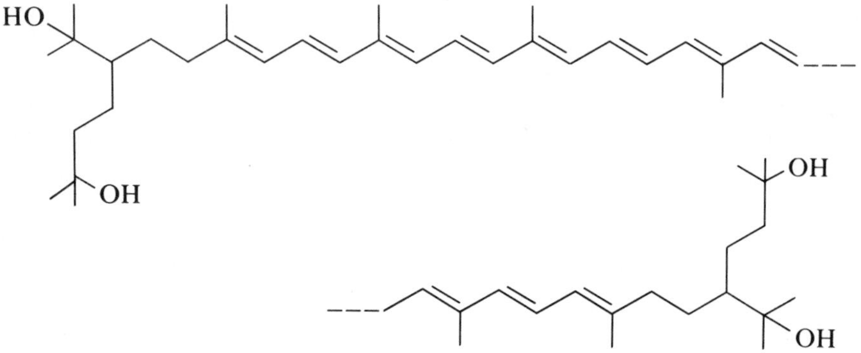

(115)

carotenoid skeleton. Another C-50 carotenoid, dehydrogenans P439, was found in the non-photosynthetic bacterium *Flavobacterium dehydrogenans Arnandi* (Liaaen-Jensen, 1967b, Liaaen-Jensen and Hertzberg, 1968). Ozonolysis showed the absence of an isopropylidene group and acetylation indicated two hydroxyl groupings which were proved to be primary by oxidation to a dialdehyde. The C-50 skeleton was deduced from consideration of the mass spectral fragmentation leading to M-140 (see Section V E for fragmentation of α- end groups). If this was due to the breakdown of an α- end group, then the extra isoprenoid units must be attached to the 2 and

2 positions of ε-carotene (6). This led to structure (116) being assigned to dehydrogenans P439. The stereochemistry of the terminal double bonds was determined by comparing the chemical shift of the aldehyde proton (0.6τ) of the oxidation product, with analogous systems. This showed the hydroxymethyl grouping to be *trans* to the alkyl grouping. (Schwieter and Liaaen-Jensen, 1969), since the alternative configuration of the double bond would require the aldehyde proton in the oxidation product to resonate at lower field.

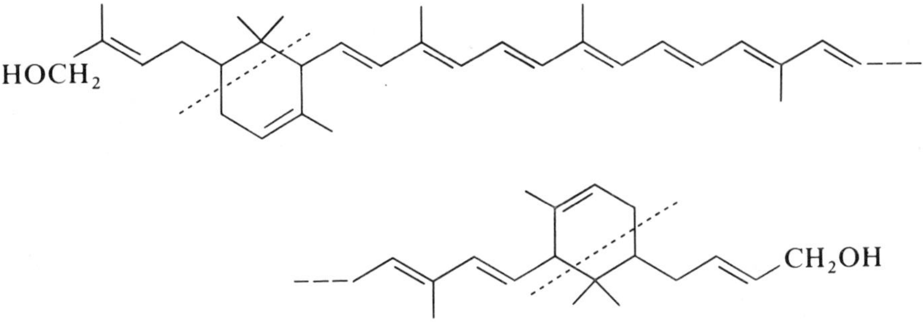

(116)

VII. Vitamin A

Vitamin A, (117) is necessary for normal growth and vision in animals and, as the structure suggests, may be considered to be a degraded carotenoid. Steenbock *et al.* (1921) showed that carotene isolated from plant sources was vitamin A-active; a finding which was later confirmed by Moore (1929) who showed that the feeding of carotene to rats resulted in the appearance of vitamin A in the liver. Other carotenoids exhibited varying degrees of efficiency as vitamin A precursors (Table 3) showing the great dependence of biological activity on the β- end group common to Vitamin A and β-carotene. Since carotenoids are only synthesized *de novo* by higher plants and protista, and since vitamin A is derived from carotenoids, then the vitamin A present in animals is ultimately derived from these sources.

Two general mechanisms for degrading β-carotene (2) to vitamin A have been discussed by Goodwin (1963). The first step is stepwise degradation from one end to vitamin A-aldehyde (118) via the apocarotenals (17) and (18). Alternatively, fission of the 15,15′-double bond would afford vitamin A aldehyde (118) directly. Tracer evidence from feeding [15,15′-^{14}C]-β-carotene (2) to rats seems to support the latter view. Wendler *et al.* (1950) oxidized β-carotene (2) *in vitro*, using hydrogen peroxide and osmium

TABLE 3

CAROTENOID STRUCTURE AND VITAMIN A ACTIVITY

Structure		Activity
α-Carotene	(1)	53%
β-Carotene	(2)	100%
γ-Carotene	(3)	43%
Lycopene	(4)	0

tetroxide, and obtained vitamin A aldehyde (118). The discovery of vitamin A has been covered by Moore (1957) and the successful syntheses follow closely on the methods of carotenoid chemistry (Isler and Schudel, 1963a).

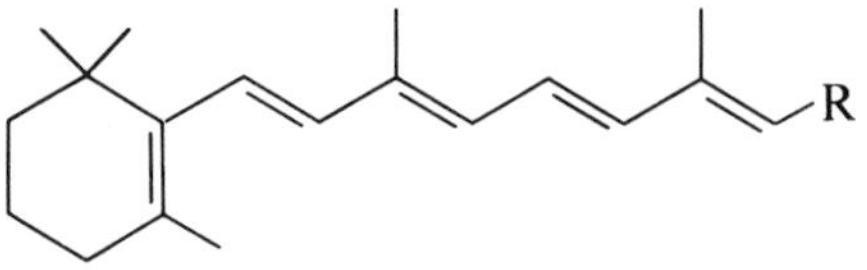

R = CH_2OH (117)
R = CHO (118)

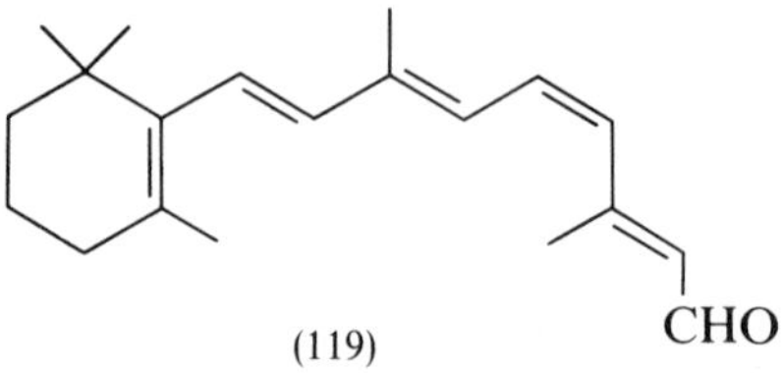

(119)

Retinene, a sterioisomer of (118), is the prosthetic group of rhodopsin, the photosensitive pigment for scotopic or dimlight vision (Wald, 1943) and is formed in the retina by reduction of vitamin A by the enzyme *alcohol dehydrogenase*. It was found (Hubbard and Wald, 1953) all-*trans* vitamin A aldehyde (118) would not combine with the protein, opsin unless it was exposed to light. Irradiation, of course, caused stereomutation of the double bonds. After careful work (See Morton and Pitt, 1957) it was discovered that only the 11-*cis* isomer (119) would unite with opsin to give rhodopsin (Blatz *et al.*, 1968, 1969).

It has been found recently that (Blatz *et al.*, 1968, 1969) 5,6-dihydroretinal after irradiation can combine with visual protein to give a new visual chromophore (λ463 mμ). The stereoisomers involved were the 9-*cis* and 11-*cis* forms. Blatz *et al.* attempted to correlate structural variation with opsin activity.

Vitamin A_2 (120) occurs mainly in fresh water fish, but is also present in small amounts in marine fish. The corresponding retinene$_2$ is the component of the visual pigments, porphyropsin, in fresh water fish. Vitamin A_2 (120) can be prepared from vitamin A_1 by allylic bromination of the acetate with N-bromosuccinimide, followed by dehydrohalogenation and hydrolysis.

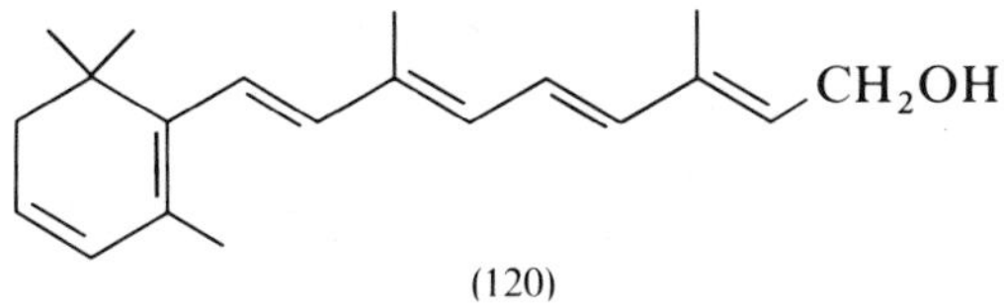

(120)

REFERENCES

Aasen, A. J. and Liaaen-Jensen, S. (1966a). *Acta Chem. Scand.* **20**, 1970.

Aasen, A. J. and Liaaen-Jensen, S. (1966b). *Acta Chem. Scand.* **20**, 811.

Aasen, A. J. and Liaaen-Jensen, S. (1967a). *Acta Chem. Scand.* **21**, 371.

Aasen, A. J. and Liaaen-Jensen, S. (1967b). *Acta Chem. Scand.* **21**, 2185.

Aasen, A. J. and Liaaen-Jensen, S. (1967c). *Acta Chem. Scand.* **21**, 970.

Aitzetmuller, K., Svec, W. A., Katz, J. J. and Strain, H. H. (1968). *Chem. Commun.* 32.

Baldas, J., Porter, Q. N., Cholnky, L., Szabolcs, J. and Weedon, B. C. L. (1966). *Chem. Commun.* 852.

Barber, M. S., Davis, J. B., Jackman, L. M. and Weedon, B. C. L. (1960). *J. Chem. Soc.* 2870.

Barber, M. S., Jackman, L. M., Warren, C. K. and Weedon, B. C. L. (1961). *J. Chem. Soc.* 4019.

Barber, M. S., Jackman, L. M., Manchand, P. S. and Weedon, B. C. L. (1966). *J. Chem. Soc.* (*C*) 2167.

Blatz, P. E. and Pippert, D. L. (1968), *J. Amer. Chem. Soc.* **90**, 1296.

Blatz, P. E., Balasubramaniyan, P. and Balasubramaniyan, V. (1968). *J. Amer. Chem. Soc.* **90**, 3282.

Blatz, P. E., Balasubramaniyan, P., Balasubramaniyan, V. and Dewhurst, P. B. (1969). *J. Amer. Chem. Soc.* **91**, 5930.

Bohlmann, F. (1953). *Chem. Ber.* **86**, 63, 657.

Bonnett, R., Mallams, A. K., Spark, A. A., Tee, J. L., Weedon, B. C. L. and McCormick, A. (1969). *J. Chem. Soc.* 429.

Brown, B. O. and Weedon, B. C. L. (1968). *Chem. Commun.* 382.

Burnett, J. H. (1965). "Chemistry and Biochemistry of Plant Pigments" (T. W. Goodwin, ed.), p. 381, Academic Press, London and New York.

Caglioti, L., Cainelli, B., Camerino, B., Mondelli, R., Prieto, A., Quilico, A., Salvatori, T. and Selva, A. (1966). *Tetrahedron, Suppl.* **7**, 175.

Cambell, S. A., Mallams, A. K., Waight, E. S., Weedon, B. C. L., Barbier, M., Lederer, E. and Salaque, A. (1967). *Chem. Commun.* 941.

Cooper, R. D. G., Jackman, L. M., and Weedon, B. C. L. (1962). *Proc. Chem. Soc.* 215.

Cooper, R. D. G., Davis, J. B. and Weedon, B. C. L. (1963). *J. Chem. Soc.* 5637.

Cheeseman, D. F., Zagalsky, P. F. and Ceccaldi, H. J. (1966). *Proc. Roy. Soc.* **B.164**, 130.

Cholnoky, L., Szabolcs, J. and Waight, E. S. (1968). *Tetrahedron Lett.* 1931.

Dale, J. (1954). *Acta Chem. Scand.* **8**, 1235.

Davies, B. H. (1961a). *Biochem. J.* **80**, 48p.

Davies, B. H. (1961b). *Phytochemistry* **1**, 25.

Davies, B. H. (1965). *In* "Chemistry and Biochemistry of Plant Pigments" (T. W. Goodwin, ed.), p. 489, Academic Press, London and New York.

Davis, J. B., Jackman, L. M., Siddons, P. T. and Weedon, B. C. L. (1966). *J. Chem. Soc.* (*C*) 2154.
Donohue, H. V., Lowry, L. K., Chichester, C. O. and Yokoyama, H. (1966). *Chem. Commun.* 807.
Dutton, H. J., Manning, W. M. and Duggar, B. M. (1943). *J. Phys. Chem.* **47**, 308.
Egger, K., Dabbagh, A. G. and Nitsche, N. (1969). *Tetrahedron Lett.* 2995.
Emerson, R. and Fox, D. L. (1940). *Proc. Roy. Soc.* **B.128**, 275.
Entschel, R. and Karrer, P. (1959). *Helv. Chim. Acta* **42**, 466.
Entschel, R. and Karrer, P. (1960). *Helv. Chim. Acta* **43**, 89.
Enzell, C. R., Francis, G. W. and Liaaen-Jensen, S. (1968). *Acta Chem. Scand.* **22**, 1054.
Eugster, C. H., Buchecker, R., Tscharner, Ch., Uhde, G. and Ohloff, G. (1969). *Helv. Chim. Acta* **52**, 1729.
Faigle, H. and Karrer, P. (1961). *Helv. Chim. Acta* **44**, 1904.
Gillam, A. E. and El Ridi, M. S. (1935). *Nature* (*London*) **136**, 914.
Gillam, A. E. and El Ridi, M. S. (1936). *Biochem. J.* **30**, 1735.
Goodwin, T. W. (1955). *In* "Modern Methods of Plant Analysis" (K. Paech and M. V. Tracey, eds.), Vol. III, p. 272, Springer, Heidelberg.
Goodwin, T. W. (1963). "The Biosynthesis of Vitamins and Related Compounds", p. 286, Academic Press, London and New York.
Goodwin, T. W. (1965). *In* "Chemistry and Biochemistry of Plant Pigments" (T. W. Goodwin, ed.), p. 127, Academic Press, London and New York.
Hertzberg, S. and Liaaen-Jensen, S. (1967). *Acta Chem. Scand.* **21**, 15.
Hertzberg, S. and Liaaen-Jensen, S. (1968). *Acta Chem. Scand.* **22**, 1714.
Holzel, R., Leftwick, A. P. and Weedon, B. C. L. (1969). *Chem. Commun.* 128.
Hora, J., Toube, T. P. and Weedon, B. C. L. (1970). *J. Chem. Soc.* (*C*) 241.
Hubbard, R. and Wald, G. (1953). *J. Gen. Physiol.* **36**, 415.
Inhoffen, H. H., Pommer, H. and Bohlmann, F. (1950a). *Liebigs Ann.* **569**, 237.
Inhoffen, H. H., Pommer, H. and Westphal, F. (1950b). *Liebigs Ann.* **570**, 69.
Inhoffen, H. H., Bohlmann, F., Bartram, K., Rummert, G. and Pommer, H. (1950c). *Liebigs Ann.* **570**, 54.
Isler, O. and Schudel, P. (1963a). *Adv. Org. Chem.* **4**, 115, Interscience Publishers, New York.
Isler, O. and Schudel, P. (1963b). "Carotine and Carotinoide", Vol. 9, p. 54, Steinkopff, Darmstadt.
Jensen, A. (1966). *Acta Chem. Scand.* **20**, 1728.
Jones, E. R. H., Lee, H. H. and Whiting, M. C. (1960). *J. Chem. Soc.* 3483.
Karrer, P. and Helfenstein, A. (1929a). *Helv. Chim. Acta* **12**, 1142.
Karrer, P. and Bachmann, W. E. (1929b). *Helv. Chim. Acta.* **12**, 285.
Karrer, P. Helfenstein, A., Wehrli, H. and Wettstein, A. (1930). *Helv. Chim. Acta* **13**, 1084.
Karrer, P. and Morf, R. (1931a). *Helv. Chim. Acta* **14**, 1033.
Karrer, P., Helfenstein, A., Pieper, B. and Wettstein, A., (1931b). *Helv. Chim. Acta.* **14**, 435.
Karrer, P. and Solmssen, U. (1937). *Helv. Chim. Acta* **20**, 1396.
Karrer, P., Konig, H. and Solmssen, U. (1938). *Helv. Chim. Acta* **21**, 445.
Karrer, P. and Kramer, H. (1944). *Helv. Chim. Acta* **27**, 1301.
Karrer, P. and Jucker, E. (1945). *Helv. Chim. Acta* **28**, 300.
Karrer, P. and Jucker, E. (1948). "Carotinoide", Birkhauser, Basle.
Karrer, P. and Eugster, C. H. (1950). *Helv. Chim. Acta* **33**, 1172.

Kelly, M. and Liaaen-Jensen, S. (1967). *Acta Chem. Scand.* **21**, 2578.
Kleinig, H., Nitsche, H. and Egger, K. (1969). *Tetrahedron Lett.* 5139.
Kuhn, R. and Ehmann, L. (1929). *Helv. Chim. Acta* **12**, 904.
Kuhn, R. and Lederer, E. (1931a). *Chem. Ber.* **64**, 1354.
Kuhn, R. and L'Orsa, F. (1931b). *Chem. Ber.* **64**, 1732.
Kuhn, R. and Grundmann, Ch. (1932a). *Chem. Ber.* **65**, 898, 1880.
Kuhn, R. and Winterstein, A. (1932b). *Chem. Ber.* **65**, 1873.
Kuhn, R. and Roth, H. (1933a). *Chem. Ber.* **66**, 1274.
Kuhn, R. and Winterstein, A. (1933b). *Chem. Ber.* **66**, 429.
Kuhn, R. and Möller, E. F. (1934). *Angew. Chem.* **47**, 145.
Kuhn, R. and Brockmann, H. (1935). *Liebigs Ann.* **516**, 98, 123.
Leftwick, A. and Weedon, B. C. L. (1966). *Acta Chem. Scand.* **20**, 1195.
Leftwick, A. and Weedon, B. C. L. (1967). *Chem. Commun.* 49.
Liaaen-Jensen, S. (1960). *Acta Chem. Scand.* **14**, 950.
Liaaen-Jensen, S. (1962a). "The Constitution of Some Bacterial Carotenoids and their Bearings on Biosynthetic Problems", Bruns Trondheim.
Liaaen-Jensen, S. (1962b). *Kgl. Norske Videnskab. Selskabs. Skrifter* **8**, 5.
Liaaen-Jensen, S., Hegge, E. and Jackman, L. M. (1964). *Acta Chem. Scand.* **18**, 1703.
Liaaen-Jensen, S. (1965). *Acta Chem. Scand.* **19**, 1025.
Liaaen-Jensen, S. (1967a). *Acta Chem. Scand.* **21**, 961.
Liaaen-Jensen, S. (1967b). *Acta Chem. Scand.* **21**, 1972.
Lunde, K. and Zechmeister, L. (1955). *J. Amer. Chem. Soc.* **77**, 1647.
Mallams, A. K. (1967). *Chem. Commun.* 485.
Mallams, A. K., Waight, E. S., Weedon, B. C. L., Chapman, D. J., Haxo, F. T., Goodwin, T. W. and Thomas, D. M. (1967b). *Chem. Commun.* 301.
Markham, M. C. and Liaaen-Jensen, S. (1968). *Phytochemistry* **7**, 839.
McCormick, A. and Liaaen-Jensen, S. (1966). *Acta Chem. Scand.* **20**, 1989.
Milas, N. A., Davis, P., Belič, I. and Fleŏ, D. (1950). *J. Amer. Chem. Soc.* **72**, 4844.
Moore, T. (1929). *Biochem. J.* **33**, 318.
Moore, T. (1957). "Vitamin A". Elsevier, New York.
Morton, R. A. and Pitt, G. A. J. (1957). *Fortschr. Chem. Org. Naturst.* **14**, 244.
Nietsche, H., Egger, K. and Dabbagh, A. G. (1969). *Tetrahedron Lett.* 2999.
Norgard, S. and Liaaen-Jensen, S. (1969). *Acta Chem. Scand.* **23**, 1463.
Pauling, L. (1939). *Fortschr. Chem. Org. Naturst.* **3**, 203.
Petracek, F. J. and Zechmeister, L. (1956). *J. Amer. Chem. Soc.* **78**, 1427.
Rabourn, W. J., Quackenbush, F. W. and Porter, J. W. (1954). *Arch. Biochem. Biophys.* **48**, 267.
Rabourn, W. J. and Quackenbush, F. W. (1956). *Arch. Biochem. Biophys.* **61**, 111.
Schwieter, U., Bollinger, H. R., Chopard-Dit-Jean, L. H., Englert, G., Kofler, M., Konig, A., v. Planta, C., Ruegg, R., Vetter, W. and Isler, O. (1965). *Chimia* **19**, 294.
Schwieter, U. and Liaaen-Jensen, S. (1969). *Acta Chem. Scand.* **23**, 1057.
Schneider, D. F. and Weedon, B. C. L. (1967). *J. Chem. Soc.* (*C*) 1686.
Steenbock, H., Snell, M. T. and Bontwell, P. W. (1921). *J. Biol. Chem.* **47**, 303.
Tswett, M. (1903). *Proc. Warsaw. Soc. Nat. Sci. Biol. Sec.* **14**, Min. 6.
Wackenroder, H. W. F. (1831). *Giegers Magazin Pharm.* **33**, 141.
Wald, G. (1943). *Vitam. and Horm.* **1**, 195.
Williams, R. J. H., Britton, G. and Goodwin, T. W. (1967). *Biochem. J.* **105**, 99.
Yamaguchi, M. (1960). *Bull. Chem. Soc.* (*Japan*) **33**, 1560.
Yamamoto, H. Y., Yokoyama, H. and Boettger, H. (1969). *J. Org. Chem.* **34**, 4207.
Yokoyama, H. and White, M. J. (1968). *Phytochemistry* **7**, 1031.

Zechmeister, L. (1960). *Prog. Chem. Org. Nat. Prod.* **18**, 223.
Zechmeister, L. (1962). "*Cis-trans* Isomeric Carotenoids, Vitamin A and Arylpolyenes". Springer-Verlag, Vienna.
Zechmeister, L., v. Cholnoky, L. and Vrabély, V. (1928). *Chem. Ber.* **61**, 566.
Zechmeister, L. and Tuzon, P. (1938). *Biochem. J.* **32**, 1305.

7

BIOGENESIS OF TERPENES

D. V. BANTHORPE and B. V. CHARLWOOD

University College, London

I. Fundamental Patterns

A. Introduction

Most living species can synthesize terpenoids, although terpenes as distinct from steroids only accumulate in many higher plants and microorganisms and a few animals. The biosynthetic pattern is shown in Diagram I and the situation is of almost limitless structural variation of a few precursors (Sandermann, 1962).

The hemiterpenes are essential building bricks but they do not accumulate whereas other classes can comprise several per cent of the body weight in favourable cases. In addition to the pathways shown, a few C_{25}, C_{35} and C_{50} compounds, and several classes of partly terpenoid origin (isoprenoid quinones, terpene glycosides, indole alkaloids etc.) are known. The detailed investigation of the complex biosynthetic patterns reveals the remarkable accuracy with which chemical theory can predict the course of these biological processes. Enzymes exploit the inherent reactivities of their substrates and biosynthetic schemes can invariably be dissected into steps involving elimination, addition, rearrangement etc. that are controlled by the electronic and stereochemical factors found in non-biological systems. Even the reactivity of apparently non-activated carbon atoms can usually be rationalized in terms of conformational and electronic changes imposed by the formation of the substrate-enzyme link. Knowledge of these pathways has not only revealed the underlying pattern for the varied and frequently bizarre structures that are found, but provides guide-lines for the elucidation of new structures and for the design of biogenetic-type syntheses *in vitro*.

Modern methods of structure-determination and the advent of radioisotopic techniques have enabled much information concerning terpene biosynthesis to be accumulated in the past two decades. Although only a small fraction of the field has been investigated, over one thousand papers have appeared on these topics and so the present review, which covers the literature to January, 1970, cannot be exhaustive. Excellent accounts describe the historical development (Richards and Hendrickson, 1964; Nicholas, 1967a) or give specific details (Clayton, 1965a,b; Waller, 1969). A useful distinction can be drawn between the terms biogenesis and biosynthesis although they are often used synonymously; the former applies to speculative ideas, often derived from information concerning simple precursors, but the latter refers to detailed schemes that have some experimental justification (Stermitz and Rapoport, 1961).

B. The Isoprene Rule

The earliest successful attempts to rationalize the structures of terpenes were the rules proposed by Wallach in 1887 and elaborated by Robinson some thirty years later. These investigators suggested that terpenes were derived from isoprene (or more generally isopentane) units condensed head to tail. Simple examples are in Diagram 2 where limonene (1) and camphor (2) are dissected into their structural components. Many higher terpenes

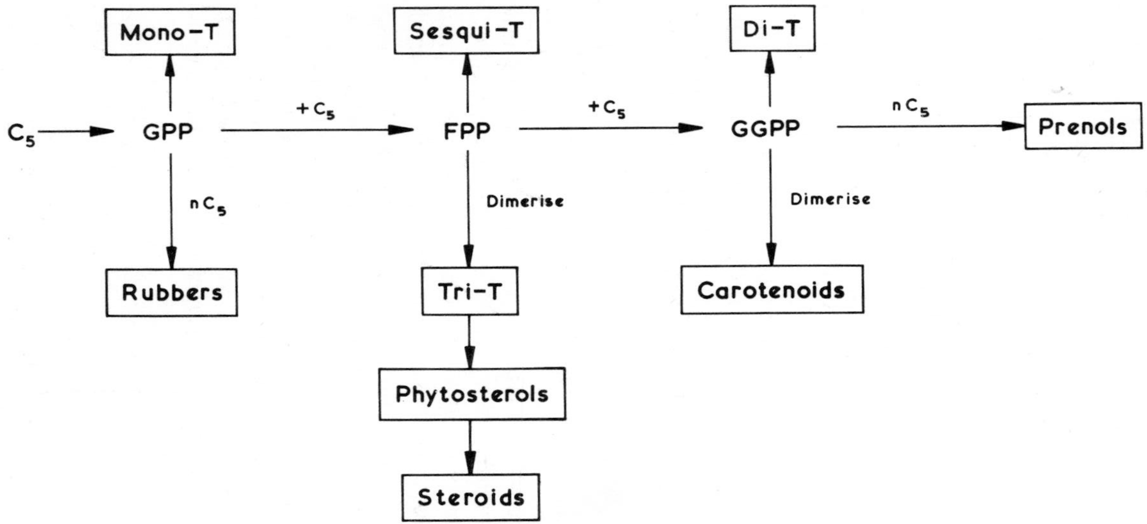

DIAGRAM 1.

T = Terpene. GPP, FPP and GGPP are the pyrophosphates of geraniol, farnesol, and geranylgeraniol respectively.

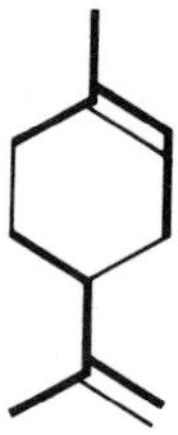

(1)

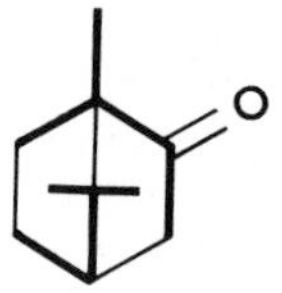

(2)

DIAGRAM 2.

were subsequently found not to obey these generalizations, but all could be accommodated by a "biogenetic" rule (Ruzicka *et al.*, 1953; Ruzicka, 1959) which states that naturally occurring terpenes are derived directly, or by way of stereospecific dimerizations, cyclizations and rearrangements, from the acyclic precursors geraniol (C_{10}), farnesol (C_{15}), geranylgeraniol (C_{20}) and squalene (C_{30}). This rule implies common pathways of biosynthesis and proposals for "irregular" biogenetic routes are to be regarded with reserve.

C. METHODS AND LIMITATIONS OF BIOSYNTHETIC STUDIES

Early ideas on terpenoid biosynthesis were based on studies of precursor relationships in higher plants (Nicholas, 1967a) but most of the detailed biochemistry since 1950 has been concerned with the synthesis of steroids in mammalian liver and in yeast. However, common patterns for routes to steroids and terpenes were soon established. Probable precursors of steroids that had been isotopically labelled were incubated with tissue slices or homogenates, the percentage incorporations into products were measured, isotope-dilution analyses were applied to characterize intermediates and the importance of additives such as coenzyme A, ATP and metal ions was explored. Similar investigations were carried out with cell-free systems and with fractions from the crude homogenates that were capable of sustaining parts of the reaction sequence.

Certain confusing and sometimes conflicting results arose in the pioneering studies, and although most can now be understood as arising from complex interconversions of precursors, it is important to appreciate the limitations of these methods of study. For plants, carbon dioxide is the only precursor that can be fed under physiological conditions: all others necessarily perturb the system and may evoke unnatural patterns of metabolism, particularly in the stages immediately after feeding and before the excess of precursor is degraded by salvage mechanisms. Although such feeding experiments may indicate a possible pathway, additional criteria are necessary to decide whether the pathway is obligatory (Davis, 1955; Swain, 1965). Ideally, tracer-containing precursors should be fed at the highest available specific activities in the lowest feasible concentrations and obligate precursors are expected to be efficiently incorporated into the end-products. Postulated biogenetic schemes should account for groups of concurrently formed products and should invoke reactions feasible on the principles of physical organic chemistry. Such reactions should also be energetically possible for the organisms concerned and should lead to an experimentally verified pattern of tracer in products biosynthesized from precursors labelled at specific positions. In addition, postulated intermediates and enzymic activities should be detected and characterized. For higher plants, the sequential synthesis of related compounds during the growing season should be consistent with the scheme proposed; and for microorganisms the use of species with blocked biosynthetic pathways (caused by either mutations or the application of growth-inhibitors) should lead to the expected results.

These criteria have never all been applied to any one example, but their existence emphasizes that it is not sufficient to isolate a radioactive product after feeding a radioactive precursor and to claim to have characterized a normal pathway. For example: recovery of a radioactive product after feeding a plant with [^{14}C]-mevalonic acid (see Section I D) is no evidence that the product is of terpenoid origin or nature for, although direct pathways for return of tracer from mevalonic acid to the acetate pool appear to be uncommon (but are not unknown), the same is not true for some of the labelled terpenes that are initially formed. An additional obvious, but nevertheless sometimes overlooked, technical point is that the isolated product must be highly purified, preferably by recrystallization to constant specific activity. Contamination with traces of more heavily labelled impurities can invalidate any conclusions.

D. The Characterization of "Active Isoprene"

Early attempts to characterize the unit ("active isoprene") that was believed to condense, according to the Isoprene Rule, showed that acetate and certain C_5 and C_6 acid radicals acted as precursors for cholesterol to the extent of a few percent incorporation of their tracer but with extensive randomization of their carbon atoms (Nicholas, 1967a). Similar observations were made for terpenes in plants (Sandermann and Stockmann, 1958). The key discovery (Tavormina *et al.*, 1956) was that mevalonic acid (MVA, 6), a compound isolated from brewers' solubles, was an acetate-replacing factor for certain bacteria and was incorporated almost quantitatively into cholesterol in cell-free systems from yeast and liver with concomitant loss of carbon dioxide.

The biosynthetic route to MVA and the early steps in steroid biosynthesis were subsequently worked out by Bloch and Lynen and their research groups, and the relationships between steroids and terpenes became apparent.

Acetate ion, in the form of acetyl-coenzyme A derived mainly from carbohydrate and fat metabolism, is generally considered to be converted into MVA by the steps shown in Diagram 3 (Clayton, 1965a) where, for clarity, only the absolute stereochemistry of the lactone (7) derived from R-MVA at physiological pH is shown. The sole known biological role of R-MVA is in steroid and terpenoid synthesis; the S-isomer is inactive. The thiol esters of acetic, acetoacetic (3) and β-hydroxy-β-methylglutaric acids (4) are normally interconvertible and the synthesis of MVA is controlled by the virtually irreversible reductions with NADPH. The occurrence of free mevaldehyde as an obligatory species on the path to MVA is doubtful, as addition of isotopically-normal compound to a cell-free system that synthesized MVA, neither reduced the incorporation of tracer from acetate into the product nor led to passage of label into re-isolated additive (Ferguson *et al.*, 1959). Mevaldehyde is either bonded to a carrier enzyme (cf. 5) or occurs as a hemithioacetal of coenzyme A that cannot exchange to any extent with the added free form. Free β-hydroxy-β-methylglutaric acid (HMG) may similarly not exist on the direct route to MVA although its metabolism has been demonstrated in both plants and animals. MVA is usually too reactive to accumulate *in vivo* but it has been isolated from liver and other tissue (Knauss *et al.*, 1959).

The condensing enzyme from yeast will accept acetoacetyl-acyl carrier protein from *E. coli* in addition to (3) to form HMG-acyl carrier protein (Rudney *et al.*, 1966) and this direct link with fatty acid metabolism may be significant in the control of terpene synthesis. Other pathways to MVA have been proposed but are generally considered to be of minor importance. One such pathway involves condensation of acetyl-coenzyme A with malonyl-acyl carrier protein to form acetoacetyl-carrier protein that is structurally elaborated without loss of the protein moeity, such that there is no scrambling of inactive additives with radioactive material on the [^{14}C]-tagged synthetic route (Brodie *et al.*, 1964). Another proposed route is for a biotin-dependent carboxylation of β,β-dimethylacrylate, derived from leucine, to form HMG-coenzyme A; this is suggested by the findings that leucine greatly stimulated carotenogenesis in moulds and was sometimes incorporated in high yield (Goodwin, 1965a).

MVA is converted into C_5-compounds by the route shown in Diagram 4. The first step is catalysed by MVA-kinase and all four nucleoside triphosphates can act as phosphoryl donors (Waller, 1969). Magnesium ion is required for the activation of this enzyme in yeast and bacterial systems but manganese ion is more effective for plant and mammalian systems. Phospho-

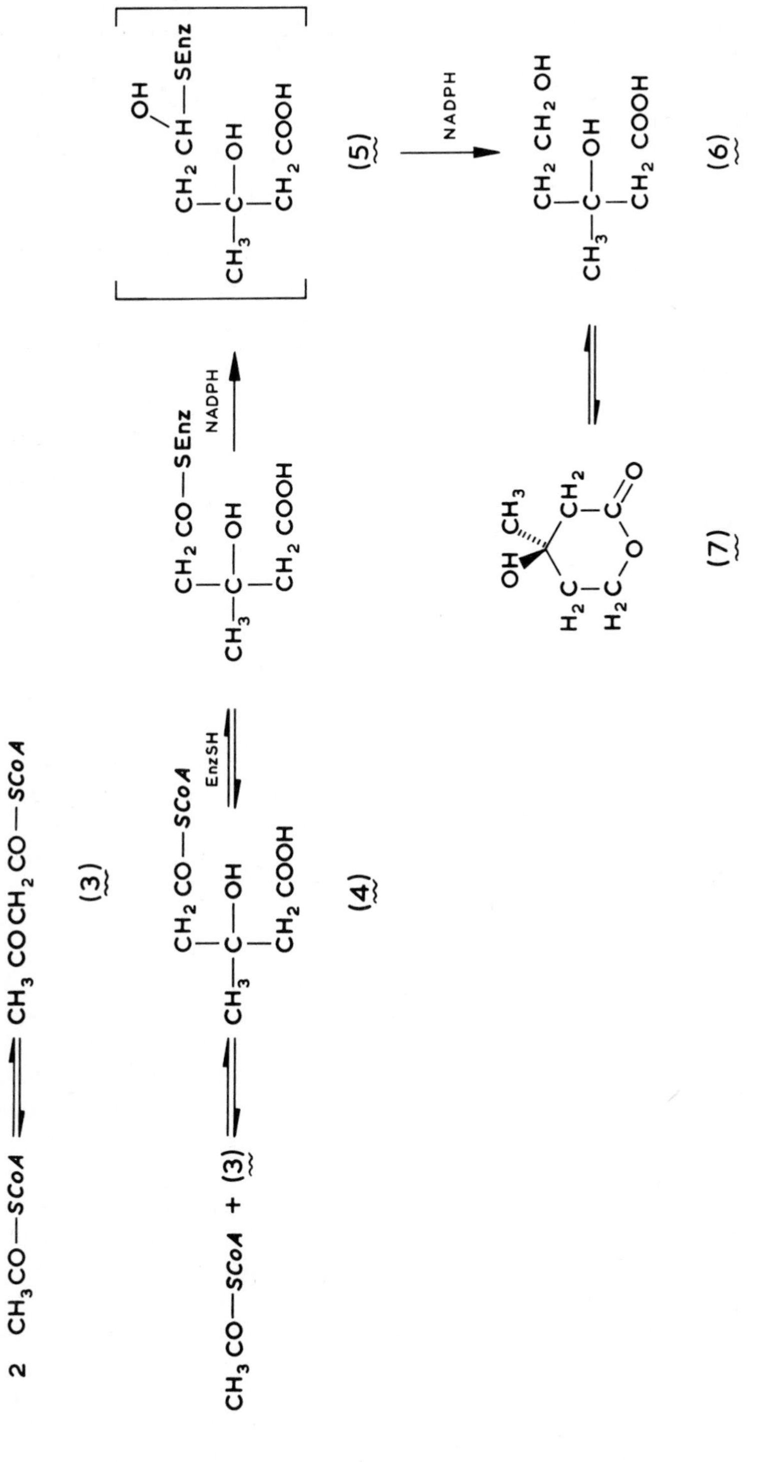

DIAGRAM 3.

CoASH = Coenzyme A. NADPH = Reduced form of the Coenzyme NADP. Enz–SH = carrier enzyme with thiol group at active site.

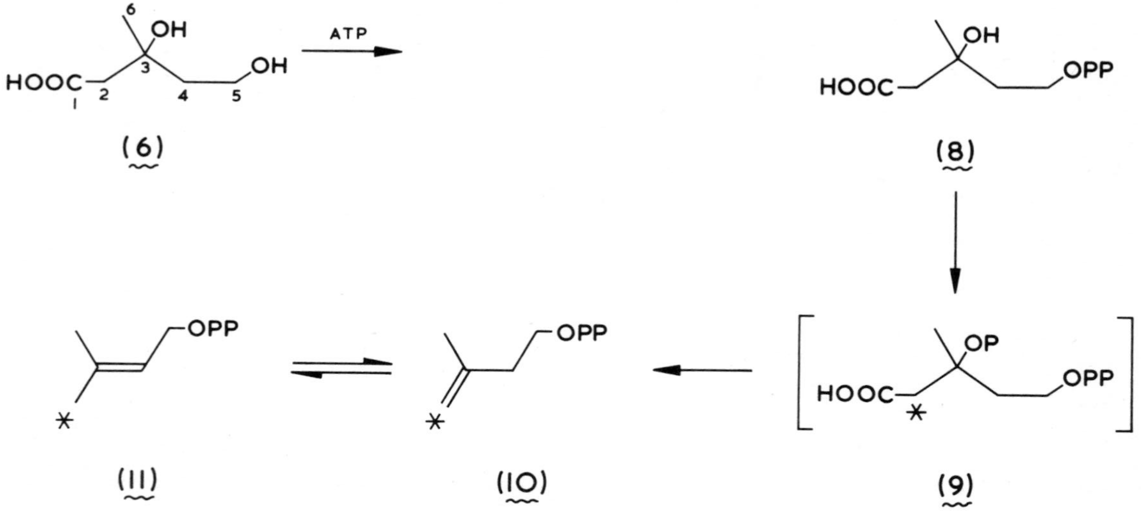

DIAGRAM 4.

P and PP represent phosphate and pyrophosphate groups respectively.

MVA kinase which catalyses the formation of (8) also requires the same co-factors. The intermediate (9) has not been isolated but is inferred from the stoichiometric and synchronous formation of ATP together with the elimination product Δ^3-isopentenyl pyrophosphate (IPP, 10), and the carboxyl group has been shown to be lost at this step by studies with [1-^{14}C]-MVA. An isomerase that is easily inhibited by iodoacetamide and so probably possesses a thiol group at the active site converts (10) into the more stable 3,3-dimethylallyl pyrophosphate (DMAPP, 11) in a reaction that probably involves addition of the thiol group across the double bond of (10) followed by elimination. A radioactive compound that is acid-stable and bonded to protein has been isolated at this stage and was probably the adduct of the enzyme with (10). The isomerization is such that the marked carbon of (10) takes the *trans*-orientation with respect to the ester group as shown in (11), and this non-equivalence of the *gem*-dimethyls of (11) has been proved by detailed studies on higher terpenes (Sections V B, VI B) and by an ingenious degradation of the terpenoid part of the fungal product mycelianamide (12, Diagram 5). In the crucial step of the degradation of the last product, part of the molecule that had been labelled with [2-^{14}C]-MVA was fed to a rabbit and the stereospecifically oxidized product in which all the tracer resided in the carboxyl group was isolated from the animal's urine (Birch *et al.*, 1958, 1962).

The interconversion of IPP and DMAPP is followed by a condensation by a formal combined SN2-E2 process (cf. Section I E) to give geranyl pyrophosphate (GPP, 13) and neryl pyrophosphate (NPP, 14). Thus the chemical significance of the interconversion of the two C_5 units is that a molecule with a relatively unreactive ester group and a nucleophilic double bond is converted into a highly reactive allylic ester. IPP and DMAPP taken together may be regarded as "active isoprene", the biological equivalent of the isoprene unit.

IPP can further condense with GPP to give farnesyl pyrophosphate (FPP, 15) in the all *trans* form which in turn can either dimerize in a tail to tail fashion to give squalene (16) or add another C_5 unit to form geranylgeranyl pyrophosphate (GGPP, 17); the latter can dimerize to carotenoids. The whole family of terpenes can be obtained by structural elaboration of these fundamental, ubiquitous, but usually rapidly turned-over intermediates.

It is not clear whether individual transferases catalyse the additions of the C_5 units or if one enzyme functions in all the reactions. A partially purified enzyme from pumpkin catalysed the formation of GPP and FPP but not the subsequent step to GGPP, and similar results have been found for other systems (Ogura *et al.*, 1969). For most preparations the enzymic activities for the formation of GPP and FPP run parallel throughout the various stages of the purification. Enzyme preparations from certain micro-

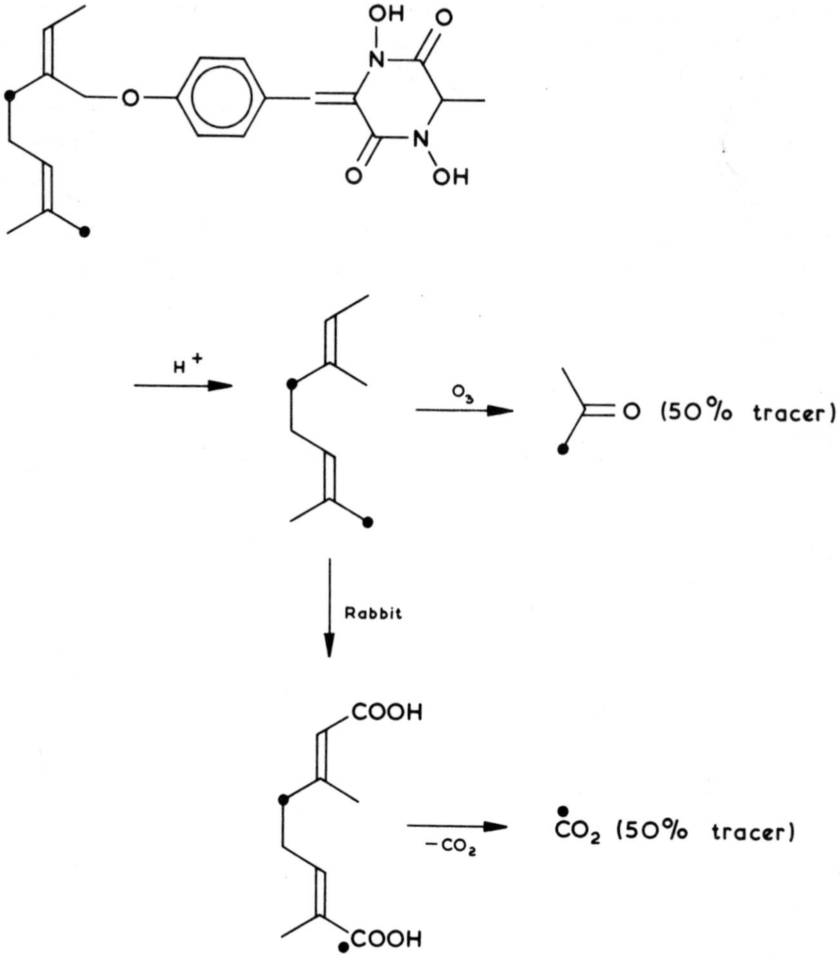

(12)

DIAGRAM 5.

Dots indicate the position of tracer here, and in subsequent diagrams.

organisms catalyse all three steps but a mixture of enzymes may have been present. Although the sequence of additions is sufficiently similar for one enzyme to be involved, there could be obvious biological advantages in using a system of enzymes that controlled the sizes of the pools of the different pyrophosphates.

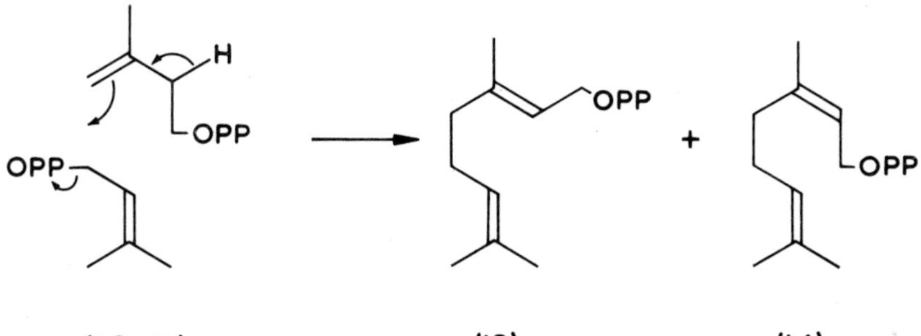

(10, 11) (13) (14)

DIAGRAM 6.

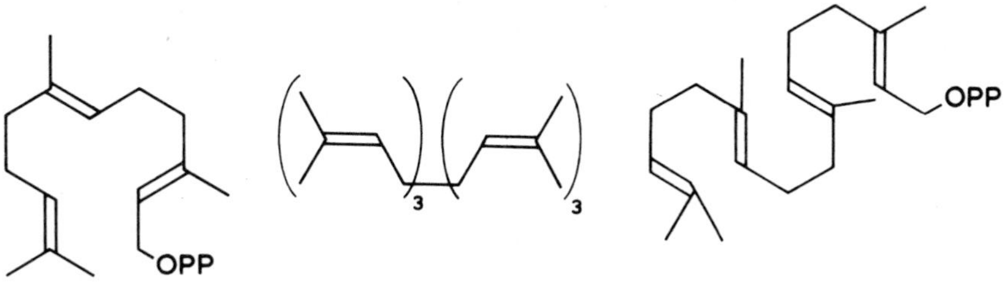

(15) (16) (17)

DIAGRAM 7.

E. STEREOCHEMICAL DETAILS

Some recent approaches have elucidated the stereochemistry of the condensations and dimerization in a most elegant manner using precursors doubly-labelled with ^{14}C and ^{3}H, or less usually, with ^{14}C and ^{2}H (Cornforth, 1968, 1969). Incubation of either [4R-4-$^{3}H_{1}$]-MVA (18) or its [4S] isomer (19) together with [2-^{14}C] MVA in liver homogenates and comparison of the ratio of tracer, $^{3}H:^{14}C$, in the starting materials and products enabled far-reaching deductions to be drawn. Squalene biosynthesized from the [4R] isomer had the same isotope ratio as the starting material, whereas that from the [4S] isomer contained no ^{3}H. Thus, the hydrogen originally at the [4S] position of MVA was stereospecifically lost in both the condensations of the C_5 units and those of the C_5 units with the C_{10} units. Moreover, the same hydrogen must have been lost in the formation of DMAPP from IPP (for DMAPP residues are retained at the terminals of the squalene chain.) If loss of hydrogen in these steps had been non-stereospecific the ratio $^{3}H:^{14}C$ in the products would have been half that in the starting materials. Stereospecific losses of [4S] hydrogen have also been

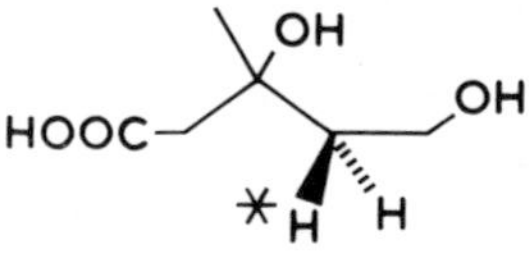
(18)

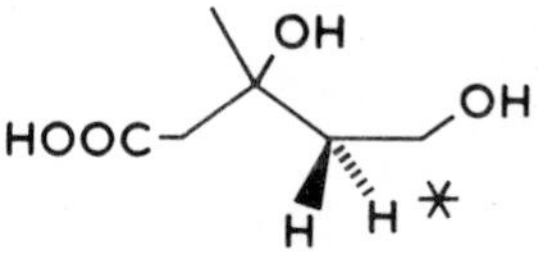
(19)

DIAGRAM 8.

demonstrated for the formation, *inter alia*, of geraniol and nerol (Le Patourel, 1970) and of FPP, squalene, β-carotene and the side chains of ubiquinones in plants (Goodwin and Williams, 1965a,b, 1966). All these compounds except nerol (see Section II B) have *trans*-olefin linkages, but the opposite stereospecificity ([4R] lost, [4S] retained) occurred for rubber which has *cis*-olefin linkages (see Section VIII A).

The condensation of two molecules of FPP to form squalene is formally the construction of a whole from two halves, but the actual situation is more complex since squalene biosynthesized from [1-$^2H_2$]-FPP by a liver preparation contained only three tracer atoms per molecule. This asymmetric labelling proved that the redistribution of hydrogen occurred at a stage subsequent to the formation of FPP rather than in any kind of exchange process of this molecule. More detailed investigations showed the intruding protium to have originated from the [β-4S-] position of NADPH and not from water.

The stereochemistry and tracer-content of succinic and laevulinic acids resulting from ozonolysis of squalene biosynthesized from [5R-5-$^2H_1$] MVA were worked out by standard chemical methods and were found to be consistent with the pattern in Diagram 9 (Cornforth, 1969). Here the precursor was so chosen that the tracer was not lost at the centre of the molecule and it was concluded from the absolute configuration of the succinic acid that the formation of the central bond A resulted in retention of configuration at the terminus of one moeity and inversion at the other. Moreover, the chain extension to form FPP was deduced to have involved inversion at C_1 of both DMAPP and GPP with a stereochemistry akin to that of the classical S_N2 mechanism. This latter result, taken in association with the known geometry of proton-loss in the process of chain extension, required the incoming alkyl group to be on the same side of the IPP molecule that was extending the chain as was the departing proton. This is equivalent to an S_N2 reaction at C_4 of IPP with retention of configuration and was considered unlikely. Consequently a two-step process has been proposed (Cornforth, 1968) whereby an *anti*-1,2-addition is followed by an *anti*-1,2-elimination in which an intervening electron-donating group X adds to C_3 of IPP to permit acceptable stereo-electronic transactions to be carried out at this

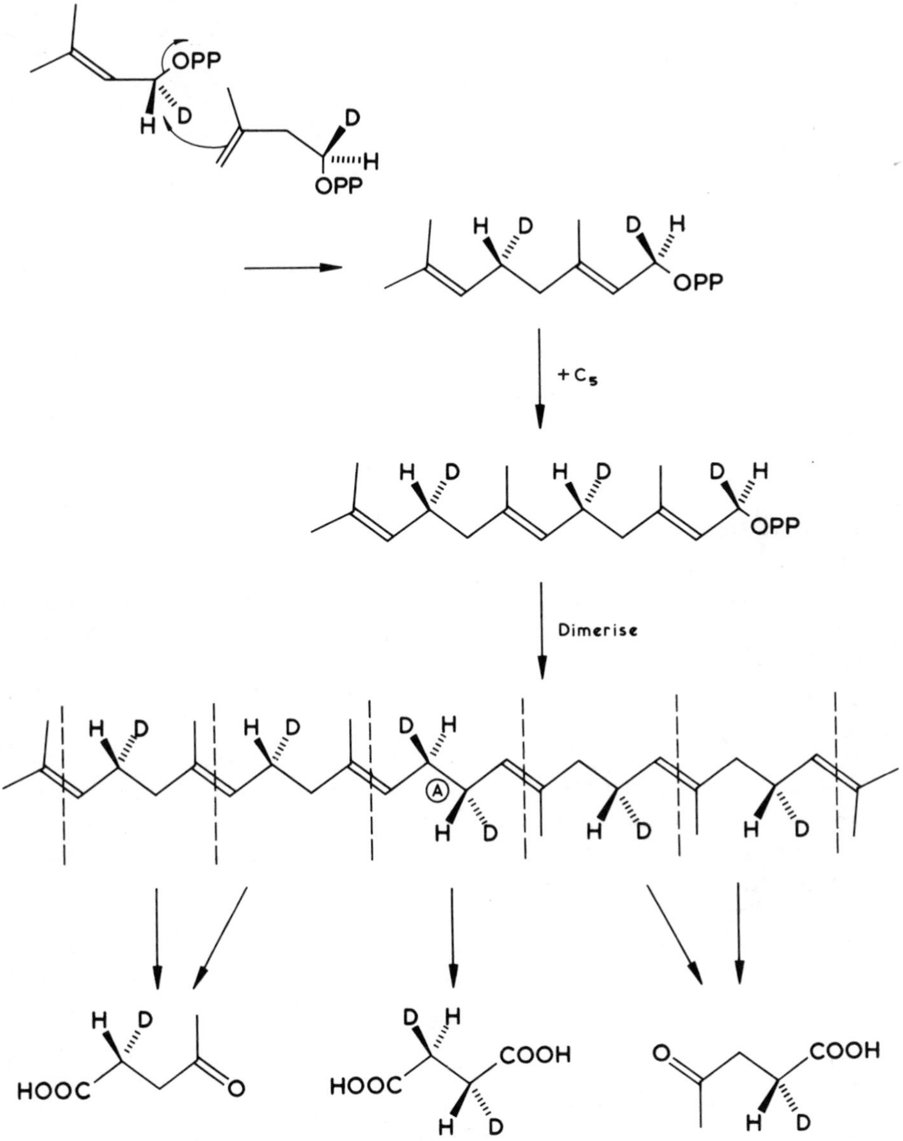

DIAGRAM 9.
D represents Deuterium (^2H).

atom (Diagram 10). As yet, X is unspecified: it may be a group such as a water molecule bonded to an enzyme, or perhaps an oxygen atom of the pyrophosphate group of the IPP moeity.

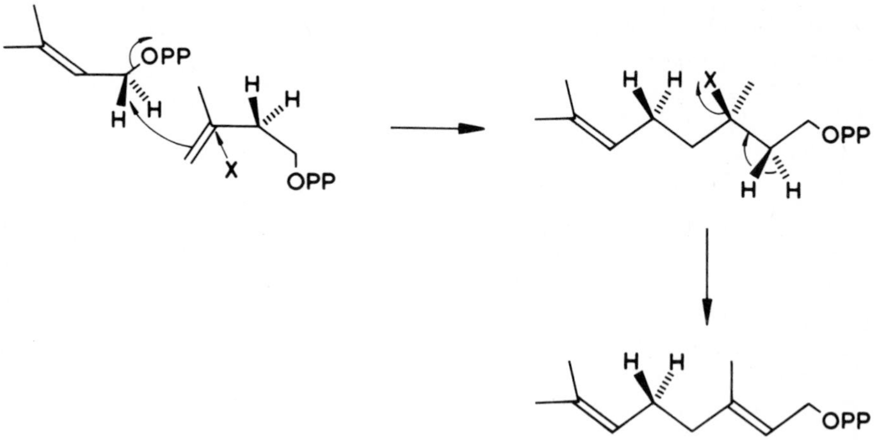

DIAGRAM 10.

II. Monoterpenes

Few tracer studies have been made for this family but attractive hypothetical schemes of biogenesis have been proposed (Ruzicka *et al.*, 1953; Loomis, 1967).

A. HYPOTHESES

Acyclic monoterpenes undoubtedly arise from chemical modification of GPP or NPP whereas cyclic members of the class are generally considered to be derived from NPP or, less probably, from the isomeric linaloyl ester (20); it is not stereochemically possible for GPP to cyclize directly. It is probable that NPP initially forms a species (21), that can lead to α-terpineol or can undergo hydride shift to form (22), the progenitor of terpinen-4-ol.

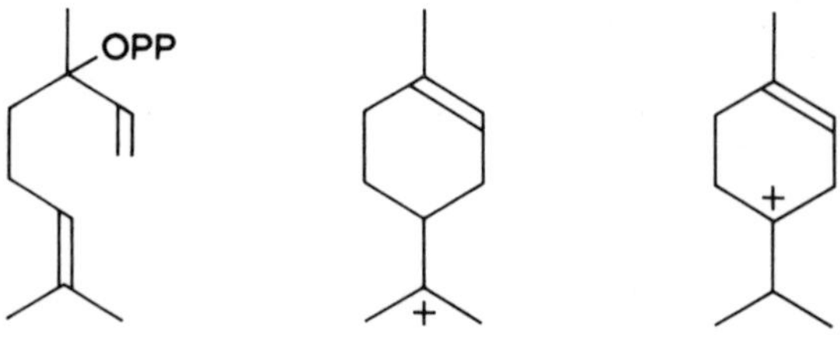

(20) (21) (22)

DIAGRAM 11.

Dehydration of these alcohols or elimination of a proton from the corresponding carbonium ions leads to menthadienes. In this and the following discussions classical carbonium ions will be used to represent biochemically and structurally equivalent species. Thus (21) may represent a phosphate ester, an alcohol or an ion, free or bonded to protein: the actual details of structure are not known for any case.

Bicyclic skeletons can be hypothetically derived by cyclization, controlled by either electronic (Markownikoff) or steric factors, within monocyclic frameworks. Pinane and bornane skeletons (23 and 24) and, by a more unusual cyclization, carene skeletons (25), can be thus formed from (21); and (22) leads to the thujane skeleton (26). Isobornanes, e.g. fenchanes, isocamphanes, can be derived from (23) and (24) by Wagner-Meerwein, and Nametkin shifts for which there are numerous analogues *in vitro*. These basic skeletons, and those of the acyclics and monocyclics, may subsequently be modified in numerous ways by the introduction of functional groups, e.g.

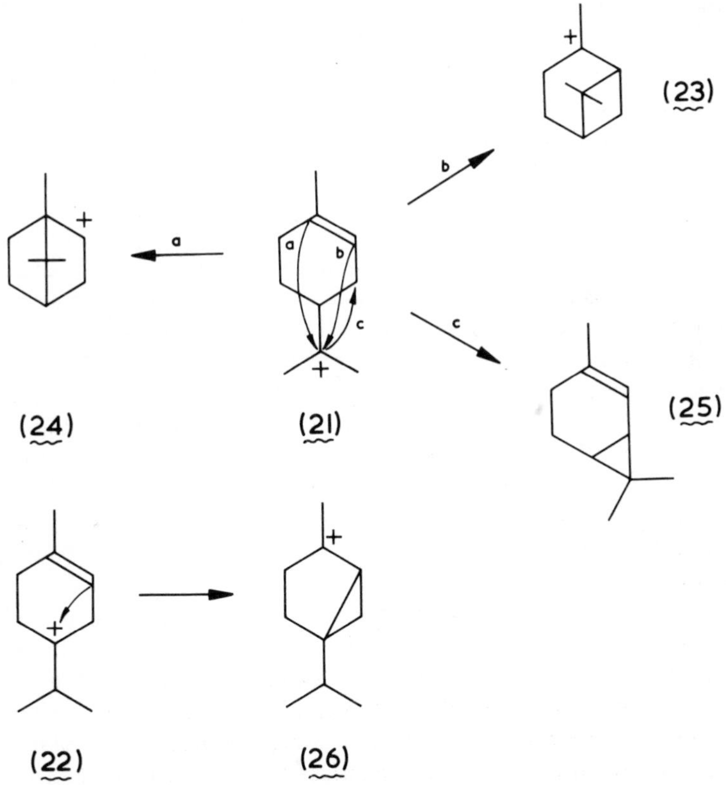

DIAGRAM 12.

γ-oxidation of α-pinene (27) could lead to a series of myrtenyl and verbenyl derivatives (Diagram 13) that occur together in certain plants. A good example

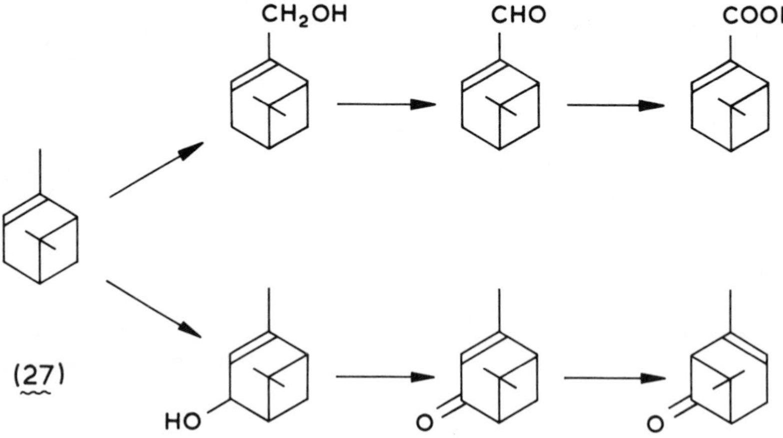

DIAGRAM 13.

of secondary transformations is the scheme in Diagram 14 that was deduced from phytochemical and tracer studies on *Mentha piperita* (Reitsema, 1958; Hefendehl, 1967; Burbott and Loomis, 1967): the detailed optical and stereo-isomerism is ignored in this diagram.

Other, less well-defined biogenetic schemes have been proposed. Radical-induced cyclizations of acyclics to bicyclics (Diagram 15) were suggested (Ruzicka *et al.*, 1953) and similar direct cyclizations were supported by thermodynamic arguments (Gascoigne, 1958), but although photochemically-promoted bicyclizations have been demonstrated *in vitro* (Crowley, 1965; Cookson, 1968), these views have received little support. In order to account for the stereospecificity of the cyclizations any radical must be protein-bound. A detailed investigation of the time-dependence of incorporation of tracer into the monoterpenes of *Tanacetum vulgare* has ruled out direct bicyclization of C_5 units or of acyclics to thujane derivatives and was consistent with the ionic route *via* NPP and monocyclic structures (Banthorpe and Wirz-Justice, 1969).

Details of the cyclizations of acyclics to monocyclics are also obscure. Epoxides, which extensively occur in nature (Cross, 1960), or sulphonium salts may be involved. Both classes are implicated in the construction of rings in higher terpenes (Sections V, VI) and interesting model systems for the synthesis of monoterpenes *in vitro* using sulphur ylides have been described (Blackburn *et al.*, 1969). Investigation of the importance of such routes *in vivo* must await the advent of suitable cell-free systems.

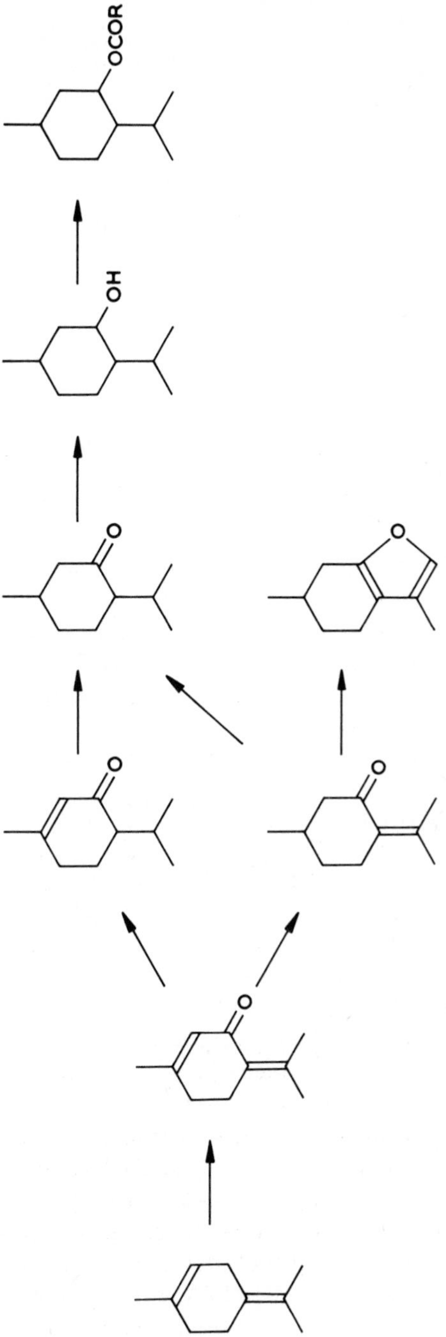

DIAGRAM 14.

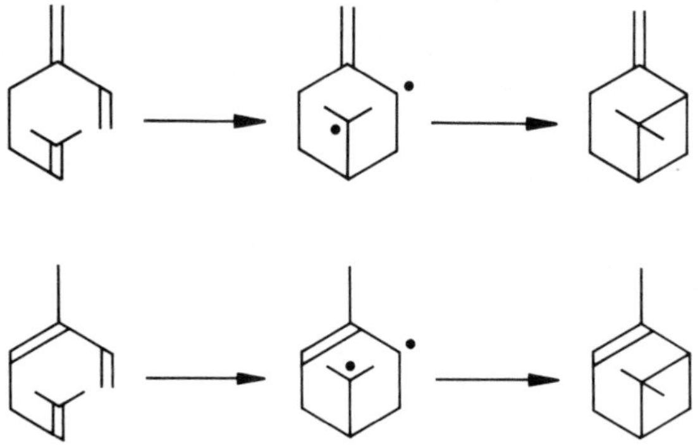

DIAGRAM 15.

Some monoterpenes, e.g. linalool and camphor, occur in plant oils as either practically pure (+) or (−)-isomers or as mixtures, occasionally racemic (Plouvier, 1966), and enzymes producing each optical isomer can presumably coexist. Although it has not been generally demonstrated, the construction of the skeletons of mono- and higher terpenes is almost certainly under enzymic control, but some of the secondary reactions may be acid or photochemically-induced processes, or even artefacts of extraction procedures.

B. Tracer Studies

Monoterpenes of *Mentha piperita* were rapidly labelled after exposure to [14C]-carbon dioxide but not after feeding [2-^{14}C]-MVA, although the latter was an efficient precursor of carotenoids (Battaile and Loomis, 1961). This and the observation of rapid evolution of labelled carbon dioxide after feeding [2-^{14}C]-MVA in other species, was considered to signify that MVA was degraded *in situ* to carbon dioxide which was fixed into monoterpenes (Battu and Youngken, 1966). However, more refined studies on *M. piperita* have shown MVA to be significantly incorporated into the monoterpenes (Hefendehl, 1966), and measurements of the course of incorporation of acetate, carbon dioxide and MVA into *M. piperita* and *Tanacetum vulgare* both in the dark and light ruled out degradation of mevalonate to C_1 or C_2 units and subsequent incorporation of these (Hefendehl *et al.*, 1967; Banthorpe and Wirz-Justice, 1969). A particularly direct demonstration of the unimportance of carbon dioxide derived from MVA as a contributor to the synthesis of the monoterpene pool of *Artemisia annua* was the negligible incorporation of tracer from [1-^{14}C]-MVA, although extensive degradation to radioactive carbon dioxide (that was liberated) did occur; very little degradation of [2-^{14}C]-MVA to carbon dioxide occurred under the conditions of these experiments (Charlwood 1970).

Uptake of [2-^{14}C]-MVA into the monoterpenes of several other species of plants (Loomis, 1967) and a few of insects (Meinwald *et al.*, 1966) has been reported. Although the incorporations were invariably very poor (ca. 0.01 to 0.1 %) this substrate is generally regarded as an obligate precursor. In contrast is the 10.8 % uptake of (racemic) [2-^{14}C]-MVA into monoterpene glucosides in the petals of a rose species (Francis and O'Connell, 1969). Variations in efficiencies of incorporation must depend on the species, the type of tissue used and environmental conditions, and apparent inconsistencies that are probably attributable to such factors have been discussed (Banthorpe and Wirz-Justice, 1969).

Numerous reports of incorporation of tracer from other labelled compounds are available (Nicholas, 1967a; Loomis, 1967; Waller, 1969). Often tracer from glucose or glycine is more rapidly transferred to the terpene pool than that from either acetate or MVA; this may be due to more efficient translation to, and penetration of, the appropriate metabolic sites by the former followed by degradation and incorporation of the fragments.

More subtle information about biosynthetic routes comes from the use of specifically-labelled precursors and degradation to find the site of tracer in products. A fundamental point is the inter-relationship of GPP and NPP, the parents of the higher terpenes and cyclic monoterpenes respectively. NPP could be formed either by direct coupling of IPP and DMAPP or by isomerisation of preformed GPP. The former route is attractive in requiring a specific route in plants that is generally deficient in animals (which do not generally synthesize cyclic monoterpenes), but experiments using [4R-4-$^{3}H_{1}$]-MVA and its [4S] isomer have ruled this out as a general occurrence (Le Patourel, 1970). Incorporation of each isomer in turn in admixture with [2-^{14}C]-MVA into various species of plants and isolation of geraniol, nerol (both free and as their glucosides) and α-pinene, indicated that the [4S] hydrogen was stereospecifically lost in all cases whereas the direct *cis*-condensation of C_5 units would have resulted in loss of the [4R] hydrogen. The formation of monocyclic terpenes from the solvolysis of various phosphate esters of nerol differed from the open chain products solely formed from similar geraniol derivatives (Cramer and Rittersdorf, 1967; Haley *et al.*, 1969) and this led to the suggestion that conversion of GPP into NPP or linaloyl pyrophosphate followed by cyclization was unlikely *in vivo*; but such extrapolations from *in vitro* systems are obviously inconsequential.

Incorporation of [2-^{14}C]-MVA into thujane derivatives produced by several plant species gave the unexpected result that tracer was located essentially in that part of the skeleton derived from IPP, e.g. for thujone (28, Diagram 16) the tracer was almost entirely at the carbonyl carbon atom as predicted by Ruzicka's hypothesis, but only a few percent was in the "bottom" half of the molecule (derived from DMAPP) and this small amount could be attributable to degradation of MVA to acetate and incorporation of this (Banthorpe and Turnbull, 1966; Mann, 1970). This labelling pattern

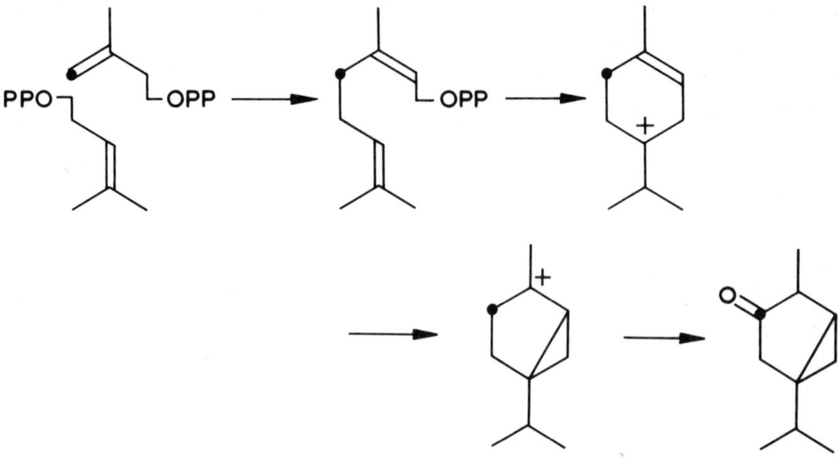

(28)

DIAGRAM 16.

differed entirely from that deduced from previous incomplete degradations of thujone (Sandermann and Schweers, 1962a) and from an analogous pattern that had been reported for α-pinene (Sandermann and Schweers, 1962b). Asymmetric labelling into that part of the molecule derived from IPP also occurred for both optical isomers of camphor (Banthorpe and Baxendale, 1968), for pulegone and artemisia ketone (Charlwood, 1970) and for the sesquiterpenes tutin and coriamyrtin (Biollaz and Arigoni, 1969); the phenomenon appears general for monoterpenes in a wide range of feeding conditions. The unlabelled "bottom" halves of these molecules could be derived from DMAPP that is either present in a large metabolic pool or is of non-mevalonoid origin; this could condense with labelled IPP before the latter could isomerize. Alternatively, a compartmentation effect (Section X), could be invoked, in which radioactive IPP but not the derived DMAPP could penetrate to the biosynthetic site, to react with endogenous DMAPP. If the DMAPP is of non-mevalonoid origin, an attractive possibility is that it is derived from degradation of the monoterpene pool. Interesting pathways can be devised, based on known pathways of microbial degradation of monoterpenes (Loomis, 1967), whereby this could happen, and tracer studies on the fate of monoterpenes injected into higher plants indicate that enzymic pathways are available for degradation to C_2 and C_5 units (Justice, 1967; Banthorpe and Wirz-Justice, 1969).

Tracer has been shown to be located in that part of the skeleton derived from DMAPP only for the terpenoid parts of pyrethrins (29), (Crowley *et al*., 1962) and neryl and geranyl β-glucosides (Francis and Allcock, 1969; Le Patourel, 1970). All are conjugated monoterpenes, all are biosynthesized in petals (as opposed to leaves for all other non-conjugated monoterpenes that have been studied from this aspect), and all show very high (ca. 10%) incorporation of [2-^{14}C]-MVA with the tracer divided almost equally between the IPP and DMAPP moieties.

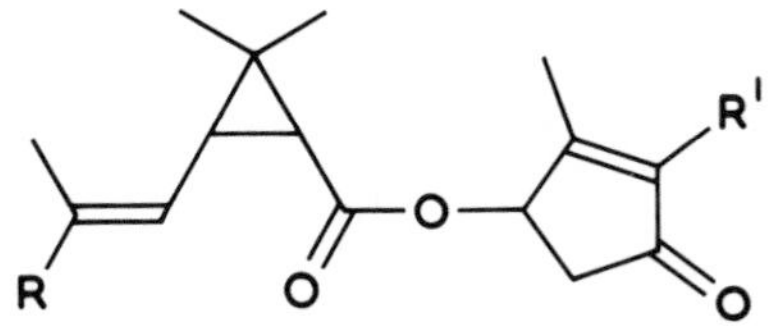

R CO_2Me
R' $CH_2CH{=}CHCH{=}CH_2$

(29)

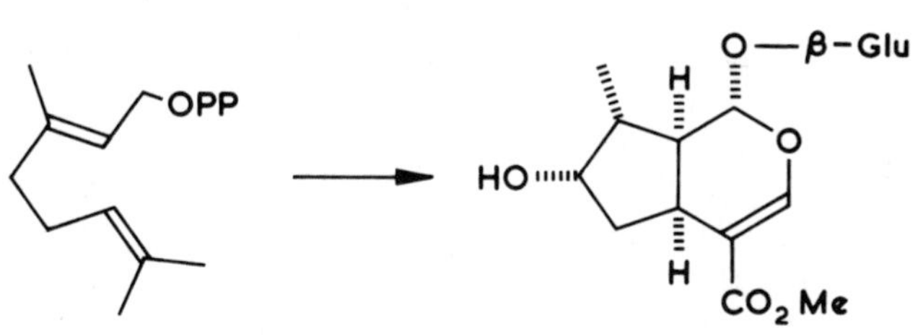

(31) (30)

Diagram 17.

Studies on unconjugated leaf monoterpenes, e.g. limonene (Sandermann and Bruns, 1962a), β-phellandrene (Bruns, 1964), carvone (Sandermann and Bruns, 1965), citronellal and cineole (Birch *et al*., 1959) have led to the assumption of equal labelling in these moieties on the basis of incomplete or ambiguous degradations but the results obtained are also consistent with a pattern of asymmetrical labelling. An unusual conclusion (Sandermann and Stockmann, 1958) is that two molecules of β,β-dimethylacrylic acid are incorporated intact into pulegone in *Mentha pulegium*; however, degradation was incomplete, the derived products were neither satisfactorily purified nor identified and the significance of the work is not clear. Under similar conditions a pattern of asymmetric labelling was obtained on feeding with [2-^{14}C]-MVA (Charlwood, 1970). In contrast to these results, fungal products, e.g. mycelianamide (12) and many higher terpenes from plants have been proved to possess a labelling pattern resulting from equilibration of tracer between the C_5 units.

It is attractive to speculate that the geranyl and neryl glucosides are either storage compounds or precursors for free terpenes, but their labelling pattern is different from that of most monoterpenes that have been examined, and their apparent nonoccurrence in most plant species, may indicate that they occupy a side path of terpene metabolism. Glucosides of cyclopentane terpenoids are known to be the precursors of the non-tryptamine part of certain indole alkaloids (Battersby and Gregory, 1968; Taylor and Battersby, 1969), and some such compounds, e.g., loganin (30) and related glucosides, have been shown both by direct incorporation of specifically labelled precursor (Brechbühler-Bader *et al.*, 1968) and by studies with [4R-4-$^{3}H_1$]-

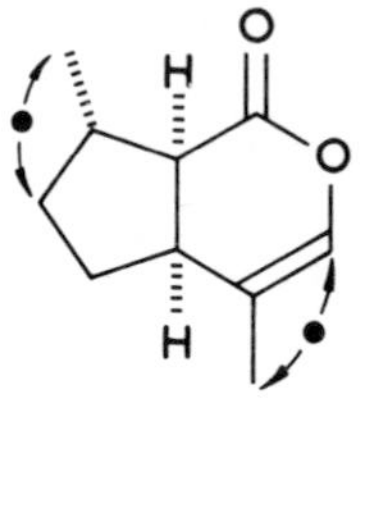

(32)

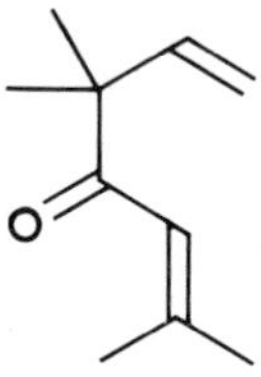

(33)

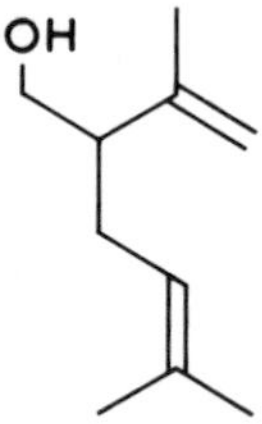

(34)

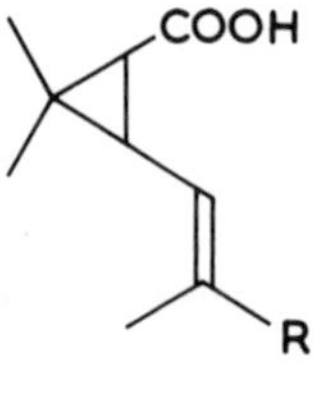

(35)

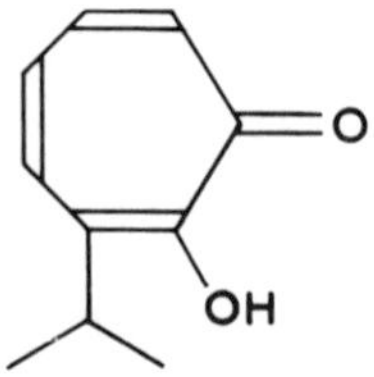

(36)

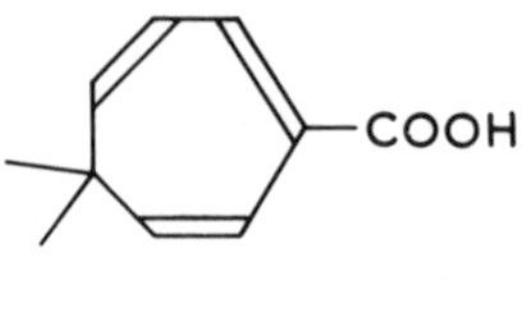

(37)

DIAGRAM 18.

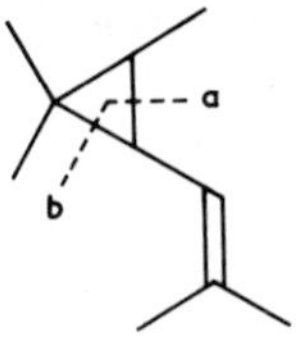

(38)

DIAGRAM 18 (contd.).

MVA (Guaranaccia *et al.*, 1969) to be derived directly from geraniol (31). [2-^{14}C]-MVA is also incorporated into nepetalactone (32), which has the terpenoid skeleton of loganin, in positions derived from both IPP and DMAPP although with considerable randomization between the methyl groups shown (Regnier *et al.*, 1968). A similar situation is found in the related glycosides plumieride (Yeowell and Schmid, 1964) and verbenalin (Horodysky *et al.*, 1969).

C. APPARENT EXCEPTIONS TO THE ISOPRENE RULE

Certain monoterpenes, e.g. artemisia ketone (33), lavandulol (34), chrysanthemic acid (35, R = Me), α-thujaplicin (36) and thujic acid (37) appear not to be formed by head to tail fusion of isoprene units but in view of the wide success of the biogenetic isoprene rule in accommodating the diverse skeletons of higher terpenes, it seems likely that all are derived from GPP. A route to (35) consistent with the labelling pattern (Crowley *et al.*, 1962; Godin *et al.*, 1963) is by fission of a carane ring (Diagram 19), and the skeletons of (33) and

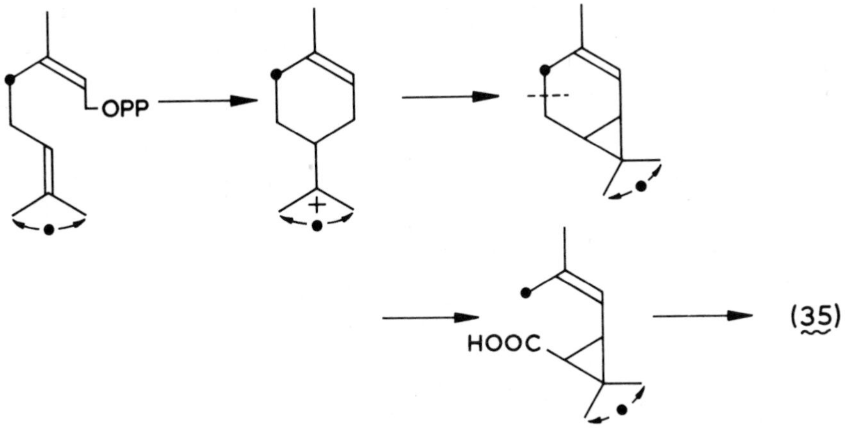

DIAGRAM 19.

(34) could arise from subsequent bond breakage at (a) and (b) in the skeleton (Bates and Feld, 1967; Crombie *et al.*, 1967) of (38). However, tracer from [2-^{14}C]-MVA is located almost exclusively in the *gem*-dimethyl groups in the top half of structure (33) and this, together with the position of label incorporated from [2-^{14}C]geraniol (Charlwood, 1970) rules out the carane as an intermediate and also excludes a mechanism involving coupling of equivalent C_5 units (Ruzicka *et al.*, 1953). An ion with the skeleton (38) was concluded to be a precursor of (33) but its mode of biosynthesis is quite obscure, cf. Diagram 20.

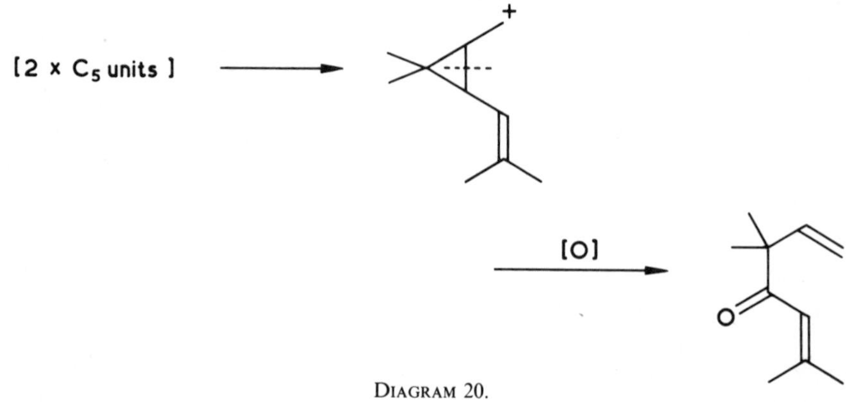

DIAGRAM 20.

No tracer studies have been made on the other compounds mentioned, but (36) and (37) and related structures may arise from a benzillic acid-like rearrangement leading to ring expansions of menthane derivative (Nicholas, 1967a), and a fission of a carane skeleton respectively. An alternative hypothetical route is via the polyacetate pathway wherein three molecules of malonyl-coenzyme A condense with DMA-coenzyme A to form a branched C_{11}-polyketide that can cyclize, decarboxylate, and be functionalized.

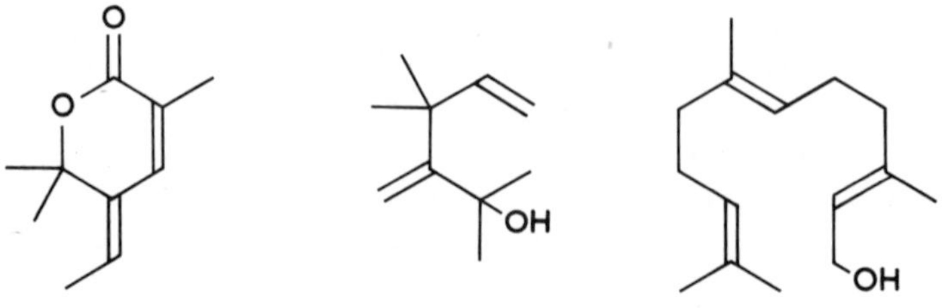

(39) (40) (41)

DIAGRAM 21.

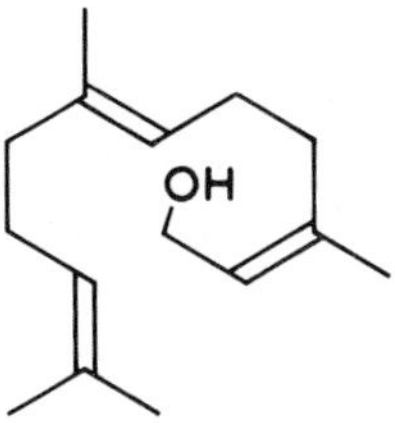

(42)

DIAGRAM 21 (contd.).

Recently the "non-isoprenoid" compounds (39) and (40) have been isolated (Bohlmann and Grenz, 1969; Hayashi *et al.*, 1968). The former represents a further variation in methods of linking C_5 units, but the latter was probably an isomer of artemisia alcohol rather than the structure proposed (Willhalm and Thomas, 1969).

III. Sesquiterpenes

Over forty different carbon skeletons are found in this class and there is much scope for biogenetic speculation, but the few tracer studies have confirmed the accepted theoretical views in a very satisfactory manner.

A. Hypotheses

All sesquiterpenes were originally (Ruzicka *et al.*, 1953) considered to be derived from cyclization and rearrangement of either *trans-trans* farnesol (41) or its *trans-cis* isomer (42). More recently the *cis-trans* isomer has been invoked to account for the stereochemical details in a few compounds, and the role of FPP rather than the free alcohol has been accepted. These ideas were mechanistically and stereochemically developed (Hendrickson 1959) whereby anchimerically-assisted ionization of pyrophosphate esters and 1,2 or 1,3-hydride shifts were formally considered to lead to carbonium ions. These ions cyclized by addition of the cationic centres to the double bonds in a manner governed either by the Markownikoff rule (electronic control) or the need to minimize strain (steric control). In general FPP (unlike triterpenes, Section VI) is not oxidized during cyclization and so the usual oxidation state of the products is that of the parent. Applications of these ideas to most classes of sesquiterpenes have been brilliantly summarized (Parker *et al.*, 1967).

Although it is convenient to visualize particular carbonium ions, these bond fissions and cyclizations are almost certainly concerted. Similarly, migrations of hydrogens or alkyl groups and β-eliminations proceed with the *anti*-stereospecificity expected for concerted sequences (Diagram 22).

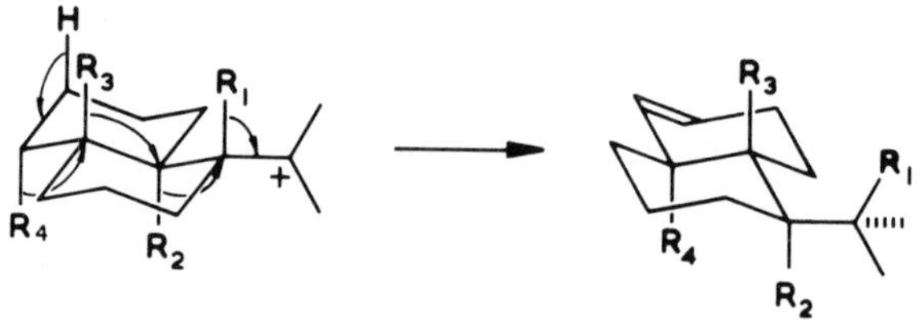

DIAGRAM 22.

Applications of these principles of mechanistic organic chemistry fully rationalize the structures found *in vivo*: presumably the acyclic substrates are specifically orientated on an enzyme surface to provide each particular stereoelectronically permitted concerted pathway. Attempts to cyclize the starting materials by treatment with acids *in vitro* usually lead to a spectrum of products most of which are not found in nature; such reactions must be multi-step with discrete carbonium ions as intermediates that, lacking the template of the enzyme surface, can equilibrate to a variety of conformations before reacting further. Similar stereoelectronic rationalizations can be applied to the other families of terpenes, and their application to triterpenes (Section VI) pre-dated that to sesquiterpenes. Monoterpenes are generally of insufficient complexity to analyse in such terms.

A few hypothetical schemes are given below. Most are due to Hendrickson (1959) and full details are available (Parker *et al*., 1967). The pyrophosphate of (41) can cyclize to either (43) or (44) (Diagram 23). Models suggest that the former is much less strained and further cyclization can lead to the eudesmane (45) and eremophilane skeletons (46) or to bulnesol (47), guaiol (48) and patchouli alcohol (49) (Diagram 24). The hydroxyl attached to the ring may

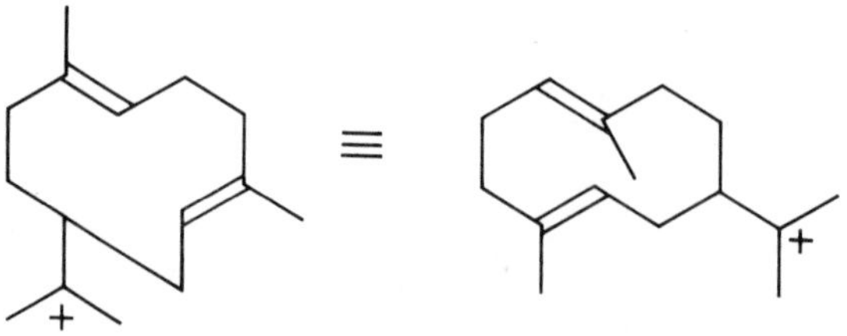

(43)

DIAGRAM 23.

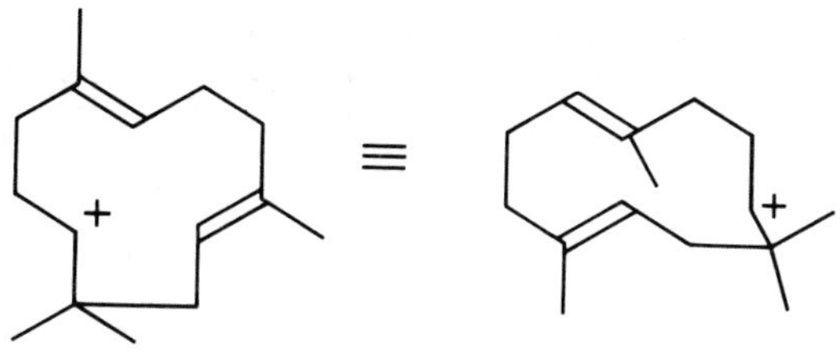

(44)

DIAGRAM 23 (contd.).

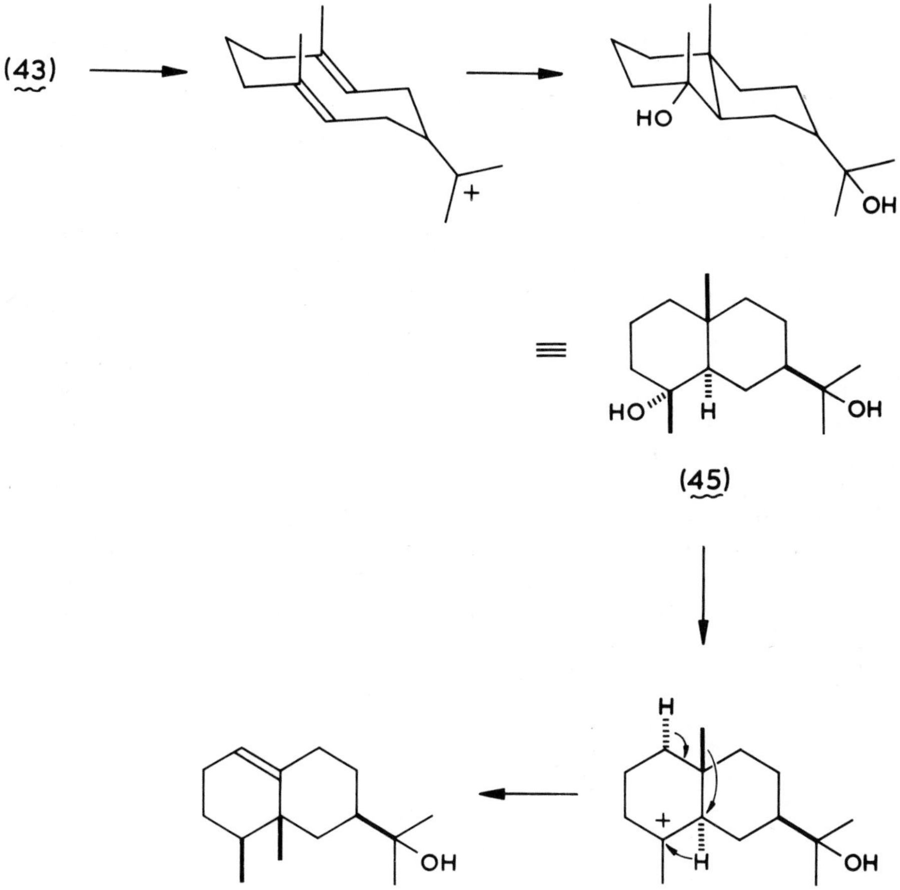

(46)

DIAGRAM 24.

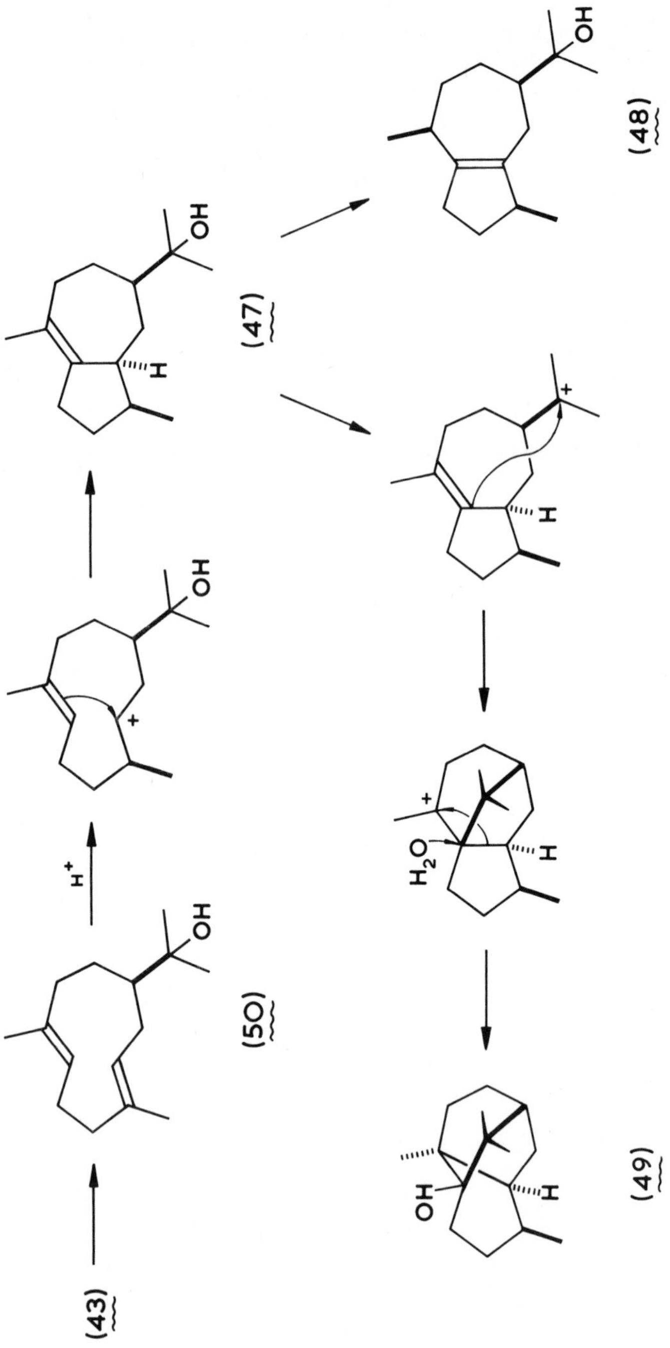

DIAGRAM 24 (contd.).

be formed from an epoxide group (see Section V) and certain sesquiterpenes of these types have been prepared *in vitro* by acid-catalysed cyclization of epoxides of cyclodecadienes (Brown and Sutherland, 1968). An alternative route to [5.3.0] bicyclo-decanes involves similar cyclizations of *cis-trans* FPP (Parker *et al.*, 1967). Cope rearrangement of (50) could lead to elemol (51) and its relative (Diagram 25), although such products may be artefacts of the isolation procedures.

(50)
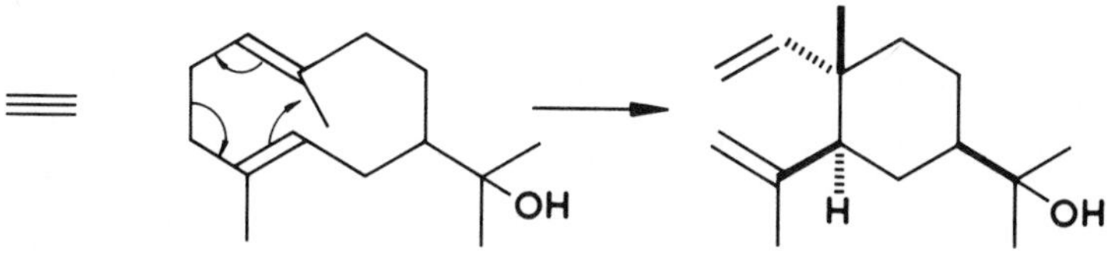

(51)

DIAGRAM 25.

Cyclization of the pyrophosphate of (42) is predicted to lead preferentially to the less strained product (52) rather than (53); and it was argued (Hendrickson 1959), that (52) would not further cyclize to give bicyclic skeletons of the sort just described as a hydrogen at C_1 (Diagram 26) is

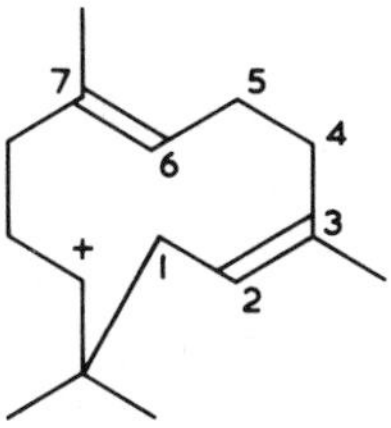

(52)

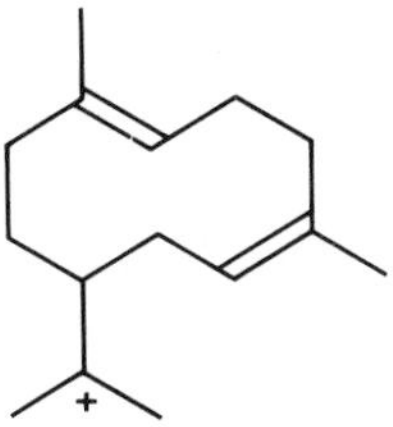

(53)

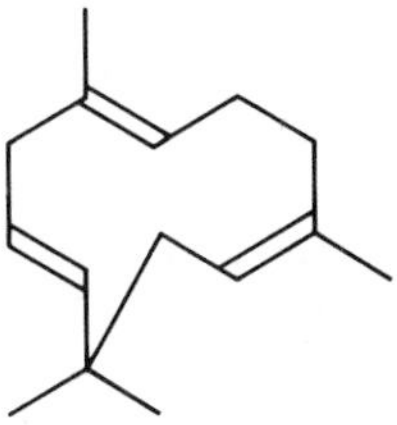

(54)

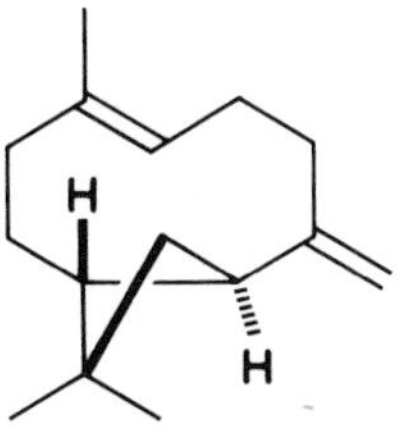

(55)

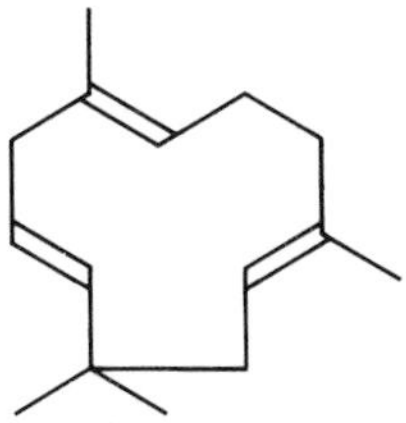

(56)

DIAGRAM 26.

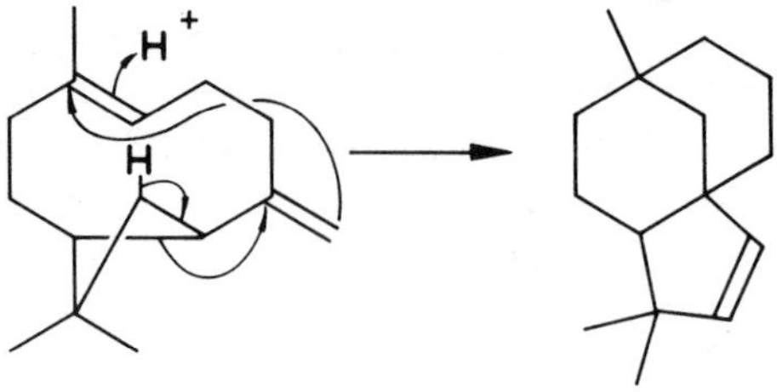

(57)

DIAGRAM 26 (contd.).

situated inside the ring and blocks attack of the cationic centre on the C_6 end of the C_6–C_7 double bond. Consequently, proton loss from (52) was supposed to lead to humulene (54), whilst attack of the cationic centre on the C_2–C_3 double bond would lead to the co-occurring caryophyllene (55). The latter in turn can undergo further cyclization to clovene (57). Caryophyllene undoubtedly has the structure shown, but this neat theoretical analysis was marred by crystallographic investigations (McPhail *et al.*, 1964) which showed that humulene and certain related compounds had the all-*trans* structure (56). Consequently, a common precursor of caryophyllene and humulene seems unlikely.

A naturally-occurring isomer of caryophyllene has a *cis*-double bond in the ring and so may be derived from *cis-cis* FPP. Another reasonable

(52)

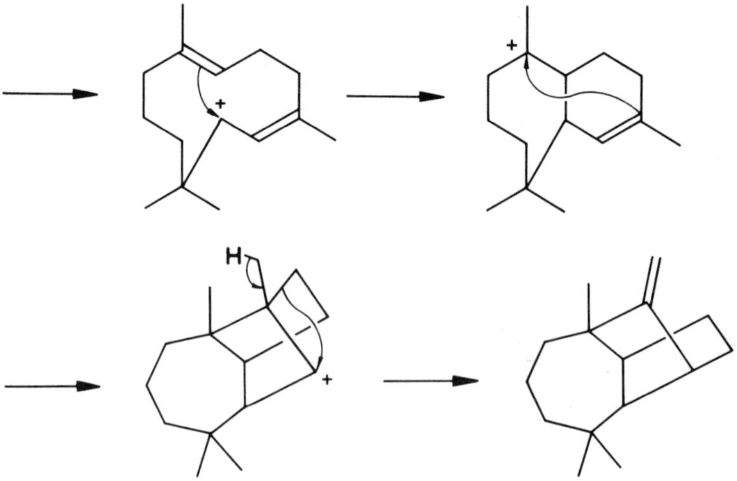

(58)

DIAGRAM 27.

(53)

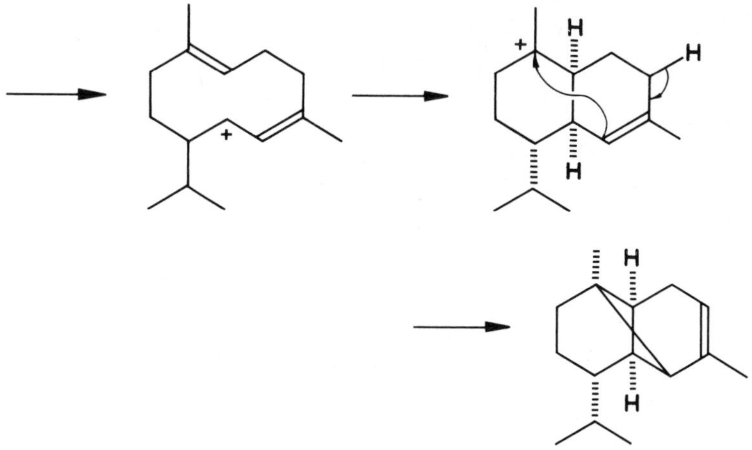

(59)

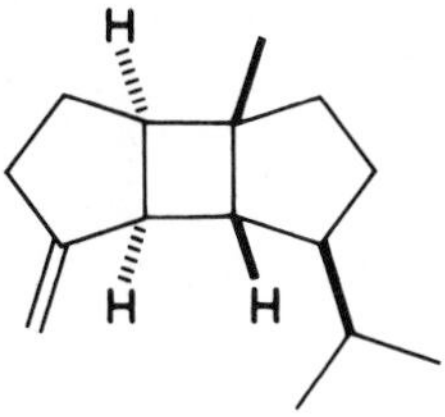

(60)

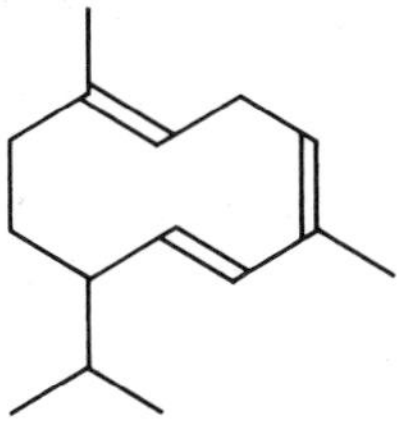

(61)

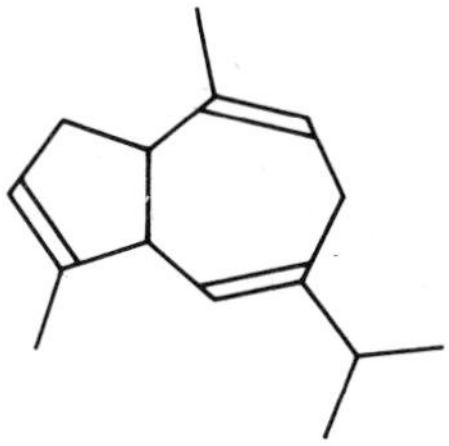

(62)

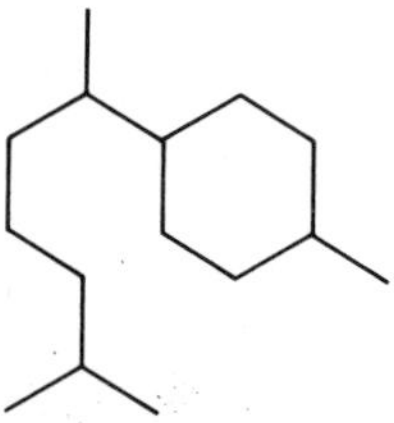

(63)

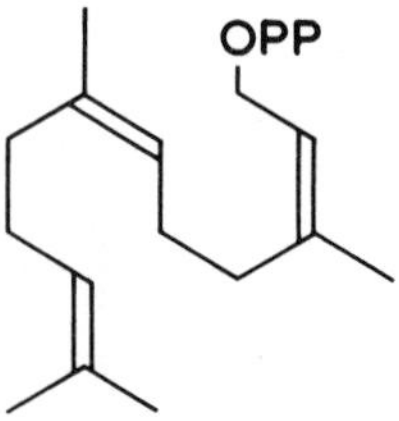

(64)

DIAGRAM 27 (contd.).

biogenetic route from (52) is to longifolene (58), Diagram 27, and modifications lead to longipinene and longicyclene in which the camphene moiety has undergone Wagner–Meerwein rearrangement.

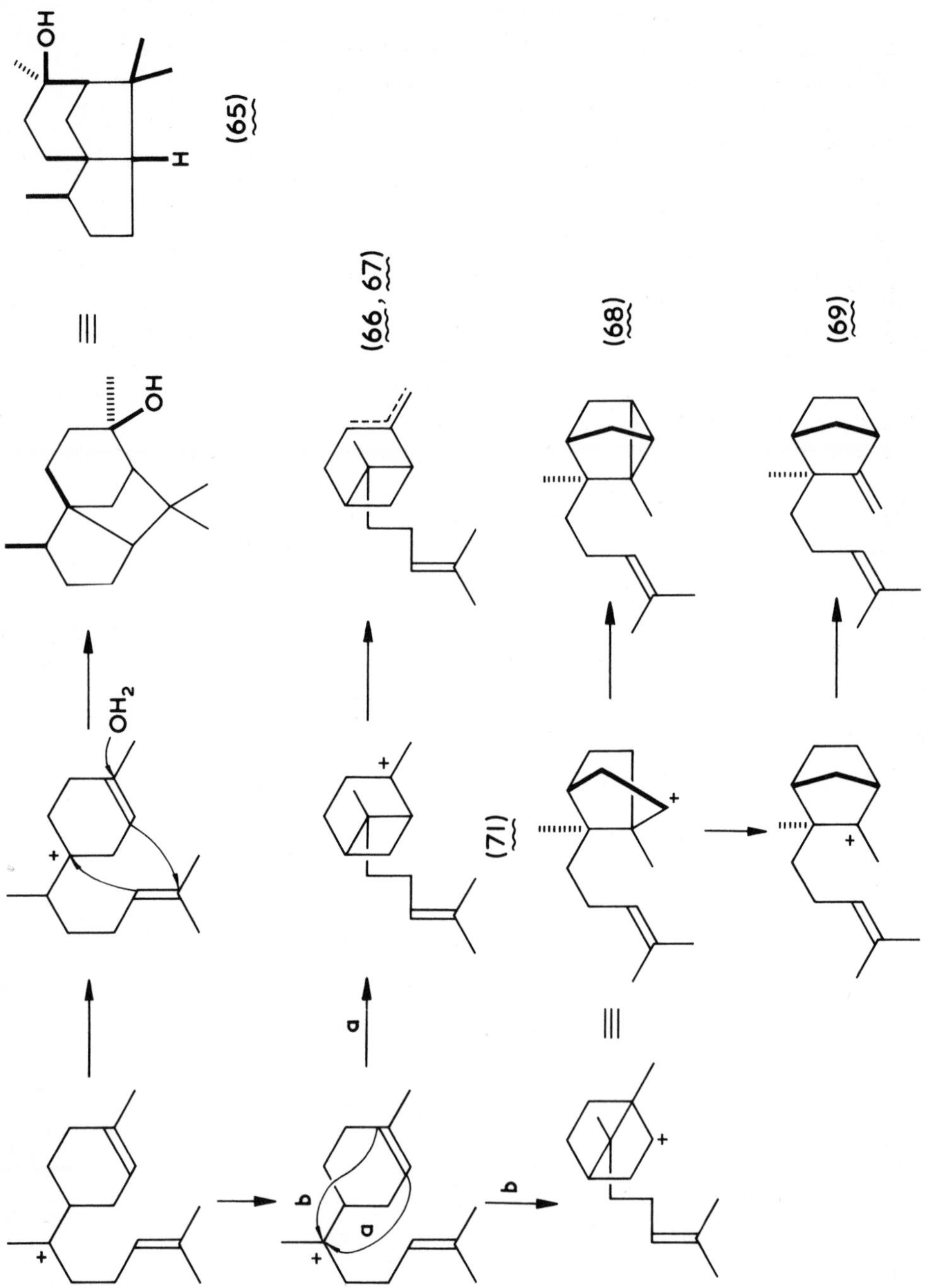

DIAGRAM 28.

Despite its strained structure, (53) has been proposed as the precursor for a number of skeletal types, although reasonable alternative mechanisms are possible. One proposal is the route to copaene (59) and an analogous pathway may lead to helminthosporal (see Section III B). Alternatively, (59) and *β*-bourbonene (60) may be derived from pigment-sensitized photocyclization of the trienes (61) or (62), (Brown 1968; Porter *et al.*, 1967).

Not all sesquiterpenes are constructed from the frameworks so far described. Bisabolenes with the skeleton (63) may result from cyclization of either (42) or (64). The former precursor seems more likely, as there is little evidence for the occurrence of the other, but the two possible routes should give characteristic labelling patterns in products biosynthesized from [2-^{14}C]-MVA (Ruzicka, 1963). The initial product of cyclization may either be elaborated into cedrol (65) (Diagram 28), or converted by electronically and sterically-favoured cyclizations into α- and *β*-bergamotenes (66, 67) and α- and *β*-santalenes (68, 69). The antibiotic fumagillin (70) may arise by oxidative fission of the four-membered ring of the intermediate (71), (Parker *et al.*, 1967). Compounds with the cadinane skeleton (72) are probably derived from (53) by appropriate hydride shifts and ring-closures.

(71)

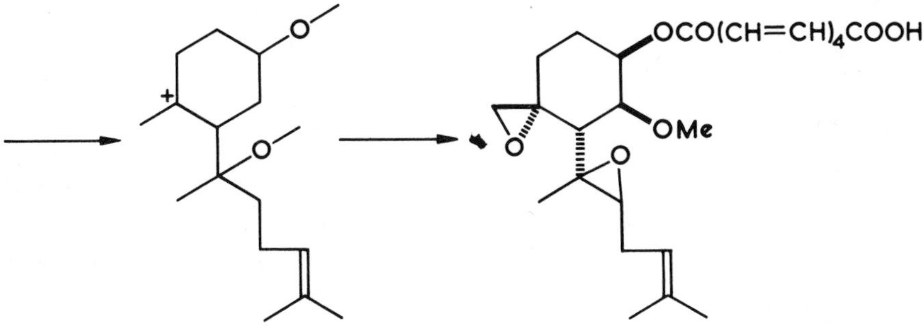

(70)

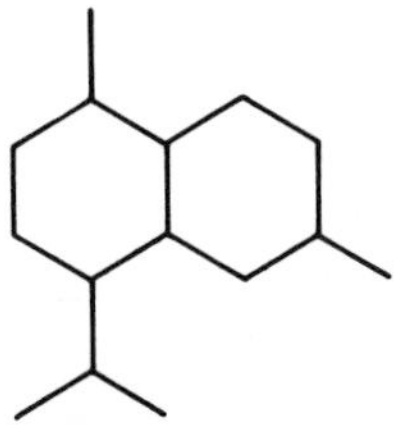

(72)

DIAGRAM 29.

The structures of several less common groups have been rationalized biogenetically. Iresin (73) and its derivatives are believed to be formed by oxidatively-induced cyclization, such as is common for di- and tri-terpenes, that leads to an antipodal ($5\beta:10\alpha$ rather than the more usual $5\alpha:10\beta$) configuration at the A:B ring junction (Djerassi *et al.*, 1959). Drimenol (74) has a similar skeleton (Appel *et al.*, 1959) but now cyclization is initiated by protonation and the A:B stereochemistry is normal. Iresin-like skeletons

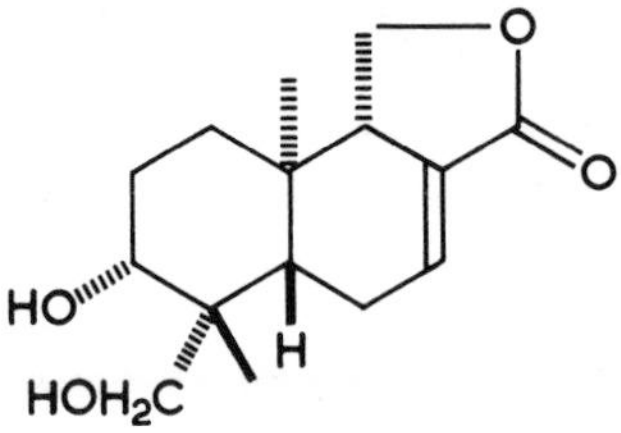

(73)

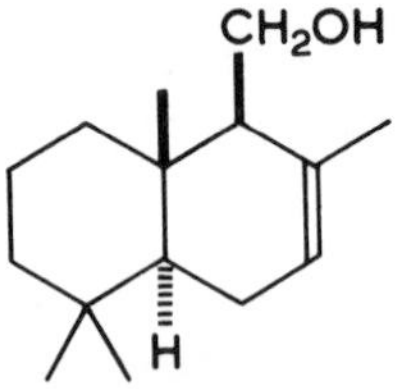

(74)

DIAGRAM 30.

can be synthesized by treatment of the terminal epoxides of *trans-trans* farnesyl acetate with boron trifuoride or mineral acids (van Tamelen *et al.*, 1963, 1966) but it is uncertain whether the natural products are true sesquiterpenes or are degraded di or triterpenes. The unusual compounds humbertiol (75) and calacone (76) can be derived by cyclization of (64) or by addition of a C_5 unit to a pre-formed monoterpene respectively (Parker *et al.*, 1967). The C_{14}-compound (77) is probably a degraded sesquiterpene (Maheshwari *et al.*, 1963) as compounds of the type (78) have been isolated from the same plants. Another interesting sesquiterpene is picrotoxin (79) which is believed to be synthesized by the route outlined (Parker *et al.*, 1967).

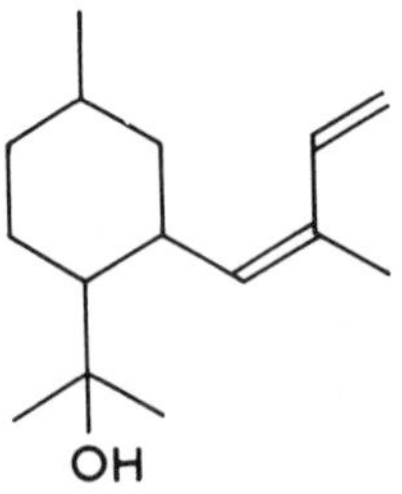

(75)

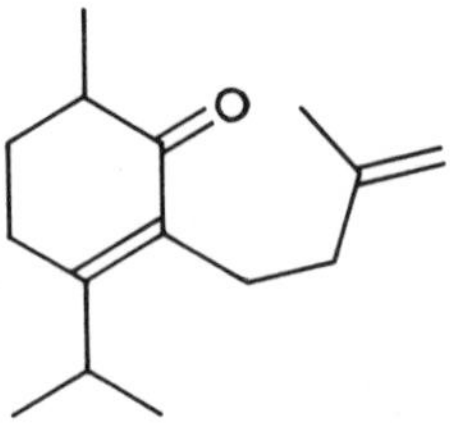

(76)

DIAGRAM 31.

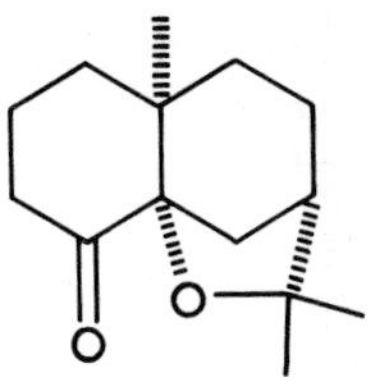

(77)

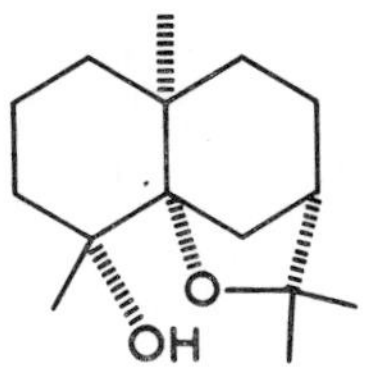

(78)

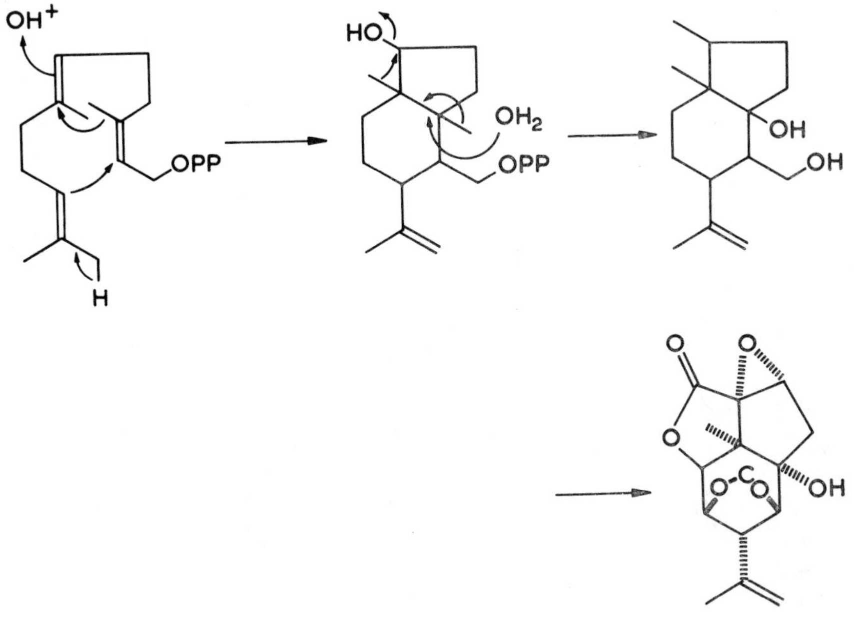

(79)

DIAGRAM 31 (contd.).

B. TRACER STUDIES

No more elegant unifying pattern for an apparently loosely related group of compounds has been made in the whole of chemistry than that in the hypotheses briefly outlined above, but few experimental studies have been made.

Degradation of trichothecin (80) biosynthesized from [2-^{14}C]-MVA indicated that bisabolene (81) derived from *trans-trans* FPP was precursor (Jones and Lowe, 1960; Godtfredsen and Vangedal, 1965) with stereospecific migration of one of the *gem*-dimethyl groups. The latter observation reinforces conclusions derived from certain di- and tri-terpenes that these methyls do not become stereochemically equivalent during the formation of FPP and subsequent biosynthetic sequences.

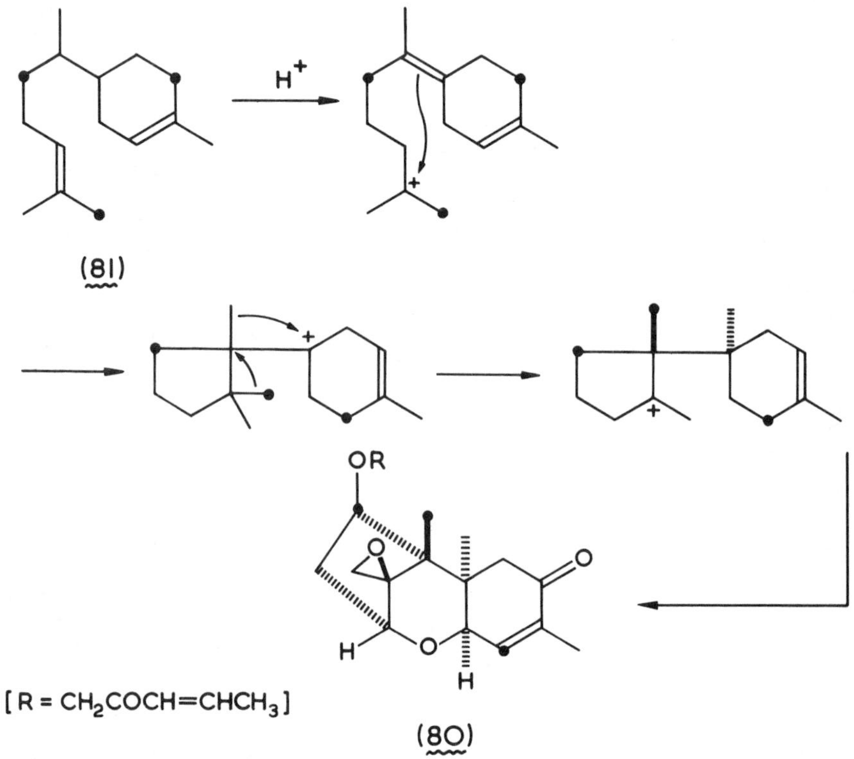

DIAGRAM 32.

In another study (Soucek, 1962) elucidation of the labelling pattern in carotol (82) biosynthesized from [2-^{14}C]-acetate allowed a decision between two possible routes of cyclization; the proven route (Diagram 33) involved cyclization of *trans-cis* FPP. Incomplete degradation of the fungal toxin helminthosporal (83), that is probably formed by breakdown of an acetal during extraction (de Mayo *et al.*, 1965), was in accord with the scheme in Diagram 34 (de Mayo *et al.*, 1962). The unsaturated aldehyde group contained 38 % of tracer from [2-^{14}C]-MVA and the intermediate sativene (84) was also isolated (de Mayo and Williams, 1965).

A detailed degradation scheme was consistent with a route from humulene (or its structural equivalent) to the mould product marasmic acid (85) (Dugan *et al.*, 1966); the labelling pattern for product biosynthesized from

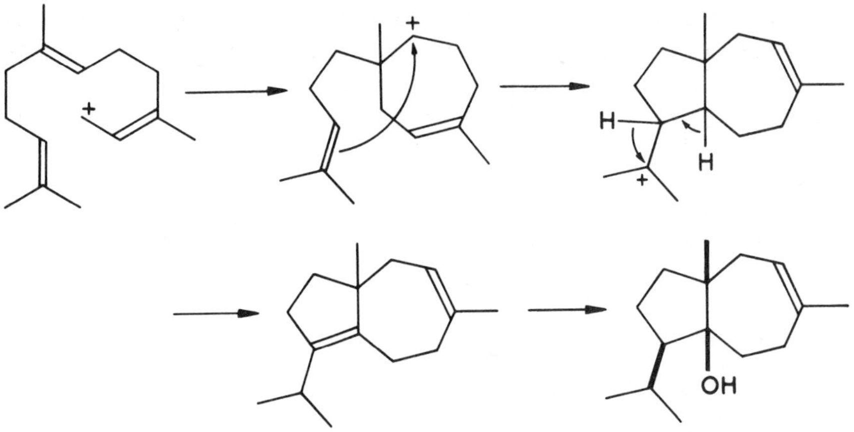

(82)

DIAGRAM 33.

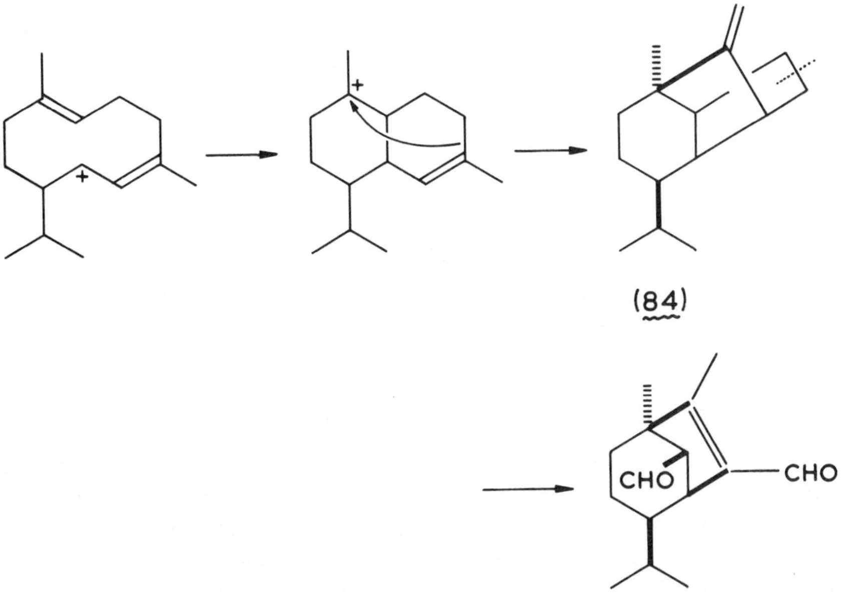

(84)

(83)

DIAGRAM 34.

[2-^{14}C]-MVA is in Diagram 35. Other detailed degradations have elucidated the routes to coriamyrtin (86, R = H) and tutin (86, R = OH) in higher plants (Biollaz and Arigoni, 1969; Corbella *et al.*, 1969) and feeding of presumed precursors labelled with ^{14}C suggested that eudesmol (87) was a precursor of

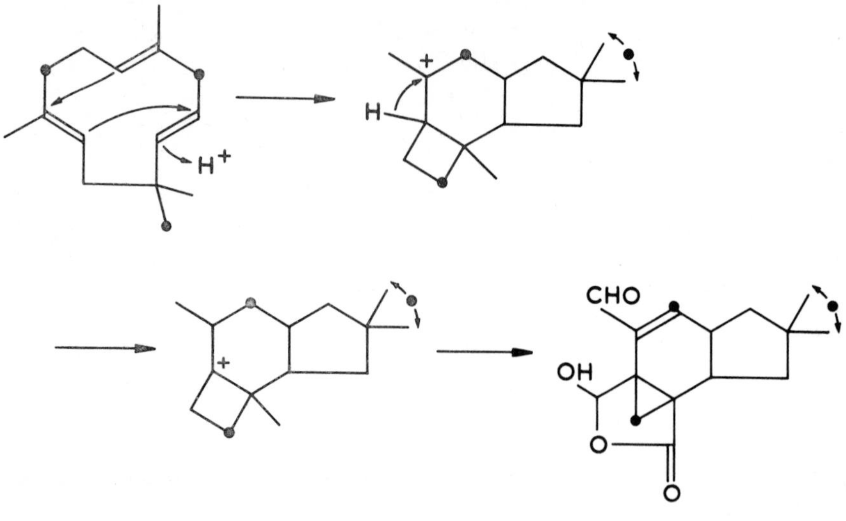

(85)

DIAGRAM 35.

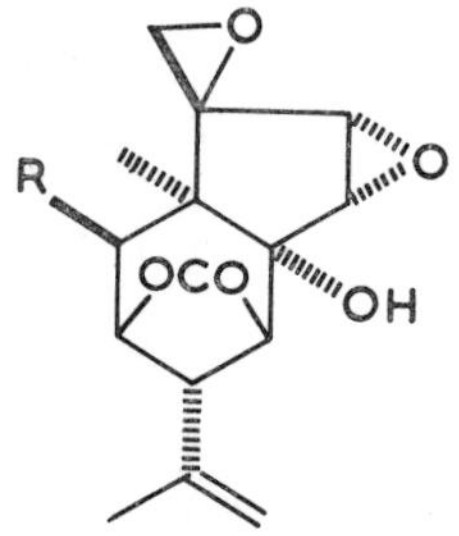

(86)

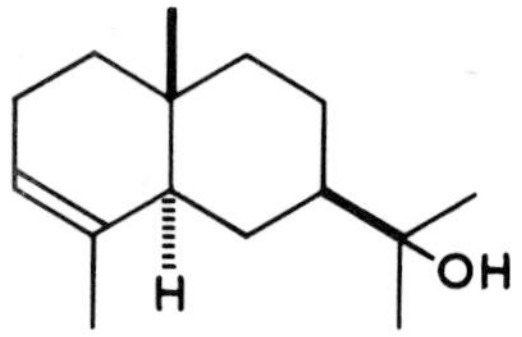

(87)

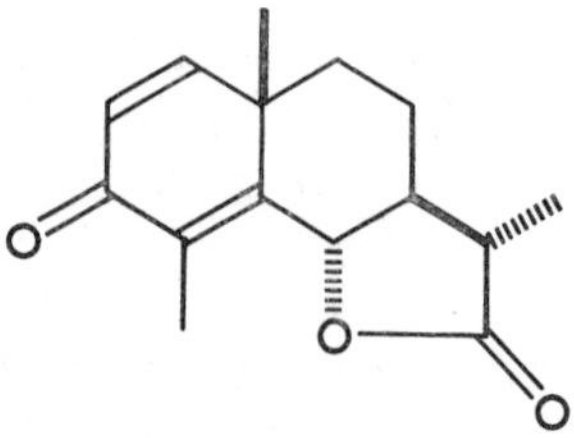

(88)

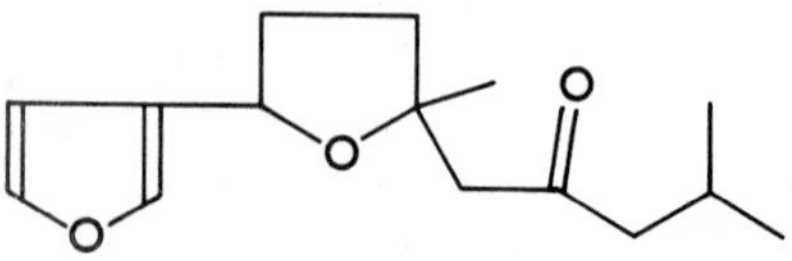

(89)

DIAGRAM 36.

santonin (88) and also that formation of the lactone ring in the latter preceded the construction of the diene group (Barton *et al.*, 1968). An interesting point was that FPP was not incorporated into (88) and farnesol was only so in minute traces, presumably owing to difficulty of translocation to the appropriate metabolic sites.

Perfunctory studies have been made on tracer incorporation into longifolene (58) (Sandermann and Bruns, 1962b,c) and ipomeamarone (89) found in sweet potatoes that had become infected with a fungus (Akazawa *et al.*, 1962). The latter was considered to be entirely of mevalonoid origin but no degradation studies were reported.

IV. Diterpenes

The majority of diterpenes are cyclic with no C_3-OH group and with the $5\alpha:10\beta$ "normal" configuration at the A:B ring junction that implies the stereospecific cyclization shown in Diagram 37. A multitude of hypothetical

DIAGRAM 37.

routes have been proposed (Ruzicka *et al.*, 1953; Hanson, 1968) and often acid-catalysed cyclizations *in vitro* parallel the proposed schemes (Oehlschlager and Ourisson, 1967).

A. Hypotheses

GGPP (17) is considered to be the progenitor of this class and some, e.g. phytol (90), are directly formed; but most, e.g. sclareol (91) and manool (92) are best derived from the isomer of GGPP (93). Formation of the allylic ion (94) permits further cyclization to rimuene and abietic acid (95, 96) and the former can again cyclize to compounds such as phyllocladene and isophyllocladene (97, 98). Penetrating stereochemical analyses (Wenkert, 1955; Wenkert and Chamberlin, 1959; Scott *et al.*, 1964) suggest that the formation of the last two compounds occur by the migrations shown rather than the shift of the angular methyl group.

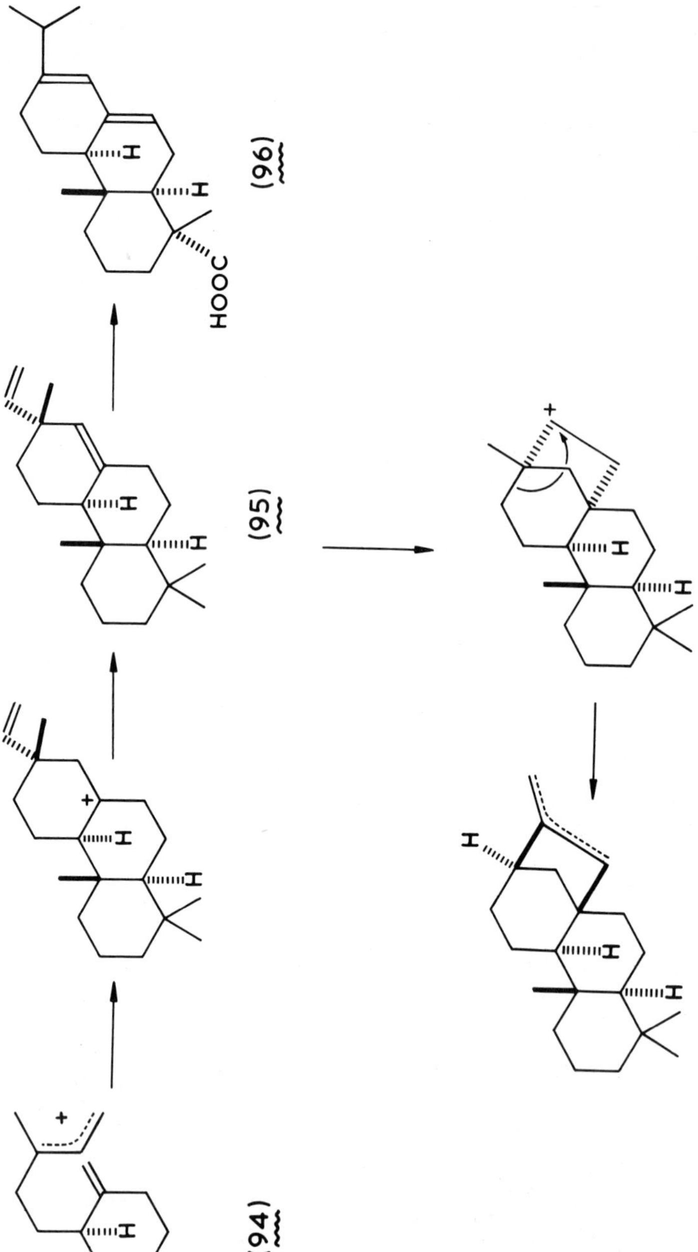

(97, 98)

DIAGRAM 39.

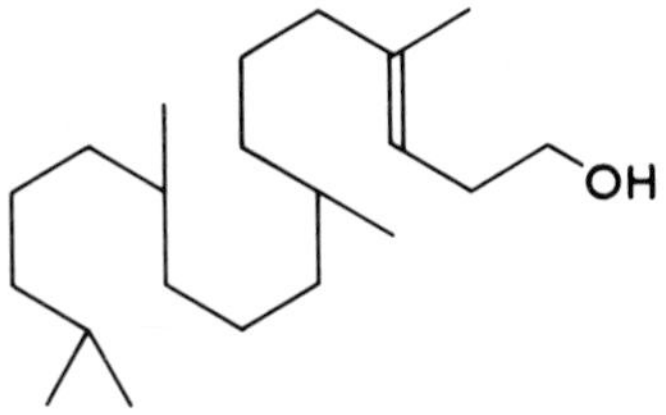

(90)

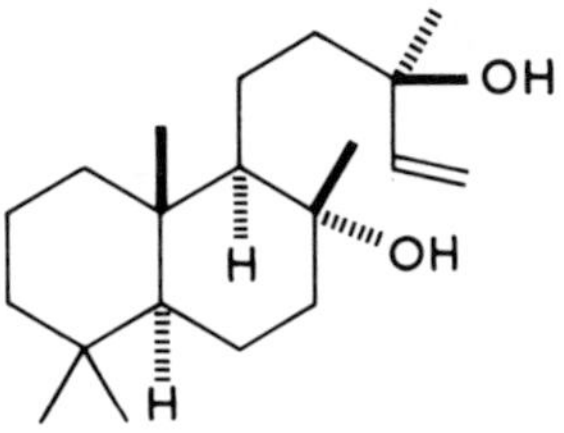

(91)

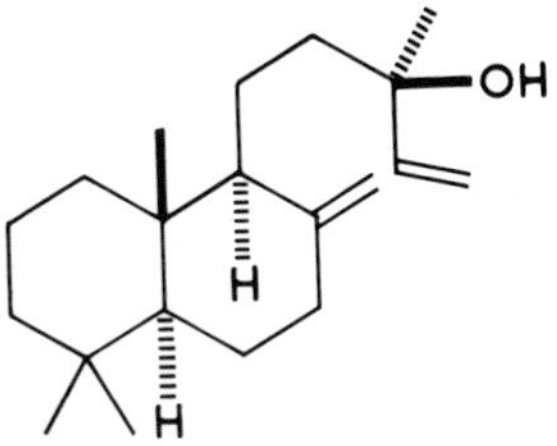

(92)

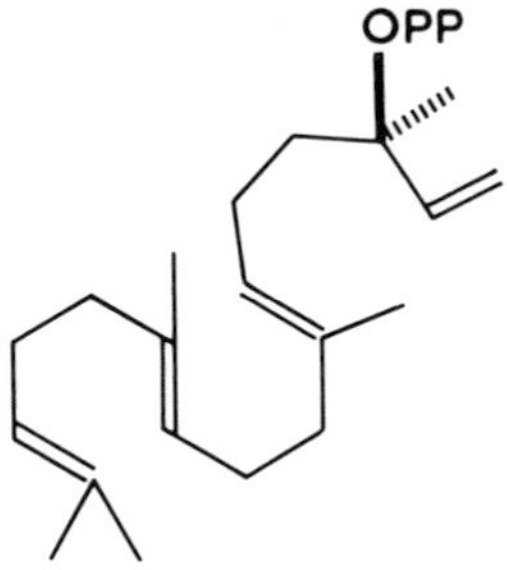

(93)

DIAGRAM 38.

More exotic structures are represented by the bitter principle columbin (99), that is believed to be formed by a backbone rearrangement, and the macrocyclic compound cembrene (100) that is probably derived from GGPP with a *cis*-olefin linkage (Dauben *et al.*, 1965).

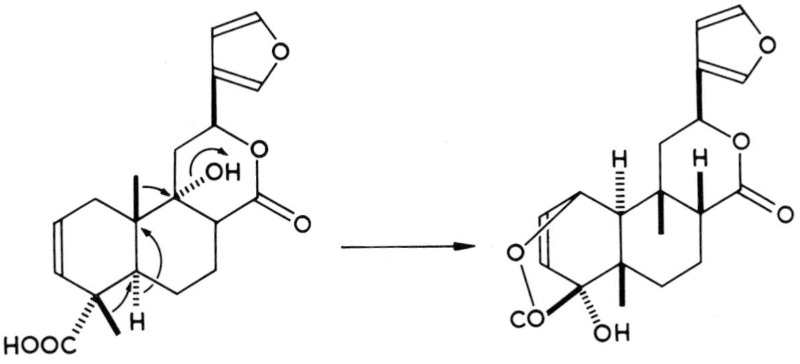

(99)

DIAGRAM 40.

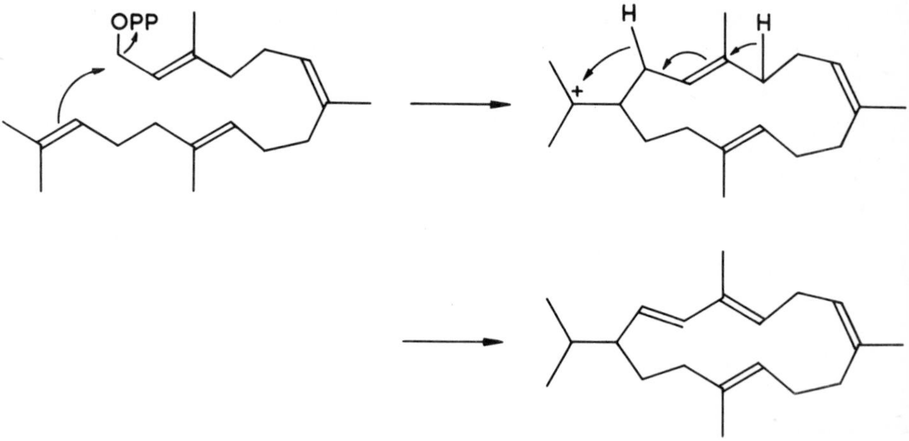

(100)

Diagram 40 (contd.).

B. Tracer Studies

Several important tracer studies have been made on members of this class and the hypothetical routes have been amply confirmed. Location of four equally labelled positions in the mould product rosenolactone (101) biosynthesized from [2-^{14}C]-MVA confirmed a route from GGPP via (102)

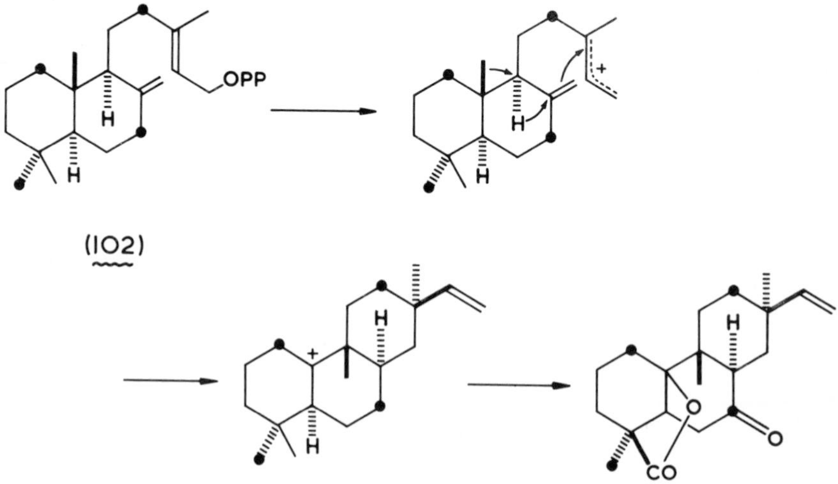

(102)

(101)

Diagram 41.

with no randomization of tracer in the *gem*-dimethyl groups (Britt and Arigoni, 1958; Birch *et al.*, 1959). The expected labelling pattern from [^{14}C]-acetate was also obtained and details have been elucidated by determining the location of tracer in the lactone after feeding with supposed intermediates labelled with ^{14}C (Hanson and Achilladelis, 1967; Achilladelis and Hanson, 1969a,b).

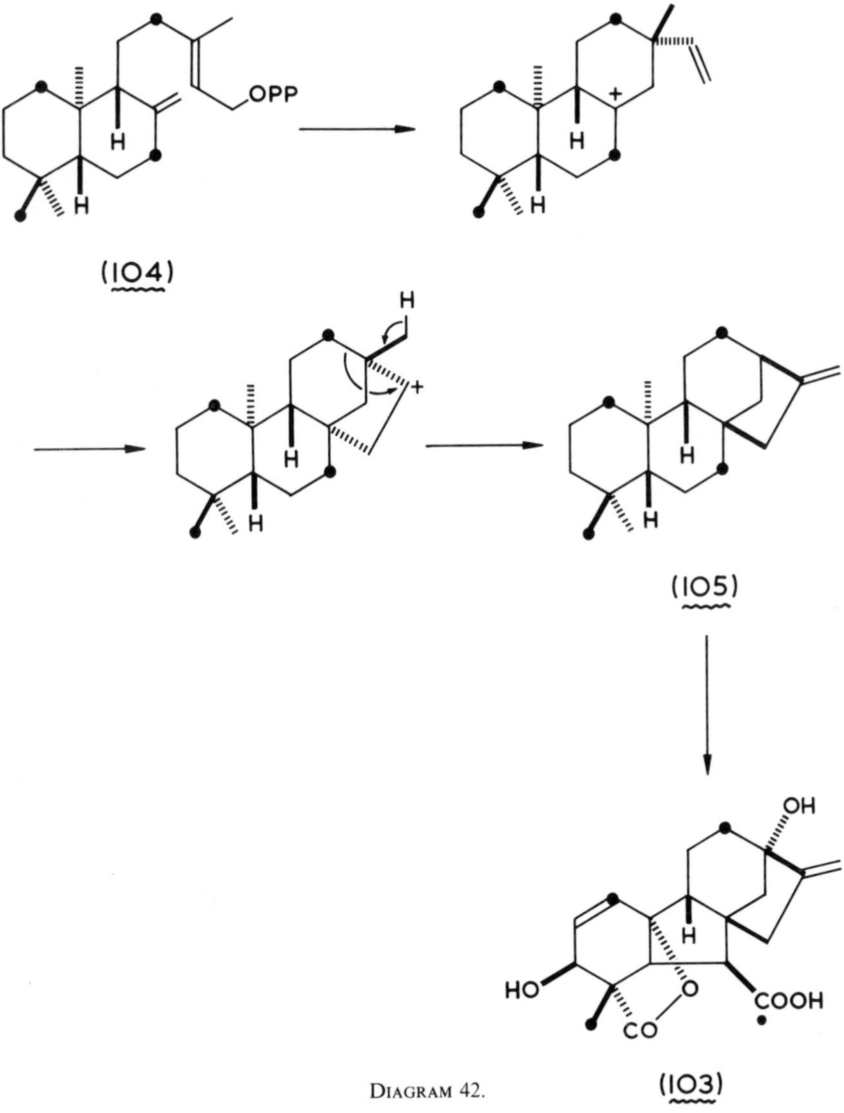

DIAGRAM 42.

The related gibberellins of fungal origin, e.g. gibberellic acid (103) also gave a labelling pattern from [2-^{14}C]-MVA that was completely in accord with predictions (Birch *et al.*, 1959), although the initial precursor (104) is now enantiomeric with (102); after the formation of kaurene (105) the methyl at the A:B ring-junction is lost. The theoretical proposals for the construction of the D-ring that were outlined in the previous section were confirmed by the pattern of tracer in this ring when [1-^{14}C]- and [2-^{14}C]-acetate were used as precursors: the results were only consistent with the scheme in Diagram 42, i.e. a Wagner–Meerwein skeletal rearrangement rather than a 1,2 methyl shift. Kaurene and other assumed precursors singly or doubly labelled with ^{14}C and ^{3}H at specific sites were incorporated into (103) without loss or redistribution of tracer (Cross *et al.*, 1963; Hanson and White, 1968). The whole subject has been reviewed (Cross, 1968) and specific details clarified (Hanson *et al.*, 1968; Cross *et al.*, 1968; Cross and Stewart, 1968). An interesting minor detail is that the hydroxyl at C_3 must be unrelated to the initiation of cyclization.

Pleuromutilins (106, R = H, $COCH_2OH$) biosynthesized from either [2-^{14}C]-MVA or [2-^{14}C]-acetate give a tracer pattern consistent with the route shown (Arigoni, 1962; Birch *et al.*, 1963), and partial degradation of marrubiin (107) formed from the same precursors also gave the expected distribution (Breccia and Badiello, 1967). However the pattern in glaucarubolone (108) and related compounds was consistent with their being degraded triterpenes (Moron *et al.*, 1966). [^{14}C]-β-β-Dimethylacrylic acid transferred tracer to the resin acids of pines (Sandermann and Stockmann, 1957) but it was not determined whether the C_5 unit was directly incorporated.

V. Sterols

These compounds, which have the characteristic steroid skeleton, could be classed as triterpenes but it is convenient to discuss them separately.

A. Sterols

More is known concerning the biosynthesis of cholesterol than of any other terpenoid, and much of this information, accrued from cell-free systems of mammalian tissue and yeast, has provided the background for theories of the routes to the various terpene families.

The stereochemistry of the condensation of two molecules of FPP to squalene is given in Section I E. Two attractive mechanisms have been proposed for this reaction (Clayton, 1965a): one envisages reaction of FPP

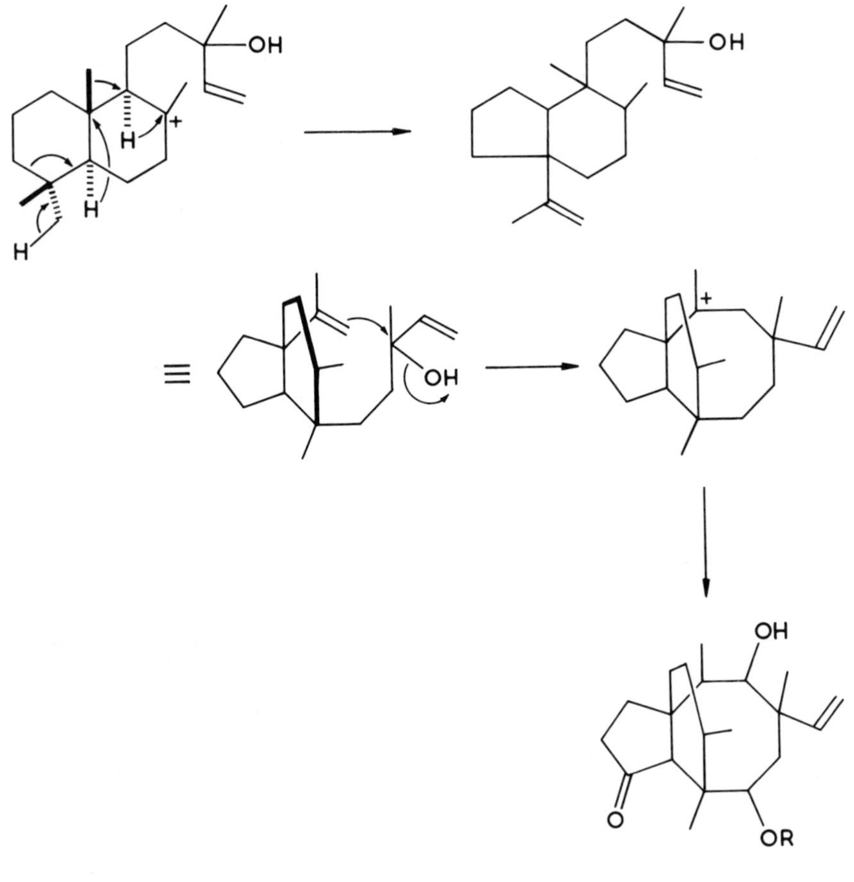

(106)

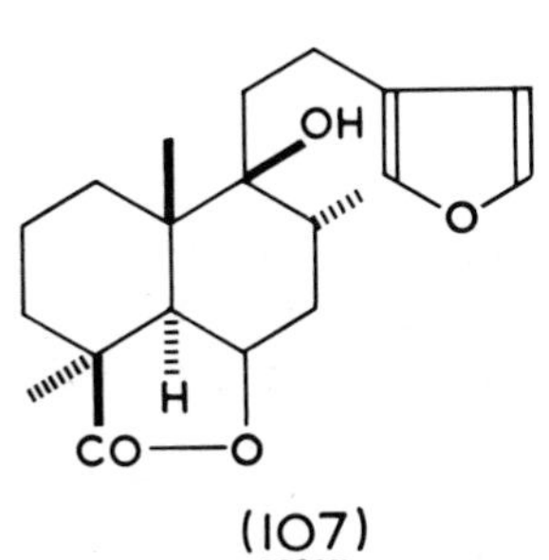

(107)

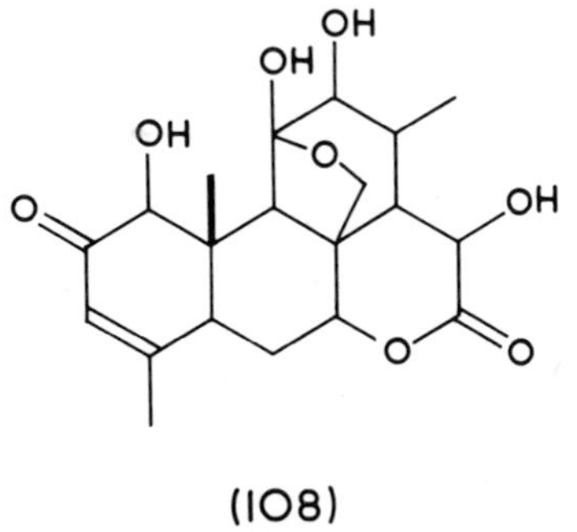

(108)

DIAGRAM 43.

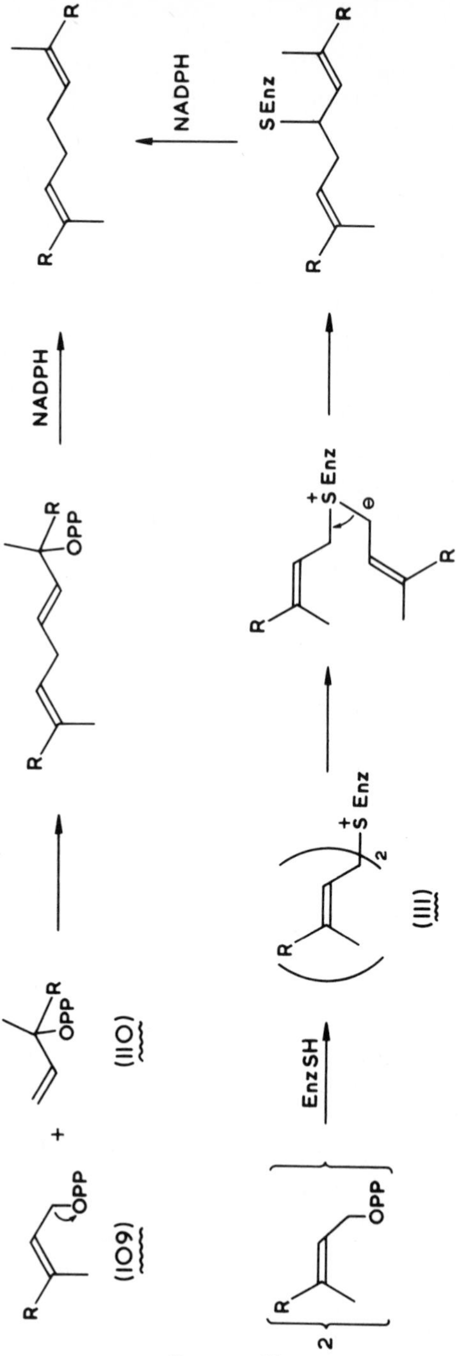

DIAGRAM 44.

(109) with its allylic isomer (110) and the intervention of hydride ion from NADPH, and the other requires a Stevens rearrangement within a sulphonium complex involving an enzyme (111) (Cornforth *et al.*, 1966). The latter route is consistent with the formation of protein-bonded squalene during reaction (Krishna *et al.*, 1964). Both routes have a choice of internal detail, but a rather complex sequence of events is likely as thiamine is implicated in an obscure manner (Risinger and Durst, 1968) and intermediates with cyclopropyl rings or cyclic phosphate groups have been identified (Rilling and Epstein, 1969; Popjak *et al.*, 1969a, Rilling, 1970; Epstein and Rilling, 1970).

A biogenetic relationship between squalene and cholesterol had long been suspected and the implication of lanosterol (112) and the determination of the labelling patterns in squalene and cholesterol (113) formed from [^{14}C]-acetate in cell-free systems led to the proposal of the route in Diagram 45 (Woodward and Bloch, 1953; Dauben *et al.*, 1953; Eschenmoser *et al.*, 1955). Squalene was supposed to be held on the enzyme surface in a chair-boat-chair-boat-unfolded (c-b-c-b-u) conformation and to undergo concerted cyclization and rearrangement (dissected into two discrete steps for clarity in Diagram 45) to lanosterol that was in turn functionalized to cholesterol. Much evidence supported this mechanism: thus, no intermediates could be detected between squalene 2,3-oxide (see below) and lanosterol; two 1,2-methyl shifts in (114) rather than a single 1,3-shift have been proved by tracer methods (Maudgal *et al.*, 1958; Cornforth *et al.*, 1958); and migration of the appropriate hydrogens to the C_{17} and C_{20} positions have been demonstrated (Cornforth *et al.*, 1965a; Caspi and Mulheirn, 1969). An "X-group" mechanism (see Section I E) has been proposed to explain certain stereochemical details of this and other cyclizations (Cornforth, 1969) although no direct evidence for this has been adduced.

More recent work has clarified the nature of the electrophilic attack that initiates cyclization. A rat liver preparation converted squalene 2,3-oxide into lanosterol almost quantitatively under anaerobic conditions, and the epoxide could be isolated by isotopic dilution when squalene was metabolized aerobically to cholesterol (van Tamelen *et al.*, 1966; Corey *et al.*, 1966). Consequently, cyclization undoubtedly is triggered by the protonation and opening of the epoxide ring (Diagram 46), such that the 4α-methyl group is derived from the C_2 of MVA (Moss and Nicolaidis, 1969). Structurally modified squalene oxides, e.g. with an entire C_5 unit missing from the non-oxidized end, have been cyclized to analogues of lanosterol and various triterpenes in cell-free systems from plants and animals, and the nature of the enzymes involved has been explored (Corey *et al.*, 1968a,b,c, 1969; van Tamelen *et al.*, 1968; Sharpless and van Tamelen, 1969; Anderson *et al.*, 1969).

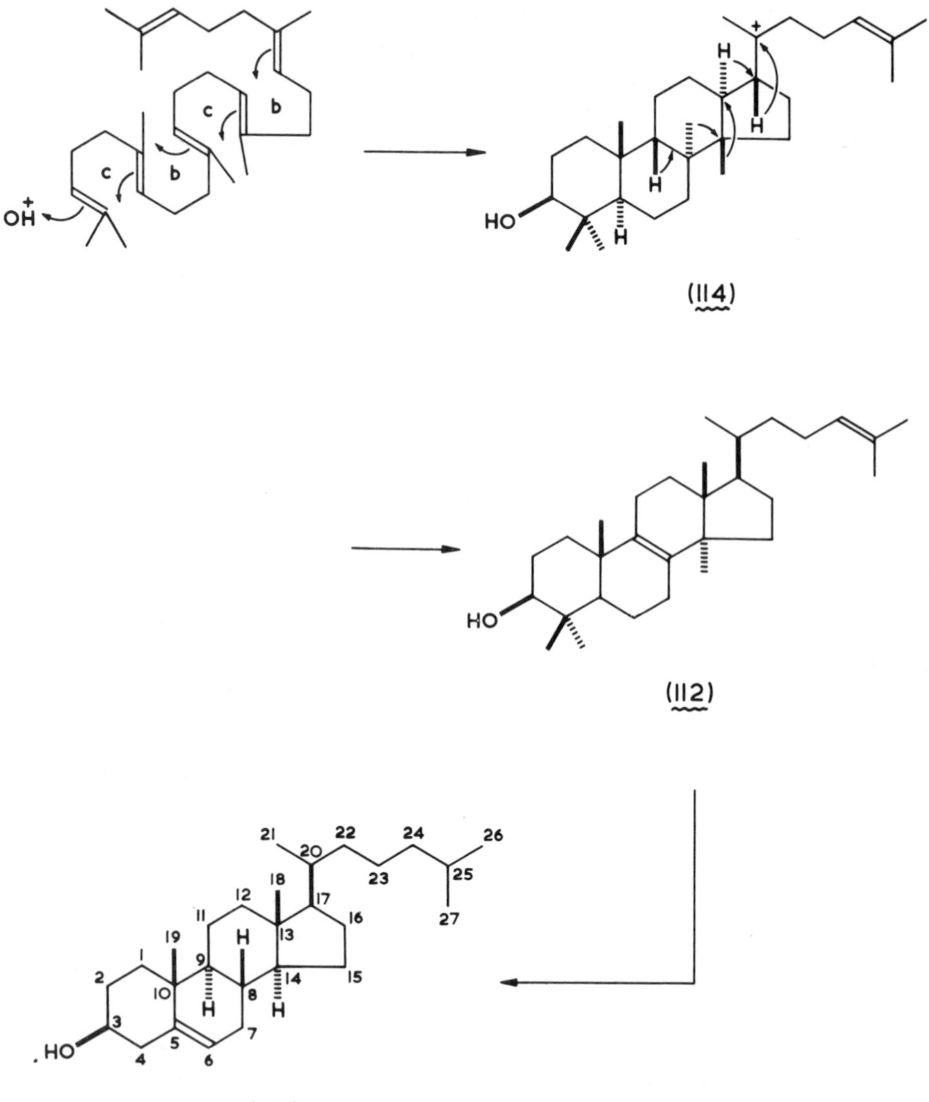

DIAGRAM 45.

An ingenious experiment was devised to determine whether squalene is both synthesized and cyclized within an enzyme complex or whether desorption from the first system is followed by cyclization at another site. When isotopically normal FPP is coupled in the presence of [^{3}H]-NADPH,

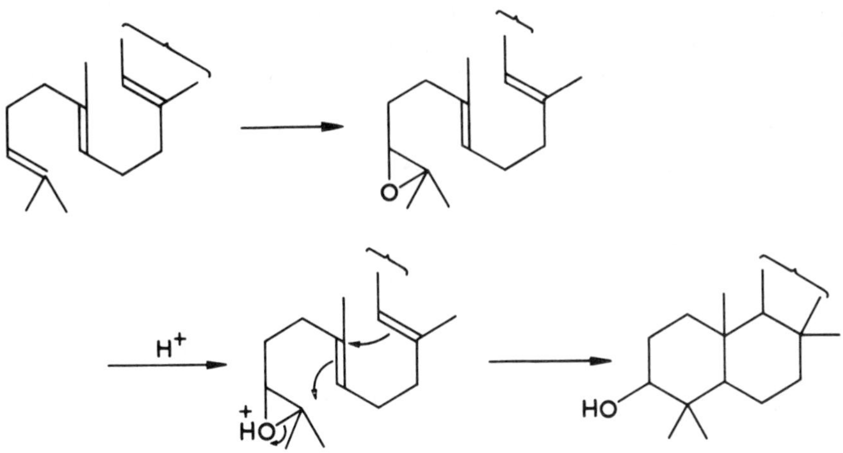

DIAGRAM 46.

the squalene will be labelled in one half of the molecule but not in the other (Section I E): if the first possibility applies, the asymmetric molecule will be passed on in a spatially defined manner to be oxidized and cyclized from one particular end; whereas, for the second route either end may be oxidized to initiate cyclization. In consequence, the distribution of tracer between C_{11} and C_{12} in the cholesterol produced is a direct index of the randomness of oxidation and hence of mechanism. Cholestrol synthesized by pig liver homogenates showed an inequality of labelling but owing to ambiguities of the experimental technique, no decision could be drawn (Etemadi *et al.*, 1969).

Several pathways have been suggested for the conversion of lanosterol into cholesterol, but although each step is fairly well understood, the exact sequence, if this is invariant, is uncertain. However the C_{14}-methyl group is probably removed before the *gem*-dimethyls (Clayton, 1965b) by means of the sequence: hydroxymethyl-aldehyde-acid, followed by decarboxylation. Cholesterol is undoubtedly the precursor for other steroids, sex hormones, vitamins D etc. (Clayton, 1965b).

B. PHYTOSTEROLS

The pathway to compounds such as β-sitosterol (115), stigmasterol (116) and ergosterol (117) that occur in plants was considered to involve the same intermediates as the routes to sterols in yeast and animals (Clayton, 1965b), but more recent views (Goad, 1967; Lederer, 1969) demur in certain respects. The central role of squalene is unquestioned; and although it is rapidly turned-over (Nicholas, 1962), it has been isolated from higher plants and tissue cultures (Anderson and Porter, 1962; Benveniste *et al.*, 1964) as the

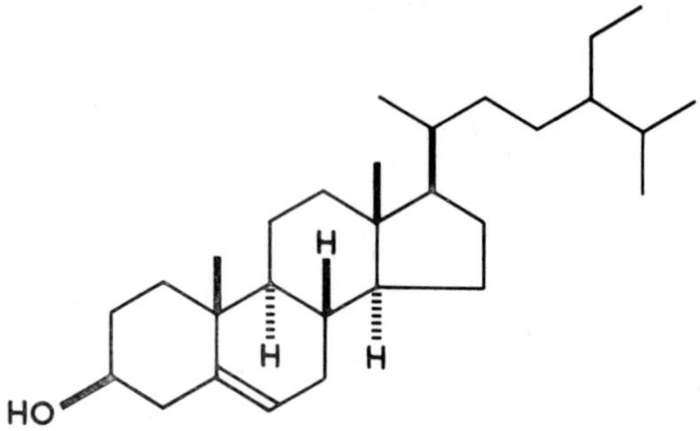

(115)

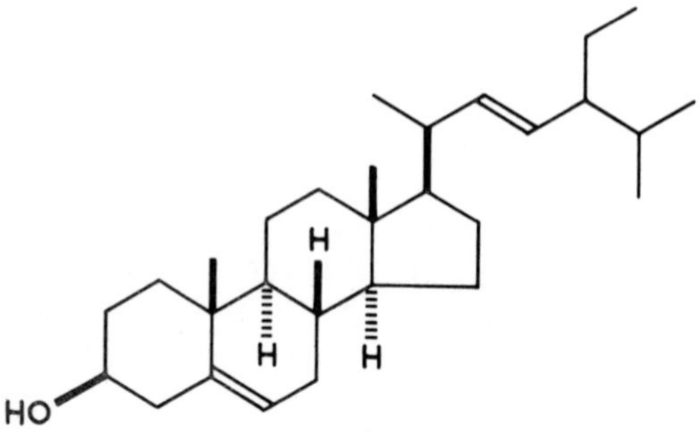

(116)

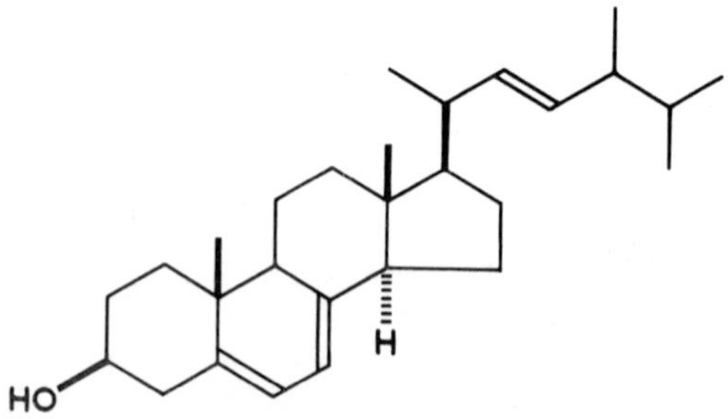

(117)

DIAGRAM 47.

all-*trans* isomer (Beeler *et al.*, 1963) with the same labelling pattern as that formed in animals (Capstack *et al.*, 1965). However, although lanosterol has been isolated from microorganisms and a few plants (Bazzano *et al.*, 1964), it is not generally found in the latter and evidence has accumulated that it is not on the direct route to phytosterols (Goad, 1967). Cycloartenol (118) which occurs widely and was rapidly labelled with tracer from acetate or MVA in plants and tissue cultures (Benveniste *et al.*, 1966) under conditions where lanosterol could not be detected (Goad and Goodwin, 1966, 1967) appears to take its place. Squalene 2,3-oxide was converted into cycloartenol but not into lanosterol *in vivo* and in tissue cultures (Eppenberger *et al.*, 1969; Rees *et al.*, 1968) and cycloartenol could be further converted *in vivo* into other phytosterols (Hewlins *et al.* 1969; Hall *et al.* 1969). The presence of lanosterol in latex from certain species was the result of enzymic modification of cycloartenol (Ponsinet and Ourisson, 1968).

Cycloartenol can be hypothetically derived from the c-b-c-b-u conformation of squalene *via* the intermediate (119): the hydrogen at C_9 migrates to C_8 instead of being lost and a 9β:19 cyclopropane ring is constructed by proton loss from C_{19}. Evidence for the hydrogen migration was provided by feeding experiments using equal specific activities of ^3H and ^{14}C in a mixture of [4R-4-$^3H_1$]-MVA and [2-^{14}C]-MVA: here the C_9 hydrogen in (119) was labelled but was expelled in the formation of lanosterol in liver tissue to give a product with an isotope ratio (^3H:^{14}C) of 5:6, but the ratio

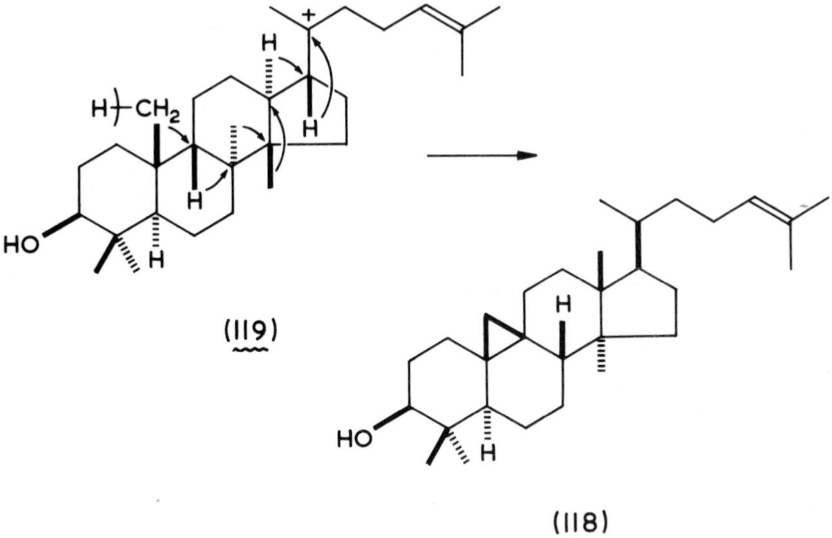

(119)

(118)

DIAGRAM 48.

in cycloartenol produced in potato leaves remained at 6:6. This result also eliminated the hypothesis of formation of cycloartenol from lanosterol (Rees, Goad and Goodwin, 1968a,b, 1969). The mechanism in Diagram 48 appears to contradict the usually rigid stereoelectronic requirements for rearrangement in that the migrating C_{10}-linked methyl group is on the same face of the molecule as the hydrogen at C_9 which also migrates; an intermediate of the "X-group" type (Section I E) may intervene to resolve this apparent anomaly.

Few dissenting views (cf. Bernard and Reid, 1967) have been raised to these mechanistic proposals, but it is not known if the small amounts of cholesterol found in some plants and algae have lanosterol or cycloartenol as their immediate precursor (Jacobsohn and Frey 1967). Certainly, some species appear capable of transforming [^{14}C]-lanosterol into phytosterols (Hewlins *et al.*, 1969).

Many reports of tracer uptake from acetate or MVA into various phytosterols are available (Goad, 1967) and acetate was incorporated into β-sitosterol and digitoxigenin with the expected labelling pattern (Battersby and Parry, 1964; Gros and Leete, 1965). Use of [4R-4-$^3H_1$]-MVA showed that the stereochemistry of elimination of hydrogen in the conversion of IPP into DMAPP was the same in plant and animal tissue (Goad and Goodwin, 1965) and the [4R-^{3}H] was retained in squalene, α- and β-amyrin and phytol (Rees *et al.*, 1966, 1968; Wellburn *et al.*, 1966).

The sequence from cycloartenol to other phytosterols is obscure although precursor relationships have been explored (Bennett *et al.*, 1963; Johnson *et al.*, 1964; Waters and Johnson, 1965; Baisted, 1969). Although proposed intermediates have been detected (Lederer, 1964; Goad, 1967), there is no evidence for their participation in any well defined pathway, and because of the prodigality of enzyme patterns practically every possible compound between cycloartenol and the phytosterols may eventually be isolated. The C-alkylation of the side chain of ergosterol etc. can probably occur at the C_{27}, C_{30} or any intermediate structural level (Goad and Goodwin, 1967; Lederer, 1969) by a route of methyl transfer from S-adenosyl-methionine (Goad and Goodwin, 1969) and migration of hydrogen from C_{24} and C_{25} (Diagram 49). Further alkylation and reduction of the methylene sterols formed can subsequently occur (Goad *et al.*, 1969; Baisted 1969; van Aller *et al.*, 1968), and tracer studies indicate similar routes to triterpenes such as cyclolaudenol with a 25-methylene group (Ghisalberti *et al.*, 1969). The C_{22}–C_{23} double bond in the side chain is apparently introduced after the alkylation steps by desaturation reactions (Lederer, 1969; Akhtar *et al.*, 1969). The second C-alkylation to form an ethyl group in the side chain probably occurs after partial dealkylation of the skeleton (at C_{14} or C_4): there is only one report of a C_{32} ($C_{30} + C_2$) compound (Lederer, 1969).

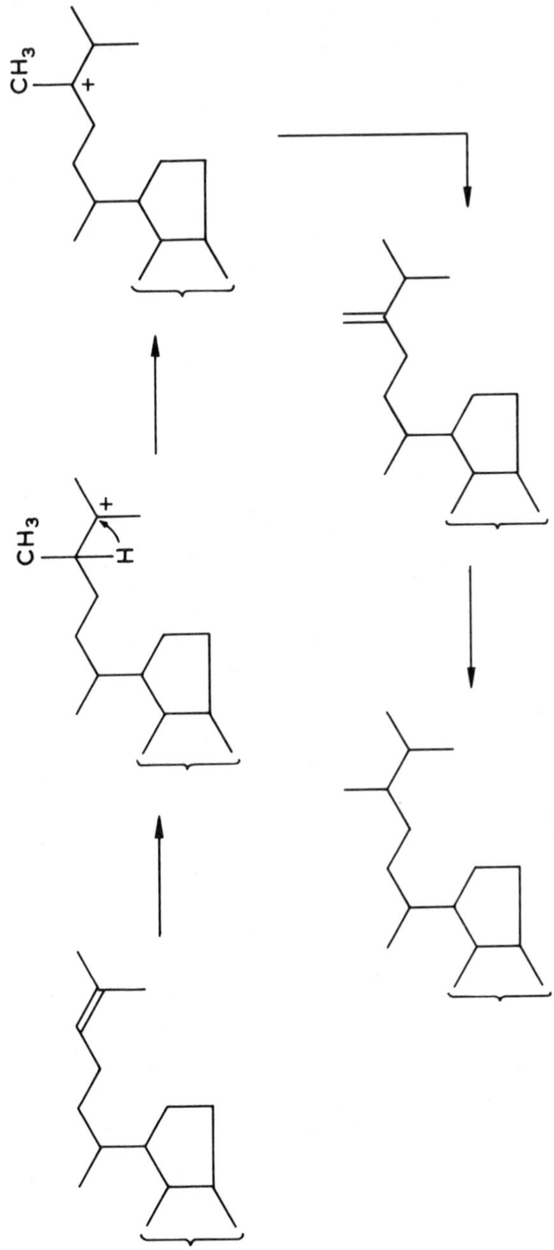

DIAGRAM 49.

The result of this plethora of pathways is the formation of compounds such as 24-methylene-lophenol (120) with an unusual 4α-methyl group, and macdougallin (121), found on a species of cactus in which, contrary to

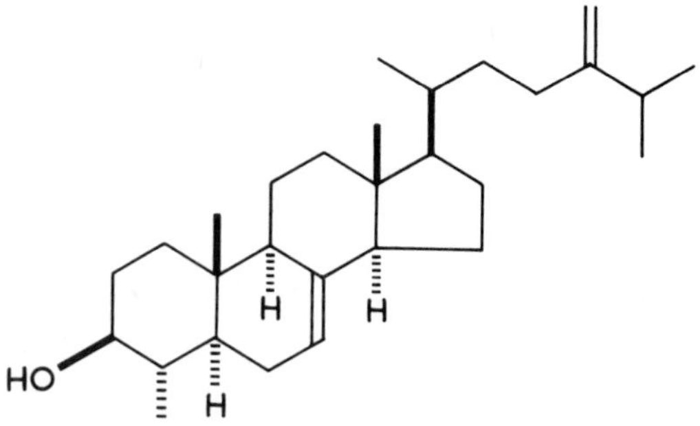

(120)

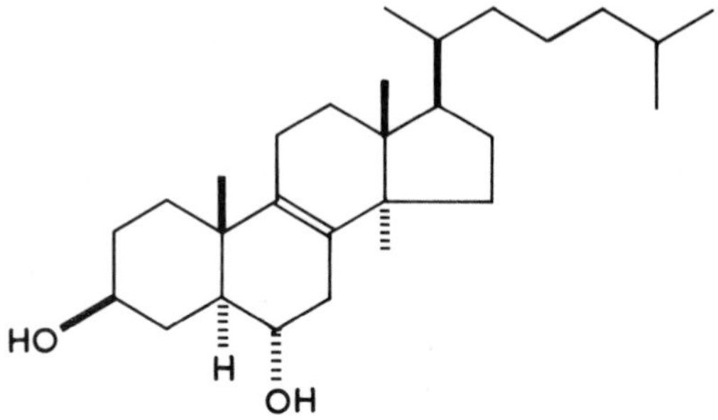

(121)

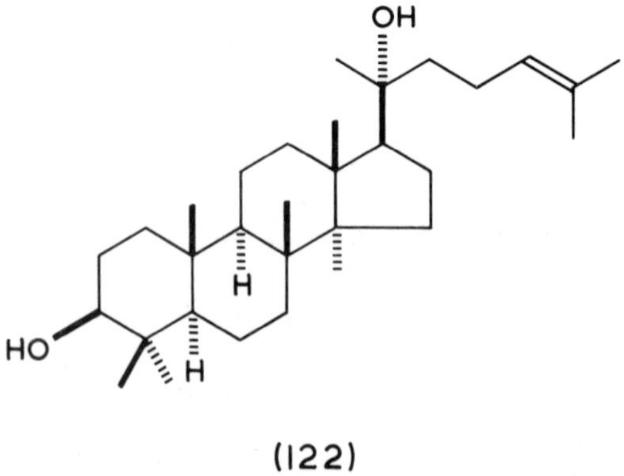

(122)

Diagram 50.

the situation in mammalian systems, removal of the C_{14}-methyl group has not preceeded loss of the *gem*-dimethyls (Djerassi *et al.*, 1963). Other compounds have a C_{14}-methyl together with one-C_4 methyl group (Goad, 1967), but oxidation and elimination of one of the *gem*-C_4-methyls apparently must be complete before the second is attacked. Use of doubly labelled-MVA revealed that the 4α-methyl of cycloartenol is derived from C_2 of MVA and that this group is removed in the conversion into the *nor*-methyl sterol (Ghisalberti *et al.*, 1969b).

VI. Triterpenes

A. Hypotheses

Soon after the proposed mechanism for the conversion of squalene into lanosterol came a brilliant hypothesis for the derivation of the various families of triterpenes from the same parent (Eschenmoser *et al.*, 1955) that was the forerunner of analogous proposals (that are described in previous sections) for sesqui and diterpenes. The stereochemistry of each family could be rationalized, with no exceptions, by allowing each of the 5 segments of all-*trans* squalene to be held in potential chair or boat conformations on the enzyme surface or to be unfolded, and applying the usual stereoelectronic requirements for cyclization. Thus the c-c-c-b-u form leads via a smooth cyclization to dammerenediol (122) and pentacyclics such as α-amyrin (123), β-amyrin (124) and lupeol (125) arise from the same conformation of precursor but with a pause after the construction of the steroid skeleton (126) that permits further rearrangements. Pentacyclic compounds such as hydroxyhopanone (127) can be derived from a c-c-c-c-c conformation of squalene without rearrangement, and cyclization can start from each end leaving a potential ring unenclosed at the centre; thus, c-c-u-c-c squalene can lead to onocerin (128) that can cyclize to serrantenediol (129). Initiation of cyclization by attack of a proton, rather than by the epoxide route, can lead to ambrein (130) from the c-c-u-u-c conformer.

These routes are impressive in their simplicity and generality, but certain other proposals do exist (Richards and Hendrickson, 1964). Large numbers of triterpenes are also formed by secondary transformations of the preformed skeletons; thus, nyctanthic acid (131) contains the β-amyrin skeleton opened oxidatively (Arigoni *et al.*, 1960a) and limonin (132) (Arigoni *et al.*, 1960b) contains a much adapted skeleton. Another elegant modification is the presumed conversion of β-amyrin into the enol of friedelin (133) via a backbone rearrangement (Corey and Ursprung, 1956).

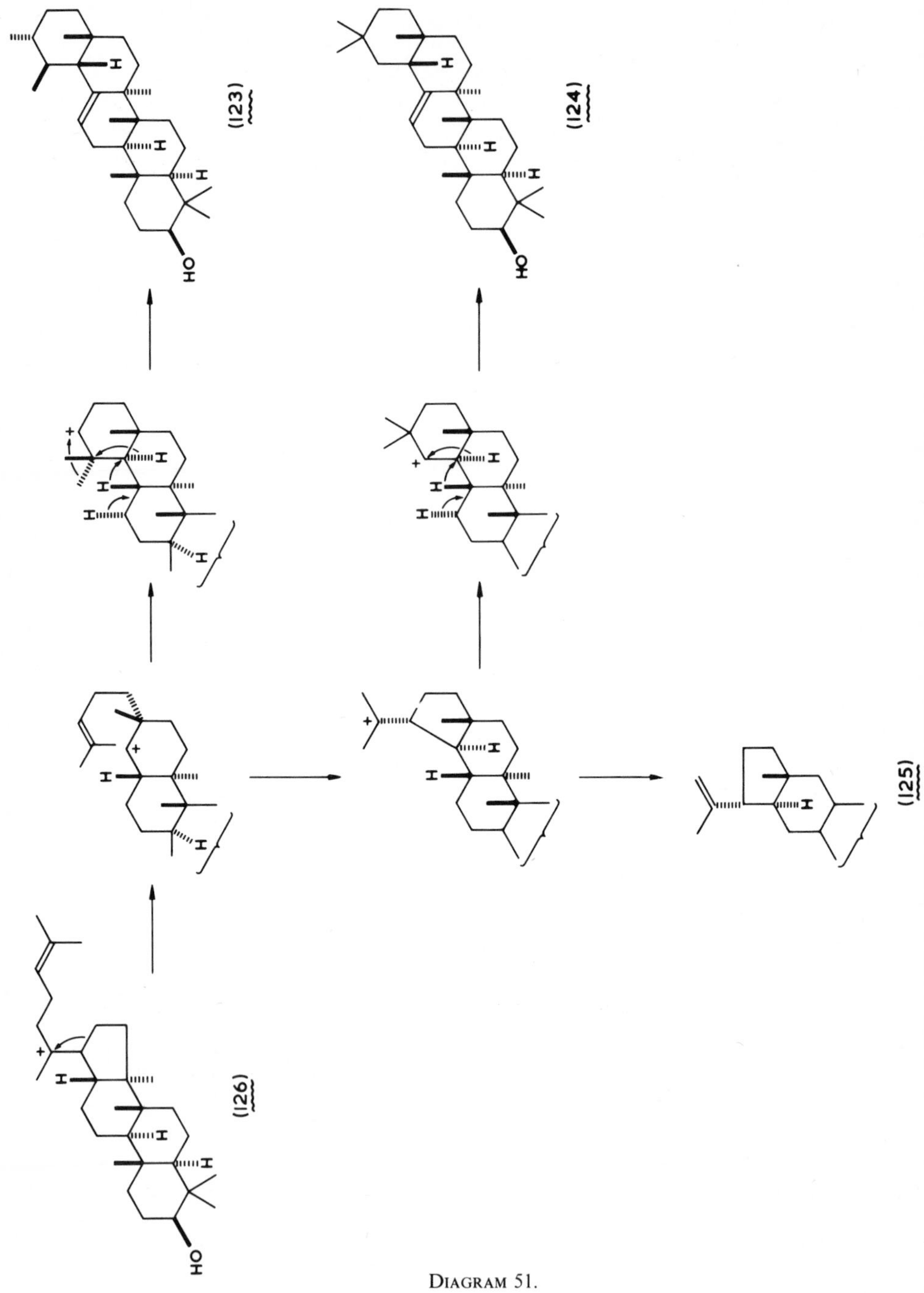

DIAGRAM 51.

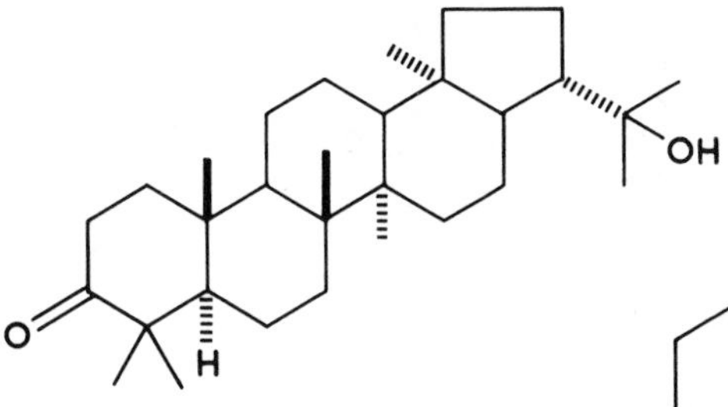

(127)

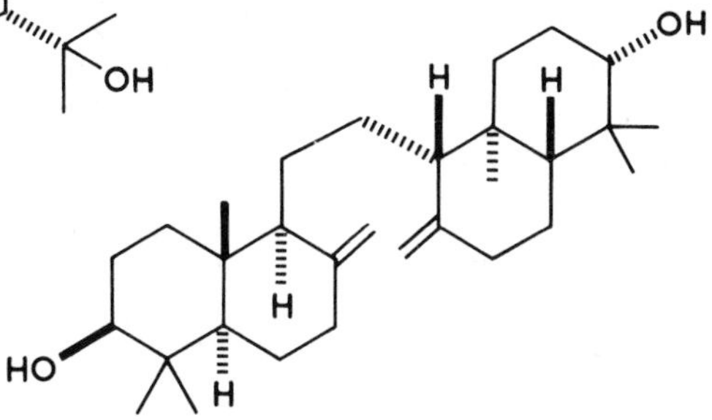

(128)

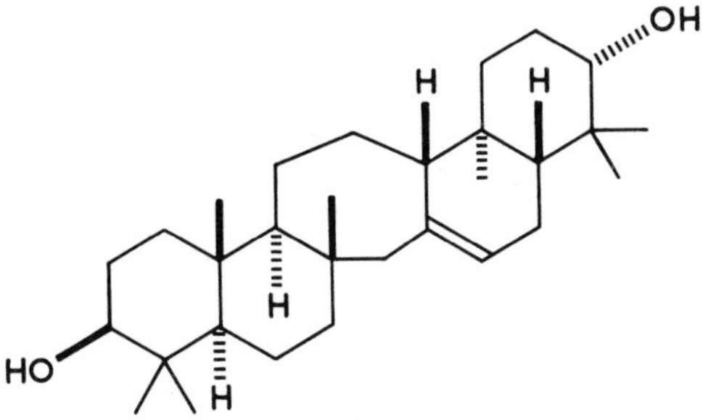

(129)

DIAGRAM 52.

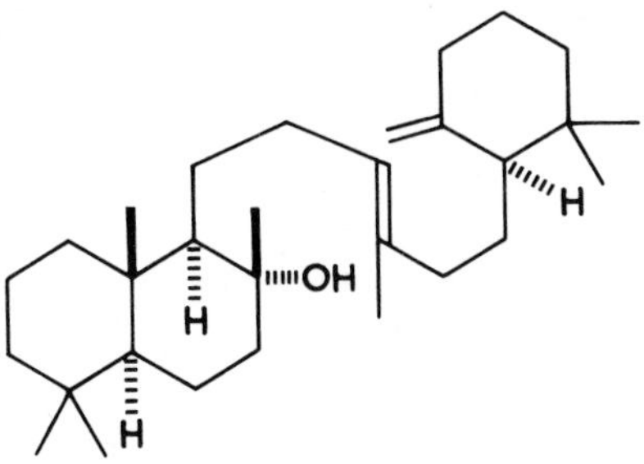

(130)

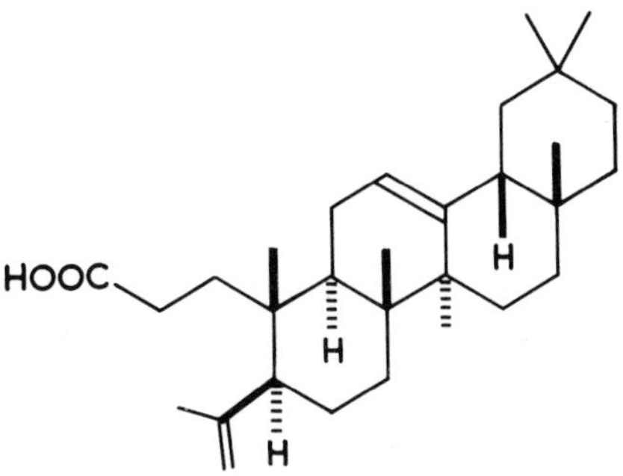

(131)

DIAGRAM 53.

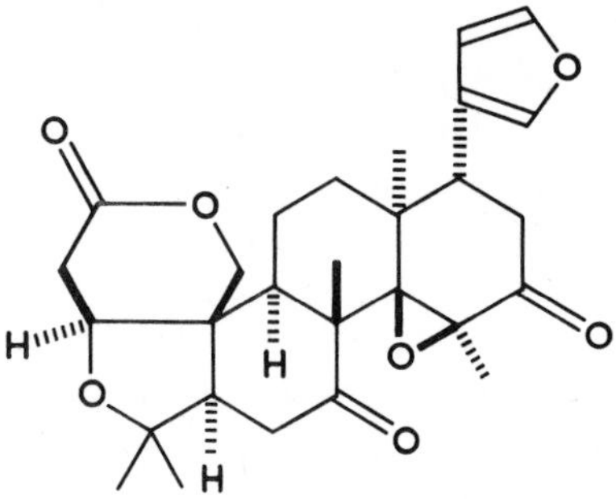

(132)

DIAGRAM 53 (contd.).

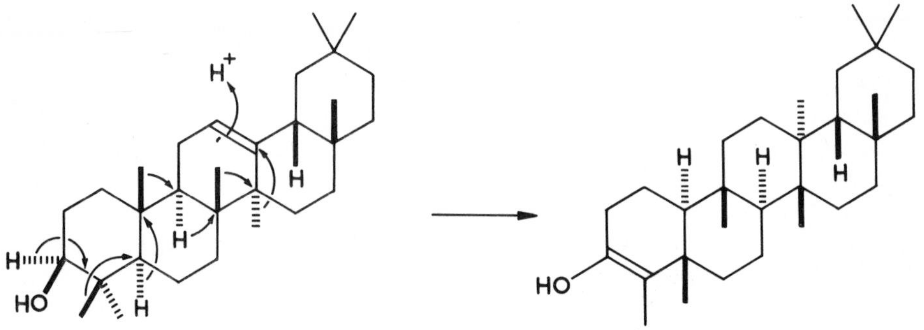

(133)

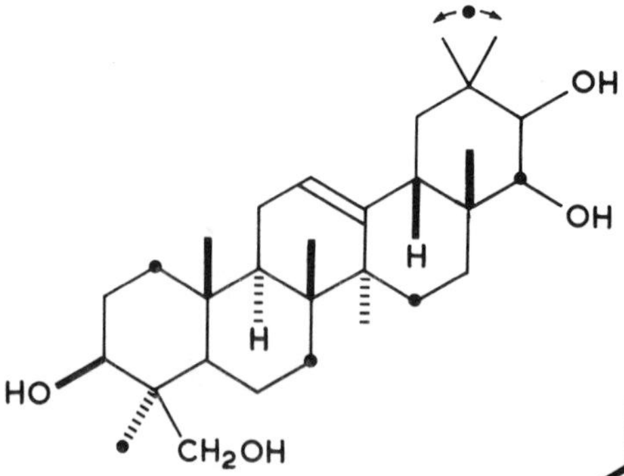

(134)

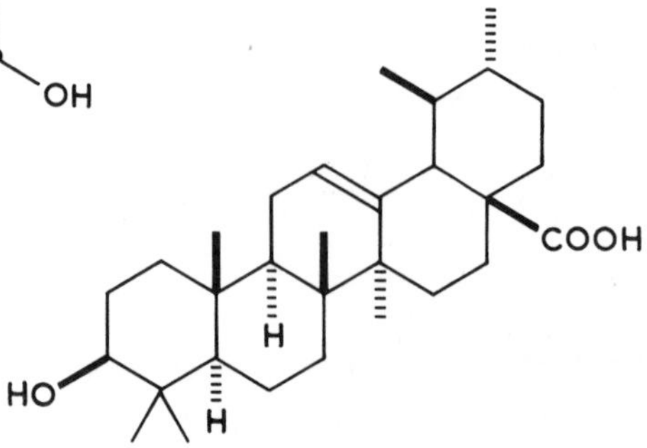

(135)

DIAGRAM 54.

B. Tracer Studies

The most detailed investigation has been carried out on soyasapogenol A (134). Biosynthesis from [2-^{14}C]-MVA gave the tracer pattern as expected (Arigoni, 1959) which revealed the non-equivalence of the *gem*-dimethyl groups of ring A. Similar results were obtained from the isomeric soyasapogenol D (Arigoni, 1958). The *gem*-dimethyls distal to the site of oxidation in lupeol (125) similarly retained their stereochemical individuality (Ruzicka, 1963), and the tracer pattern in ring E of this compound formed from [^{14}C]-MVA and acetate was also consistent with the hypothetical route (Gugliel-metti, 1967). Partial degradation of the mould product eburicoic acid also gave the expected tracer pattern (Dauben *et al.*, 1957; Lawrie *et al.*, 1967), but decarboxylation of ursolic acid (135) gave the unexpected result that the removed group was not labelled by acetate or by MVA (Nicholas 1967b). Fern-9-ene (Barton *et al.*, 1969) and tetrahymanol (Battersby and Gregory, 1968) have been shown to be cyclized by proton-addition, rather than via an epoxide, but the latter route has been demonstrated for certain triterpene antibiotics (Godtfredsen *et al.*, 1968). The plant pigment gossypol (136) is of mevalonoid origin (Heinstein *et al.*, 1962) but probably is derived from a cyclized, aromatized sesquiterpene that is oxidized to a naphthol derivative and dimerized. The compound (137) could be isolated from cell-free systems in which oxidative coupling was inhibited and this implies that *cis-trans* FPP is the precursor (Waller, 1969).

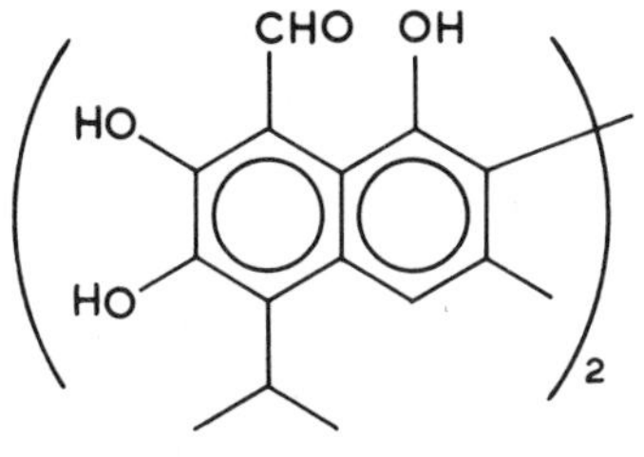

(136)

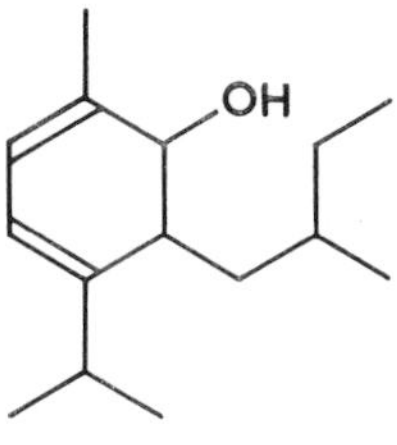

(137)

Diagram 55.

VII. Carotenoids

Carotenoids are found mainly in the photosynthetic organelles of plants and are constructed by tail to tail coupling of C_{20} units with cyclization almost invariably restricted to the ends of the chain. Their biosynthesis is well understood (Goodwin, 1965a; Porter and Anderson, 1967; Porter 1969).

A. Carotenes

Early studies revealed the incorporation of [^{14}C]-acetate and MVA with no randomization of tracer into carotenoids of plants and numerous protista (Goodwin *et al.*, 1965a) and although HMG was reported to be incorporated (Chichester *et al.*, 1959) the substrate used may have been isotopically impure. GGPP is undoubtedly the precursor of the class and coupling, which does not require NADPH (Charlton *et al.*, 1967), probably leads to phytoene (138) with three conjugated double bonds at the centre of the molecule. Coupling analogous to that forming squalene would give lycopersene (139) with a central single bond, and although the latter has been

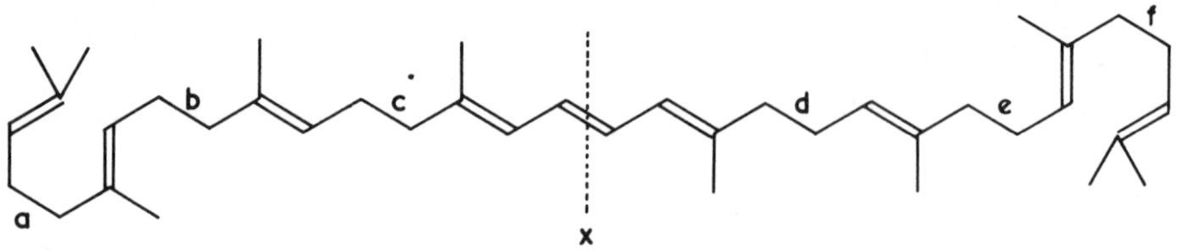

(138)

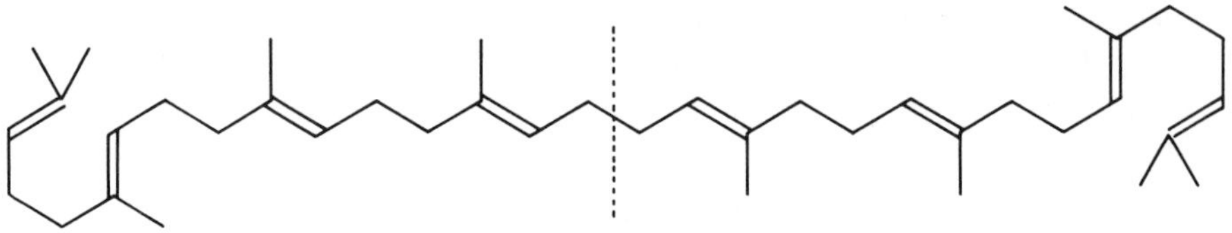

(139)

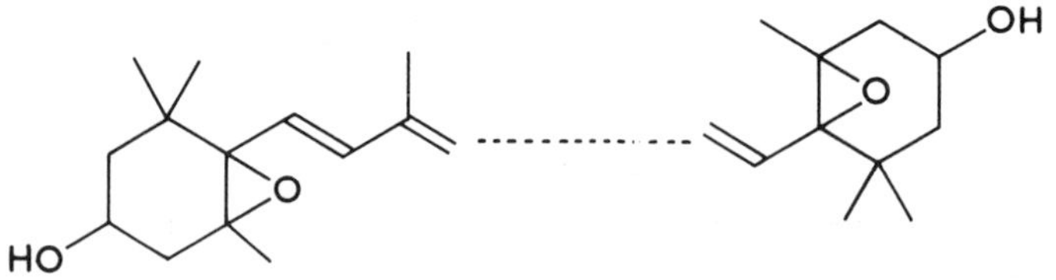

(142)

Diagram 56.

detected in moulds and higher plants (Grob and Boschetti, 1962; Nusbaum-Cassuto and Villoutreix, 1965) there is much evidence from tracer and analytical investigations that phytoene is the primary product (Rilling, 1962;

Goodwin and Williams, 1965a,b, 1966; Goodwin, 1965a). Phytoene has the important property that the central triad of double bonds prevents folding on an enzyme surface and consequent extensive cyclization whereas lycopersene would readily lead to polycyclic products. Naturally-occurring phytoene (138) probably has the central bond in the *cis*-configuration (Goodwin, 1969).

Phytoene is probably converted into more unsaturated derivatives by loss of pairs of hydrogen atoms alternately to the left and right of the central triad in (138). In higher plants, *trans* double bonds are introduced at c,d,b and e in turn (Goodwin, 1965b), but a different sequence may occur in bacteria (Davies and Holmes 1969). This order is based on the occurrence of carotenoids in tomato and chlorella mutants and the order of labelling from [^{14}C]-acetate or MVA. The *trans*-orientation of these introduced double bonds is confirmed by the stereospecific loss of [4S-^{3}H] when phytoene and related compounds were biosynthesised *in vivo* from [4S-4-$^{3}H_{1}$]-MVA (Goodwin and Williams, 1966) and the loss of two [S-$^{3}H_{1}$] from [5S-5-$^{3}H_{1}$]-MVA implied that the central bond was *cis* (Williams *et al.*, 1966). The latter result was confirmed for the synthesis of phytoene in a cell-free system from bean leaves (Buggy *et al.*, 1969).

Carotenes such as neurosporene and lycopene, of different degrees of unsaturation, can undergo terminal cyclization to lead to a variety of isomers (Goodwin, 1965b) and detailed schemes for the sequences have been proposed (Decker and Uehleke, 1961; Davies *et al.*, 1963; Wells *et al.*, 1964; Goodwin 1965a,b). The logical route to α- and β-ionone rings (140, 141) is as

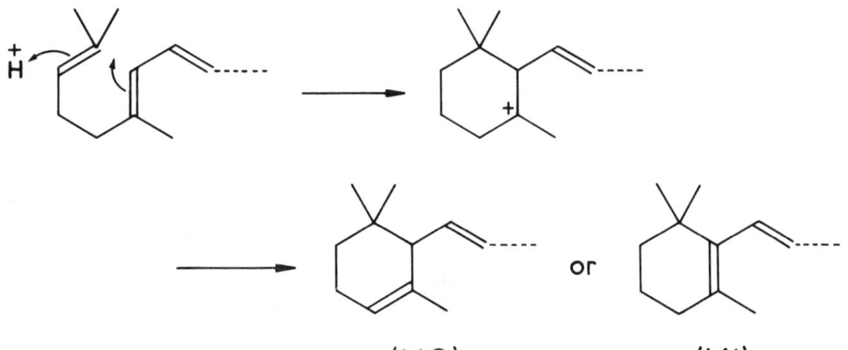

(140) (141)

DIAGRAM 57.

shown, and this requires that β-ionone rings biosynthesized from [2-^{14}C] and [4R-4-$^{3}H_{1}$]-MVA would lose labelled hydrogen whereas α-ionone rings would not. Consistently with this, several species formed phytoene with an isomer ratio ($^3H:^{14}C$) of 8:8 which dropped to 7:8 in α-carotene which has one α- and one β-ionone ring and to 6:8 in β-carotene with two β-ionone residues (Goodwin and Williams, 1965a,b). Analogous results were found for other carotenoids formed from differently isotopically labelled MVA (Williams *et al.*, 1967a,b) and the possibility that the α-ionone ring was derived by isomerization of its β-isomer was definitely excluded.

B. Xanthophylls

These, a typical example of which is violaxanthin (142), are oxygenated derivatives believed to be derived from α- and β-carotenes and lycopene (Shneour, 1962; Goodwin, 1965a,b); a report that xanthophylls are reduced to carotenes by isolated chloroplasts (Costes 1963) requires confirmation. Tracer studies using [2-^{14}C] and [5R-$^3H_1$]-MVA showed that the hydroxyl groups in (142) and related compounds were introduced by stereospecific replacement of the R-hydrogen of the carotene precursor; this rules out the intermediacy of a 3-keto group and confirms that xanthophylls are not precursors of carotenes, for the latter retained tracer at C_3 (Goodwin, 1969).

VIII. Other Terpenoids

A. Rubber

Rubber consists of head to tail polymers of 500 to 5000 *cis*-linked isoprene residues, whereas gutta percha typically has 100 *trans*-linked residues. Ability to make these polymers is scattered in about 1% of all plant species but the commercially-used species are physiological freaks that have been selected for their ability to channel most of their mevalonoid products into rubber. [^{14}C]-MVA and IPP are incorporated without scrambling of tracer, and chain extension occurs on the surface of existing insoluble rubber particles; initiation of a new chain in the absence of these particles requires the addition of DMAPP as starter (Archer *et al.*, 1963). The IPP-polymerase has been studied (Chesterton and Kekwick, 1966) but little is known about the mechanism of the chain extension, the number of polymerases involved or the factors that control cessation of polymerization (Bonner, 1967). The [4S] hydrogen of MVA is retained in the formation of the *cis*-linked polymer whereas the FPP simultaneously formed in the specialized latex cells has the same pattern as squalene, i.e. the [4S] hydrogen is lost (Archer *et al.*, 1966).

B. Prenols

The best studied of these polyisoprenoid alcohols is solanesol (143), a major component of tobacco leaves. Little uptake of tracer from [^{14}C]-acetate or MVA occurred into this compound (Nicholas, 1967a; Reid, 1961)

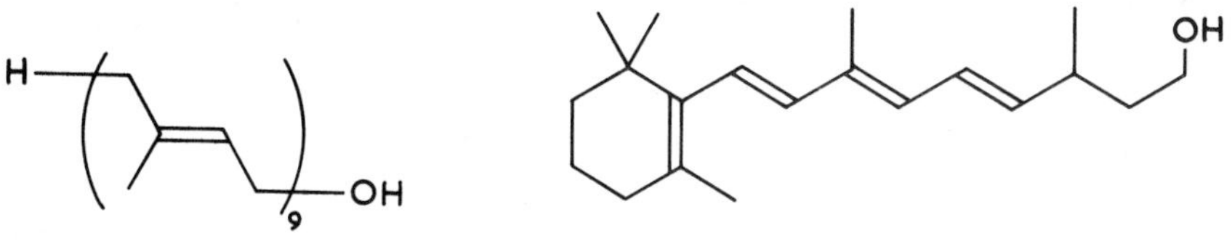

(143) (144)

Diagram 58.

and its turnover is presumably slow. Other prenols with up to 13 or 22 isoprene residues occur in non-photosynthetic plant tissue (Hemming, 1967, 1969) or liver (Dunphy *et al.*, 1967) and some of the former type have both *cis*- and *trans*-olefin linkages and give the expected isotope ratios when biosynthesized from doubly-labelled MVA (Gough and Hemming, 1967). Farnesyl-farnesyl pyrophosphate, the precursor of the side chain of ubiquinone-6, has been isolated but the pyrophosphates of most of the other known prenols have not been detected in plants.

C. Miscellaneous

Ubiquinones 6 to 10, plastoquinone and the vitamins K have isoprenoid side chains of from 4 to 10 condensed units that have been proved to be of mevalonoid origin (Threlfall, 1967). Certain of these side chains have been shown to be synthesized in the chloroplast (Griffiths *et al.*, 1968) and to have an isotope ratio when biosynthesized from doubly-labelled MVA consistent with an all-*trans* pattern of linkages (Dada *et al.*, 1968). The side chain, as the pyrophosphate, probably condenses with the phenol moeity and the product is reduced to a quinone; the tocopherols indeed retain the phenol ring in a modified form.

Other compounds of terpenoid origin are vitamin A (144) that is formed from *β*-carotene in the intestinal wall (Chichester and Nakayama, 1967); trisporic acids that are also degraded carotenoids (Austin *et al.*, 1969); and fungal sesterterpenoids (C_{25}) that are believed to be derived from geranyl-farnesyl pyrophosphate (Canonica *et al.*, 1966). Terpene alkaloids (Waller, 1969), cannabis and hop constituents, and certain mould metabolites are of partial mevalonoid origin (Crout, 1969). The last three classes are formed by C- or O-alkylation of polyketide rings by prenyl pyrophosphates.

IX. Biological Significance

A few groups of terpenes play a known biological role. Carotenoids are associated with photosynthesis (Goodwin, 1961; Burnett, 1965); sterols and prenols play a structural role in membranes; gibberellins, dormin and abscissin are plant hormones (Cornforth *et al.*, 1965a); phytol is an essential component of chlorophyll; prenols are components of essential quinones; and monoterpenes can play rather undefined roles in inhibiting competing plants and bacteria (Muller, 1965) and repelling predators (Goodwin, 1967). However there appears no obvious role for most terpenes and it is obviously unrealistic to seek specific functions for such a wide variety of compounds. Any biological significance must pertain either to the class as a whole or to the main groups within the class, not to individual compounds.

The classical view is that terpenes are, apart from the exceptions noted above, metabolic wastage; inert and slowly formed ballast of a prolific metabolic scheme (Sandermann, 1962) that are only synthesized in phylogenetically recent plants that have the means for storing and discarding these unwanted oddities. Recently, on the other hand, evidence has accumulated that these compounds are by no means inert and are usually rapidly synthesized even in young tissues. Analysis of the interconversions *in vivo* after feeding [^{14}C]-labelled carbon dioxide, MVA or terpenes has shown that in many plants tracer rapidly passed through the monoterpene pool within a matter of hours without a change in the size of the latter (Sukhov, 1958; Banthorpe and Wirz-Justice, 1969; Loomis, 1967; Francis and O'Connell, 1969; Nicholas, 1962; Battaile and Loomis, 1961; Reitsema, 1958; Attaway and Buslig, 1969) and a similar situation, although usually with a slower turn-over, exists for di- and tri-terpenes (Breccia and Badiello, 1967; Nicholas, 1964; Aexel *et al.*, 1967, Kasprzyk and Wojciechowski, 1969). Tracer from administered monoterpenes is also efficiently transferred to higher terpenes (Baisted, 1967) and amino-acids and sugars (Banthorpe and Wirz-Justice, 1969); and the monoterpene pool may act as a reservoir of material to maintain the respiratory coenzymes in a reduced form (Burbott and Loomis, 1967, 1969).

There is, thus, evidence that many terpenes can play a dynamic role in metabolism. Two theories have been proposed to account for the complexity of the compounds. One is that the vast majority have no biological function but are by-products of an evolving network from which routes to essential plant hormones and carotenoids have been selected (Goodwin, 1967): on this basis all but the few essential terpenes will disappear in the future course of evolution. The other directs attention to the activity of biosynthesis rather than the nature of the products: plants and micro-organisms are considered to produce terpenes during periods of dormancy (either of the whole plant or of localized tissue) to keep the enzyme systems in an active state. Specific intermediates that define a whole group, e.g. GPP, FPP, are subjected to relatively nonspecific interconversions and the diversity of products is beneficial in preventing the accumulation of possible inhibitory or toxic products (Bu'lock, 1965).

X. Regulation of Metabolism

Particular terpenes occur rather selectively within certain genera and species, and crossing experiments (Wildman *et al.*, 1946; Alston, 1966; Hanover, 1966) show that the distribution is under fairly rigid genetic control. There certainly appears to be much scope for genetic and chemical analysis of hybrids of different subspecies that differ in terpene content although being indistinguishable morphologically (Loomis, 1967).

A decisive factor governing the pattern of terpene metabolism is the ease of access of precursors to the synthetic sites. These sites are unknown, but storage and perhaps synthesis of monoterpenes and others largely occurs in oil glands that are modified epidermal hairs (Loomis, 1967; Amelunxen, 1964, 1965). The low incorporations, typically $<0.1\%$, of MVA into higher plants may reflect the inaccessibility of these sites. Moulds and fungi are less structurally segmented and consistently incorporate MVA into terpenoids in high yields (5 to 10%). Low incorporation of MVA could also be a consequence of its inability to intervene in a particulate enzyme system that converts acetate to terpene by a sequence of protein-bonded intermediates; such a system would not be easily isolated in an intact functional form and this may be why preparation of cell-free systems for terpene synthesis is notoriously difficult (Section XI).

The most detailed rationalization of factors governing terpene synthesis invokes intra- or extra-chloroplastic compartmentation (Goodwin, 1965a), whereby (a) enzymes responsible for the initial reactions occur both within and without the chloroplast, whereas those for the synthesis of chloroplastic terpenes (phytol, carotenes, ubiquinones) are inside and those for the steroids occur outside, and (b) the chloroplast membrane is impermeable to MVA. In agreement with this, [^{14}C]-MVA was incorporated into sterols but not into the chloroplastic terpenes in many species of higher plants, whereas [^{14}C]-carbon dioxide showed the reverse behaviour (Treharne *et al.*, 1966; Wieckowski and Goodwin, 1967; Griffiths *et al.*, 1968) and evidence for the localization of the MVA-activating enzymes was put forward (Rogers *et al.*, 1966). Observations that [^{14}C]-acetate or carbon dioxide labelled the diterpenes of certain plants but not the triterpenes although MVA gave the opposite pattern (Nicholas, 1964; Ruddat *et al.*, 1965) may have a similar explanation; and the location of tracer from MVA in the sub-cellular particles of lettuce was held to eliminate the chloroplasts as sites of sterol synthesis (Nicholas, 1967a).

Not all the evidence is in favour of the compartmentation effect. Barley cultivated in the presence of [^{14}C]-acetate accumulated tracer mainly in the non-phytol part of chlorophyll, whereas MVA mainly went into phytol (Fischer *et al.*, 1962), and [^{14}C]-geranylgeraniol and geranylinalool were incorporated into phytol in maize in high yields (Costes, 1964). However, moulds and non-proliferating plant tissue such as tomato fruit that do not contain chloroplasts utilize MVA very efficiently for carotenoid synthesis.

Considerable progress has been made in the elucidation of the feedback controls governing steroid synthesis in mammals and micro-organisms (Clayton 1965b; Bucher *et al.*, 1959), e.g. farnesoic acid (derived from FPP) and cholesterol can both inhibit the reductase that converts HMG-coenzyme A into MVA. Such pathways are undoubtedly present in plants.

XI. Cell Free Systems

Most of the biochemistry of the initial steps in steroid synthesis was elucidated using homogenates of liver tissue and yeast although usually only purified extracts of enzymes were prepared. The soluble fraction could generally support the route to FPP or GGPP but after this stage, when the substrate becomes water-insoluble, the enzymes become particle-bound; this has the great advantage that the substrate can be circulated through the system with none of the delays and risks attendant on dissociation from the catalytic surface.

Few studies on non-plant enzymes that produce mono- and higher terpenes have been made (Popjak, 1959), although recent investigations on the inhibition of MVA-kinase by terpenyl phosphate esters (Dorsey and Porter, 1968) and of the effects of analogues of GPP on prenyltransferase (Popjak *et al.*, 1969b,c) point the way for future developments.

Most of the work on synthesis of terpenes in plants has involved feeding either whole plants or tissue slices. Cell-free systems appear to be much more difficult to prepare from plants than from liver and yeast, probably due both to the ordered nature of the enzyme systems which do not survive the extraction and to the presence of phenols which complex with the enzymes. However, recently methods have been developed (Loomis and Battaile, 1966) to overcome these difficulties and future studies will, no doubt, be able to define the biosynthetic pathways free of such complications as compartmentation effects.

MVA-kinase has been purified from several plant sources (Loomis and Battaile, 1963; Rogers *et al.*, 1966; Loomis, 1967; Gray and Kekwick, 1969), but few other studies have been made. A preparation from peas incorporated MVA into MVAP, MVAPP, IPP, DMAPP, GPP and FPP albeit with less than 1% efficiency (Pollard *et al.*, 1966). Others, from pine seedlings incorporated 0.1% MVA into limonene (Valenzuela *et al.*, 1966), or 7% into GPP and NPP (Beytia *et al.*, 1969) and phospholipids stimulated this incorporation (George-Nascimento *et al.*, 1969). A cell-free system from mint converted pulegone into menthone and menthols (Battaile *et al.*, 1968).

Few investigations on sesquiterpene biosynthesis have been made save incidentally when FPP has been co-formed with other types of terpenes. GGPP synthetase has been isolated from carrots and pig liver and has approximately the same characteristics from each source (Nandi and Porter, 1964) and the corresponding bacterial enzyme has been isolated in a good state of purity (Kandutsch *et al.*, 1964). GGPP synthetase has also been demonstrated in extracts of various roots and seeds (Oster and West, 1968) and MVA and GGPP were converted to kaurene and other diterpenes in similar preparations (Anderson and Moore, 1967; Graebe, 1968; Shechter and West, 1969).

Squalene was formed from MVA or GPP in extracts of peas, carrots and tomatoes (Capstack *et al.*, 1962; Beeler *et al.*, 1963; Van Aller and Nes, 1968) and has been converted into β-amyrin in a preparation from peas (Nes *et al.*, 1965; Capstack *et al.*, 1965). MVA was converted into di- and triterpenes in similar preparations (Baisted *et al.*, 1962; Nicholas, 1962).

Numerous studies have been made on the conversion of MVA, IPP, acetate and FPP into carotenoids in particulate systems from microorganisms and tomatoes (Goodwin, 1965b; Suzue and Porter, 1969; Kushwaha *et al.*, 1969; Lee and Chichester, 1969). Isolation of chloroplasts by non-aqueous procedures is advantageous as often aqueous methods remove much lipid material and destroy the structural integrity of the organelle (Charlton *et al.*, 1967). In a recent study, lycopene was converted in 10% yield into β-carotene by a preparation from bean chloroplasts (Hill and Rogers, 1969).

Another type of approach, as yet almost unexplored, is the study of tissue cultures whereby root, leaf or stem tissues are excised and grown in sterile media under defined conditions such that growth, development or reversion to a primitive state can be controlled. This technique overcomes many of the difficulties found in whole plants without the loss of organization inherent in cell-free systems. Biochemical studies have been reported *inter alia* on cultures of tobacco (cf. Section V) and *Tanacetum vulgare* (Justice, 1967). The latter study showed that monoterpene metabolism in culture was significantly different from that in the intact plant.

Tissue cultures would be ideal for the study of hormonal control of metabolism and of differentiation and development but as yet no definitive studies appear to have been made on these lines.

REFERENCES

Achilladelis, B. and Hanson, J. R. (1969a). *Chem. Commun.* 488.

Achilladelis, B. and Hanson, J. R. (1969b). *J. Chem. Soc.* **(C)**, 2010.

Aexel, R., Evans, S., Kelley, M. and Nicholas, H. J. (1967). *Phytochemistry* **6**, 511.

Akazawa, T., Uritani, I. and Akazawa, Y. (1962). *Arch. Biochem. Biophys.* **99**, 52.

Akhtar, M., Brooks, W. A. and Watkinson, I. A. (1969). *Biochem. J.* **115**, 135.

Alston, R. E. (1966). *In* "Comparative Phytochemistry" (Swain, T. W. ed.), p. 33, Academic Press, London and New York.

Amelunxen, F. (1964). *Planta Med.* **12**, 121.

Amelunxen, F. (1965). *Planta Med.* **13**, 457.

Anderson, D. G. and Porter, J. W. (1962). *Arch. Biochem. Biophys.* **97**, 509.

Anderson, J. D. and Moore, T. C. (1967). *Plant Physiol.* **42**, 1527.

Anderson, R. J., Hanzlik, R. P., Sharpless, K. B., van Tamelen, E. E. and Clayton, R. B. (1969). *Chem. Commun.* 53.

Appel, H. H., Brooks, C. J. W. and Overton, K. H. (1959). *J. Chem. Soc. London* 3322.

Archer, B. L., Audley, B. G., Cockbain, E. G. and McSweeney, G. P. (1963). *Biochem. J.* **89**, 565.

Archer, B. L., Barnard, D., Cockbain, E. G., Cornforth, J. W., Cornforth, R. H. and Popjak, G. (1966). *Proc. Roy. Soc. London* **163B**, 519.

Arigoni, D. (1958). *Experientia* **14**, 153.

Arigoni, D., Barton, D. H. R., Corey, E. J., and Jeger, O., Caglioti, L., Dev. S., Ferrini, Symposium), p. 231, Churchill, London.

Arigoni, D. (1962). *Gazz. Chim. Ital.* **92**, 884.

Arigoni, D., Barton, D. H. R., Bernasconi, R., Djerassi, C., Mills, J. S. and Wolff, R. E. (1960a). *J. Chem. Soc. London* 1900.

Arigoni, D., Barton, D. H. R., Corey, E. J., and Jeger, O., Caglioti, L. Dev. S., Ferrini, P. G., Glazier, E. R., Melera, A., Pradhan, S. K., Schaffner, K., Sternhell, S., Templeton, J. F. and Tobinaga, S. (1960b). *Experientia* **16**, 41.

Attaway, J. A. and Buslig, B. S. (1969). *Phytochemistry* **8**, 1671.

Austin, D. G., Bu'lock, J. D. and Winstanley, D. J. (1969). *Biochem. J.* **113**, 34P.

Baisted, D. J. (1967). *Phytochemistry* **6**, 93.

Baisted, D. J. (1969). *Phytochemistry* **8**, 1697.

Baisted, D. J., Capstack, E. and Nes, W. R. (1962). *Biochemistry* **1**, 537.

Banthorpe, D. V. and Baxendale, D. (1968). *Chem. Commun.* 1553.

Banthorpe, D. V. and Turnbull, K. W. (1966). *Chem. Commun.* 177.

Banthorpe, D. V. and Wirz-Justice, A. (1969). *J. Chem. Soc.* **(C)**, 541.

Barton, D. H. R., Gosden, A. F., Mellows, G. and Widdowson, D. A. (1969). *Chem. Commun.* 184.

Barton, D. H. R., Moss, G. P. and Whittle, J. A. (1968). *J. Chem. Soc.* **(C)**, 1813.

Bates, R. B., and Feld, D. (1967). *Tetrahedron Lett.* 4875.

Battaile, J., Burbott, A. J. and Loomis, W. D. (1968). *Phytochemistry* **7**, 1159.

Battaile, J. and Loomis, W. D. (1961). *Biochim. Biophys. Acta* **51**, 545.

Battersby, A. R. and Gregory, B. (1968). *Chem. Commun.* 134.
Battersby, A. R. and Parry, G. V. (1964). *Tetrahedron Lett.* 787.
Battu, R. G. and Youngken, H. W. (1966). *Lloydia* **29**, 360.
Bazzano, G., Hamilton, J. G., Miller, O. N. and Hansen, J. H. (1964). *Fed. Proc.* **23**, 425.
Beeler, D. A., Anderson, D. G. and Porter, J. W. (1963). *Arch. Biochem. Biophys.* **102**, 26.
Bennett, R. D., Heftmann, E., Preston, W. H. and Haun, J. R. (1963). *Arch. Biochem. Biophys.* **103**, 74.
Benveniste, P., Durr, A., Hirth, L. and Ourisson, G. (1964). *Compt. Rend.* **259**, 2005.
Benveniste, P., Hirth, L. and Ourisson, G. (1966). *Phytochemistry* **5**, 45.
Bernard, M. J. and Reid, W. W. (1967). *Chem. Ind.* (*London*) 997.
Beytia, E., Valenzuela, P. and Cori, O. (1969). *Arch. Biochem. Biophys.* **129**, 346.
Biollaz, M. and Arigoni, D. (1969). *Chem. Commun.* 633.
Birch, A. J., Boulter, D., Fryer, R. I., Thompson, P. J. and Willis, J. L. (1959). *Tetrahedron Lett.* No. 3, 1.
Birch, A. J., Cameron, D. W., Holzapfel, C. W. and Rickards, R. W. (1963). *Chem. Ind.* (*London*) 374.
Birch, A. J., English, R. J., Massy-Westropp, R. A. and Smith, H. (1958). *J. Chem. Soc. London* 369.
Birch, A. J., Kocor, M., Sheppard, N. and Winter, J. (1962). *J. Chem. Soc. London* 1502.
Birch, A. J., Richards, R. W., Smith, H., Harris, A. and Whalley, W. B. (1959). *Tetrahedron* **7**, 241.
Blackburn, G. M., Ollis, W. D., Smith, C. and Sutherland, I. O. (1969). *Chem. Commun.* 99.
Bohlmann, F. and Grenz, M. (1969). *Tetrahedron Lett.* 2413.
Bonner, J. (1967). *In* "Biogenesis of Natural Compounds" (Bernfeld, P. ed.), 2nd Ed., p. 941, Pergamon Press, Oxford.
Breccia, A. and Badiello, R. (1967). *Z. Naturforsch.* **22B**, 44.
Brechbühler-Bader, S., Coscia, C. J., Loew, P., Von Szczepanski, C. H. and Arigoni, D. (1968). *Chem. Commun.* 136.
Britt, J. J. and Arigoni, D. (1958). *Proc. Chem. Soc. London* 224.
Brodie, J. D., Wasson, G. and Porter, J. W. (1964). *J. Biol. Chem.* **239**, 1346.
Brown, E. D. and Sutherland, J. K. (1968). *Chem. Commun.* 1060.
Brown, M. (1968). *J. Org. Chem.* **33**, 162.
Bruns, K. (1964). "Dissertation Universität Hamburg", cited by Weissmann, G. (1966). *In* "Comparative Phytochemistry" (Swain, T. ed.), p. 97, Academic Press, London and New York.
Bucher, N. L. R., McGarrahan, K., Gould, E. and Loud, A. V. (1959). *J. Biol. Chem.* **234**, 262.
Buggy, M. J., Britton, G. and Goodwin, T. W. (1969). *Biochem. J.* **114**, 641.
Bullock, J. D. (1965). "The Biosynthesis of Natural Products," p. 9, McGraw-Hill, London.
Burbott, A. J. and Loomis, W. D. (1967). *Plant Physiol.* **42**, 20.
Burbott, A. J. and Loomis, W. D. (1969). *Plant Physiol.* **44**, 173.
Burnett, J. H. (1965). *In* "Chemistry and Biochemistry of Plant Pigments" (Goodwin, T. W. ed.), p. 381, Academic Press, London and New York.
Canonica, L., Fiecchi, A., Kienle, M. G., Ranzi, B. M. and Scala, A. (1966). *Tetrahedron Lett.* 3035.
Capstack, E., Baisted, D. J., Newschwander, W. W., Blondin, G., Rosin, N. L. and Nes, W. R. (1962). *Biochemistry* **1**, 1178.

Capstack, E., Rosin, N., Blondin, G. A. and Nes, W. R. (1965). *J. Biol. Chem.* **240**, 3258.
Caspi, E. and Mulheirn, L. J. (1969). *Chem. Commun.* 1423.
Charlton, J. M., Treharne, K. J. and Goodwin, T. W. (1967). *Biochem. J.* **105**, 205.
Charlwood, B. V. (1970). Ph.D. Thesis, London.
Chesterton, C. J. and Kekwick, R. G. O. (1966). *Biochem. J.* **100**, 56P.
Chichester, C. O. and Nakayama, T. O. M. (1967). *In* "Biogenesis of Natural Compounds" (Bernfeld, P. ed.), 2nd Ed., p. 641, Pergamon Press, Oxford.
Chichester, C. O., Yokoyama, H., Nakayama, T. O. M., Lukton, A. and Mackinney, G. (1959). *J. Biol. Chem.* **234**, 598.
Clayton, R. B. (1965a). *Quart. Rev.* **19**, 168.
Clayton, R. B. (1965b). *Quart. Rev.* **19**, 201.
Cookson, R. C. (1968). *Quart. Rev.* **22**, 423.
Corbella, A., Gariboldi, P., Jommi, G. and Scolastico, C. (1969). *Chem. Commun.* 634.
Corey, E. G. and Gross, S. K. (1968b). *J. Amer. Chem. Soc.* **90**, 5045.
Corey, E. J., Lin, K. and Jautelat, M. (1968a). *J. Amer. Chem. Soc.* **90**, 2724.
Corey, E. J., Lin, K. and Yamamoto, H. (1969). *J. Amer. Chem. Soc.* **91**, 2132.
Corey, E. J., De Montellano, P. R. O. and Yamamoto, H. (1968c). *J. Amer. Chem. Soc.* **90**, 6254.
Corey, E. J., Rossey, W. E. and De Montellano, P. R. O. (1966). *J. Amer. Chem. Soc.* **88**, 4750.
Corey, E. J. and Ursprung, J. J. (1956). *J. Amer. Chem. Soc.* **78**, 5041.
Cornforth, J. W. (1968). *Angew. Chem. Int. Ed.* **7**, 903.
Cornforth, J. W. (1969). *Quart. Rev.* **23**, 125.
Cornforth, J. W., Cornforth, R. H., Donninger, C. and Popjak, G. (1966). *Proc. Roy. Soc. (London)* **163B**, 492.
Cornforth, J. W., Cornforth, R. H., Donninger, C., Popjak, G., Shimizu, Y., Ichii, S., Forchielli, E. and Caspi, E. (1965a). *J. Amer. Chem. Soc.* **87**, 3224.
Cornforth, J. W., Cornforth, R. H., Pelter, A., Horning, M. G. and Popjak, G. (1958). *Proc. Chem. Soc. (London)* 112.
Cornforth, J. W., Milborrow, B. V. Ryback, G. and Wareing, P. F. (1965). *Nature (London)* **205**, 1269.
Costes, C. (1963). *Compt. Rend.* **256**, 3535.
Costes, C. (1964). 6th International Biochem. Congress, New York, Abstract, p. 569.
Cramer, F. and Rittersdorf, W. (1967). *Tetrahedron* **23**, 3015.
Crombie, L., Houghton, R. P. and Woods, D. K. (1967). *Tetrahedron Lett.* 4553.
Cross, A. D. (1960). *Quart. Rev.* **14**, 317.
Cross, B. E. (1968). *Prog. Phytochem.* **1**, 195.
Cross, B. E., Galt, R. H. B. and Hanson, J. R. (1963). *J. Chem. Soc. London* 2944.
Cross, B. E., Norton, K. and Stewart, J. C. (1968). *J. Chem. Soc. London (C)* 1054.
Cross, B. E. and Stewart, J. C. (1968). *Tetrahedron Lett.* 5195.
Crout, D. H. G. (1969). *Topics Carbocyclic Chem.* **1**, 63.
Crowley, K. J. (1965). *Tetrahedron Lett.* 2863.
Crowley, M. P., Godin, P. J., Inglis, H. S., Snarey, M. and Thain, E. M. (1962). *Biochim. Biophys. Acta* **60**, 312.
Dada, O. A., Threlfall, D. R. and Whistance, G. R. (1968). *Eur. J. Biochem.* **4**, 329.
Dauben, W. G., Abraham, S., Hotta, S., Chaikoff, I. L., Bradlow, H. L. and Soloway, A. H. (1953). *J. Amer. Chem. Soc.* **75**, 3038.
Dauben, W. G., Fonken, G. J. and Boswell, G. A. (1957). *J. Amer. Chem. Soc.* **79**, 1000.

Dauben, W. G., Thiessen, W. E. and Resnick, P. R. (1965). *J. Org. Chem.* **30**, 1693.
Davies, B. H. and Holmes, E. A. (1969). *Biochem. J.* **113**, 33P.
Davies, B. H., Villoutreix, J., Williams, R. J. H. and Goodwin, T. W. (1963). *Biochem. J.* **89**, 96P.
Davis, B. D. (1955). *Advan. Enzymol.* **16**, 247.
Decker, K. and Uehleke, U. (1961). *Z. Physiol. Chem.* **323**, 61.
de Mayo, P., Robinson, J. R., Spencer, E. Y. and White, R. W. (1962). *Experientia* **18**, 359.
de Mayo, P. and Williams, R. E. (1965). *J. Amer. Chem. Soc.* **87**, 3275.
de Mayo, P. Williams, R. E. and Spencer, E. Y. (1965). *Can. J. Chem.* **43**, 1357.
Djerassi, C., Cais, M. and Mitscher, L. A. (1959). *J. Amer. Chem. Soc.* **81**, 2386.
Djerassi, C., Knight, J. C. and Wilkinson, D. I. (1963). *J. Amer. Chem. Soc.* **85**, 835.
Dorsey, J. K. and Porter, J. W. (1968). *J. Biol. Chem.* **243**, 4667.
Dugan, J. J., de Mayo, P., Nisbet, M., Robinson, J. R. and Anchel, M. (1966). *J. Amer. Chem. Soc.* **88**, 2838.
Dunphy, P. J., Kerr, J. D., Pennock, J. F., Whittle, K. J. and Feeney, J. (1967). *Biochim. Biophys. Acta* **136**, 136.
Eppenberger, U., Hirth, L. and Ourisson, G. (1969). *Eur. J. Biochem.* **8**, 180.
Epstein, W. W. and Rilling, H. C. (1970). *J. Biol. Chem.* **245**, 4597.
Eschenmoser, A., Ruzicka, L., Jeger, O. and Arigoni, D. (1955). *Helv. Chim. Acta* **38**, 1890.
Etemadi, A. H., Popjak, G. and Cornforth, J. W. (1969). *Biochem. J.* **111**, 445.
Ferguson, J. J., Durr, I. F. and Rudney, H. (1959). *Proc. Nat. Acad. Sci. U.S.A.* **45**, 499.
Fischer, F. G., Märkl, G., Hönel, H. and Rüdiger, W. (1962). *Ann.* **657**, 199.
Francis, M. J. O. and Allcock, C. (1969). *Phytochemistry* **8**, 1339.
Francis, M. J. O. and O'Connell, M. (1969). *Phytochemistry* **8**, 1705.
Gascoigne, R. M. (1958). *J. Chem. Soc. London* 876.
George-Nascimento, C., Beytia, E., Aedo, A. R. and Cori, O. (1969). *Arch. Biochem. Biophys.* **132**, 470.
Ghisalberti, E. L., de Souza, N. J., Rees, H. H., Goad, L. J. and Goodwin, T. W. (1969a). *Chem. Commun.* 1401.
Ghisalberti, E. L., de Souza, N. J., Rees, H. H., Goad, L. J. and Goodwin, T. W. (1969b). *Chem. Commun.* 1403.
Goad, L. J. (1967). *In* "Terpenoids in Plants" (Pridham, J. B. ed.), p. 159, Academic Press, London and New York.
Goad, L. J., Gibbons, G. F., Bolger, L. M., Rees, H. H. and Goodwin, T. W. (1969). *Biochem. J.* **114**, 885.
Goad, L. J. and Goodwin, T. W. (1965). *Biochem. J.* **96**, 79P.
Goad, L. J. and Goodwin, T. W. (1966). *Biochem. J.* **99**, 735.
Goad, L. J. and Goodwin, T. W. (1967). *Eur. J. Biochem.* **1**, 357.
Goad, L. J. and Goodwin, T. W. (1969). *Eur. J. Biochem.* **7**, 502.
Godin, P. J., Inglis, H. S., Snarey, M. and Thain, E. M. (1963). *J. Chem. Soc. London* 5878.
Godtfredsen, W. O., Lorck, H., van Tamelen, E. E., Willett, J. D. and Clayton, R. B. (1968). *J. Amer. Chem. Soc.* **90**, 208.
Godtfredsen, W. O. and Vangedal, S. (1965). *Acta Chem. Scand.* **19**, 1088.
Goodwin, T. W. (1961). *Ann. Rev. Plant. Physiol.* **12**, 219.
Goodwin, T. W. (1965a). *In* "Biosynthetic Pathways in Higher Plants" (Pridham, J. B. and Swain, T. eds.), p. 37, Academic Press, London and New York.

Goodwin, T. W. (1965b). *In* "Chemistry and Biochemistry of Plant Pigments" (Goodwin, T. W. ed.), p. 143, Academic Press, London and New York.
Goodwin, T. W. (1967). *In* "Terpenoids in Plants" (Pridham, J. B. ed.), p. 1, Academic Press, London and New York.
Goodwin, T. W. (1969). *In* "Perspectives in Phytochemistry" (Harborne, J. B. and Swain, T. eds.), p. 75, Academic Press, London and New York.
Goodwin, T. W. and Williams, R. J. H. (1965a). *Biochem. J.* **94**, 5C.
Goodwin, T. W. and Williams, R. J. H. (1965b). *Biochem. J.* **97**, 28C.
Goodwin, T. W. and Williams, R. J. H. (1966). *Proc. Roy. Soc. London* **163B**, 515.
Gough, D. P. and Hemming, F. W. (1967). *Biochem. J.* **105**, 10C.
Graebe, J. E. (1968). *Phytochemistry* **7**, 2003.
Gray, J. C. and Kekwick, R. G. O. (1969). *Biochem. J.* **113**, 37P.
Griffiths, W. T., Threlfall, D. R. and Goodwin, T. W. (1968). *Eur. J. Biochem.* **5**, 124.
Grob, E. C. and Boschetti, A. (1962). *Chimia* **16**, 15.
Gros, E. G. and Leete, E. (1965). *J. Amer. Chem. Soc.* **87**, 3479.
Guarnaccia, R., Botta, L. and Coscia, C. J. (1969). *J. Amer. Chem. Soc.* **91**, 204.
Guglielmetti, L. (1967). Quoted in "Biogenesis of Natural Compounds" (Bernfeld, P. ed.), 2nd Ed., p. 889, Pergamon Press, Oxford.
Haley, R. C., Miller, J. A. and Wood, H. C. S. (1969). *J. Chem. Soc.* **(C)**, 264.
Hall, J., Smith, A. R. H., Goad, L. J. and Goodwin, T. W. (1969). *Biochem. J.* **112**, 129.
Hanover, J. W. (1966). *Heredity* **21**, 73.
Hanson, J. R. (1968). *Perfum. Essent. Oil. Rec.* **59**, 802.
Hanson, J. R. and Achilladelis, B. (1967). *Tetrahedron Lett.* 1295.
Hanson, J. R., Hough, A. and White, A. F. (1968). *Chem. Commun.* 467.
Hanson, J. R. and White, A. F. (1968). *Chem. Commun.* 1689.
Hayashi, S., Yano, K. and Matsuura, T. (1968). *Tetrahedron Lett.* 6241.
Hefendehl, F. W. (1966). *Planta Med.* **14**, 66.
Hefendehl, F. W. (1967). *Planta Med.* **15**, 121.
Hefendehl, F. W., Underhill, E. W. and Von Rudloff, E. (1967). *Phytochemistry* **6**, 823.
Heinstein, P. F., Smith, F. H. and Tove, S. B. (1962). *J. Biol. Chem.* **237**, 2643.
Heinstein, P. F., Herman, D. L., Tove, S. B. and Smith, F. H. (1970). *J. Biol. Chem.* **245**, 4658.
Hemming, F. W. (1967). *In* "Terpenoids in Plants" (Pridham, J. E. ed.), p. 223, Academic Press, London and New York.
Hemming, F. W. (1969). *Biochem. J.* **113**, 23P.
Hendrickson, J. B. (1959). *Tetrahedron* **7**, 82.
Hewlins, M. J. E., Ehrhardt, J. D., Hirth, L. and Ourisson, G. (1969). *Eur. J. Biochem.* **8**, 184.
Hill, H. M. and Rogers, L. J. (1969). *Biochem. J.* **113**, 31P.
Horodysky, A. G., Waller, G. R. and Eisenbraun, E. J. (1969). *J. Biol. Chem.* **244**, 3110.
Jacobsohn, G. M. and Frey, M. J. (1967). *J. Amer. Chem. Soc.* **89**, 3338.
Johnson, D. F., Heftmann, E. and Houghland, G. V. C. (1964). *Arch. Biochem. Biophys.* **104**, 102.
Jones, E. R. H. and Lowe, G. (1960). *J. Chem. Soc. London* 3959.
Justice, A. M. (1967). Ph.D. Thesis, London.
Kandutsch, A. A., Paulus, H., Levin, E., and Bloch, K. (1964). *J. Biol. Chem.* **239**, 2507.
Kasprzyk, Z. and Wojciechowski, Z. (1969). *Phytochemistry* **8**, 1921.

Knauss, H. J., Porter, J. W. and Wasson, G. (1959). *J. Biol. Chem.* **234**, 2835.
Krishna, G., Feldbruegge, D. H. and Porter, J. W. (1964). *Biochem. Biophys. Res. Commun.* **14**, 363.
Kushwaha, S. C., Subbarayan, C., Beeler, D. A. and Porter, J. W. (1969). *J. Biol. Chem.* **244**, 3635.
Lawrie, W., McLean, J., Pauson, P. L. and Watson, J. (1967). *J. Chem. Soc.* **(C)**, 2002.
Lederer, E. (1964). *Biochem. J.* **93**, 449.
Lederer, E. (1969). *Quart. Rev.* **23**, 453.
Lee, T. C. and Chichester, C. O. (1969). *Phytochemistry* **8**, 603.
Le Patourel, G. N. J. (1970). Ph.D. Thesis, London.
Loomis, W. D. (1967). *In* "Terpenoids in Plants" (Pridham, J. B. ed.), p. 59, Academic Press, London and New York.
Loomis, W. D. and Battaile, J. (1963). *Biochim. Biophys. Acta* **67**, 54.
Loomis, W. D. and Battaile, J. (1966). *Phytochemistry* **5**, 423.
Maheshwari, M. L., Varma, K. R. and Bhattacharyya, S. C. (1963). *Tetrahedron*, **19**, 1519.
Mann, J. (1970). Ph.D. Thesis, London.
Maudgal, R. K., Tchen, T. T. and Bloch, K. (1958). *J. Amer. Chem. Soc.* **80**, 2589.
McPhail, A. T., Reed, R. I. and Sim, G. A. (1964). *Chem. Ind.* (*London*) 976.
Meinwald, J., Happ, G. M., Labows, J. and Eisner, T. (1966). *Science, N.Y.* **151**, 79.
Moron, J., Rondest, J. and Polonsky, J. (1966). *Experientia* **22**, 511.
Moss, G. P. and Nicolaidis, S. A. (1969). *Chem. Commun.* 1072.
Muller, W. H. (1965). *Botan. Gazz.* (*Chicago*) **126**, 195.
Nanoi, D. L. and Porter, J. W. (1964). *Arch. Biochem. Biophys.* **105**, 7.
Nes, W. R., Capstack, E. and Blondin, G. A. (1965). *Fed. Proc.* **24**, 660.
Nicholas, H. J. (1962). *J. Biol. Chem.* **237**, 1485.
Nicholas, H. J. (1964). *Biochim. Biophys. Acta* **84**, 80.
Nicholas, H. J. (1967a). *In* "Biogenesis of Natural Compounds" (Bernfeld, P., ed.), 2nd Ed., p. 829, Pergamon Press, Oxford.
Nicholas, H. J. (1967b). *Phytochemistry* **6**, 371.
Nusbaum-Cassuto, E. and Villoutreix, J. (1965). *Compt. Rend.* **260**, 1013.
Oehlschlager, A. C. and Ourisson, G. (1967). *In* "Terpenoids in Plants" (Pridham, J. B. ed.), p. 83, Academic Press, London and New York.
Ogura, K., Koyama, T. and Seto, S. (1969). *Biochem. Biophys. Res. Commun.* **35**, 875.
Oster, M. O. and West, C. A. (1968). *Arch. Biochem. Biophys.* **127**, 112.
Parker, W., Roberts, J. S. and Ramage, R. (1967). *Quart. Rev.* **21**, 331.
Plouvier, V. (1966). *Phytochemistry* **5**, 955.
Pollard, C. J., Bonner, J., Haagen-Smit, A. J., and Nimmo, C. C. (1966). *Plant Physiol.* **41**, 66.
Ponsinet, G. and Ourisson, G. (1968). *Phytochemistry* **7**, 757.
Popjak, G. (1959). *Tetrahedron Lett.* No. 19, 19.
Popjak, G., Edmond, J., Clifford, K. and Williams, V. (1969a). *J. Biol. Chem.* **244**, 1897.
Popjak, G., Holloway, P. W., Bond, R. P. M. and Roberts, M. (1969b). *Biochem. J.* **111**, 333.
Popjak, G., Rabinowitz, J. L. and Baron, J. M. (1969c). *Biochem. J.* **113**, 861.
Porter, J. W. and Anderson, D. G. (1967). *Annu. Rev. Plant Physiol.* **18**, 197.
Porter, J. W. (1969). *Pure Appl. Chem.* **20**, 449.
Rees, H. H., Britton, G. and Goodwin, T. W. (1968). *Biochem. J.* **106**, 659.
Rees, H. H., Goad, L. J. and Goodwin, T. W. (1968a). *Tetrahedron Lett.* 723.
Rees, H. H., Goad, L. J. and Goodwin, T. W. (1968b). *Biochem. J.* **107**, 417.

Rees, H. H., Goad, L. J. and Goodwin, T. W. (1969). *Biochim. Biophys. Acta* **176**, 892.
Rees, H. H., Mercer, G. I. and Goodwin, T. W. (1966). *Biochem. J.* **99**, 726.
Regnier, F. E., Waller, G. R., Eisenbraun, E. J. and Auda, H. (1968). *Phytochemistry* **7**, 221.
Reid, W. W. (1961). *Chem. Ind.* (*London*) 1489.
Reitsema, R. H. (1958). *J. Amer. Pharm. Ass. Sci. Ed.* **47**, 267.
Richards, J. H. and Hendrickson, J. B. (1964). "The Biosynthesis of Steroids, Terpenes and Acetogenins," W. A. Benjamin, New York and Amsterdam.
Rilling, H. C. (1962). *Biochim. Biophys. Acta* **65**, 156.
Rilling, H. C. and Epstein, W. W. (1969). *J. Amer. Chem. Soc.* **91**, 1041.
Rilling, H. C. (1970). *J. Lipid Res.* **11**, 480.
Risinger, G. E. and Durst, H. D. (1968). *Tetrahedron Lett.* 3133.
Rogers, L. J., Shah, S. P. J. and Goodwin, T. W. (1966). *Biochem. J.* **99**, 381.
Ruddat, M., Heftmann, E. and Lang, A. (1965). *Arch. Biochem. Biophys.* **110**, 496.
Rudney, H., Stewart, P. R., Majerus, P. W. and Vagelos, P. R. (1966). *J. Biol. Chem.* **241**, 1226.
Ruzicka, L. (1959). *Proc. Chem. Soc. London* 341.
Ruzicka, L. (1963). *Pure and Appl. Chem.* **6**, 493.
Ruzicka, L., Eschenmoser, A. and Heusser, H. (1953). *Experientia* **9**, 357.
Sandermann, W. (1962). *In* "Comparative Biochemistry" (Florkin, M. and Mason, H. S., eds), Vol. III, pp. 503, 591, Academic Press, London and New York.
Sandermann, W. and Bruns, K. (1962a). *Naturwiss*, **49**, 258.
Sandermann, W. and Bruns, K. (1962b). *Chem. Ber.* **95**, 1863.
Sandermann, W. and Bruns, K. (1962c). *Tetrahedron Lett.* 261.
Sandermann, W. and Bruns, K. (1965). *Planta Med.* **13**, 364.
Sandermann, W. and Schweers, W. (1962a). *Tetrahedron Lett.* 257.
Sandermann, W. and Schweers, W. (1962b). *Tetrahedron Lett.* 259.
Sandermann, W. and Stockmann, H. (1957). *Fette, Seifen, Anstrichm.* **59**, 852.
Sandermann, W. and Stockmann, H. (1958). *Chem. Ber.* **91**, 930.
Scott, A. I., McCapra, F., Comer, F., Sutherland, S. A., Young, D. W., Sim, G. A. and Ferguson, G. (1964). *Tetrahedron* **20**, 1339.
Sharpless, K. B. and van Tamelen, E. E. (1969). *J. Amer. Chem. Soc.* **91**, 1848.
Shechter, I. and West, C. A. (1969). *J. Biol. Chem.* **244**, 3200.
Shneour, E. A. (1962). *Biochim. Biophys. Acta* **65**, 510.
Soucek, M. (1962). *Collect. Czech. Chem. Commun.* **27**, 2929.
Stermitz, F. R. and Rapoport, H. (1961). *J. Amer. Chem. Soc.* **83**, 4045.
Sukhov, G. V. (1958). Radioisotopes in Scientific Research (First Int. Conference), Vol. 4, p. 535, Pergamon Press, New York.
Suzue, G. and Porter, J. W. (1969). *Biochim. Biophys. Acta* **176**, 653.
Swain, T. (1965). *In* "Biosynthetic Pathways in Higher Plants" (Pridham, J. B. and Swain, T., eds), p. 9, Academic Press, London and New York.
Tavormina, P. A., Gibbs, M. H. and Huff, J. W. (1956). *J. Amer. Chem. Soc.* **78**, 4498.
Taylor, W. I. and Battersby, A. R. (1969). "Cyclopentanoid Terpene Derivatives", M. Dekker, New York.
Threlfall, D. R. (1967). *In* "Terpenoids in Plants" (Pridham, J. B. ed.), p. 191, Academic Press, London and New York.
Treharne, K. J., Mercer, E. I. and Goodwin, T. W. (1966). *Biochem. J.* **99**, 239.
Valenzuela, P., Beytia, E., Cori, D. and Yudelevich, A. (1966). *Arch. Biochem. Biophys.* **113**, 536.
van Aller, R. T., Chikamatsu, H., de Souza, N. J., John, J. P. and Nes, W. R. (1968). *Biochem. Biophys. Res. Commun.* **31**, 842.

van Aller, R. T. and Nes, W. R. (1968). *Phytochemistry* **7**, 85.
van Tamelen, E. E. and Coates, R. M. (1966). *Chem. Commun.* 413.
van Tamelen, E. E., Hanzlik, R. P., Sharpless, K. B., Clayton, R. B., Richter, W. J. and Burlingame, A. L. (1968). *J. Amer. Chem. Soc.* **90**, 3284.
van Tamelen, E. E., Storni, A., Hessler, E. J. and Schwartz, M. (1963). *J. Amer. Chem. Soc.* **85**, 3295.
van Tamelen, E. E., Willett, J. D., Clayton, R. B. and Lord, K. E. (1966). *J. Amer. Chem. Soc.* **88**, 4752.
Waller, G. R. (1969). *Progr. Chem. Fats Other Lipids* **10**, 151.
Waters, J. A. and Johnson, D. F. (1965). *Arch. Biochem. Biophys.* **112**, 387.
Wellburn, A. R., Stone, K. J. and Hemming, F. W. (1966). *Biochem. J.* **100**, 23C.
Wells, L. W., Schelble, W. J. and Porter, J. W. (1964). *Fed. Proc.* **23**, 426.
Wenkert, E. (1955). *Chem. Ind.* (*London*) 282.
Wenkert, E. and Chamberlin, J. W. (1959). *J. Amer. Chem. Soc.* **81**, 688.
Wieckowski, S. and Goodwin, T. W. (1967). *Biochem. J.* **105**, 89.
Wildman, S. G., Abegg, F. A., Elder, J. A. and Hendricks, S. B. (1946). *Arch. Biochem.* **10**, 141.
Willhalm, B. and Thomas, A. F. (1969). *Chem. Commun.* 1380.
Williams, R. J. H., Britton, G., Charlton, J. M. and Goodwin, T. W. (1967a). *Biochem. J.* **104**, 767.
Williams, R. J. H., Britton, G. and Goodwin, T. W. (1966). *Biochem. J.* **101**, 7P
Williams, R. J. H., Britton, G. and Goodwin, T. W. (1967b). *Biochem. J.* **105**, 99.
Woodward, R. B. and Bloch, K. (1953). *J. Amer. Chem. Soc.* **75**, 2023.
Yeowell, D. A. and Schmid, H. (1964). *Experientia* **20**, 250.

AUTHOR INDEX

Numbers in *italics* refer to pages on which references are listed at the end of each chapter. Those numbers in parentheses refer to Chapter 5 in which the system of numbered references has been used.

C

D

H

N

O

P

Q

R

S

T

Subject Index

In this index, references to substituted organic compounds will be found under the name of the parent compound; esters are indexed under the name of the terpene moiety. Prefixes such as "*cis*", "*trans*", "*endo*" ar ! "*exo*" are ignored in alphabetization; the prefixes "allo", "apo", "epi", "iso" and "neo" are treated in each case as a part of the name concerned.

J

K